# Technology for Creativity and Innovation:

## Tools, Techniques and Applications

Anabela Mesquita
*ISCAP/IPP, Portugal*

Information Science REFERENCE

**INFORMATION SCIENCE REFERENCE**
Hershey · New York

Senior Editorial Director: Kristin Klinger
Director of Book Publications: Julia Mosemann
Editorial Director: Lindsay Johnston
Acquisitions Editor: Erika Carter
Development Editor: Joel Gamon
Production Coordinator: Jamie Snavely
Typesetters: Keith Glazewski & Natalie Pronio
Cover Design: Nick Newcomer

Published in the United States of America by
Information Science Reference (an imprint of IGI Global)
701 E. Chocolate Avenue
Hershey PA 17033
Tel: 717-533-8845
Fax: 717-533-8661
E-mail: cust@igi-global.com
Web site: http://www.igi-global.com



Library of Congress Cataloging-in-Publication Data

Technology for creativity and innovation : tools, techniques and applications / Anabela Mesquita, editor.
p. cm.
Includes bibliographical references and index.
Summary: "This book provides empirical research findings and best practices on creativity and innovation in business, organizational, and social environments"--Provided by publisher.
ISBN 978-1-60960-519-3 (hbk.) -- ISBN 978-1-60960-520-9 (ebook) 1. Creative ability in business. 2. Creative thinking. 3. Diffusion of innovations. 4. Technological innovations. I. Sarmento, Anabela.
HD53.T434 2011
658.4'063--dc22

2010051811

British Cataloguing in Publication Data
A Cataloguing in Publication record for this book is available from the British Library.

*To Pedro Ivo and Nuno Miguel*

# Table of Contents

# Detailed Table of Contents

This chapter examines how specific technologies may influence performance at each stage of the creative process, and provides specific recommendations for how technology may be used to facilitate the development of creative solutions. This research is justified by the fact that in our fast-paced world, organizations must continually innovate in order to stay competitive. At the same time, technology is continually advancing, and tools to facilitate work are frequently changing. This forces organizations to stay abreast of current technologies, and also puts pressure on employees to use the technologies available to them in order to devise innovative solutions that further accomplish the organization's goals. To date, there has been little research on how such technologies may best be used to facilitate such creative performance. The present chapter addresses this gap by integrating a model of the creative process from the psychology literature with technology literature from engineering and information technology.



This chapter presents the findings of a study aimed at investigating how the fit between the problem-solving style of a product development team and the cognitive environment induced by the perceived problem complexity affects the amount and type of knowledge generated. It is assumed that organizational knowledge is created as a by-product of collective creative technical problem-solving, when people work together to deal with unfamiliar and unexpected situations. Two major outcomes emerge from the analysis of experimental data: (1) different cognitive environment patterns are more conducive than others to organizational learning; (2) there is some fit between the cognitive environment pattern and the team technical problem-solving style, as some cognitive practices and social behaviours adopted during technical problem-solving are more effective than others in certain cognitive environments. Particularly, practices and behaviours that are more associated to creativity have a not negligible weight in the generation of knowledge. Ninety-one cases of technical problem-solving occurred during product innovation within the 35 small firms were studied.



This chapter investigates some recent developments and proposes a conceptual framework for creative participation as a personality-driven interface between creativity and innovation. To retain competitiveness, succeed, and flourish, organizations are forced to continuously innovate. This drive for innovation is not solely limited to product/process innovation but more profoundly relates to a continuous process of improving how organizations work internally, requiring a constant stream of ideas and suggestions from motivated employees. Under the assumption that employees' intrinsic willingness to contribute novel ideas and solutions requires a set of personal characteristics and necessary skills that might well be unique to each organizational unit. The chapter explores personal characteristics associated with creativity, innovation and innovative behavior. Various studies on the correlation between creativity and personality types are also reviewed. The chapter provides a discussion of solutions and future development together with recommendations for the future research.



This chapter describes and explores social relationship patterns associated with outstanding innovation. In doing so, the authors draw upon the findings of 16 in-depth interviews with award-winning Australian innovators from science & technology and the creative industries. The interviews covered topics relating to various influences on individual innovation capacity and career development. Authors found that for all of the participants, innovation was a highly social process. Although each had been recognised individually for their innovative success, none worked in isolation. The ability to generate innovative outcomes was grounded in certain types of interaction and collaboration. Authors outline the distinctive features of the social relationships which seem to be important to innovation, and ask which 'social network capabilities' might underlie the ability to create an optimal pattern of interpersonal relationships. Authors discuss the implications of these findings for universities, which we argue play a key role in the development of nascent innovators



This chapter intends to clarify some issues about creativity and innovation methods, since it is believed that the term is often misunderstood. For the authors, neither creativity nor innovation is guided by a method. There are only methods against methods that can help the extraordinary individual to step faster and easier into a state of mind that is conducive to creativity, but which has no effect on whether the creative output becomes an innovation. In order to support this claim, the authors outline three major reasons that seem to be responsible for making people believe that such methods for creativity and innovation exist. Then authors present their understanding of creativity and continue with a discussion on the systemic character of creativity and innovation. Finally, the authors show that there are no methods for creativity, but methods against non-creativity by explaining in particular how one of these methods against non-creativity works. What the authors outline here is a necessarily one-sided and partial view. Their aim is not to convince the readers that they are right but to make them think by presenting one possible consistent approach.

## Section 2
## Techniques for Creativity



This chapter focuses on the development of organizational creativity, using the CPS methodology, aiming at demonstrating its effectiveness in using the individual and team divergent thinking improvement in identifying organizational problems. A study was undertaken using problem solving teams in seven companies, in which each individual was submitted to a pre-post test in attitudes towards divergent thinking and asked to express the evaluation of the method. All the information reported in the sessions was recorded. The results indicate a change in attitude favourable to divergent thinking, the provision of a professional, efficient method of organizing knowledge in such a way that can help individuals find original solutions to problems, and an important way to lead teams to creativity and innovation, according to companies' different orientations.



This chapter describes a workshop concept for small groups that aims at the qualification for creative business-idea generation in interdisciplinary teams of academics. The chapter's aim is to provide a theory-based and application-oriented description of the workshop, including a first-hand report on implementation. The chapter starts with a description of theoretical contributions and underlying research results to illustrate the workshop's framework. Next is a description of the aims, methods and target group follows, as well as the organizational settings and training methods. Then the chapter will focus on the realization of the workshop. Finally, authors discuss the workshop concept's scope of application.



This chapter discusses the important role played by idea generation techniques in the innovation process. Until recently, the space of techniques has been unstructured, and no clear guidelines have been available for the selection of an appropriate technique for a given innovation goal. Authors use an engineering approach to study and develop idea generation techniques with the aim of obtaining more structured and rigorous guidelines for generating ideas. One objective of this approach was to identify and understand the fundamental mental principles underlying an idea generation technique. In this chapter, the authors show how three such principles suffice to cover a large range of published idea generation techniques and suggest how these can be used to improve the utility of idea generation within the innovation process.

This chapter aims to demonstrate how PAOL - Unit for Innovation in Education, a project from IS-CAP - School of Accounting and Administration of Oporto - Institute Polytechnic of Oporto, Portugal - prompted new educational initiatives and new learning scenarios at a Higher Education Institution. Furthermore, authors intend to demonstrate PAOL's lines of intervention through an extensive analysis based on the 6 years of experience that this unit has in the educational technology field; a project that began small but that, due to the force of innovation, has progressively conquered new adepts. Therefore the unit described in this chapter relates all these factors, as a whole, capable of attaining changes that influence mentalities and methodologies, overcoming cultural and technical barriers. This case study can serve as a catalyst, potentiating the creation of new multi-faceted projects in the scope of Web technologies in higher education teaching-learning processes



This chapter elaborates on some trends that will govern creativity and innovation. Authors argue that the onset of the 21st century is marked by deep psychological and sociological transformations affecting every scale of human endeavour, ranging from individual to crowd behaviour. Deep and central to these transformations is the penetration of digital communication and computer technology into modern day life. Above all, this new and evolving technological landscape has opened exciting new possibilities to drive creative behaviour, organizational creativity and innovation through computer-mediated interactions. Such opportunities are met with equal challenges that need to be addressed in order to harness the full potential of massively distributed creative collaborations. This chapter will elaborate on the underlying trends that give rise to these opportunities and challenges and to what extent these trends will govern creativity and innovation in areas of organizational life such as business, education, science and design in the next 10 to 30 years.

# Foreword

The recursive nature of Inquiry as a process of turning back on itself can be transformative by the tension it generates between describing the process and exemplifying it. One might wonder what role each one of us plays when 'opportunities' present themselves as possibilities that eventually become realities. Although much confusion has been generated in sorting out what is science from what is technology, conceiving science as applied technology and technology as applied science has only muddled matters more. On the other hand, a useful distinction[i] has been offered that defines Science as being "concerned with processes that seek out the meaning of the natural world by 'inquiry', 'discovering what is', 'exploring', and using 'the scientific method'; and technology as being concerned with such processes that we use to alter/change the natural world such as 'invention', 'innovation', practical problem solving, and design." I highlighted another useful distinction[ii] elsewhere between "creativity" as the generation of new ideas, and "innovation" as the successful exploitation of new ideas, and that the most fundamental technologies available for inquiry are creativity and innovation.

As editor of Technology for Creativity and Innovation: Tools, Techniques and Applications, Anabela Mesquita has demonstrated her creative and innovative intentions in this book as a form of communities of practice. Dr. Mesquita is a professor of the School of Accounting and Administration of the Polytechnic Institute of Porto (ISCAP), one of Portugal's largest and most prestigious public Polytechnic Institutes (IPP). She is also the Principal Editor of the International Journal of Technology and Human Interaction, an interdisciplinary platform for leading research that addresses issues of human and technology interaction. She has transformed these opportunities and conveyed a challenge that has pulled together contributors from different fields to demonstrate their conception of what it means to relate to technology. Through their differences, she created a community of individuals willing to share their creativity and innovation through their relationship to and with technology.

Focusing on the relevancy of the demand for academic work, empirical research and best practices found in business areas of technologies, Dr. Mesquita's book, Technology for Creativity and Innovation: Tools, Techniques and Application, is a definite asset to anyone interested in creativity, innovation and technology, particularly educators, professionals and researchers working in these fields.

With the range of knowledge, experience, creativity and innovation presented here, readers will find their way toward greater confidence and competence relating to technology in their own creative and innovative ways.

*Mario Spiler*
*Barrie, Ontario, Canada*

## ENDNOTES

i William E. Dugger, Jr., Senior Fellow International Technology Education Association (ITEA) & Emeritus Professor, Virginia Tech, USA. "The Perspective of Technology Education", the Second International Symposium on Educational Cooperation for Industrial Technology Education, Kariya City, Japan, 2000.

ii Mario Spiler, "Creative inquiry and the discourse of creativity", unpublished manuscript, 2009; and Mario Spiler, "The Epistemology of Creative Inquiry", unpublished chapter, 2010.

***Mario Spiler,*** *holds a BA in Philosophy and Sociology from York University, Toronto and a Master's degree in Clinical Social Worker from Wilfred Laurier University, Kitchener-Waterloo. He attained a Master's Practitioner Certificate in Neuro-Linguistic Programming. He has been practicing clinical social work for 25 years. His research interests include Epistemology, Metaphor and the structure of subjective experiences and transactional dynamics of Transformation, Learning, Change and therapeutic interactions. He is a psychotherapist and consultant in private practice, Mapping Insights, in Barrie and Toronto, providing psychotherapeutic, as well as clinical training, research and consultancy services. He has taught Social Work at both the Bachelor and Masters levels at Laurentian University and Windsor University. He is currently enrolled and working toward a Certificate in Clinical Hypnosis through The Canadian Society of Clinical Hypnosis, Ontario Division.*

# Preface

## TECHNOLOGY FOR CREATIVITY AND INNOVATION: TOOLS, TECHNIQUES AND APPLICATIONS

The year 2009 was considered the European Year of Creativity and Innovation (http://create2009.europa.eu/about_the_year.html). The objective of this decision was to "raise awareness of the importance of creativity and innovation for personal, social and economic development; to disseminate good practices; to stimulate education and research and to promote policy debate on related issues". It is widely accepted that in order to cope with the rapid pace of change while being competitive, companies (and people) must be innovative, create new knowledge and have new ideas constantly. As a matter of fact innovation constitutes an important issue in organizational management in order to be competitive today. It is considered as the most important drive for business (Darso, 2008), providing economic value and "social prosperity through benefits to the individual and society" (European Parliament, 2008). Innovation is often coined with creativity, which can be described as a

*process of playing with ideas, thoughts, and possibilities. It is often part of an innovation process, but whereas creativity is inspired activity, innovation is more about the strategic overview in order to create an output that will be used and bought by the customers and clients (Darso, 2008).*

In order to be creative and innovative, it is necessary to help people to develop some competences (usually designated as soft skills - creativity, innovation, collaboration, communication, and critical thinking, among others). These are important because, as Chandler and Grzyb (2005:2) refer: "If we are creative, if we are skilled at innovation, we can come up with new ways of approaching situations that have changed".

Although some persons may be born more creative than others, it is possible to help those less creative to improve their innovative capacities and competences. It is in this context that this book appears. We decided to put together contributions from several scientific areas and from experts covering almost all corners of the world.

This book is divided in five sections: (1) The Process of Creativity and Innovation, (2) Techniques for Creativity, (3) Tools for Creativity, (4) Best Practices and (5) Conclusion.

## SECTION 1: THE PROCESS OF CREATIVITY AND INNOVATION

Creativity comes from the Latin term *creo* – to create, make. During the Christian period, *creatio* designated God's act *Ex nihilo*, "creation from nothing". *Creatio* thus meant something different than *facere* – to make – and did not apply to human activity. Throughout history this perspective has changed, but very often it was coined with only gifted people. Today it is widespread recognized that creativity should be seen as something that exists in a wider range of professions and people (Andriopoulos and Dawson, 2009: 21).

In the last century some researchers contributed to the development of creativity and to the creative thought. Here, we only present the contribution of 3 of them. For example, Wallas, in his work *Art of Thought* (1926) presents one of the first models of the creative process. It consisted of five stages, namely: (1) preparation – period when an individual may refine his/her goals in response to a particular issue or question or where relevant material from a wide range of primary and secondary sources is collected. The reason is that one needs first to prepare oneself with the relevant skills, knowledge, and abilities in order to be able to refine the problem at hand; (2) incubation – process of subconscious data processing, (3) intimation – the creative person gets a "feeling" that a solution is on its way, (4) illumination –when someone suddenly becomes aware of a core answer to the problem (sometimes it is very difficult to distinguish intimation and illumination, and as such, intimation is designated as a sub stage of illumination); and (5) verification – this period corresponds to the translation of a new idea into a doable solution. Since this early work, other studies followed focused on the process of creative problem solving. For instance, Basadur *et al* (1982) proposed a three-stage model comprising problem finding, problem solving (generating as many ideas as possible) and solution implementation. And this approach is more sophisticated than the one proposed by Wallas (1926) as it not only distinguishes between the behaviours that occur in creative problem solving but also is concerned with the thought processes involved at each stage.

Finally, we would like to refer to the Amabile's (1983) work. She suggests that while innovation begins with creative ideas, creativity by individuals and teams is a starting point for innovation: the first is a necessity but not sufficient condition for the second. This idea led her to develop her five-stage model and to the identification of key components of creativity at certain stages of the creative process. The first stage is problem or task presentation; the second stage concerns preparation and at this moment the creative person develops or reactivates a data store/background relevant to the problem or opportunity identified. The stage that follows is response generation, and here the individual comes up with a set of possible ideas to the issue in question. The fourth stage is response validation and it refers to the process through which new ideas are verified and validated. The last stage concerns the assessment of the outcome based on tests performed in the previous stage.

Due to the importance of this topic we decided to prepare a set of papers dealing precisely with it. In this section of the book, readers will find studies concerning the identification of practices leading to creativity in high technology industries, an example of how to integrate technology in the creative design process and how development teams learn when they carry on innovation. Aware that there are factors that enable or prevent creativity and innovation, we also decided to include some studies dealing with issues such as the personality traits of individuals and their motivation to contribute to novel ideas and solutions; the role of social relationships and the existence of methods that go against methods to promote creativity and innovation.

**Chapter 1**, *A Benchmarking Study on Organizational Creativity Practices in High Technology Industries,* by Fernando Sousa and Ileana Monteiro, aims to provide a benchmarking list of initiatives that deal with the development of corporate or organizational creativity and innovation in the emerging sectors of bio-technology, nano-technologies, information and communication technologies, and eco-innovation, together with companies of other sectors, perceived as good examples of organizational innovation. Twenty-two interviews made by telephone addressing specific strategies in three domains: creative management, creative people management, and creativity management were conducted to top management in these organizations. Results indicate that high technology organizations, dependent upon financial support, do not seem to concentrate on corporate innovation, and do not have alternatives to budget reductions made in R&D, due to the present financial crisis, in order to innovate. The remaining companies provided several suggestions as to the way corporate innovation systems can be built and sustained within the framework of the future European innovation policies, devoted to workforce development, the service sector and the SMEs.

**Chapter 2**, *Integrating Technology with the Creative Design Process,* by Joshua Fairchild, Scott Cassidy, Liliya Cushenbery and Samuel T. Hunter, examines how specific technologies may influence performance at each stage of the creative process, and provides specific recommendations for how technology may be used to facilitate the development of creative solutions. This research is justified by the fact that in our fast-paced world, organizations must continually innovate in order to stay competitive. At the same time, technology is continually advancing, and tools to facilitate work are frequently changing. This forces organizations to stay abreast of current technologies, and also puts pressure on employees to use the technologies available to them in order to devise innovative solutions that further accomplish the organization's goals. To date, there has been little research on how such technologies may best be used to facilitate such creative performance. The present chapter addresses this gap by integrating a model of the creative process from the psychology literature with technology literature from engineering and information technology.

**Chapter 3**, *Problem-Solving Style, Problem Complexity and Knowledge Generation: How Product Development Teams Learn When They Carry on Innovation,* by Corrado lo Storto, presents the findings of a study aimed at investigating how the fit between the problem-solving style of a product development team and the cognitive environment induced by the perceived problem complexity affects the amount and type of knowledge generated. It is assumed that organizational knowledge is created as a by-product of collective creative technical problem-solving, when people work together to deal with unfamiliar and unexpected situations. Two major outcomes emerge from the analysis of experimental data: (1) different cognitive environment patterns are more conducive than others to organizational learning; (2) there is some fit between the cognitive environment pattern and the team technical problem-solving style, as some cognitive practices and social behaviours adopted during technical problem-solving are more effective than others in certain cognitive environments. Particularly, practices and behaviours that are more associated to creativity have a not negligible weight in the generation of knowledge. Ninety-one cases of technical problem-solving occurred during product innovation within the 35 small firms were studied.

**Chapter 4**, *Towards a Theoretical Framework for Creative Participation: How Personal Characteristics Influence Employees' Willingness to Contribute Ideas,* by Natalya Sergeeva and Milan Radosavljevic, investigates some recent developments and proposes a conceptual framework for creative participation as a personality-driven interface between creativity and innovation. To retain competitiveness, succeed, and flourish, organizations are forced to continuously innovate. This drive for innovation is not solely limited to product/process innovation but more profoundly relates to a continuous process of improving how organizations work internally, requiring a constant stream of ideas and suggestions from motivated employees. Under the assumption that employees' intrinsic willingness to contribute novel ideas and solutions requires a set of personal characteristics and necessary skills that might well be unique to each organizational unit. The chapter explores personal characteristics associated with creativity, innovation and innovative behavior. Various studies on the correlation between creativity and personality types are also reviewed. The chapter provides a discussion of solutions and future development together with recommendations for the future research.

**Chapter 5**, *Cultivating Innovation through Social Relationships: A Qualitative Study of Outstanding Australian Innovators in Science and Technology and the Creative Industries,* by Ruth Bridgstock, Dr Shane Dawson and Professor Greg Hearn, describes and explores social relationship patterns associated with outstanding innovation. In doing so, the authors draw upon the findings of 16 in-depth interviews with award-winning Australian innovators from science & technology and the creative industries. The interviews covered topics relating to various influences on individual innovation capacity and career development. Authors found that for all of the participants, innovation was a highly social process. Although each had been recognised individually for their innovative success, none worked in isolation. The ability to generate innovative outcomes was grounded in certain types of interaction and collaboration. Authors outline the distinctive features of the social relationships which seem to be important to innovation, and ask which 'social network capabilities' might underlie the ability to create an optimal pattern of interpersonal relationships. Authors discuss the implications of these findings for universities, which we argue play a key role in the development of nascent innovators.

**Chapter 6**, *Methods against Methods,* by Marc Stierand and Viktor Dörfler, intends to clarify some issues about creativity and innovation methods, since it is believed that the term is often misunderstood. For the authors, neither creativity nor innovation is guided by a method. There are only methods against methods that can help the extraordinary individual to step faster and easier into a state of mind that is conducive to creativity, but which has no effect on whether the creative output becomes an innovation. In order to support this claim, the authors outline three major reasons that seem to be responsible for making people believe that such methods for creativity and innovation exist. Then authors present their understanding of creativity and continue with a discussion on the systemic character of creativity and innovation. Finally, the authors show that there are no methods for creativity, but methods against non-creativity by explaining in particular how one of these methods against non-creativity works. What the authors outline here is a necessarily one-sided and partial view. Their aim is not to convince the readers that they are right but to make them think by presenting one possible consistent approach.

## SECTION 2: TECHNIQUES FOR CREATIVITY

All creativity processes start with a problem that requires a solution, which means that the first task is to correctly define the problem. The accuracy of this problem definition will affect the quantity and quality of ideas generated in the subsequent stages. The next step is generating as many ideas as possible because one assumes that the probability of coming up with a good idea increases with the number of ideas. The question now is to know if there are techniques that can help the individual to solve a problem or to increase the number of ideas generated. In this part of the book readers will find three studies that bring some light to this discussion.

**Chapter 7**, *Methods to Improve Creativity and Innovation: The Effectiveness of Creative Problem Solving,* by Fernando Sousa, Ileana Monteiro and René Pellissier, focuses on the development of organizational creativity, using the CPS methodology, aiming at demonstrating its effectiveness in using the individual and team divergent thinking improvement in identifying organizational problems. A study was undertaken using problem solving teams in seven companies, in which each individual was submitted to a pre-post test in attitudes towards divergent thinking and asked to express the evaluation of the method. All the information reported in the sessions was recorded. The results indicate a change in attitude favourable to divergent thinking, the provision of a professional, efficient method of organizing knowledge in such a way that can help individuals find original solutions to problems, and an important way to lead teams to creativity and innovation, according to companies' different orientations.

**Chapter 8**, *An Interdisciplinary Workshop for Business-Idea Generation,* by Astrid Lange, describes a workshop concept for small groups that aims at the qualification for creative business-idea generation in interdisciplinary teams of academics. The chapter's aim is to provide a theory-based and application-oriented description of the workshop, including a first-hand report on implementation. The chapter starts with a description of theoretical contributions and underlying research results to illustrate the workshop's framework. Next is a description of the aims, methods and target group follows, as well as the organizational settings and training methods. Then the chapter will focus on the realization of the workshop. Finally, authors discuss the workshop concept's scope of application.

**Chapter 9**, *The Structure of Idea Generation Techniques: Three Rules for Generating Goal-Oriented Ideas,* by Stefan Werner Knoll and Graham Horton, discusses the important role played by idea generation techniques in the innovation process. Until recently, the space of techniques has been unstructured, and no clear guidelines have been available for the selection of an appropriate technique for a given innovation goal. Authors use an engineering approach to study and develop idea generation techniques with the aim of obtaining more structured and rigorous guidelines for generating ideas. One objective of this approach was to identify and understand the fundamental mental principles underlying an idea generation technique. In this chapter, the authors show how three such principles suffice to cover a large range of published idea generation techniques and suggest how these can be used to improve the utility of idea generation within the innovation process.

## SECTION 3: TOOLS FOR CREATIVITY

In creativity and innovation, although the individual is important, usually he / she is part of a team which supports, nurtures and often delivers his / her vision and goals. The importance of the team leads us to organize a section in this book dealing with the tools to promote creativity but used in the context of a group. Here, readers will find studies concerning the use of brainstorming supported by a system for problem solving, the development of a tool to assess the organizational culture towards learning (the support of creativity and innovation) and the use of technology in knowledge management

**Chapter 10**, *Experience with Self-Guiding Group Support Systems for Creative Problem Solving Tasks,* by Gwendolyn L. Kolfschoten and Calvin Lee, explores the use of brainstorming to improve creativity. Brainstorming can be supported with Group Support Systems (GSS). However, GSS are most successful when offered in combination with facilitation, or at least training. Unfortunately, facilitation or training will impose a barrier to use such systems. In this chapter, authors evaluated the use of a GSS for a multi-step, creative, problem solving task. The groups using this GSS got no training, had no GSS experience and got no support, other than a 1 page log-in instruction. With this limited instruction and no training all participating groups handed in a report with the results of their brainstorm, using the tool. In this chapter, authors will report the process, the way it is embedded in the tool, and will report the results of our exploratory questionnaire among the participants.

**Chapter 11**, *The OLC Questionnaire: A Measure to Assess an Organization's Cultural Orientation towards Learning,* by Teresa Rebelo, and A. Duarte Gomes, is centred on the psychometric qualities of the OLC questionnaire, which has the objective of measuring the orientation of organizational culture towards learning – a kind of culture that promotes creativity and innovation in organizations. Hence, it includes description and discussion of its conception, assessment of content validity, and the main construct validity studies already carried out. Its bi-dimensionality in terms of internal integration and external adaptation processes and its potentialities for research and intervention are also discussed, as well as future research directions to continue its journey of validation.

**Chapter 12**, *Knowledge Management and Innovation: Mapping the Use of Technology in Organizations,* by Leonor Cardoso and A. Duarte Gomes, explores the role of technology in organizational knowledge management, inasmuch as it provides new forms for holding and exchanging information and knowledge in intra and inter-organizational contexts. The potential of technological means is therefore emphasized as tools supporting the various organizational processes related to knowledge, and questions arise concerning their comparative or relative relevance for increased innovation and creativity. In a sample of 1275 individuals belonging to 50 Portuguese organizations, the use of technology plays an important part in terms of the organizational processes related to knowledge management, but this is limited above all to those which are formally instituted and based on knowledge of a mainly explicit nature. In addition, this study highlights the importance of management of organizational processes related to tacit knowledge, which emerges essentially from processes of social and discursive interaction involving organizational actors.

## SECTION 4: BEST PRACTICES

In this section we present three examples of best practice on how to use technology to promote creativity and innovation. The first one presents the use of digital storytelling in primary schools classrooms, the second one deals with the use of technology as a creative management tool at the BBC. The last chapter describes a b-learning case in a Higher Education Institution.

**Chapter 13**, *Exploring Alternative Assessments to Support Digital Storytelling for Creative Thinking in Primary School Classrooms,* by Lee Yong Tay, Siew Khiaw Lim and Cher Ping Lim, documents the use of digital storytelling as a teaching approach to facilitate the learning of creative thinking among students (aged 7 and 8) in a primary school setting. A constructive teaching approach is adopted to allow students to create their own digital stories based on an authentic experience and expression of their thoughts. The focus of this chapter is to show how a shift from traditional classroom assessment to more flexible, alternative assessment format facilitates higher level thought processes (e.g., creative thinking) and range of skills. Several issues and challenges related to the use of alternative assessment in digital storytelling are explored and discussed. Findings suggest that digital storytelling may be effectively used as an approach to foster creative thinking. The authors also suggest that refinements to the assessment process are needed to make it more formative in nature.

**Chapter 14**, *Creative Management, Technology and the BBC,* by Nicholas Nicoli, argues that the past decade has witnessed increased attempts by managers, scholars and policy-makers to stimulate the creativity of organisations. The practice of stimulating organisational creativity has led to a paradigm shift known as creative management, the focus of which is to use these practices to achieve competitive advantages. Such creative stimulation can come in a variety of forms. These include identifying and influencing environmental conditions that can increase the chances of creating new and significant products or services. In order to stimulate creativity, current creative management literature proposes the use of technology as a disseminator of knowledge and ideas. This chapter offers a literature review of creative management and technology use for creativity. It next introduces a case study of how technology is used as a creative management tool at the BBC. The findings of the study indicate that although the BBC's yearly revenues are under sustained pressure, the organisation has invested heavily in technology in order to maintain its high creative standing. In conclusion, supported by the findings of the case study, this chapter corroborates and further advocates the use of technology as a significant component of creative management practices.

**Chapter 15**, *Sustainable Blended-Learning in HEI: Developing and Implementing Multi-Level Interventions,* by Paula Peres, Sandra Ribeiro, Célia Tavares, Luciana Oliveira and Manuel Silva, aims to demonstrate how PAOL - Unit for Innovation in Education, a project from ISCAP - School of Accounting and Administration of Oporto - Institute Polytechnic of Oporto, Portugal - prompted new educational initiatives and new learning scenarios at a Higher Education Institution. Furthermore, authors intend to demonstrate PAOL's lines of intervention through an extensive analysis based on the 6 years of experience that this unit has in the educational technology field; a project that began small but that, due to the force of innovation, has progressively conquered new adepts. Therefore the unit described in this chapter relates all these factors, as a whole, capable of attaining changes that influence mentalities and methodologies, overcoming cultural and technical barriers. This case study can serve as a catalyst, potentiating the creation of new multi-faceted projects in the scope of Web technologies in higher education teaching-learning processes.

## SECTION 5: CONCLUSION

It is very difficult to forsee the future. Anyway, we decided to present here a perspective of two authors on how the future will look in the field of creativity.

**Chapter 16,** *The Opportunities and Challenges of Technology Driven Creative Collaborations,* by Diego E. Uribe Larach and John F. Cabra, elaborates on some trends that will govern creativity and innovation. Authors argue that the onset of the 21st century is marked by deep psychological and sociological transformations affecting every scale of human endeavour, ranging from individual to crowd behaviour. Deep and central to these transformations is the penetration of digital communication and computer technology into modern day life. Above all, this new and evolving technological landscape has opened exciting new possibilities to drive creative behaviour, organizational creativity and innovation through computer-mediated interactions. Such opportunities are met with equal challenges that need to be addressed in order to harness the full potential of massively distributed creative collaborations. This chapter will elaborate on the underlying trends that give rise to these opportunities and challenges and to what extent these trends will govern creativity and innovation in areas of organizational life such as business, education, science and design in the next 10 to 30 years.

A final word just to say that we hope that the reader enjoys reading this book as much as we enjoyed preparing it.

## REFERENCES

Amabile, T. M. (1983).*The social psychology of creativity*. New York: Springer-Verlag.

Andriopoulos, C. & Dawson, P. (2009). *Managing change, creativity and innovation*. London: Sage Pubs Ltd.

Basadur M. & Graen, G.B. (1982), Training in creative problem solving: Effects on ideation and problem finding in an applied research organization. *Journal of Applied Psychology, 71*, 612-617.

Chandler, R. & Grzyb, J. E. (2005). *Creativity and innovation – a view from impact factory.* Retrieved from http://www.impactfactory.com/gate/free/creativityandinnovation.pdf

Darso, L. (2008). *Creativity and innovation. Francesca Pagliuca interviews Lotte Darso.* Retrieved from http://uninews.unicredit.it/en/articles/page.php?id=9284

European Parliament (2008). *2009 to be designated European Year of Creativity and Innovation – press release*. Retrieved from http://www.europarl.europa.eu/news/expert/infopress_page/037-37791-266-09-39-906-20080922IPR37790-22-09-2008-2008-false/default_en.htm

Wallas, G. (1926). *Art of thought*.

# Acknowledgment

As Editor, I would like to acknowledge the help of those involved in the collation and review of this book, without whose unstinting support the project could have not been completed. A further special note goes to all the staff of IGI Global whose contribution throughout the whole process has been invaluable. Thanks to all those who provided constructive and comprehensive reviews.

Special thanks go to Mario Spiler for having accepted the invitation to write the foreword of this book and to Alexandra Albuquerque for her valuable comments and for editing the editor's contribution.

Thank you all.

*Anabela Mesquita*

Section 1

# The Process of Creativity and Innovation

# Chapter 1
# A Benchmarking Study on Organizational Creativity Practices in High Technology Industries[1]

**Fernando Sousa**
*INUAF, CIEO, Portugal*

**Ileana Monteiro**
*University of Algarve, CIEO, Portugal*

## ABSTRACT

*The aim of this chapter is to provide a benchmarking list of initiatives that deal with the development of corporate or organizational creativity and innovation in the emerging sectors of bio-technology, nano-technologies, information and communication technologies, and eco-innovation, together with companies of other sectors, perceived as good examples of organizational innovation.*

*Twenty two interviews were conducted with top management in these organizations. The interviews were made by telephone addressing specific strategies in three domains: creative management, creative people management, and creativity management. Results indicate that high technology organizations, dependent upon financial support, do not seem to concentrate on corporate innovation, and do not have alternatives to budget reductions made in R&D, due to the present financial crisis, in order to innovate. The remaining companies provided several suggestions as to the way corporate innovation systems can be built and sustained within the framework of the future European innovation policies, devoted to workforce development, the service sector and the SMEs.*

DOI: 10.4018/978-1-60960-519-3.ch001

## INTRODUCTION

The idea for this chapter followed a report from a contract with the *Gers Chamber of Commerce*, as partner of the European project (Interreg IV B SUDOE) "CREA BUSINESS IDEA," linked to the belief that high technology institutions, in specialized countries or regions, researching for sophisticated products, were likely to have sound policies and practices in order to lure and incentivise the best talent to produce the best products.

Thus, the initial objective of this study was to produce a list of organizational creativity best practices, drawn from the above examples, and to identify the required skills to adapt these best practices to the existing SMEs. This was done, firstly, by analysing current practices of organizations, from the technology and emerging sectors of biotechnology and bio-medicine, nano-technologies, information and communication technologies, eco-innovation, and the Irish "soft landing" policy.

As the interviews progressed, this perspective proved to be wrong, as it became clear that leading (with respect to technology) industries, laboratories and universities were highly dependent upon R&D financing and did not possess alternatives to the lack of funding, due to the present crisis, and the subsequent reduction of state and private support to research. These industries, especially those supported by public funds, did not seem to be able to develop practices of corporate innovation, so that we could learn from them. We, then, decided to take the benchmarking study to companies that had been recommended as good examples of corporate creativity and innovation, no matter the sector in which they operated.

The suggestions were made by experienced consultants and academics related with EACI – European Association for Creativity and Innovation – with whom we are connect through the association we represent. Thus, the objective remained connected to corporate creativity benchmarking but aimed at corporate good examples, irrespectively of the sector or activity in which companies operated.

The striking example was the way Ireland changed its "soft landing" policy of providing financial support to attract talented people to work in Ireland, into a policy aimed at a workforce development orientation, taking advantage of the existing personnel. This example also made us turn to the analysis of the European politics of innovation, as its development was closely linked with our research.

## THE EU APPROACH TO INNOVATION

Innovative performance and evaluation are extensively reported in various EU and other institutional documents, which provide several analyses of international examples (e.g. *Étude sur les bonnes pratiques de dix pôles de compétitivité étrangers*, from DGCIS, 2009; *Assessing Community innovation policies in the period 2005-2009*, from EU Commission, 2009; *Les clusters américains,* from DGE, 2008*; Best practices in innovation policies,* from Tekes Institute, 2005; *European innovation scoreboard, 2009*, from EU Commission, 2010).

The 3$^{rd}$ edition of the Oslo Manual has included considerations about other types of innovation besides product and process, namely marketing and organizational innovation. Nevertheless this last definition (*the implementation of a new organizational method in the organization's business practices, workplace organization or external relations)* is still far from allowing a quantitative analysis of data, thus making it difficult to gain further insights leading to improved success rates. One of the reasons seems to be the wide spectrum of what might be designated as an example of "organizational innovation." Also, creativity appears connected with arts or creative industries, namely with design, or other indexes, as in the works of researchers like Richard Florida (Florida & Tinagli, 2004) the quality of the educational system, the desire of people to express themselves (artistically), or the openness of a society towards different countries and cultures. Therefore, measures have to rely on the so-called *proxy indicators*, which only indirectly measure creativity, thereby creating possible errors in measuring 'true' performance.

According with Arundel, Bordoy and Kanerva (2007), R&D is not the only method of innovating. Other methods include technology adoption, incremental changes, imitation, and combining existing knowledge in new ways. With the possible exception of technology adoption, all of these methods require creative effort on the part of the organization's employees and consequently will develop the organization's in-house innovative capabilities. These capabilities are likely to lead to productivity improvements, improved competitiveness, and to new or improved products and processes that could have wider impacts on the economy.

As present policies are based mainly on funding strategies, which cannot cope well with the financial crisis, the EU overall Lisbon Strategy, post-2010, and those of other countries, are changing from technology-based priorities to encompass other priorities, especially service innovation (see the *European Services Innovation Memorandum*) and workforce development (e.g. *Reviewing Community innovation policy in a changing world*, from EU Commission, 2009; *Challenges for EU support to innovation in services – Fostering new markets and jobs through innovation*, from EU Commission, 2009). The case of Ireland reflects this tendency, especially with respect to workforce development (see *Innovation in Ireland*, from the Department of Enterprise, Trade & Employment, 2007; and *Irish Workplaces - A strategy for change, innovation and partnership 2007–2010*, from The National Centre for Partnership and Performance, 2007). Ireland has created a €6 million fund to support workplace innovation, in order to develop the role of employee participation and workplace partnership in the SMEs (see *Workplace Innovation Fund*, from the NCCP, with information available in the National Workplace Strategy Website). To them, the term "workplace innovation" means the adoption of new workplace practices, structures and relationships, exactly as we designate "corporate" or "organizational" creativity and innovation.

According with Commins (2004), the Irish approach comes from previous EU programs, where some significant innovation management initiatives for SMEs have been delivered, namely:

- MINT – (Innovation Management Tools, 1996)
- TEMAGUIDE (Technology Management and Innovation in Organizations, 1998)
- TIPPS (Transnational Innovation Pilot, 1998)

The MINT program concerned the tools and methodologies used by consultants and advisers working with SMEs to assist them in managing innovation and involved 17 countries; TEMAGUIDE dealt with Innovation Management and the methodologies and tools of technology management as an important sub-set; A further programme was the Transnational Innovation Pilot Programmes (TIPPS) – An approach to continuous improvement. Major findings of the MINT programme mention aspects like the fact that *innovation management is about people issues, culture, communication and organisation and business process issues as much as technology*; and that *the more powerful mechanism for SMEs to learn is to do it from their own sector which enhances the importance of using examples of good practice to improve their own innovative performance*.

The designation of 2009 as the European Year of Creativity and Innovation seems to have contributed to a wider perception of innovation. The 2009 Scoreboard, for example, specifies as to the "neglected indicators" that *(...) the activities of organizations that innovate without performing R&D are of interest to policy.*

It follows that a new approach to innovation is needed and should include creativity and innovation together and connected with business orientation, as in the Irish example, separating it from artistic, educational or social tolerance aspects of creativity, as discussed in the following sections from a theoretical perspective. Nevertheless, an interaction between art and business, as a means to fulfil the integration of education, technology, research, business, entrepreneurship, creativity and innovation, can be made. A new culture of entrepreneurial education and innovation is needed and art can play an important role in facilitating creativity and innovation. As in the words of Koïtchiro Matsuura, Director-General of UNESCO, *creativity, imagination and the ability to adapt competencies which are developed through the Arts education are as important as the technological and scientific skills.*

Even though the discussion expressed in the European documents stress the need to go beyond R&D innovation, it seems important to discuss the concepts of creativity and innovation indicated in the Oslo Manual, as well as its connections to what seems to be mandatory to its development – the links between the workforce and the organization, or organizational commitment.

## CREATIVITY, INNOVATION AND COMMITMENT

While innovation concerns the processes of implementation, relying mainly on organizational communication and power (specifically with regards to production, adoption, implementation, diffusion, or commercialisation of ideas) (Spence, 1994), creativity remains exclusive to the relationship established between the creator and his product, where not even originality and usefulness are important, but only the notion of "trying to do better." The latter is connected to cognitive and emotional processes taking place at the individual level (Sousa, 2008).

If we relate creativity to problem definition, and innovation to decision implementation, the latter requires a series of problem definitions, in order to carry out a decision or an idea, thereby making it difficult to separate these concepts at an organizational level. In fact, when we move from the individual level to team and organizational levels, creativity and innovation become more and more difficult to separate, so that we must agree with Basadur (1997), when he says there is no difference between organizational (corporate) creativity and innovation. Therefore, the moment we move to other levels besides the individual, we will use these terms (creativity and innovation) as synonyms, and we refer to organizational creativity as *system devoted to enhance creativity in organizations,* thus using the definition proposed by Basadur (1997), not at all within the logic of the Oslo Manual.

As to the several approaches to identify types of innovation, either by separating the adoption of products and processes from their development (Cebon, Newton & Noble, 1999) or, in a more classical way, product and process innovation (Adams, 2006), most authors agree that innovativeness (or organizational innovation), is a third important type of innovation, which represents *the potential of the workforce to promote changes to the benefit of the organization*. It is only through developing and sustaining a creative workforce that the organization will succeed in maintaining the necessary potential to overcome difficult problems and situations that cannot be solved through investments in the adoption of technology or R&D financing, only (Cebon, Newton & Noble, 1999).

This creative workforce potential is both the ability to retain creative managers and employees (McAdam, 2006) and to provide an environment where each one feels free and capable of contributing to organizational success. Aspects like increasing job complexity, employee empowerment and time demands, together with low organizational controls (decision making, information flow and reward systems), are said to enhance employee creativity (Adams, 2006). However, more elements are necessary in order to make people willing and able to contribute to organizational effectiveness. For instance, supportive leadership (aka sponsorship), knowledge acquisition, and teamwork procedures favouring creativity can ensure success (Unsworth, 2005). Creative people (either managers or employees), are committed to their work and the organization, and thus may identify important issues, provided that top management values their work and commits resources to implement their ideas. In fact, according to a Gallup Management Journal (GMJ) survey (Harter, Schmidt & Keyes, 2003), engaged employees are more likely to "think outside of the box" and produce creative ideas than disengaged people; they also are more receptive to new ideas. Their research concludes that engaged people tend to find and suggest new ways to improve their work and business processes, which may lead to the assumption that creative people have a deeper understanding of the organizational processes, through being in a privileged position to identify, define and find organizational problems.

To a certain extent, all of this can be achieved by elevating the importance of creativity in the organization and by providing a system through which individual potential may be channelled into profitable innovation. What is required is the freedom to create content and process skills to be able to engender a supportive human environment (peers and team leader), as Rodrigues, Franco & Santos (2006) mention. Nevertheless, the issues surrounding the potential of an organization to innovate are still in its infancy, although Mclean (2005) and Puccio, Firestien, Coyle & Masucci (2006), and especially Basadur (1997, 2000), did some empirical research. The major challenges are to define criteria to evaluate the impact of organizational innovation on process and product innovation (Wolfe, 1994).

Let us see, then, what the literature brings us in order to raise company creativity and transform it into profitable innovation.

## BENCHMARKING ON CREATIVITY AND INNOVATION

Management conditions favouring creativity are extensively reported in the literature (Monteiro & Sousa, 2008; Sousa, 2004; Sousa & Andrade, 2007) and have to do mainly with the processes described in the methodology. As to the transformation of creativity into profitable innovation, it was only possible to find a few studies describing this, and so we tried to benchmark already listed practices, only to find that they were not reported.

According to Barker (2003), benchmarking is *the process of identifying the best practice in relation to products and processes, both within an industry and outside it, with the object of using this as a guide and reference point for improving the practice of one's own organization*. Benchmarking can take place within an organization, in relation to direct competitors, or in relation to organizations in totally different fields. According to Bandow (2009), two types of benchmarking can be distinguished: general benchmarking and best practice benchmarking. While the first one stands for the mere comparison of an organization's key data or management ratios in order to identify areas for improvement, the second one represents the comparison of respective best performers in order to learn and adopt their best practices. The latter is the intention of this study.

Several guides and indexes in innovation benchmarking are available in different sources, especially under a quantitative designation (e.g. Atkinson & Andes, 2009). As to innovation, these indexes are important to define criteria and ratings for the chosen units. However, as to creativity, the quantitative approach does not apply because the aim of this study is to describe processes to develop and evaluate creativity, not end results.

What we found was that existent indexes of creativity were based mainly on Richard Florida's (Florida & Tinagli, 2004) conceptions, which do not apply to this investigation. We searched for everything we could find about organizational creativity, to discover that the best sources were the web sites of many companies specialized in this field, together with a few books and articles.

Robinson & Stern (1998), for example, provided us with one of the best works about our subject, which we think should be considered a classic, just like Schumpeter (1934), Kanter (1983), Peters & Waterman (1982), or Collins & Porras (1994). The authors begin by clarifying that a company can be called creative when its employees do something new and potentially useful, without being directly shown or taught, and that corporate creativity systems began in the late XIX century. According to these authors, the Scottish shipbuilder William Denny was responsible for the first suggestion system, back in 1883, followed by companies like Kodak, NCR and Ford, and even Lenine's *rationalization proposal system*. Nevertheless, the one that they consider really representative of the passive idea suggestion system, was the Japanese *Keisen Teian*, or *continuous improvement proposal*. This system began after WW II, when Japan was almost destroyed and subjected to US Forces occupation. In fact, a series of coincidences favoured the success of the system (e.g., poor supervision due the lack of senior managers, an apprentice-master way of training, the need for everybody to participate in the rebuilding of the country, and upper management posts being occupied by young people), but the main reason seems to have been the adoption of the existent US Air Force's *Training Within Industries (TWI)* system. The Air Force had hired Paul Torrance to study survival competencies for the pilots, and he concluded that creativity was a critical component of the solution.

The authors consider that the basic elements of corporate creativity are *alignment, self-initiated creativity, unofficial creativity, serendipity, diverse stimuli, and within-company communication.* Considering the alignment between management objectives and employees' interests and actions, Robinson & Stern (1998) report the story of the Soviet system, which once had the biggest concentration of skilled engineers in the world; (even at the Politburo, 80% of its members were engineers). In fact, the Lenine's rationalization proposal institutionalized quotas for ideas, which led managers and employees to duplicate unworkable and silly ideas just to satisfy the audits, and managers to steal ideas from their employees, in order to reach the quotas. These ideas had to be technical, or they would not even be submitted. The system collapsed in 1940, when more than half of the workforce had earned the title of "Stakanovite" (those who had exceeded the idea quota in 30%), which provided special rights and material advantages.

As in the Soviet system, the failure of large organizations to innovate in the U.S. was the widening of the communication gap between management and employees, and the reward system aggregated to idea presentation. The authors report data showing that U.S. companies rewarded each idea, on average, a hundred times more than Japan did, while the net savings per employee suggestion, till 1995, was five hundred times smaller in the US. Shapiro (2001) also agrees that U.S. reward systems first contributed to the fall of the idea suggestion systems, by reducing employee commitment and the alignment with company objectives, together with automation, routinization, and mechanization of work. Shapiro states that *companies seem to have forgotten that innovation is carried out by people and for people.* As success comes from an uncompromising commitment to the organization's people, companies that cannot transform themselves into collaborative organizations will suffer in today's innovation economy. As in the words of Kao (1989), established companies must investigate sophisticated intrinsic or non cash rewards to stimulate internal entrepreneurship. Within this line of thinking, it is interesting to report the strategic statement that Collins & Porras (1994) quote from Johnson &

Johnson, *(...) we are in the business of preserving and improving human life. All of our actions must be measured by our success in achieving this goal*. "Human" meaning, by order of priority: costumers, employees, management, communities and stockholders.

The *Handbook of organizational creativity* (Zhou & Shalley, 2008), together with works from other authors (Hage, 1999; Hagardon, 1999; 2000; Lam, 2004) reports reasons for employees not to be creative, and possible solutions that management may bring to improve their creativity, either individually (Shalley, 2008) as in group (Paulus, 2008). Also, Perry-Smith (2008) debates whether communities of practice are privileged means to have groups addressing specific kinds of problems on a permanent basis, as well as the importance of links between groups. These links are extensively discussed by Sawyer (2007), who develops all possible aspects of collaboration between groups, and analyses the most creative webs as the ones in which good connections exist among the teams, but in which the teams still enjoy independence and autonomy.

All these assumptions were expected to arise among the innovative industries present in this study, in order to understand how the above concepts work in reality. So this is what we expected from the technology sector - a list of organizational creativity best practices, in order to adapt the required skills to implement these best practices for training existing SMEs. These were the objectives of this study.

## RESEARCH METHODOLOGY

According to information reporting on the most innovative technological sectors and countries, the Gers Chamber of Commerce provided indications (Table 1) as to which kind of organizations should

*Table 1. Cases indicated to be reported on benchmarking in creative practices by Gers Chamber of Commerce*

| | Bio-tech and bio-medicine | Nano-technologies | ICT | Attractivity of creative people and businesses | Eco-innovation |
|---|---|---|---|---|---|
| **USA** | Biotechnologies in San José and North Carolina. | | | | |
| **Scandinavia** | | Sweden: University as place for creativity: case of the Lund University and its research in nano technologies. | Finland: Big organizations triggering creativity: NOKIA and TIC clusters. | | Sweden: Eco-Construction, éco-urbanism and sustainable management of energy. Western Harbour, city of Malmo Bo01. |
| **Israel** | Best practices in the innovation funding in biotech. | Public private cooperation to back the nano-technology sector. | | | |
| **Ireland** | | | | Attracting creative people: soft landing. The Irish strategy | |
| **Japan** | | Last evolutions in RFID | Ubiquitous society: link with local development. | | Sustainable energy management Example: Program « 0 waste » City of Kamikatsu |

be included in the research. The technological sectors were the following:

## Bio-Technology and Bio-Medicine

Biotechnology is technology based on biology, agriculture, food science, and medicine. Biotechnology has applications in four major industrial areas, including: health care (medical), crop production and agriculture, non food (industrial) uses of crops and other products (e.g. biodegradable plastics, vegetable oil, biofuels), and environmental uses. Biomedicine, also known as theoretical medicine, is a term that comprises the knowledge and research which is more or less in common with the field of human medicine, veterinary medicine, odontology and fundamental biosciences such as biochemistry, chemistry, biology, histology, genetics, embryology, anatomy, physiology, pathology, biomedical engineering, zoology, botanic and microbiology.

## Nanotechnology

Nanotechnology, shortened to "nanotech", is the study of the control of matter on an atomic and molecular scale. Generally nanotechnology deals with structures with the size of 100 nanometers or smaller, and involves developing materials or devices within that size. One nanometer (nm) is one billionth, or $10^{-9}$, of a meter. Nanotechnology has the potential to create many new materials and devices with wide-ranging applications, such as in medicine, electronics, and energy production. Its wide range applications are in medicine (diagnostics, drug delivery, tissue engineering), chemistry and environment, energy, information and communication, heavy industry (aerospace, construction, refineries, vehicle manufacturers), consumer goods (foods, nano-foods, household, optics, textiles, cosmetics, agriculture).

## TIC: Information and Communications Technology

ICT (information and communications technology - or technologies) is an umbrella term that includes any communication device or application, encompassing: radio, television, cellular phones, computer and network hardware and software, satellite systems and so on, as well as the various services and applications associated with them, such as videoconferencing and distance learning.

## Eco-Innovation

Eco-innovation is the production, assimilation or exploitation of a product, production, process, service, management or business method that is novel to the organization (developing or adopting it) and which results, throughout its life cycle, in a reduction of environmental risk, pollution and other negative impacts of resources use (including energy use), compared with relevant alternatives. Eco-innovation is a broad concept and is reflected in a large number of technical fields. One important example of a technology which is pivotal for tackling climate change is the generation and transformation of energy. Examples: renewable energy, water and waste management, motor vehicle abatement technologies. Two interviews were conducted: the western harbour of Malmo and the city of Kamikatzu.

In each cell attempts were made to meet the table requirements and the purpose of the report, including previous case study, establishing contacts, interviews and report writing. The approach was personalized, following personal contacts with institutions related with science, industry and innovation (e.g. Company Association for Innovation – COTEC; Innovation Europe Network; Technological Innovation Office/National Science and Technology Foundation – FCT; Innovation Agency - ADI); laboratories and universities; e-mail addresses taken from the internet (e.g. organizations, national technological or science institutions, professional and industrial associations), and e-mails or telephone contacts given by colleagues in the field.

More than 100 personalized messages were sent and 32 responses were received. Of the 32, twenty two telephone interviews were conducted, lasting for an average of 45 min. From the 15 questionnaires sent by e-mail, after prior agreement, none of the subjects replied.

The interviews were conducted by members of the research team to top management subjects or scientists in the target organization, addressing specific strategies in three domains: creative management (leader selection, orientation and training in order to bring creative contributions out of teams and individuals), creative people management (general and specific orientations as to hiring, training and retaining highly creative employees), and creativity management (existent systems and conditions for team work and the transformation of individual and team creativity into profitable corporate innovation).

## RESULTS

For each section a general description of the sector and country-sector will be made, followed by the results obtained from the interviews and related documents.

Except for Ireland, it was not possible to fully analyse each cell, according with the project's requirements. Nevertheless, all sectors were covered. Later it was agreed that, besides the deadline extension, the cells were no longer mandatory and all efforts should be made to bring in more relevant cases. Due to participation in conferences, it was possible to obtain a series of contacts, resulting in ten additional interviews. The report now includes 21 cases, instead of the originally required ten.

## Bio-Technology and Bio-Medicine

Four cases are described: North Carolina Biotechnology Center, Bühler factories, Unilever (Port Sunlight Research Center, UK) and Pfizer.

The U.S. research center is a private non-profit state- funded organization, devoted to support research in organizations and universities. The impact of the economic crisis is causing serious problems and reductions in staff are imminent.

Bühler, situated in Zurich, is a global technology partner for the food industry, chemical processing and die casting. As to the management of new ideas, idea generation and idea development require different team compositions. The client presents the need and the teams produce the solution. Besides applied research, there is a small group of 20 scientists who conduct fundamental research, based on top management objectives. The cycle is: initial brainstorming - database search - hypothesis suggestion - team gathering to staff the development program – management submission and budgeting – periodical (36-month) reports to management to gain more resources, in 4-5 year projects. As to creativity skills and competencies, the company president has been the main source of ideas and research support and the company believes that there are no specially gifted scientists. Each researcher works concurrently on 2-3 projects and salary is less important than autonomy and productive working conditions. There is no designation of extra time (for creative work), except in production. The company considers that the worst thing would be to restrict people to work only on their specialities and supports city sport and leisure clubs where all employees can meet. There are no closed offices and internal networks are preferred due to company product secrecy.

Unilever deals with nutrition, hygiene and personal care. Research staff conduct meetings to know "what Unilever is going to do 20 years from now?". First the idea has to be sold to the project manager and, if accepted, a complete document will be subjected to top decision. Creative individuals may negotiate their salary and each person is motivated to work in another project and to propose an original one. Meetings are held in a central bar, where other people are invited, and all employees are encouraged to participate in external conferences and meetings. There are co-operation agreements with external institutions, individuals experts, consultants, and networks of specialists, under a confidentiality obligation.

Pfizer, in Belgium, adopted another method to think out of the box one year ago, after training with a steering committee. People are organized in training teams and everybody can propose new projects. Bad ideas are not punished, and people are encouraged to take risks. There is a pipeline of projects top-down, and an annual prize for the top company project, while individuals receive a symbolic award for every new idea. It is more about recognition than money.

## Nanotechnology

Also four cases are described: Lund University, The Nanometer Structure Consortium – Lund Nano Lab; Department of Physical Electronics, School of EE, Faculty of Engineering, Tel-Aviv University; Faculty of Science and Engineering, Department of Chemistry, Waseda University, Japan; and the consultant Mark Raison, with experience on Japanese organizations, from Yellowideas, Barcelona.

In the case of Lund University, people seem less aware of creativity techniques. Projects are submitted, selected and funded like normal research projects. Five years ago it was declared a top priority for the university, to leverage newly acquired national funding to attract the best masters students. They act by collaboration, meeting people at the university and competing for projects. There is no specific project structure. Projects are at risk as funds are drastically reduced due to the economic crisis.

In the Israeli case, the present economic crisis has also reduced the amount of funding available and a new policy shall be drafted for the future. As to creativity practices, if an individual proposes a good idea, he or she applies and can obtain money easily. There is a good technology transfer mechanism and a sort of supermarket of mechanisms for garnering project support. The base is fundamental research and there is a selection committee every year to select projects, from several sectors in the society. The university cannot pay as much as a private company, but attracts people by providing an interesting life and can grant special contracts for 20-30% less pay than in an Israeli company, in order to retain staff at the University. Organizations can come to the University and there is a small number of start-ups (before they had 2000; now it is 200) and a venture capital system.

The only Japanese scientist interviewed reported that his university collaborates with industry in physical chemistry projects, and that industry also sends employees to work with university scientists. He admitted that it is very difficult to garner industrial projects or funding due to industry secrecy and confidentiality restrictions.

Mark Raison disclosed that Japanese managers have difficulty accepting creativity techniques, but if they agree on the steps, then projects have great performance. With managers that have not been exposed to western culture, this is much more difficult. They are focused 100% per cent on conventional solutions, making it difficult for employees to propose something different from the boss. One possible solution is to have them submit ideas anonymously (silent brainstorms). It is necessary for Japanese employees to tell a story to gain buy-in and support for their ideas. Thinking and working as a united group is the reference; very well organized and aligned; novelty and uniqueness are eschewed. There is conflict avoidance and no debate and women's opinions are typically not sought out, so that diversity of thought is low. In R&D departments, there are typically all Japanese employees; there is no diversity. The lifelong contract between the company and the employee no longer applies and so risk-taking suffers. The current economic crisis portends the end of the Japanese era of creativity contribution.

## TIC: Information and Communications Technology

From the ICT sector, four cases are reported: the Finnish Hermia Ltd, and the Portuguese YDreams.

Hermia Living Labs integrates users: it is a discussion forum involving users in every step of the project, and they have a topic for each year. One example – the objective of allowing elderly people to live in their homes later in their lives - they invited the elderly, their families, nurses, technology organizations and researchers to a one day workshop. They generate ideas and prioritize and select a set of ideas that could be developed by organizations, using forecasting activities like *Scenario Planning* – watching what's going on world-wide (economic and socio-political trends), asking for cooperation in the projects or trying to create joint R&D projects. They also act as de facto start ups, giving training to entrepreneurs, helping them to build business plans, to get money and link them with recipient customer organizations inside the country or abroad. This is an important objective: to generate and help new entrepreneurs and surface new business ideas. Other activities are the organization of events and the fostering of open innovation, involving technology specialists in social networking activities to cooperatively generate ideas. Many organizations have lots of ideas never implemented or used, and they can challenge university students to develop them. Once an idea is produced, the students own the project and are supported to commercialize it. Tekes Institute (government innovation funding organization) provides funds for organizations and universities to implement these ideation projects. They also have an innovation mill where old ideas can be reanalysed, rejuvenated and eventually implemented.

António Câmara mentioned that there are too many ideas in a company. He considered that the key is to recruit people who think "out of the box". Brazilians, for example, merge the ability to grasp mass culture (what pleases people) with sophisticated culture (arts, humanities and technology). They have two groups of creative people: the ones in fundamental research, who have total freedom; and those of day-to-day projects. Both switch roles frequently. Teams include content and interface (design) technicians. They conduct brainstorming sessions with participants from business, financial and intellectual property disciplines. The results of these sessions are evaluated and the more productive sessions and contributors are rewarded. The salary range of participants is 5:1 and within the same category salary differentials do not exceed 1:2, whatever the evaluation. The internal network is permanent, but employees complain about the lack of communication (there is too much information and communication suffers). See the example of Coca Cola intranet in Brazil. He considers that the university is typical: however, what counts are the hidden curricula, as in the creative capacity, language mastery and tolerance for ambiguity and uncertainty among the student population.

## The Irish Case

Both reported initial cases belong to Entreprise Ireland, a state institution devoted to company development and support. It deals with indigenous SME's in Ireland. The strategy is no longer attracting creative people (the labour market has become too expensive) but to develop own organizations and companies with the people they have, either inland or overseas and, of course, attract foreign investment. And so, the concept of "soft landing" is no longer in use. The focus is on entrepreneurship and programs to help indigenous development, e.g. "leadership for growth", with Stanford university (USA) for chief executives. Also to support R&D, either within organizations, or in collaboration with universities, and to garner relationships with overseas advisors.

The second interview confirmed that the strategy is no longer attracting creative people to move to Ireland or come to Irish companies, but to improve existing organizations - workforce development - creativity and organizational innovation system suggestion to organizations. In the view of the interviewee, no country does this; it's entirely new. The thinking, practices and tools must be at the highest level; the managing director of a company must believe that innovation is both necessary and useful, and that it is not just a moment of inspiration but rather the result of a structured approach to selecting the best opportunities and developing related products or services correctly. Tools are hugely important for making judgments for a range of issues related to both the selection of the best opportunities and the adoption of related products or services in the market. (I know you don't like Rowan, but this is exactly what his Innovation Czar articles is about.) The applied innovation approach uses project – based learning to instill a creative culture in a company. This approach involves giving selected teams pieces of work to carry out and then having them make presentations some weeks later to explain their findings. The topics are deliberately chosen to raise debate and of course conflict will be evident. From this we are able to see to what extent people trust and believe in their work systems and in each other and to what extent there is hidden disagreement. Innovation will never be effective unless organizations have an ambitious culture and leadership.

The third case was Dromone Industries, which designs, manufactures and markets heavy machinery for construction and agriculture. They use idea production techniques of Brainstorming; Kepner Tregoe, Situation Appraisal, QFD – Quality Function Deployment, and Contextual Enquiry. Each project manager reports to the boss, who is in charge of integration. The emphasis is on incremental innovation. People have to see progression in the implementation of their ideas. Projects are normal work: instead of bonuses, they give educational support and travel to exhibitions. The company experienced some conflict with operations managers at the beginning of the deployment of idea production techniques. The system can be improved by adding more value to products and processes, and moving beyond purely incremental innovations.

## Eco-Innovation

Eco-innovation is the production, assimilation or exploitation of a product, production, process, service, management or business method that is novel to the organization (developing or adopting it) and which results, throughout its life cycle, in a reduction of environmental risk, pollution and other negative impacts of the use of resources (including energy use), compared with relevant alternatives. Eco-innovation is a broad concept and is reflected in a large number of technical fields. One important example of a technology which is pivotal for tackling climate change is the generation and transformation of energy. Examples: renewable energy, water and waste management, motor vehicle abatement technologies. Two interviews were made: the western harbour of Malmo, and the city of Kamikatzu.

The first interview was related with the Bo01 Area, a residential area in Malmö, Sweden, which will house 30,000 people by the completion of the project. There were several instruments used by the various teams involved in developing Bo01. At least two seem to score as keys for the project's success: planning and organisation through a quality programme for building; uniform and consensual views of the goals and vision of the area, shared by all those involved (redundant to say uniform and consensual and shared by all) in the project. The Quality Programme is a sophisticated document which deals with virtually all issues which are important in city planning. In the 1999-03-31 Version, it indexes and dedicates 69 pages to many different areas in urban planning such as buildings, light, green spaces, social life, decoration, and so on. It also puts forward the spirit and the philosophy of Bo01. Most striking, it is a document signed by 18 people, which demonstrates a strong commitment to the project.

The second interview was made to the non-pr ofit organization ZERO WASTE ACADE MY JAPAN, 3-1 Shimoyokomine, Fukuhara, Kamikatsu-cho, Katsuuragun. Dr. Paul Connett, of St. Lawrence University in New York, introduced the concept to Kamikatsu during his visit in July 2003. Kamikatsu Town made a "Zero Waste" declaration with a time limit for reducing the amount of waste disposed of by incineration and landfill to near zero by 2020. The Town actually has stopped using waste collection vehicles and has already achieved about an 80% recycling rate of waste, classified into 34 categories. Research is underway into development of new energy sources, and a Zero Waste Fund has been established, which will further promote the Zero Waste initiatives. It was not possible to collect any further data.

## Other Industries and Cases

As indicated, it was possible to obtain information from eight organizations, which were suggested by consultants because of their creativity best practices. All except one were industries, and the majority were based in the Netherlands.

The Bosh (thermotechnology) Innovation Unit, in Aveiro, is helping the company to start developing an innovation process in the competency center. The vision is applied through 3 core activities/pillars: idea generation (internal network); idea implementation (apply during Project phase); and knowledge management (tailor-made knowledge management activities). Employees are encouraged to submit ideas – fill a specific template and send it via email or using a suggestion box. There is an idea evaluation group which meets every month for an hour to evaluate ideas. The Innovation Manager is accountable for organizing and facilitating these meetings. Once approved, the ideas follow a patent application process. The employee rewards are linked with patent attribution and not only with the original idea. After approval an idea can proceed to development, in which creative problem solving is applied (CPS). A network of contacts and experts on various technical subjects was assembled and made available to every person in the organization. Any individual can use this network to solve a problem.

Rotor Company (segments for engines) represents an interesting case. Rotor produces a standard product, ranked highly in innovation business software. This product sells fine, so there is no need for product innovation. It has been the same over 50 years and the personnel have an advanced average age (more than 20 years of service). Thus, they do not have any formal system of idea generation and selection, nor of project management. All work performed is based on hiring students to develop projects following managers' requirements for new processes. The system started 3 years ago and since then the incumbent employees face new ideas from the students who come and work in projects. Directors pressure middle management to come up with new ideas to be developed by students and there is no material reward to the managers for their ideas.

At Wein Minerals Neetherlands, each day the production cell meets to come up with ideas. A problem is assigned to an interdisciplinary team that studies and proposes solutions, and is forced to think outside the box. Consultants were there at the beginning, 2 ½ years ago, and now the company maintains a team of 5 full time internal facilitators. They are used to support teams, especially the poorer performing teams and projects (manual working teams and engineers work better). There are lots of ideas, but time does not allow for more than 5 ideas to be implemented per team. The A3 method (Lean Manufacturing Techniques – A3 size paper) is used for idea selection, as well as root cause analysis and effectiveness solution checking measures. An average number of 10-15 problems are solved each month and they try to define small time range projects, presenting results every 2-4 weeks. There is no control for the smaller projects. The idea is to use the ideas of people, and so more than 60% of the personnel are involved in some kind of research project. Management defines higher level objectives on safety, quality, and how processes can be faster, better or smarter. Once a month management selects people to present the best solutions. There is no reward system but the best ideas are honoured and people feel proud of their work product. Two times a year they have a central meeting in which teams present the best ideas.

Sensata Technologies makes electronic centrals and idea generation starts with marketing people leading the projects: what is the market need? Market understanding needs improvement and more and more market evaluation is performed with engineering involvement. There are problems when the market is not well understood and misguided creativity in project teams leads them to products which are difficult to sell because they have no market.

The biggest company group in Portugal, SONAE, is the only service-based entity on our list. There are 2 facilitators and an innovation committee of 30 people, representing all areas in the company - Customer, Idea management, Networking, Culture, Governance. They develop workshops with customers, follow projects with direct impact on customers, and analyze suggestions and complaints. There is an Idea Forum, where people are asked to submit ideas or solutions for concrete challenges. In promoting corporate awards, the company rewards the best implemented ideas and the most innovative projects.

At the biggest Portuguese road construction company, BRISA, ideas are brought from different sources: internal, universities, suppliers, and clients. Project management is done at the Project Support Office, including patents and models, foresight activities, innovation portal, potential evaluation and value creation. Management is highly involved in innovation connected to value creation, using the Strategic Planning Department focalized on the organizational component. An innovation committee defines priorities and policies, and the Department of Innovation and Technology acts as a first line BRISA department. They have an integrated system of quality, innovation, research and development aimed at creating value through innovation.

These cases are reported in Table 2 and, as can be seen, all cases suggested by the Chamber of Commerce have far less creativity practices than the rest of the sample. Nevertheless, it is interesting to note that only two cases seem to demonstrate best practices in creativity outside of the R&D department. Creativity best practices, or organizational innovation, are difficult to find in current companies outside of the research department, and "Everyday Innovation" is hard to find.

## DISCUSSION AND CONCLUSION

From the interviews and reports analysed, it is possible to conclude that the present crisis is having a deep effect on national policies concerning R&D and on high technology industries, and that these industries, more than the others, are having difficulty of finding alternatives ways to generate financing. The EU and countries like Israel are changing priorities and including services and other non high technology activities and functions in innovation policies and financing. The US and Japan are losing much of their workforce capability due to cost reduction policies and a lack of organizational commitment and engagement of employees. This means, among other things, that the priority given to R&D in state institutions or big company research laboratories may be shared with innovation policies devoted to SMEs and service companies. If this new definition is accepted, innovation shall be understood as organizational, in addition to the traditional process and product-

*Table 2. Summary of characteristics of each case*

| Area | Case | List Table 1 | Idea generation systems | Idea selection systems | Idea reward | Conditions |
|---|---|---|---|---|---|---|
| Bio-Technology and bio-medicine | North Carolina Bio-technology Center<br>Bühler Factories<br>Unilever R&D<br>Pfizer | Yes<br>No<br>No<br>No | No<br>Restricted to R&D Dep.<br>Project management<br>"Think out of the box" method in teams | No<br>Project management<br>Project management<br>Top-down project pipeline | No<br>No extra pay<br>Best scientists negotiate salary<br>Symbolic prize for top company project | ?<br>Creative environment<br>Creative environment<br>Creative environment |
| Nanotechnology | Lund University<br>Tel-Aviv University<br>Waseda University | Yes<br>Yes<br>Yes | No<br>No<br>No. Conventional | Project management<br>Project management<br>Project management | No<br>Better conditions<br>No | Funds are being reduced<br>Interesting life<br>University-Industry collaboration.<br>End of the Japanese era of creative contributions |
| ITC – Informatio and Communications Technology | Hermia Living Labs<br>YDreams | No<br>No | User integration. Scenario planning<br>Yes, when in "ideators" group | Yes<br>Yes | Symbolic<br>Yes. Can double the salary | Creative environment<br>Creative environment |
| The Irish case | "Soft Landing" policy<br>Dromone Industries | Yes<br>No | Several. 40% employees involved | Short term projects | Educational support and travel | No longer applies. Replaced by "Workforce Development"<br>Creative environment |
| Eco-Innovation | Bo=1 Area in Malmo<br>Zero Waste Academy (Japan) | Yes<br>Yes | Project development<br>Project development | Quality Programme<br>? | ?<br>? | State programme<br>State programme |
| Other | Bosh Innovation Unit<br>Rotor Company<br>Wein Minerals<br>Sensata Technologies<br>SONAE<br>BRISA | No<br>No<br>No<br>No<br>No<br>No | Creative Problem Solving<br>Middle managers<br>Production cells – facilitators and method. More than 60% employees involved<br>Marketing dep. leads projects<br>Idea Forum<br>R&D Dep. | Project management<br>?<br>Interdisciplinary team<br>Too many unmarketable projects<br>2 facilitators and 1 innovation committee. Workshops with costumers<br>Project Support Office and Strategic Planning Department | Patent reward<br>No<br>Symbolic<br>?<br>Corporate awards<br>? | Creative environment<br>Students are hired to run projects<br>Creative environment<br>?<br>?<br>? |

oriented types of innovation. Instead of trying to define organizational innovation as changes in the structure of the organization, which do not allow for quantitative return on investment analysis, the concept must be seen as the enactment of a dynamic system devoted to channeling individual and team creativity into profitable corporate innovation. Organizational or corporate innovation, and organizational or corporate creativity, must be seen as synonymous.

Besides R&D departments or laboratories, it is very difficult to find organizations with an institutionalized system of innovation, and from those who have innovation programs, less than 20% of the employees are included in innovation-oriented project teams. Working within a whole workforce development system, although recommended by all types of theoretical and political sources, is rare and limited to specific industries and engineering departments. It was difficult to find recently-published literary references documenting systems of collaborative management, even though it originated in the XIX Century. So we turned to consulting companies' web sites; listed at the end of this chapter.

Even though future European innovation policies will favour the service sector and the SMEs, pragmatic application remains a challenge. We have found no cue as to the primary reasons why it seems difficult for companies to engage personnel in profitable innovation projects. The need for a relationship of trust between management and employees seems to be most relevant. Power struggles seem to be the main obstacle and it seems easier to see management spending millions in technology acquisitions than hundreds of dollars in organizational project development.

The table defined by the project contractor (Table 1) does not take into consideration the effects of the recent crisis upon high technology sectors, and does not totally agree with the purpose of the investigation. It also considers as different, sectors that overlap (nanotechnology, biotechnology, biomedicine and ICT) in companies and research centres, making it difficult to separate its analysis. Also, the access to some of the cases was extremely difficult due to the sector's secrecy, confinement of the industry and culture of the country (e.g., Japan). Collecting data by telephone interviews, with people from non-English speaking countries, in order to draw complex analogies, is a difficult exercise. Nevertheless, from the interviews, relevant aspects of organizational innovation can be indicated: complementary team work in research projects, client intervention, fundamental research role, supportive environment (no punishment of errors and encouragement to take risks), social participation and intrinsic retribution system; learning opportunities, salary negotiation, open meetings; growing private participation and the university as a place for venture capital companies and start ups; wider spectrum of project selection boards, to include companies, agencies and politicians.

From the literature review and cases analysed, we concluded that besides innovation projects and problem solving methodologies, all components of an organizational innovation system must be addressed, namely: creative management (leader selection, orientation and training in order to bring creative contributions out of teams and individuals); creative people management (general and specific orientations as to hiring, training and retaining creative employees); and creativity management (existent systems and conditions for team work and the transformation of individual and team creativity into profitable corporate innovation together with HR management, especially salary systems). Also, to be effective, organizational innovation has to address power sharing, creating a climate of mutual trust between management and employees. This is not discussed in EU or national innovation strategy documentation, nor included in the literature concerning innovation.

We found that R&D is not the only method of innovating. Other methods include technology adoption, incremental changes, imitation, and combining existing knowledge in new ways.

With the possible exception of technology adoption, all of these methods require creative effort on the part of the organization's employees and consequently will develop the organization's in-house innovative capabilities. These capabilities are likely to lead to productivity improvements, competitiveness, and to new or improved products and processes that could have wider impacts on the economy.

From the investigation it is possible, then, to draw the following remarks that can be used when building a system of corporate creativity in SMEs.

## Key Factors

- From the literature and the case study analyses, it seems that the system (organizational innovation) relies on top management orientation to innovation and in project teams, supported by idea finding and problem solving methodologies, together with value and return on investment analysis. Client or market requirements seem to be the best inspiration for projects, and fluid decision making (flat hierarchy) the best guarantee that the system may work. To be effective, organizational innovation has to address power sharing and the creation of a climate of mutual trust between management and employees, together with supporting conditions (e.g. budget, tools, control measures, facilities to conduct meetings, time, and a proper reward system). Not the trust and the loyalty expected in pre-technological times, when the company provided protection and care while "trusting" the employees to provide loyalty and engagement. But, at least, the trust that is expected from a value proposition that is authentic and gets employees engaged in the immediate task. The environment (communication) provided by leadership is the key factor in corporate creativity, and collaborative management and collaborative tools (e.g. communities of practice, collaborative webs) possible solutions to overcome the difficult times of the present crisis. Companies that cannot transform themselves into collaborative organizations will suffer in today's innovation economy, as everything that is not creative will have a low cost orientation and can be outsourced. Inertia kills innovation in many organizations, and it is absolutely necessary to give employees hope, as innovation needs optimism and employees need to feel that the company is worthy of their efforts to add value.
- For corporate creativity the real power is the unexpected, and the novelty of self-initiated projects far exceed that of the projects initiated by management. Innovation cannot be predicted but just happens, somewhere between planning and improvisation, and a company cannot predict who will play a role in a creative act or what that act will be. Therefore, the company must encourage unofficial activity and provide diverse stimuli. This can be achieved by identifying stimuli and providing them to employees, rotating jobs, arranging for outside interaction and creating opportunities for employees to take their own stimuli. Asking them to run their business like their own may produce wonders. Great ideas, visionary leaders, core values, strategic planning, vision statements, and early entrepreneurial success are common myths for why companies succeed. Innovation is not for exceptional people, like the inventors; it is just discovering new ways of creating value, which once embraced by the employees, will become a way of life.
- Alignment, i.e., the degree in which the interests and actions of the employees coincide with management objectives, is a key factor. The failure of large organizations to innovate is primarily the result of

a communication gap, not a decline in ingenuity, and so the company must provide opportunities for people to meet with each other and the priority must be to make each employee responsible to requests for information or help from other employees. The connection between organizational boxes is one of the key factors; what you do inside each box is less important than how the boxes work together. Even if one of the boxes improves, that does not mean the whole system will; the effect may even be negative. Formal innovation systems must be validated by the informal organization, so that management policies and people's beliefs coincide. This means that all important decisions must be shared so that commitment is realized. There will always be continuous improvement as long as there is trust between management and employees.

- Innovation is made by people and for people, and the success comes from an uncompromising commitment to the organization and its people. Therefore, automation, routinization and mechanization are against innovation, as organizational creativity is more about commitment than about ideas, and it is not possible to be committed to a machine. A company is creative when its employees do something new and useful without being directly shown or taught. Every employee has unique knowledge, or knowledge and insights known to only a few other employees.
- Symbolic rewards (praise at company meetings, diplomas, public announcement) are more effective than material ones. Nevertheless, some compensation might be given to ideas that produce objective results, like patents. Also, learning opportunities are a sound reward to people who generate profitable ideas. If there is prize money, it should not exceed 10% of salary. It is a mistake to believe that creativity can be motivated by offering rewards. Rewards have always been the main reason to stop suggestion systems, not to initiate them.

## Idea Production and Selection Systems

- It is top management's responsibility to define innovation objectives, designate the teams, and provide time, resources, competencies, and leadership. The innovation system must be tailored to each company and not simply copied from external examples. The idea generation and selection methods chosen are important aspects of innovation, but the organization should choose appropriate ones aligned with company goals and culture and develop expertise in their use. These methods are also a first class training tool for leaders at all levels. Even though there should be individuals or groups managing the innovation system, the moment the company needs a champion to promote ideas to top management, the company has already failed.
- If possible, idea generation and idea selection and development should correspond to different teams. For example, at Buhler, with a medicament required by the market, biologists may start the research and, after discovering how the virus might be destroyed, a team of chemists leads the making of the drug, which is then delivered to a team of medical doctors to conduct tests. Also, Buhler's cycle seems to be a guide: initial brainstorming - database search - hypothesis suggestion - team gathering to make the development program – management submission and budgeting – periodical (36 months) reports to management to free more resources
- Idea producing sessions must be open and a diverse set of people should be invited to participate. Unilever's meeting at a central

bar, and Hermia's (Finland). "Innovation Mill" are examples. All forms of art should be considered in leadership training and ideation sessions. Also, Open Innovation systems are welcomed, and the marketing sector can take control of contributions.

- Teams should be multidisciplinary and members related to finance, marketing, commercial areas and intellectual property management should be invited. Also, research and production roles should shift amongst employees. If the personnel are not available or convenient to project teams, students may be put in charge, under management orientation and following internal suggestions and ideas. The selection of students should rely on expertise and in universities with a strong "hidden curricula" of creative thinking, language mastery and tolerance for uncertainty.
- Team composition should remain stable, at least during a project development, and project teams must be "visible" within the organization, together with their activities and achievements. Each project team must add this task to the project's list. Visibility is both an extra reason for the team members to comply with the planned requirements and a condition for the rest of the company to accept more easily the changes introduced by the team. Projects should have a short range (1-2 months, or 2-4 weeks), so that people may see the impact of changes or improvements. Smaller projects do not need to be subjected to systematic controls.
- Some coordination must take place, so that the innovation system becomes embedded in the organization. Also, in accordance with the company's dimension, consultants should be available to help the project teams. Even though there may exist dedicated project management, the idea for the need of a champion reveals that the system has already failed, because management needs someone to convince the top to adopt interesting ideas.
- Management control mechanisms must be a constant in every important project, so that continuous feedback is given to management and to teams on costs and impact.
- Passive systems, from the simple suggestion boxes to sophisticated idea management. software, generate only some of potential for the unexpected. Management has to take into consideration that each idea takes roughly four hours of work to analyze and organize a project. That is why some sort of active system may be preferable or, at least, work side by side with a passive system, provided it is accessible to all employees, is easy to use, and there is strong follow-through possibilities.

The overall conclusion was that each company represents a specific case on innovation, as the interviews did not allow us to detect major similarities, as we originally thought we would find. Therefore, evidence indicates that innovation in companies does not align with standardized models of execution, but only with the attempts that the company makes to develop its organization. As these attempts become consistent and reliable, the company is likely to build its own innovation system, different from any other company. As a rule of thumb, everyone should ask which activities account for 80% of one's time but only for 20% of the accomplishments, and focus the creative energies toward how it might be reduced, delayed or even stopped.

Best practices can only be fully analysed through observation and participation, which should be considered for the future. The case of Ireland, eventually complemented with others from Denmark and Finland, deserves a deeper analysis, as suggestions for future research.

## REFERENCES

Adams, R. (2006). Innovation measurement: A review. *International Journal of Management Reviews*, *8*(1), 21–47. doi:10.1111/j.1468-2370.2006.00119.x

Arundel, A., Bordoy, C. & Kanerva, M. (2008). *Neglected innovators: How do innovative organizations that do not perform R&D innovate?* Pro Inno Europe Report: Innometrics, March 31.

Atkinson, R., & Andes, S. (2009). *The Atlantic Century: Benchmarking EU & US innovation and competitiveness*. Washington: The Information and Innovation Technology Foundation.

Bandow, R. (2009). *Balance scorecard and benchmarking*. Berlin: Fraunhofer-Gesellschaft.

Basadur, M. S. (1997). Organizational development interventions for enhancing creativity in the workplace. *The Journal of Creative Behavior*, *31*(1), 59–73.

Basadur, M. S. (2000). *The economic, social and psychological outcomes of implementing a deliberate process of organizational creativity*. Working paper n° 100. McMaster University, Management of Innovation and New Technology Research Center.

Cebon, P., Newton, P., & Noble, P. (1999). *Innovation in organizations: Towards a framework for indicator development*. Melbourne Business School Working Paper #99-9, September.

Collins, J., & Porras, J. (1994). *Build to last*. New York: HarperCollins Publishers.

Commins, E. J. (2004). *The role of the Enterprise Ireland Technologist in facilitating Innovation Management in Irish SMEs*. Unpublished Master of Science Dissertation in Technology Management. Dublin: University College Dublin.

European Commission. (1996). *Innovation Management Tools–a review of selected methodologies (MINT)*. EIMS publication No 30.

European Commission. (1998). *A guide to Technology Management and innovation for organizations*. TEAMGUIDE.

Florida, R., & Tinagli, I. (2004). *Europe in the Creative Age*. London: Demos.

Hagardon, A. (1999). Group cognition and creativity in organizations. *Research on Managing Groups and Teams*, *2*, 137–155.

Hagardon, A. (2000). *Building an innovation factory. Harvard Business Review*. May-June.

Hage, J. (1999). Organizational innovation and organizational change. *Annual Review of Sociology*, *25*, 597–622. doi:10.1146/annurev.soc.25.1.597

Harter, J., Schmidt, F., & Keyes, L. (2003). *Well-being in the workplace and its relationship with business outcomes: A review of the Gallup studies*. Washington, D.C.: American Psychological Association.

Kanter, R. M. (1983). *The change masters*. London: Routledge.

Kao, J. (1989). *Entrepreneurship, creativity, & organization*. New York: Prentice-Hall.

Lam, A. (2004). *Organizational innovation*. MPRA Paper No. 11539.

McAdam, R., & McClelland, J. (2002). Sources of new product ideas and creativity practices in the UK textile industry. *Technovation*, *22*, 113–121. doi:10.1016/S0166-4972(01)00002-5

McLean, L. D. (2005). Organizational culture's influence on creativity and innovation: A review of the literature and implications for human resource development. *Advances in Developing Human Resources*, *7*, 226–246. doi:10.1177/1523422305274528

Mendonça, S. (2008). Brave old world: Accounting for 'high-tech' knowledge in 'low-tech' industries. *Research Policy*, *38*, 470–482. doi:10.1016/j.respol.2008.10.018

Monteiro, I., & Sousa, F. (2008). A liderança inovadora na hotelaria algarvia. *Revista Portuguesa e Brasileira de Gestão*, *7*(2), 68–78.

OECD. (2005). *Oslo Manual-guidelines for collecting and interpreting innovation data* (3rd ed.). Paris.

Paulus, P. (2008). Fostering creativity in groups and teams. In Zhou, J., & Shalley, C. (Eds.), *Handbook of organizational creativity*. New York: Lawrence Erlbaum Associates.

Perry-Smith, J. (2008). When being social facilitates creativity. In Zhou, J., & Shalley, C. (Eds.), *Handbook of organizational creativity*. New York: Lawrence Erlbaum Associates.

Peters, T., & Waterman, R. (1982). *In search of excellence: Lessons from Americas best run companies*. New York: Warner Books.

Puccio, G. J., Firestien, R. L., Coyle, C., & Masucci, C. (2006). A review of the effectiveness of CPS training: A focus on workplace issues. *Creativity and Innovation Management*, *15*(1), 19–33. doi:10.1111/j.1467-8691.2006.00366.x

Robinson, A. (1998). *Corporate creativity: How innovation & improvement actually happen*. San Francisco: Berret Koehler Publishers.

Rodriguez, E., Franco, T., & Santos, M. (2006). *Nature and antecedents of organizational commitment: Considerations for human resources management*. ISEG Working Papers, 12, 75-86.

Sawyer, K. (2007). *Group genius: The creative power of collaboration*. New York: Basic Books.

Schumpeter, J. A. (1934). *Invention and economic growth*. Cambridge: Harvard University Press.

Shalley, C. (2008). Creating roles. In Zhou, J., & Shalley, C. (Eds.), *Handbook of organizational creativity*. New York: Lawrence Erlbaum Associates.

Shapiro, S. M. (2001). *24/7 Innovation: A blueprint for surviving and thriving in an age of change*. New York: McGraw-Hill.

Sousa, F. (2004). Innovación, creatividad y liderazgo organizacional. *Creatividad y Sociedad, 5*, 11–17.

Sousa, F. (2008). Still the elusive definition of creativity. *International Journal of Psychology: A Biopsychosocial Approach, 2*, 55-82.

Sousa, F., & Andrade, C. (2007). Factores organizacionais e pessoais como determinantes da percepção do clima organizacional. *Inuaf-Studia*, *10*, 219–235.

Spence, W. R. (1994). *Innovation: The communication of change in ideas, practices and products*. London: Chapman & Hall.

Unsworth, K. L. (2005). Creative requirement: A neglected construct in the study of employee creativity? *Group & Organization Management*, *30*, 541–560. doi:10.1177/1059601104267607

Wolfe, R. A. (1994). Organizational innovation: Review, critique and suggested research directions. *Journal of Management Studies*, *31*, 405–431. doi:10.1111/j.1467-6486.1994.tb00624.x

Zhou, J., & Shalley, C. (Eds.). (2008). *Handbook of organizational creativity*. New York: Lawrence Erlbaum Associates.

## ADDITIONAL READING

H2i Institute http://www.h2iinstitute.com

Accenture http://www.accenture.com/

Action Technologies www.actiontech.com

American Supplier Institute www.asiusa.com

Andrew Hagardon http://andrewhargadon.com/

Angela Edwards http://www.managedinnovation.com/

Basadur, M. S. (1999). *Simplex: A flight to creativity*. The Creative Education Foundation.

Bob Barancik http://www.creativeledgestudio.com/05institute/

Boden, M. A. (Ed.). (1994). *Dimensions of creativity*. London: MIT Press.

Brian Glassman www.TechRD.com/blog

Bruce LaDuke http://www.instantinnovation.com/

Burns, J. M. (1978). *Leadership*. New York: Harper & Row.

Burns, T., & Stalker, G. M. (1996). *The management of innovation*. New York: Oxford University Press.

C-Tech Innovation http://www.ctechinnovation.com

Center for Creative Emergence http://www.creativeemergence.com

Center for Creative Leadership http://www.ccl.org/leadership/index.aspx

Chasm Group http://www.chasmgroup.com/

Chris Harris http://chrisharrisfutures.blogspot.com/

Creative Emergence http://www.creativeemergence.com/index.html

Creativity and Innovation Management. http://www.wiley.com/bw/journal.asp?ref=0963-1690

Creativity Research Journal. http://www.informaworld.com/smpp/title~content=t775653635~db=all

Creax http://www.creax.com/index.htm

Csikszentmihalyi, M. (1996). *Creativity: Flow and the psychology of discovery and invention*. New York: HarperCollins.

Dave Pollard http://blogs.salon.com/0002007/

Design of experiments http://www.isixsigma.com/tt/doe/

Destination innovation http://www.destination-innovation.com/

DMFA Guide http://www.engineeringtalk.com/guides/dfma.html

Doblin www.doblin.com

Documentos da Comissão Europeia sobre inovação http://www.proinno-europe.eu/; http://www.proinno-europe.eu/index.cfm?fuseaction=page.display&topicID=89&parentID=0

Edward De Bono www.edwdebono.com

Entrepreneurs Grow On TRIZ http://www.ideationtriz.com/stream/ideation1.rm

European Journal of Innovation Management. http://info.emeraldinsight.com/products/journals/journals.htm?id=ejim

Failure Modes and Effects Analysis (FMEA) http://www.npd-solutions.com/fmea.html

Fast Company http://www.fastcompany.com

Front End of Innovation http://frontendofinnovation.blogspot.com/

Gardner, H. (2006). *Changing minds*. New York: Basic Books.

Glover, J. A., Ronning, R. R., & Reynolds, C. R. (Eds.). (1989). *Handbook of Creativity* (pp. 3–33). New York: Plenum Press.

Gregory Zhykov http://www.conceptdraw.com/en/

Group Cognition and Creativity in Organizations. http://www.sciencedirect.com

Halpern, D. F. (1996). *Thought and knowledge: An introduction to critical thinking*. Mahwah, NJ: Lawrence Erlbaum Associates, Publishers.

Hobie Swan http://www.mindjet.com/

Howard Smith http://www.csc.com

Idea Champions http://www.ideachampions.com/index.shtml

Idea Company http://www.ideacompany.nl/

Idea Connection http://www.ideaconnection.com/

Ideas to Go http://www.ideastogo.com/

Ideation International www.ideationtriz.com

Ideo http://www.ideo.com/

Imaginatik http://www.imaginatik.com/

Industrial Research Institute http://www.iriweb.org/

Innocentive www.innocentive.com

Innosight www.innosight.com

Innovare http://www.innovare-inc.com/

Innovaro http://www.innovaro.com/index.html

Innovation as a Learning Process http://vimeo.com/3475327

Innovation Coach www.InnovationCoach.com

Innovation Economy Conference http://www.theinnovationeconomy.org/Videos/

Innovation on My Mind http://innovationonmy-mind.blogspot.com/

Innovation Point http://www.innovation-point.com/index.htm

Inogate http://www.inogate.com/

International Center for Studies in Creativity http://www.buffalostate.edu/creativity/

International Journal of Entrepreneurship and Innovation Management. http://www.inderscience.com/browse/index.php?journalCODE=ijeim

International Journal of Innovation Management. http://www.worldscinet.com/ijim/

Jack Hipple http://www.innovation-triz.com/

Jason Womack http://www.womackcompany.com/

Jeffrey Baumgartner www.jpb.com

Jim Canterucci http://www.mypersonalbrilliance.com

Joel Orr http://cyonresearch.com/

Journal of Creative Behavior. http://www.creativ-eeducationfoundation.org/?page_id=47

Journal of Product Innovation Management. http://www.wiley.com/bw/journal.asp?ref=0737-6782

Joyce Wycoff http://www.innonet.org/

Kahn, A., Castellieon, K. B., & Griffin, A. (Eds.). (2005). *The PDMA handbook of new product development* (2nd ed.). Hoboken, NJ: John Wiley & Sons.

Keith Sawyer http://keithsawyer.wordpress.com/

Kes Sampanthar http://www.metamemes.com/

Kindling http://www.kindlingapp.com/

Mario Morales http://www.aurainteractiva.com/site/content/index.htm

Mark Raison www.yellowideas.com

Mark Turrel http://www.imaginatik.com/

Meldrum Duncan http://www.whatifinnovation.com/default

Michael Bungay Stanier http://www.boxofcray-ons.biz/

Min Basadur www.basadur.com

MIT Engineering System http://sdm.mit.edu/

Open Innovation. Ben Franklin's spin http://www.youtube.com/watch?v=e01lFztpsoY Pascal Jarry http://www.solidcreativity.com/

Paul Sloane http://www.destination-innovation.com/

PDMA http://www.pdma.org/

PRTM http://www.prtm.com/

Quality Function Online (QFD) http://www.qfdonline.com/

Real Innovation http://www.realinnovation.com/

Redefining Innovation http://www.hulu.com/watch/50991/the-business-of-innovation-redefining-innovation

Robert's Rules of Innovation http://www.robertsrulesofinnovation.com/

Rogers, E. M. (1983). *Diffusion of innovations* (3rd ed.). New York: The Free Press.

Sagentia http://www.sagentia.com/

Sandy, S. (Ed.). (2005). *The IAF handbook of group facilitation: Best practices from the leading organization in facilitation*. New York: Jossey Bass Publishers.

Schuman, S. (Ed.). (2005). *The IAF handbook of group facilitation. Danvers*. Jossey-Bass.

Solid Creativity. http://www.solidcreativity.com/references.php.

Stage Gate International http://www.stage-gate.com/knowledge.php

Stefan Lindegaard http://www.15inno.com/

Strategos www.strategos.com

Synectics www.synecticsworld.com

Systematic Innovation. TRIZ http://www.systematic-innovation.com/

Systematic Inventive Thinking. ASIT http://www.start2think.com/

The Creative Leadership Forum http://www.thecreativeleadershipforum.com/

The Creative Problem Solving Group http://www.cpsb.com/

The International Association of Facilitators (IAF) http://www.iaf-world.org/i4a/pages/index.cfm?pageid=1

The International Journal for Innovation Research, Commercialization, Policy Analysis and Best Practice: http://www.innovation-enterprise.com/

Thompson, L., & Choi, H. (Eds.). (2006). *Creativity and innovation in organizational teams*. New York: Lawrence Erlbaum Associates.

Trainmore http://www.trainmor-knowmore.eu/root.en.aspx

Value Analysis/Value Engineering http://creatingminds.org/tools/value_engineering.htm

VanGundy, A. B. (1987). *Creative problem solving*. New York: Quorum Books.

West, M. A., & Farr, J. L. (Eds.). (1990). *Innovation and creativity at work: Psychological and organizational strategies*. Chichester: Wiley.

## KEY TERMS AND DEFINITIONS

**Alignment:** The degree in which the interests and actions of the employees coincide with management objectives.

**Benchmarking:** The process of identifying the best practice in relation to products and processes, both within an industry and outside it, with the object of using this as a guide and reference point for improving the practice of one's own organization.

**Creativity:** Individual cognitive and emotional processes that take place between the individual and the product that is created.

**Creative Workforce Potential:** Both the ability to retain creative managers and employees, and to provide an environment where each one feels free and willing to contribute to organizational success.

**Corporate Innovation:** A system devoted to enhance creativity in organizations.

**Innovation:** Discovering new ways of creating value.

**Innovativeness:** The potential of the workforce to promote changes to the benefit of the organization.

**Organizational Creativity:** A system devoted to enhance creativity in organizations.

**Organizational Innovation:** A system devoted to enhance creativity in organizations.

## ENDNOTES

1 We would like to thank Maria Thompson, from Motorola Company, for her technical advice and the review of the manuscript. Also, we thank our colleagues from EACI, especially Dr. Hans Van Meer, and the managers we interviewed, especially Dr. Edward Commins, from Entreprise Ireland, without whose contribution this work would not have been possible

2 The non-profit association APGICO (Portuguese Association for Creativity and Innovation) was created in 2007 and aims at developing knowledge and experience in the area of creativity and innovation management in organizations, helping to create conditions for competitiveness of companies and effectiveness in organizations.

# Chapter 2
# Integrating Technology with the Creative Design Process

**Joshua Fairchild**
*The Pennsylvania State University, USA*

**Scott Cassidy**
*The Pennsylvania State University, USA*

**Liliya Cushenbery**
*The Pennsylvania State University, USA*

**Samuel T. Hunter**
*The Pennsylvania State University, USA*

## ABSTRACT

*In our fast-paced world, it is necessary for organizations to continually innovate in order to stay competitive. At the same time, technology is continually advancing, and tools to facilitate work are frequently changing. This forces organizations to stay abreast of current technologies, and also puts pressure on employees to utilize the technologies available to them in order to devise innovative solutions that further the organization's goals. To date, there has been little research on how such technologies may best be used to facilitate such creative performance. The present chapter addresses this gap by integrating a model of the creative process from the psychology literature with technology literature from engineering and information technology. This chapter examines how specific technologies may influence performance at each stage of the creative process, and provides specific recommendations for how technology may be used to facilitate the development of creative solutions.*

DOI: 10.4018/978-1-60960-519-3.ch002

## INTRODUCTION

We live in a continually advancing, fast-paced world. Computers and technology have become ingrained in our society and pervade every aspect of our lives. They shape how we interact with our friends, family, and colleagues, how we share information, receive entertainment, and do our work. In everyday life, there is a strong pressure to remain up to date with the latest technology. This pressure is only stronger in the business world, where organizations must continue to develop and adopt new technologies to keep pace with their peers and remain competitive (DeFillipi, Grabher, & Jones, 2007). Similarly, this competition provides a driving force for organizations' own continual need for innovation; in many industries, companies must continually innovate or lose their share of the ever-narrowing market.

As technology continues to advance, the demand that organizations "innovate or die" will only grow stronger, a pressure that passes on to employees. As such, the future of organizational innovation will in large part be dependent on how employees interact with technology. Indeed, Edmonds and Candy (2002) note that "the use of complex tools, such as computers, forms a significant part of the context in which the conditions for creativity exist" (pp. 94). Therefore, as organizations move into the 21st Century, it is becoming increasingly vital that they stay up-to-date on field-relevant technologies, and ever-developing new tools that can enhance creative performance and keep them abreast of the competition.

Given the ingrained relationship between technology and creativity in the modern workplace and the necessity of both for organizational survival, it is essential to understand how technology implementation functions in the workplace, and how it may support creativity. In order to understand this, the present chapter addresses several key components associated with technology used to facilitate innovation: (1) the relevant technologies, their use and implementation, (2) the psychological aspects of the creative process, and (3) the social framework in which the technology is implemented and the creative process occurs. This chapter seeks to build an integrative framework to describe how specific technologies may be used to facilitate the creative design process, as well as some of the potential pitfalls associated with relying on technology. Such a framework provides valuable insight into how organizations may harness both their employees' abilities and specific technologies in order to maintain a competitive edge through innovation.

## THE EMERGENCE OF PROCESS MODELS IN THE STUDY OF CREATIVITY

In order to understand creative performance at work, as well as how it may be influenced by technology, is crucial to consider that creativity is not a set outcome, but rather a multistage process, composed of interlinked steps, with different social and cognitive processes operating at each stage. In order to understand how technology may influence creativity, it will be necessary to examine each step individually, considering what psychological processes are active, what specific technologies may be effective, and how they may best be implemented.

For much of its early study, creativity was thought to occur in a "black box," with the steps leading up to production of a creative product being thought of as unobservable (Ward, Smith, & Finke, 2009). As such, examinations of creativity focused primarily on initial inputs and situational factors that might influence creativity, and finished products, without consideration of how such inputs *lead* to creative outcomes. Such intervening steps were largely thought to be unobservable.

An early attempt to examine creativity as more of a process than simply an input-outcome relationship was a basic model by Dewey (1910). This model proposed a simple stage-based conceptualization of problem solving that has frequently been identified as a particularly early model of the creative process. The model consisted of perceiving a particular difficulty, defining a specific problem to address, identifying potential solutions to the problem, building on proposed solutions, and then testing them (Dewey, 1910; Lubart, 2001). However, while this model has been applied to examine creative problem solving, it does not really address how creative solutions develop.

Shortly after Dewey's initial model was proposed, Wallas (1926) proposed a stage-based model of the creative process. This model focused on the development of creative solutions through "flashes of inspiration" that are thought to be central to the process of creativity. First, in the preparation stage, an individual studies the problem at hand and performs a preliminary analysis, using one's own knowledge, skills, and analytical abilities. Then, during incubation, the worker is not consciously thinking about the problem. Instead, Wallas (1926) suggests that one is subconsciously forming conceptual combinations and testing associations at this stage. During the next stage, illumination, a particularly salient combination of ideas breaks through into consciousness, resulting in an "aha" moment. Following this, during verification, this idea is refined and evaluated, undergoing further development, testing. (Wallas, 1926; Lubart, 2001).

Though more detailed than Dewey's 1910 model, this approach still does not allow for the detailed study of what occurs *during* creative activity. By emphasizing that creativity arises when subconscious combinations of ideas break into consciousness, without elaborating on how this happens this model still hides the actual creative process from further examination.

Much more recently, models have opened this "black box" and examined specific components and stages of the creative process, a development crucial to the present study. Such models present logical links between the results of one stage of creative problem solving, and subsequent stages, and examine how these elements of the process come together, and ultimately lead to creative outcomes.

In this way, the creative process can now be thought of broadly in terms of a series of successive iterations, in which ideas are conceived, considered, and revised multiple times, as designers explore problems and try to develop an effective, creative solution (Finke, Ward, & Smith, 1992, in Ward et al., 2009). Further, it is how these ideas are considered, combined, and revised that ultimately gives rise to creative performance. Specifically, generative processes typically begin with the generation of an initial pool of ideas, which at this stage are not complete potential solutions to a problem, and in fact are often untested, or incomplete. (Finke et al., 1992). Such ideas are often considered "preinventive," forming the fuel for the creative process, as they are original and appropriate to the task at hand. However, they are not themselves the creative ideas that will be directly utilized to address a problem (Ward et al., 2009).

Instead, such preinventive structures are combined with one another and explored, to see what might result from such combinations. It is this exploration and combination that forms the basis for creative thought, and it is the exploration, revision, and recombination of ideas that arises during this process that gives rise to the creative problem solving process (Ward et al., 2009).

Conceptualizing creativity as a process in such a way has general benefits for the study of creativity, by allowing for the examination of *how* creative thought, and ultimately, creative products arise. Further, and of direct relevance to the present chapter, it allows for an examination of potential interventions that may influence creative performance, and the principles on which such interventions may operate. In particular, given the prolific role that technology plays in the modern workplace, it is essential to consider how technology used by designers may influence creative performance. Considering creativity as a process in such a way is essential for such study.

However, such integration of technology implementation research with the psychological research on creative processes has thus far been largely ignored in the literature. Some recent work by Schneiderman (cf. Schneiderman, 1998, 2000, 2002) has described a process model that describes how computers may be used as part of creative performance, but to date, there is little established literature on how technology can be used to differentially influence stages of the creative design process. Of further note is that, while creative processes are well-studied within the field of psychology, the impact of technology has been largely ignored by this discipline.

In a review of eighty one studies across twenty years of literature on technology implementation at work (1984-2004), Rizzuto and Reeves (2007) noted that less than 10% of the published research came from the psychology literature. The present chapter addresses this gap in the literature by selecting a widely accepted model of the creative process from the psychology literature, and examining how various technologies may influence creative performance by acting on each distinct stage of the creative process.

## INTERPRETING TECHNOLOGY'S EFFECTS ON CREATIVE PERFORMANCE: THE CREATIVE PROCESS MODEL AS A LENS

Currently, there are numerous models that describe the process of creativity. These models vary in scope and form, but generally follow a framework that includes problem identification, idea generation, implementation, and revision (c.f., Scott & Bruce, 1994; Zaltman, Duncan, & Holbek, 1973). The framework for the present chapter is a specific process model by Mumford and colleagues (c.f., Mumford, Mobley, Uhlman, Reiter-Palmon, & Doares, 1991; Baughman & Mumford, 1995; Mumford & Baughman, 1996; Mumford, Supinski, Baughman, Costanza, & Threlfall, 1997; Mumford, Baughman, Maher, Costanza, & Supinski, 1997). This model was selected due to its specificity and relatively fine-grained division of the creative process into eight discrete stages. Such a division is useful in the present context, as it allows for consideration of the potential effects of technology at numerous points. This model provides a sound framework for examining the psychological processes that influence creative performance. As such, it is an excellent foundation for the present discussion, as it will be shown that key mechanisms for the influence of technology on creative performance are how such technologies impact the cognitions and behaviors of designers. Such effects will be discussed as they relate to each of the stages of the creative process.

Specifically, these stages begin with (1) *problem construction*, during which a particular problem is targeted. It is important to note that, in creative problem solving, this problem is usually ambiguous, and not initially clearly defined. It is up to the team to come up with a working definition of the project before proceeding. This approach is in line with earlier work by Simon (1973), which suggests that problem solving itself can be a vaguely defined, fluid process, which is in fact hampered by rigid structure. Therefore, at the outset of the creative process, it is up to the designers to discover an effective way to define the problem, in a manner that encourages further, open-minded pursuit of novel solutions.

Once the problem has been defined, the next stage is (2) *information gathering*, during which team members begin to gather the body of information that will help them address the problem. This is followed by (3) *concept selection*, during which the gathered information is pieced together, and the information that is most potentially useful is identified. Then, during (4) *conceptual combination*, this information is sorted and categorized, with novel, unexplored links between ideas and concepts being identified as potential means of addressing the problem at hand. During (5) *idea generation*, these patterns and groups of concepts are built into potential solutions for the problem. In (6) *idea evaluation*, these potential solutions are examined, revised, or replaced, until what appears to be a viable solution emerges. Then, during (7) *implementation planning*, this viable solution is fleshed out, and the team determines how best to go about enacting it. Once the proposed solution is put into place, during the (8) *monitoring* stage, the team observes how well the solution performs, and revises, updates, or replaces it as needed. An important consideration is that this process is nonlinenar, and developments in one stage often require designers to return to earlier stages. Such iterative development is a crucial part of the creative process, and can greatly impact creative performance.

The following pages will address each stage in more detail, particularly regarding how technology may be used at that point to enhance designers' creative performance. It is hoped that such a model will reveal new means of facilitating creative design via technology, which will have implications for both future academic work and organizational practice. A summary of the stages of the creative process model, as well as some specific technologies that may be useful in each stage, is presented in Table 1.

## Problem Construction

The first stage in Mumford and colleagues' (1991; 1994; 1996a; 1996b) eight-stage model of the creative process is *problem construction*. Mumford et al. (1991) suggest that any problem-solving effort in ill-defined or novel domains first requires definition of the problem(s) to be solved. This stage provides the context for the application of the subsequent steps in the creative process and, consequently, has a substantial impact on not only later stages but on creative problem solving as whole (e.g., Adelman, Gualtieri, & Stanford, 1995; Mumford et al., 1994; Reiter-Palmon, Herman, & Yammarino, 2008; Rostan, 1997; Scott, Leritz, & Mumford, 2004).

At this early juncture, it is essential that all designers develop a similar grasp of the problem, and that each has a chance to share his or her ideas and expertise. As such, technologies used to influence performance in this problem construction phase should focus on enhancing communication between designers and facilitating the visualization of potential design problems.

Several studies have demonstrated that problem construction is related to solution quality and originality in solving real-life problems, such as those often encountered by engineers, technical design specialists, and research and development (R&D) professionals (e.g., Dillion, 1982; Getzels & Csikszentmihalyi, 1975; Runco & Okuda, 1988; Smilansky, 1984). Reiter-Palmon et al. (2008) suggest that the problem construction process is influenced by past experience and familiarity with similar problems, as well as attention to cues signaling a problem. As such, it is expected that expertise and prior experience play major roles in problem construction. With regard to the engineering and R&D sectors, this certainly appears to be the case. In particular, tools that enable team members to better share knowledge and expertise will likely greatly facilitate performance in the problem construction phase.

In support of this, a recent study investigating the use of bibliometric citation techniques for mapping and visualizing data about the oral communication patterns of R&D engineers (Skovvang, Elbaek, and Hertzum, 2006) found that maps of old projects were viewed as particularly useful in the early stages of problem solving. More specifically, the authors found that R&D engineers considered such maps about who has what knowledge- which they termed *personometric* maps - to be potentially useful as tools for finding people with specific competencies. The authors concluded that face-to-face communications and communications

*Table 1. Potential benefits of technology at each creative process stage*

| Process | Description | Opportunities for Technological Benefits | Potential Technologies |
|---|---|---|---|
| *Problem Definition* | Identification of a broad, often ambiguous problem to address | Facilitating communication and personometric maps; sharing ideas | Email, smart phones, CAD software |
| | | Enhancing problem visualization | CAD software |
| *Information Gathering* | Assembly of a general body of information to be used in formulation of a solution | Gathering data from a broad range of sources; instant knowledge access | Internet/World Wide Web |
| | | Facilitating communication with colleagues and experts off-site | Email, smart phones, video conferencing, instant messaging |
| *Concept Selection* | Identification of most useful information; sorting into categories relevant to the problem | Supporting, sorting, organizing and cataloging bodies of information | Online design databases (e.g., *Wikis*), interactive, searchable design repositories |
| *Conceptual Combination* | Combination of groups of ideas in novel ways in order to address the problem at hand | Assisting designers in developing a framework for combining ideas in novel ways | Interactive, searchable design repositories |
| | | Facilitating in-person discussion that leads to development of novel idea combinations | Large format displays or other visualization tools |
| *Idea Generation* | Shaping and integration of concepts into specific potential solutions | Providing examples to support for brainstorming abstract concepts into concrete ideas | Publically-available, Internet-based resources (e.g., YouTube) |
| | | Providing tools to help designers flesh out concepts into potential solutions | CAD software |
| | | Facilitating communication among design team members | Virtual meeting software, online collaboration tools (e.g., Google Wave) |
| *Idea Evaluation* | Examination of the costs, benefits, potential merits and pitfalls of ideas, and selection of those that are likely to be successful | Providing an efficient, cost effective basis for examining and testing the viability of design ideas | Rapid prototyping |
| *Implementation Planning* | Identification of design details, and determination of how to best enact them | Developing and examining design details to determine best routes for implementation | CAD software, design repositories and databases of past projects |
| | | Enabling cross-disciplinary collaboration for implementation | CAD software |
| *Monitoring* | Systematic examination of a solution's success following implementation | Providing regular updates and feedback regarding design implementation | Hardware and software sensors, tracking hardware |
| | | Facilitating customer/user feedback on design implementation | Web-based surveys, email |

via phone, email, and other systems are critical to practical application of personometric maps.

Such communication maps facilitate the sharing of information between team members, and therefore are a critical means of effectively and efficiently addressing problem construction at the beginning of the creative process, regardless of the specific occupational domain. However, other sophisticated technological tools can be beneficial as well. One such technology widely among R&D

professional and engineers of all types is computer aided design, or CAD.

While CAD is typically thought of as part of the design process, it can also be used during problem solving because it provides alternative ways to look at parts and structures in both two- and three-dimensions. Such a visualization tool can provide designers with new insight into issues with existing designs, and can potentially help guide them in better defining a design problem at hand. Such software also speeds up and simplifies the visualization of designs, allowing designers to "play" with new ideas with increased efficiency, in turn allowing them to spend less time on detail and more time on identifying novel conceptualizations of design problems.

In a recent study, Robertson and Radcliff (2009) investigated several mechanisms by which CAD tools may influence the creative problem solving process; generally speaking, the results demonstrated that CAD is a very useful tool for communication and visualization. However, CAD-facilitated design can also result in circumscribed thinking, premature design fixation, and bounded ideation, to some extent, particularly during the early stages of the creative process. Similarly, Fallon (2000) expressed concern that current CAD tools, used improperly, can potentially have a negative impact on the engineering and design professions, concluding that typical CAD users often spend their time solving computer problems regarding how to draw, rather than solving architectural problems and that the observed gap between engineers and CAD technicians tended to isolate people rather than support a communicative team environment.

As such, CAD should be considered as a potentially supplement to the problem construction phase of creative design; it can benefit the design process by providing new ways to visualize problems, but should not take the place of face-to-face interaction and communication among designers, which is likely the most critical requirement for success in this first stage of the creative process. Once designers have successfully identified and clarified the problem at hand, they need to begin to construct a course of action to approach developing a solution. This process begins with the information gathering stage.

## Information Gathering

In this second stage of the creative process, simple, reliable access to data is crucial. However, very rarely does any individual have all of the relevant or necessary information available to him or her. Therefore, as with the first stage of the creative process, the most essential tools used for *information gathering*, appear to be those that provide efficient access to rich sources of relevant data. However, while technologies such as the Internet can certainly facilitate information gathering, the role of human interaction is likely more valuable at this stage than any particular technology.

Given their role of developing new technology themselves and the technical savvy and proclivity to work with machines typically associated with engineers and R&D professionals, it comes as a bit of a surprise that a consistent finding throughout the literature dating back to the 1960's is the critical importance of, and steadfast reliance on, human interaction as the primary source of information among professionals in these fields. In fact, in the conduct of their day-to-day work, engineers make extensive use of communications through interpersonal means as well as through information found in other sources (e.g., handbooks, internal reports) (Hertzum & Peftersen, 2000). Indeed, communication is a crucial aspect of co-operative work, and several studies have indicated that engineers typically spend 40%-66% of their time communicating, which includes gathering information and producing the results from their work (King et al., 1994).

In fact, most studies conclude that internal communication of any kind is generally more critical in engineering work than is communication with external sources (i.e., outside of the organization) (e.g., Bichteler & Ward, 1989, Gerstenfeld & Berger, 1980, Kasperson, 1978, Rosenberg, 1967; Von Seggern & Jourdain, 1996; Zipperer, 1993). More to the point, it has been demonstrated that face-to-face interactions remain essential for engineers and that R&D engineers rely on their own information and communication with their colleagues to acquire information, get trusted opinion, and as the impetus for creative discourse before all other available internal sources (Fidel & Green, 2004; Hertzum & Peftersen, 2000; Skovvang, Elbaek, & Hertzum, 2006). Therefore, at this early stage in the creative process, instead of considering technologies that may themselves drive creativity and innovation, it may be more valuable to consider technologies that facilitate communication among designers. Such technologies may include email, teleconferencing and video conferencing. However, it is essential to remember that such virtual communication should be in *supplement* to face-to-face interaction, which seems to be critical at this point in the creative process.

What is perhaps most surprising is that this phenomenon of reliance internal, face-to-face personal communication is not new, nor has it dramatically changed over the past several decades. For example, in 1967 Rosenberg presented 96 R&D professionals with three hypothetical situations and asked them to rank order eight information-gathering methods according to personal preference in each of the three situations. The resulting ranking for the engineering sample clearly demonstrated a strong preference for proximal, oral sources (see Table 2).

Several related, but more contemporary, studies have reported quite similar results. For example, Shuchman (1982) found that the combination of three internal sources of information - conversations with colleagues, consulting supervisors, and reading in-house technical reports – emerged as the most important internal resources for information gathering. Such internal sources were ranked as moderately or very important by 82% of the sampled engineers. Similarly, in a survey of 500 scientists and engineers in industrial R&D, Chakrabarti et al. (1983) reported that work groups were the most frequently used source of information followed by trade journals, handbooks, newspapers, in-firm experts, and 17 other information sources. More recently, Von Seggern and Jourdain (1996) asked aerospace engineers and scientists to indicate which information sources were used in solving a technical problem. The 228 respondents across three different work sites displayed nearly identical patterns of usage almost 30 years later (see Table 3).

As can be readily determined from the extant research, engineers and R&D professionals get

*Table 2. Information-gathering methods used in problem solving (Rosenberg, 1967)*

| Rank-ordered information gathering method |
|---|
| 1 Search your personal library |
| 2 Search material in the same building where you work, excluding your personal library |
| 3 Visit a knowledgeable person nearby (within your organization) |
| 4 Telephone a knowledgeable person who may be of help |
| 5 Consult a reference librarian |
| 6 Use a library that is not within your organization |
| 7 Write a letter requesting information from a knowledgeable person – 20 |
| 8 Visit a knowledgeable person – 20 miles away or more |
| * Adapted from Skovvang, Elbaek, and Hertzum (2006) |

*Table 3. Information sources used in problem solving (Von Seggern & Jourdain, 1996)*

| Rank-ordered information-gathering method |
|---|
| 1 Personal store of technical information 99% |
| 2 Spoke with a co-worker or people inside my organization 99% |
| 3 Spoke with colleagues outside my organization 93% |
| 4 Used literature resources found in my organization's library 88% |
| 5 Searched an electronic database in the library 72% |
| 6 Spoke with a library or technical information specialist 62% |
| * Adapted from Skovvang, Elbaek, and Hertzum (2006) |

most of their information from colleagues and other, immediate internal sources, and a number of studies have found that *the cost* associated with using an information source is the single most important determinant of its use, as opposed to *the quality* of the information. This scenario has given rise to a law of least effort whereby, when selecting among information sources, engineers act in a manner intended to minimize the expended effort, not to maximize gain.

Given the application of the law of least effort, one might conclude that with the arrival and subsequent proliferation of information gathering and communication technologies that have characterized the current 'Information Age,' that engineers and R&D professionals have embraced such tools to facilitate the acquisition of critical information, there is at least some evidence that this is the case.. Allard, Levine, and Tenopir (2008) conclude that searching technology (e.g., the Internet and similar resources) has improved, making the search for information less time consuming, and that engineers are choosing the Internet as a primary source despite the fact that this information may not be as focused, as timely, or as authoritative. It would likely often be more effective if designers instead consider technologies, such as the Internet, as *secondary* sources during the information gathering stage, supplementing information gathered through more direct, interpersonal means.

As a primary resource, technology should not be considered as a replacement for interpersonal communication, but instead as a means of facilitating it. Such tools may not necessarily change the reliance on, and important of, interpersonal communication as an information source, but that such tools may change *the way in which* this communication is transacted. In other words, colleagues and other internal sources of information will likely remain crucial sources of information, but emerging technologies (e.g., smart phones, chat rooms, blogs, instant messaging services, file transfer and video conferencing software) may provide more efficient means to access such sources of data. Once designers have gathered sufficient information, they must then determine how best to apply it to the identified problem.

## Concept Selection

Following the information gathering stage, designers likely have a large pool of raw information to work with. While such information can ultimately be useful in approaching and developing a solution to a problem, such an array of information does not itself constitute a solution. Instead, once sufficient information has been gathered, designers can begin to identify what pieces of data might be useful to the present problem, and begin sorting it into relevant groups of concepts. (Hunter et al., 1996; Mumford et al., 1991; Mumford et al., 1996).

At this *concept selection* stage, designers would likely most benefit from tools that help them catalog their ideas and sort them into coherent groups. The act of sorting information and determining which ideas to pursue requires analytical ability on the part of the designers (Sternberg, O'Hara, & Lubart, 1997), and the use of software or other technology to supplement such abilities could facilitate the design process. Of particular note here are technologies that enable designers to share and communicate their ideas with one another. As discussed in the preceding stages, facilitating such communication enables designers to share knowledge and expertise, in this case, providing additional resources to structure disparate information into more coherent concepts.

One set of tools that may facilitate concept selection are wiki-based design databases, in which designers can freely upload information, see what their peers have also provided, and construct groupings and categories that may be meaningful for later stages of the creative problem solving process. Such systems also have the benefit of being searchable, allowing designers to quickly call up potentially relevant pieces of information uploaded by themselves or their peers, reexamine them, and decide how to categorize them. Schneiderman (2007) sites such rapid exploration as being essential if a tool is to be a successful component of the creative design process. Ideally, such a system will have a search system that goes beyond simple key words, allowing users to search by individual elements, and apply and remove filters freely, with search results updating automatically (Schneiderman, 2007). Software such as this may speed up some of the more tedious aspects of this stage of the creative process, helping designers work their way toward later stages of the process more efficiently.

However, an important potential limitation here is that, by design, such tools are web-based, allowing users to access and share information with one another from anywhere in the world. While on the surface it may seem as though this would enhance productivity, by allowing teams of designers to continue working with one another even when not in proximity, over-reliance on such virtual communication can stifle the creative process. As has been emphasized previously, face-to-face communication is crucial for the success of the creative design process, and sacrificing it for virtual communication will likely hinder performance and productivity (Hertzum & Peftersen, 2000). As such, web-based technology, while an excellent tool to support creative design, is not a replacement for face-to-face interaction, where engineers can see one another, and talk through their ideas, concerns, and potential means of categorizing information in real time. As such technology continues to develop and become more prominent in the workplace, it will be important to remember this key limitation.

In essence, as designers navigate the concept selection stage, and prepare to combine and integrate concepts into coherent designs in the subsequent stages, it remains important that technology is used to *supplement* and provide a framework for the selection of concepts, not to replace legwork, discussion, and free-thinking among the designers. It is worth emphasizing that technologies such as those presented here, while potentially beneficial if used correctly, will show more utility once groups of task-relevant ideas have been established, and designers are preparing to combine them in novel ways, in order to approach a solution. Such is the case in the subsequent *conceptual combination* stage.

## Conceptual Combination

Once designers have organized the information available to them into groups of potentially relevant ideas, they can truly begin to innovate, by combining ideas in unusual or previously unexamined ways. As it is the combinations of ideas developed during this stage that later give rise to actual design solutions, this conceptual combination stage may have the most direct impact on creative performance (Hunter et al., 2006). It is important to note that the process of conceptual combination is itself abstract, involving combining disparate, broad concepts in new ways, versus using ideas to create a single coherent design (Hunter et al., 2006).

In this stage, beneficial technology should provide a platform on which engineers can quickly compose and examine numerous potential combinations of ideas, in order to determine how best to proceed with the development of a novel and effective design. Such technology should either allow users to explore how others have previously combined ideas and used them to drive designs, or more ideally, provide a framework in which users or design teams may do so themselves.

Somewhat related to the design wikis discussed previously are online, searchable design repositories. While not necessarily as readily editable by the end-user as a wiki may be, these repositories act as libraries of data, organized by type of design. Examples of such systems include a still-growing repository being developed at the University of Missouri-Rolla (UMR), in collaboration with engineers at the Pennsylvania State University, Bucknell University, and Virginia Tech, as well as the *National Design Repository,* hosted at Drexel University.

Such resources contain a sizeable number of designs, from a broad range of engineering fields, and enable users to search for designs and related information that might help them address their present problem. Technologies such as this can be useful when dealing with approaching problems similar to ones that have been solved by previous designs. Here, designers may look at what has been used previously, and apply the same system for combining ideas (Mumford, et al., 1997). In this way, using technology that displays existing designs may help engineers develop a road map for combining their own ideas into novel, but similarly successful, designs.

However, it is important to note that, at this stage, examination of the designs produced by others should be used carefully. As suggested above, design repositories may help users at this stage of the creative process to develop a framework for combining concepts into new ideas, *not* to generate new ideas directly. If used improperly at this stage, information technology such as design repositories may ultimately be detrimental to the creative design process, causing users to fixate on the types of designs created by previous users. This has been empirically demonstrated, showing that exposure to detailed example designs early in the design process resulted in less variability and innovation among designers' products (Jannson & Smith, 1991).

At this early stage of the creative process, it is important to remember that the goal is to identify combinations of *concepts* that might be used to address a given problem, not on selecting a fully-established solution. Regardless of the specific technologies being used at this stage, designers should be encouraged to focus on the combination of ideas underlying designs, and not the completed design itself. If designers dwell too much on the completed designs provided by others (or their peers), they may become fixated on certain types of solutions, unnecessarily narrowing the range of solutions they later explore, and limiting the potential for innovative outcomes.

To best avoid this pitfall, users must be able to mentally decompose designs into relevant conceptual elements from the designs they are viewing, examining the conceptual make up of these designs, rather than the overall design itself. In order to support this process, technologies utilized at this stage should ideally present designs more abstractly, in terms of broader functional models or conceptual designs, instead of finished products. Such is the case with the design repository developed at UMR, which allows users to browse designs at multiple levels of complexity, from broad functional models to much more specific designs. In this way, users can study at a basic level how previous solutions combined design concepts, and incorporate these into their own conceptual combination approach.

Clearly, such technology can be useful at this point, but it must be used with care in order to ensure it facilitates, rather than stifles, the creative process. Doing so requires "synthetic ability" (Sternberg et al., 1997), a skill which enables designers to see connections between elements and redefine problems. Indeed, in order to innovate, designers need to be able to be able to go beyond manipulating an established solution, deconstructing it and building new, useful design elements from what they learned in this deconstruction (Edmonds, Candy, Jones, & Souff, 1994) As this is a skill set that is amenable to training (Sternberg et al., 1997), it may be worthwhile to train designers in this area, better enabling them to utilize the tools available to them to extract novel, abstract combinations of design concepts.

It is also crucial to remember that design is often a social process (Warr & O'Neill, 2005), and so effectively facilitating collaboration between designers is essential for the success of the process. Indeed, in Schneiderman's (1998, 1999, 2002) model of creative performance, he indicates that social interaction, in the form of idea sharing between peers and mentors, occurs throughout the process, and is especially essential at the brainstorming stage, which is similar in content to Mumford's *conceptual combination* stage. Similarly, Hertzum and Peftersen (2000) note that engineers seeking information tend to seek out both informational documents and experts to ask, suggesting a perhaps unexpected reliance on interpersonal communication. As such, if technology can provide a framework to facilitate peer and mentor interaction and information sharing, it is expected that the designers will arrive at more novel and effective combinations of concepts, enhancing creative performance at this stage, and also providing a stronger framework for the following *idea generation* stage.

## Idea Generation

After creating new combinations of abstract concepts in the conceptual combination stage, designers in *idea generation* stage begin to integrate these concepts into coherent design ideas to address the problem at hand. This stage of the process has similar cognitive demands to those of the preceding stage (Hunter et al., 2006), but more concretely applied to the development of a specific potential product. As such, effective technologies to support this stage are those that reduce cognitive load on the designers, allowing them to focus on developing new ideas, by reducing the difficulty of combining and visualizing concepts.

One interesting and socially-pervasive instance of technology that may support the design process are publically available, Internet-based resources, such as the online video site YouTube. While typically not highly technical, and not inherently moderated or structured in a way that makes it seem immediately conducive to the research and development process, such resources are continuing to grow, and are likely to become increasingly prevalent throughout the foreseeable future. Such services house a massive and ever-growing quantity of user-provided content, and can deliver it with haste. While not currently recognized as a readily harnessed professional or academic resource, such services could provide a useful framework for technology to support the idea generation phase of the creative process.

At this point, once users have decided on the combinations of design concepts they want to pursue, examining multimedia representations of how other users have addressed such problems could be incredibly useful. It must be noted again, however, that in using such tools, users would need to guard against design fixation. Perhaps by balancing such online exploration with interpersonal and in-lab design sessions, this issue may be minimized, and the potential benefits maximized.

Additionally, this stage of the creative process can be positively influenced by tools that allow users to take broad concepts and use them to compose potential product designs with relative speed and ease. Schneiderman (2007) notes that such tools support creativity by allowing users to generate alternative designs quickly, as well as test, revise, and replace them as necessary, until a seemingly workable design is decided upon. When such tools are well designed and easy to use, they should help to minimize the cognitive complexity of visualizing a design idea, allowing users to instead focus on building upon and improving it (Schneiderman, 2007). CAD software, such as Autodesk's Inventor and AutoCAD, and Dassault Systèmes' SolidWorks, are one clear example of technology that can be incredibly useful in facilitating the idea generation stage in such a way.

Schneiderman (2007), in his recommendations for effective tools to support creativity, provides several suggestions that clearly apply to any such technology utilized during the idea generation stage. Regardless of the specific technology in question, a crucial element to consider is accessibility to new users. The interface for such a tool should be streamlined and logically laid out, so that users can find their way around quickly. This prevents frustration or performance delays where users must struggle to learn the system before they can get back to designing a product. Schneiderman suggests layering functions of the software, so that the most beginner-friendly and frequently used features are readily accessible, while less frequently used or more advanced features can be accessed as users require them. A simple analogy for such a system is a search engine such as Google, where users can either enter a simple query and click one button, or access an "advanced search" menu, with many more options (Schneiderman, Hewett, Fischer, & Jennings, 2007). At any level of complexity, the design of such a software system should emphasize allowing the user to draw out and test his or her ideas, rather than wrestling with the program.

Additionally, supporting communication and collaboration is again an important theme of technology use at the idea generation stage. While certainly essential during the earliest stages of the design process, when the team is deciding how to approach a problem, they can also facilitate performance during the present stage. This is in line with the concern that over-reliance on CAD may actual impair designers' performance (Fallon, 2000). Again, the combinations of concepts that users may be working with as they enter this stage are likely abstract and quite broad, leaving much room for disagreement and differences of opinion among designers.

If such differences can be handled constructively, then a team of designers' creative performance can be enhanced. However, if "creative differences" are left unchecked, it is likely that the team's performance will suffer, falling below the level of its individual members (Taggar, 2001). In the present context, effective technologies would be those that enable users to communicate in a manner that allows them to feel safe with one another, share ideas easily, and maintain accurate records of their ideas and interpersonal interactions. Such systems should make it easy for users to share comments, images, documents, and exchange ideas related to the design process, with as minimal a delay as possible. Google's new Wave software is an example of such a novel technology that may be used to support the design process. Such a tool could be useful in managing some of the exchanges of information that occur in the inherently social design process (Warr & O'Neill, 2005). If such conditions are met, it is expected that higher quality creative performance should result.

However, it must be reiterated that, while technologically-mediated communication can be beneficial for users seeking to exchange and develop ideas during this phase of the creative process, such systems must be considered a *supplement* to more traditional face-to-face communication, which is absolutely essential to the success of the creative design process (Fidel & Green, 2004; Hertzum & Peftersen, 2000). Such tools have potentially large benefits for project management at the idea generation stage, but should under no circumstances be considered a substitute for face-to-face, interpersonal communication.

If designers are sufficiently skilled, communicate well, and appropriately use the tools available to them, they will eventually produce a body of potential design ideas. Before any of these ideas can be put into production, their quality and potential effectiveness must be assessed. It is this procedure that is the subject of the next stage of the creative process.

## Idea Evaluation

While the *idea evaluation* stage has typically received less attention than others, it is an essential component of the creative process (Hunter et al., 2006; Lonergan, Scott, & Mumford, 2004). Paying close attention to practical constraints on the design process leads effective designers to discard designs that are likely to fail, or would be too costly to implement, while pushing forward the development of ideas that address problems in new, more efficient and effective ways. At this stage, technology can benefit the creative design process by providing designers with tools to examine potential design ideas prior to production, preferably in a manner that saves time and resources.

One set of tools already discussed previously are wikis and online design repositories. Originally discussed in the context of the concept selection and conceptual combination stages, these tools are again potentially useful during idea evaluation. At this stage, designers can again examine what others have done to address potentially similar problems. If the proposed idea is overly similar to something already in existence, then perhaps the designers could go back to their pool of ideas and select something new, or look for ways to revise their concept.

Alternatively, if a proposed design is entirely different from how others have approached a similar problem, it may be worth evaluating the cause of the discrepancy; either the design may be truly innovative, or the engineers may be too far off base, and need to revise their design idea. While using the work of others as a reference for evaluating new designs is potentially useful, it must be noted that simply being similar to, or different from, an existing design is rarely sufficient justification to judge a design a success or failure. Such over reliance on these tools could lead to discarding a valuable, novel idea on one hand, or fixating on an established design on the other.

Again, creative performance requires a measure of risk, and a willingness to try out ideas that may have been ignored or insufficiently considered by others (Sternberg et al., 1997). Therefore, while such tools may be a good starting point for idea evaluation, they should not be the sole crutch at this stage of the design process. Instead, designers should use this technology to support their own knowledge and skills with regard to designing a potential solution, instead of relying on it as the primary source of validation.

Another technology that may be quite effective in supporting the idea evaluation stage is rapid prototyping. Originally released on the commercial market in 1988, this is a form of technology where a stripped-down version of a proposed design idea is quickly developed and constructed, typically from low-cost materials, the purpose being to approximate some aspect of a proposed design or its components (Chua, Leong, & Lim, 2003). Often, this technology involves fabrication from a machine that is directed by a CAD file containing the engineer's design. As such, using rapid prototyping during the idea evaluation stage is often a natural extension of the modeling often undertaken during the idea generation stage.

Rapid prototyping has several potential benefits for creative design. First, and most generally, it enables engineers to better visualize their designs. This may point out particular strengths or weaknesses in a design that would not have been identifiable in a two-dimension representation on paper or a computer screen. It also enables users to test questions they may have about a proposed design idea by actually interacting with model. Chua et al. (2003), for instance, cite the example of designers testing a prototype of a chair to ensure that it has comfortable arm support, or examining the hinge on a prototype of a pair of folding reading glasses, to make sure it functions properly.

A key benefit of rapid prototyping technology is the potential time and financial savings it can impart on the design process. Since such prototypes can be produced quickly, and with minimal cost (von Hippel, 1994), designers are able to explore a larger number of ideas than they may have without such a system to support evaluation, increasing the likelihood that a novel idea will be found. Furthermore, if problems are found with a design, it is typically easy and inexpensive to reproduce a new model for testing, once the issues have been investigated and resolved (von Hippel, 1994).

Another crucial consideration is how technology can facilitate the social processes essential to idea evaluation. In order to critically analyze and refine an idea, designers must be willing to express their own ideas, and to use their expertise to critique the ideas of their peers. In such a context rapid prototyping is again an example of a valuable technology. The low cost of models built via rapid prototyping reduce the risk associated with experimentation; since it is faster and cheaper to correct for mistakes or errors in design in a low cost prototype than in a completed design, designers may be more willing to take risks on less conventional preliminary designs, potentially giving rise to a novel design that otherwise would have been lost (Chua et al., 2003).

Such technology also reduces the costs, and therefore the risks, associated with a failed design, likely leading designers to be more willing to suggest unconventional ideas to test. As creative performance often requires individuals to be able to identify unlikely or discarded ideas and adopt them (Sternberg et al., 1997), knowing that there is little cost to attempting to do so may facilitate such actions that can give rise to creative performance.

Such an environment where users are able to express their ideas, try them out, and also critique others is considered high in *participative safety*, a crucial element for a climate for innovation (West, 1990; Anderson & West, 1998). Similar to the concept of psychological safety proposed by Edmonson (1999), this essential factor for team creative performance enables team members to feel more comfortable taking risks within the team, and is also associated with criticism of ideas being perceived as constructive, rather than derogatory. In this way, a psychologically safe team context encourages the free exchange of ideas, and encourages members to present suggestions and critiques in order to improve the product. In practice, this should lead to the acceptance of more original ideas, and support the continued development of more novel, unconventional ideas into potential solutions in the subsequent stages of the creative process.

Ideally, in such circumstances, the use of rapid prototyping and the subsequent low cost of revising a design will encourage team members to voice any concerns, leading to the necessary design revisions and, ultimately, a design that has the potential to be effective. However this stage is reached, once such a potentially effective design has been identified, the designers need to begin planning its implementation.

## Implementation Planning

After an idea is formed and evaluated, the next phase of innovation is the implementation planning stage. In this stage, the organization makes an adoption decision for an idea and creates a plan to carry it out (Anderson & Gasteiger, 2007). This stage requires task-relevant knowledge, skills, and abilities, as well as personality characteristics such as persuasion (Mumford et al., 1991). Planning is an important stage of creativity because plans (a) help maximize limited resources, (b) help creative ideas align with broader business strategies, and (c) are important for implementation of ideas over a prescribed time period (Hunter, Friedrich, Bedell, & Mumford, 2006).

In contrast to early stages of the creative process, implementation planning focuses on all of the details of a design and how they each fit together. Here, technology becomes a vital tool for visualizing how disparate design elements fit together to address the problem in practice, in ways that align with the organization's goals.

For instance, while a sketch pad may be sufficient for early functional designs, CAD software can greatly contribute to a project planning effort (Inouye, Mitchell, & Blumenthal, 2003). In the implementation phase, details become important as potential problems are sought out. In addition, the generative processes of software such as CAD allow designers to create entire models automatically rather than building them one piece at a time (Inouye, Mitchell, & Blumenthal, 2003). Such procedures again help facilitate the integration of pieces of a solution into a complete package.

With the increasing speed and decreasing size of computer processors, computer systems are becoming ever faster and more practical to use

(Inouye, Mitchell, & Blumenthal, 2003). Given their increased processing capabilities, software that can be used to generate predictive models can be incredibly useful at this stage. Such software allows scientists and designers to generate "what if" scenarios, based on physical boundaries or human behavior, to foresee problems before they occur. Often, business plans fail because they are too rigid (Van Der Zee & De Jong, 1999) perhaps as scenarios are run through computer programs, implementation teams can broaden the boundaries of their thinking. In addition, computer systems can allow for easy access to previous data on projects including duration, cost, and manpower. Facilitating such access can facilitate the production of an efficient, cost-effective, implementation plan.

Technology can also be useful in facilitating multi-disciplinary cooperation at the implementation stage. For instance, the amount of information stored in a digital model of a building can capture design elements from several teams to highlight where system overlap can become a problem. In urban planning, teams of architects, civil planners, politicians, trades, and stakeholders can test designs in a forum that allows everyone to communicate with each other using concrete examples (Hughes & Stapleton, 2005).

Technology can also aid in more routine tasks, such as sending large documents quickly or organizing planning meetings with team members globally. Information that is stored on an internet server or transmitted by several users is unlikely likely to be lost and can be given to a new project member immediately. Meetings can be digitally recorded and all aspects of the project can be tracked. Information is quickly obtained, speeding up decision making processes. Daily planning tasks become ever more manageable. Such advances make the implementation planning stage more efficient, and enable designers and project managers to respond and adapt to changing circumstances more readily.

Additionally, as suggested by Schneiderman (2000), advancements in instant communication has enabled "more people to be more creative more of the time." In fact, organizations who do not utilize these technologies may not be able to compete with rival companies because traditional approaches may be ineffective because of how rapidly the business environment changes (Van Der Zee & De Jong, 1999).

However, it is important to note that there are potential barriers to utilizing new technologies during the implementation planning stage. There may be intentional opposition to new products and systems by their users. For example, employees may not be willing to learn new technology that is necessary for a system wide change (Anderson & Gasteiger, 2007). This resistance should be anticipated, and organizations can use online learning modules or a technical support staff available to assist in the transition. In addition, the organization should make it clear which type of process change will occur at what time and should be aware of other initiatives that are being implemented concurrently (Davenport, 1993).

Finally, the speed of information that technology allows can increase expectations and result in inadequate planning. With the complexities of planning for an ill defined task, there may be an overreliance on technology to do more than it was intended to do. Organizations should allow adequate time to prepare for possible problems and to ensure ideas are implemented correctly. Since new ideas are often highly scrutinized, implementation planning is crucial for successful innovation. However, the creative process does not end with implementation of a solution; following implementation, solutions must be monitored and evaluated for success.

## Monitoring

Once implemented, prospective solutions must be carefully monitored and evaluated, order to ensure they are working. In the event that a solution appears ineffective, designers must be ready to revise or replace it as necessary. Therefore, prompt feedback and continual information about solution performance is vital. Carson and Carson (1993) suggest that feedback about creativity in one round of performance results in improved creativity in the next round. In addition, a series of qualitative interviews with those in jobs dedicated to innovation identified a need for feedback on innovative performance (Hellstrom & Hellstrom, 2002). Thus, not only is feedback essential for continuous improvement, but is specifically asked for by the creators of innovation. At this stage of the creative process, beneficial technologies would be those that facilitate the rapid collection and dissemination of data regarding solution performance.

One such useful set of technologies are sensors or trackers that can convey detailed information about system or product is performing. For example, tracking and recording the specific ways that consumers are using a new software may yield more information than just counting the number of products sold. Developers could use such information to determine which features are used the most, and could continue to improve these features in future iterations of the product. As a further example, Hughes and Stapleton (2005) describe a software called Mixed Simulation Demo Dome, which could be used to monitor data about a newly built roller coaster ride. The software captures real time data such as safety, crowd control, staffing assignments, materials management, merchandise stocking, and technical maintenance. Feedback from all of these areas can reveal successful innovations as well as areas for improvement.

The Internet can also provide useful a venue for instant feedback from the consumers of innovative products. This could be a simple website as a form of a suggestion box, or a more complex system of surveys via product registration or by email. If implemented properly, such as by asking specific questions and prompting respondents for detailed information about which aspects of a product work well and which do not, such Internet-based surveys can provide product information that can be a vital part of the monitoring process.

On the whole, once a system is in place, and feedback is being gathered, designers should be ready to make modifications or adjustments as need be, in order to ensure that the solution put in place remains effective. Ideally, such a solution will be highly novel, and beneficial to the organization, though if revisions are necessary, the requisite social and technical processes should remain in place to facilitate this performance.

## FUTURE RESEARCH DIRECTIONS

This chapter has sought to provide insight into how technology may best be integrated into the creative process, and the authors believe that the inferences and recommendations contained here are sound. However, it must be noted that there are some limitations that may influence the generalizability, scope, and future research directions for this topic. For one, in many cases, there is limited empirical data to support conjectures, with much of the support for specific ideas or implementations coming from examination of case studies or other anecdotal evidence.

Additionally, in effort to broaden applicability, this chapter has examined the implementation of technology across types of work within engineering and design jobs. It is likely that there are task-specific processes (along with relevant technologies) within individual types of jobs. While it is unrealistic to examine every possible job individually, it must be noted that the approach taken in this chapter may exclude certain job-specific scenarios. Conversely, this chapter focused only on the impact of technology on the creative process within the fields of engineering. A broader study of technology implementation at work would likely find recommendations that generalize across a wider range of work areas, including other academic disciplines, and even artistic fields.

Additionally, for the sake of examining the potential effects of technology at each stage of the creative process, the process model was in many areas portrayed as effectively linear. *However*, as was previously mentioned, creativity is not a linear process, and the process model does not assume it to be; stages can be recursive, with later stage performance feeding back into earlier stages, potentially leading designers to revise their design and revisit earlier stages prior to arriving at a solution. Technology may influences some of these recursive relationships that may be overlooked by following this more linear approach to analysis.

Finally, the issue of potential obsolescence must be addressed. Rather than a limitation of this research specifically, this is an acknowledgement that, given constant advancement and new technological developments, any study of technology must remain up-to-date in order to retain its usefulness. As technology advances rapidly, there is always the chance that some of the specific technologies discussed in this chapter may prove to be out of date, and of decreased relevance. However, it is important to remember that even as new technologies develop, the creative process model, and the analysis of necessary conditions for each stage, will remain constant, providing a framework upon which future examination of technological implementation can be carried out.

Future research on in the area of facilitating creative performance with technology should focus on empirical examination of technology as a moderator of creative process stages. As discussed previously, much of the existing information on this specific topic is based on conjecture arising from case studies or anecdotal evidence. Empirical study would be highly beneficial in establishing further, and more detailed, recommendations for implementing technologies to support creativity.

Additionally, future research should begin to examine the effects of technology on the creative process in a non-linear manner. Since the creative process is itself not linear, with designers often revisiting earlier stages to revise their ideas or potential solutions, further attention should be given to how technology may facilitate this process, and help designers navigate such transitions. Similarly, it will be vital to examine how technology impacts creative performance *across* the creative process. A broader-level examination will help to shed light on what sort of technology, or combination of technologies, may have the most pronounced benefits across the creative process, which could help organizations plan more cost-effective interventions.

Further, given the continual advancement inherent in technology, future research should examine newer, emerging technologies with regard to their potential influence on creative performance. In particular, the coming ubiquity of mobile communications devices, such as smart phones, will likely have a lasting impact on creative work. Though briefly addressed in this chapter, it is expected that the miniaturization and increased access to information and social connectedness inherent in using such devices will have a lasting impact on creative performance.

Finally, in order to broaden the applicability of the recommendations provided in this chapter, future research should examine the impact of technology on creative performance in a broader sample of industries and content areas. As the creative process model is not industry-specific, it will still provide a crucial framework for such study; adopting this approach will allow researchers to focus on how technologies may influence creative performance in specific fields, without having to start from a blank slate.

## CONCLUSION

As has been discussed in this chapter, there are numerous ways by which technology can influence each stage of the creative process, both in terms of benefits if implemented properly, or hindrances if done poorly. Any organization seeking to remain competitive innovative should be aware of such elements, and be prepared to implement new technologies as appropriate. The technologies discussed in this chapter are also only some of many that may influence the creative process. Furthermore, as technology continues to advance, new developments will inevitably arise that will have further potential to impact the creative process. It will no doubt be vital for organizations to stay up to date with such developments, in order to maintain a competitive edge.

However, as asserted by Mumford & Hunter (2005), having more resources (such as novel technologies) is indeed helpful for innovation, but there is a limit to how much assistance resources can provide. The time necessary to learn new technology and the vast possibilities in design that technology allows may lead to a loss of focus that may hamper success in creative endeavors. Therefore, organizations must maintain sight of their goals with respect to innovation, and only adopt and implement new systems when they will have clear benefits, without impairing employee's creative performance in other ways.

Such understanding requires conceptualizing creativity as a detailed process, and not simply an input-output relationship. If creativity were only considered in terms of outcomes, it is impossible to gain insight into *how* technology can impact creative performance; it becomes just another input, filtered through a "black box." In order to identify what technologies may be useful, when, and in what ways, it is vital to consider the cognitive and behavioral mechanisms operating throughout the design process.

The eight-stage creative process model discussed in this chapter provides an excellent framework for studying the mechanisms by which various technologies can impact creative performance, which can be useful for both academic study and planning for practical implementation. It is important to note that such a model is not technology-specific. Rather, it provides a broad framework within which technology may interact with user characteristics, behaviors, cognitions, and attitudes. In this way, the designers themselves, and their interactions with one another, still drive the creative process. Again, creative design is a social process (Warr & O'Neill, 2005), and any effective attempt to study it must necessarily consider it as such.

In essence, when considering the influence of technology on the creation of new products and systems, it is merely a tool of human innovation; no matter how technology develops, it is people who will always be the driving force of the creative process.

## REFERENCES

Adelman, L., Gualtieri, J., & Stanford, S. (1995). Examining the effect of causal focus on the option generation process: An experiment using protocol analysis. *Organizational Behavior and Human Decision Processes*, *61*, 54–66. doi:10.1006/obhd.1995.1005

Allard, S., Levine, K. J., & Tenopir, C. (2008). Design engineers and technical professionals at work: Observing information usage. *Journal of the American Society for Information Science and Technology*, *60*, 443–454. doi:10.1002/asi.21004

Anderson, N., & Gasteiger, R. M. (2007). Helping creativity and innovation thrive in organizations: Functional and dysfunctional perspectives. In Langan-Fox, J., Cooper, C. L., & Klimoski, R. J. (Eds.), *Research companion to the dysfunctional workplace: Management challenges and symptoms* (pp. 422–440). Cheltenham, UK: Edward Elgar Publishing.

Anderson, N. R., & West, M. A. (1998). Measuring climate for work group innovation: Development and validation of the team climate inventory. *Journal of Organizational Behavior*, *19*, 235–258. doi:10.1002/(SICI)1099-1379(199805)19:3<235::AID-JOB837>3.0.CO;2-C

Baughman, W. A., & Mumford, M. D. (1995). Process-analytic models of creative capacities: Operations influencing the combination-and-reorganization process. *Creativity Research Journal, 8*, 37–62. doi:10.1207/s15326934crj0801_4

Bichteler, J., & Ward, D. (1989). Information-seeking behavior of geoscientists. *Special Libraries, 79*, 169–178.

Carson, P. O., & Carson, E. D. (1993). Managing creativity enhancement through goal-setting and feedback. *The Journal of Applied Psychology, 27*, 36–45.

Chen, H. L., Cannon, D., Gabrio, J., Leifer, L., Toye, G., & Bailey, T. (2005). Using Wikis and weblogs to support reflective learning in an introductory engineering design course. In Gero, J. S., & Lindermann, U. (Eds.), *Human Behaviour in Design 05* (pp. 95–105). Key Centre of Design Computing and Cognition, University of Sydney.

Chua, C. K., Leong, K. F., & Lim, C. S. (2003). *Rapid prototyping: Principals and applications* (2nd ed.). Singapore: World Scientific Publishing.

Davenport, T. H. (1993). *Process innovation: Reengineering work through information technology*. New York: Harvard Business Press.

Dewey, J. (1910). *How we think*. New York: Heath. doi:10.1037/10903-000

Dillion, J. T. (1982). Problem finding and solving. *The Journal of Creative Behavior, 16*, 97–111.

Edmonds, E., Candy, L., Jones, R., & Soufi, B. (1994). Support for collaborative design: Agents and emergence. *Communications of the ACM, 37*(7), 41–47. doi:10.1145/176789.176793

Edmonson, A. (1999). Psychological safety and learning behavior in work teams. *Administrative Science Quarterly, 44*(2), 350–383. doi:10.2307/2666999

Fallon, K.K. (2000, April 3). *Design firm principals unite to demand better technology: Leading architects examine practice, question impact of CAD tools on architecture, engineering, and construction*. [Press release]. New York: Business Wire.

Fidel, R., & Green, M. (2004). The many faces of accessibility: Engineers' perception of information sources. *Information Processing and Management: an International Journal, 40*, 563–581. doi:10.1016/S0306-4573(03)00003-7

Finke, R. A. Ward., T.B. & Smith, S.M. (1992). *Creative cognition: Theory, research, and applications*. Cambridge, MA: MIT Press.

Gerstenfeld, A., & Berger, P. (1980). An analysis of utilization differences for scientific and technical information. *Management Science, 26*, 165–179. doi:10.1287/mnsc.26.2.165

Getzels, J. W., & Csikszentmihalyi, M. (1975). From problem solving to problem finding. In Taylor, I. A., & Getzels, J. W. (Eds.), *Perspectives in creativity* (pp. 90–116). Chicago: Aldine.

Hellstrom, C., & Hellstrom, T. (2002). Highways, alleys and by-lanes: Charting the pathways for ideas and innovation in organizations. *Creativity and Innovation Management, 11*(2), 107–114. doi:10.1111/1467-8691.00242

Hertzum, H., & Pejtersen, A. M. (2000). The information-seeking practices of engineers: Searching for documents as well as for people. *Information Processing & Management, 36*, 761–778. doi:10.1016/S0306-4573(00)00011-X

Hughes, C. E., & Stapleton, C. B. (2005). *The shared imagination: Creative collaboration in augmented virtuality*. Paper presented at HCI International, Las Vegas, NV, July 2005.

Hunter, S. T., Friedrich, T. L., Bedell, K. E., & Mumford, M. D. (2006). Creative thought in real world innovation. *Serbian Journal of Management, 1*, 29–39.

Inouye, A. S., Mitchell, W. J., & Blumenthal, M. S. (2003). *Beyond productivity: Information technology, innovation, and creativity*. Washington, D.C.: National Academies Press.

Jannson, D. G., & Smith, S. M. (1991). Design fixation. *Design Studies, 12*(1), 3–11. doi:10.1016/0142-694X(91)90003-F

Kasperson, C. J. (1978). Psychology of the scientist: 37. Scientific creativity: A relationship with information channels. *Psychological Reports, 42*, 691–694.

Lonergan, D. C., Scott, G. M., & Mumford, M. D. (2004). Evaluative aspects of creative thought: Effects of idea appraisal and revision standards. *Creativity Research Journal, 16*, 231–246. doi:10.1207/s15326934crj1602&3_7

Lubart, T. I. (2001). Models of the creative process: Past, present, future. *Creativity Research Journal, 13*(3/4), 295–308. doi:10.1207/S15326934CRJ1334_07

Mumford, M. D., & Baughman, W. A. (1996). Process-based measures of creative problem-solving skills: 1. Problem construction. *Creativity Research Journal, 9*, 63–76. doi:10.1207/s15326934crj0901_6

Mumford, M. D., Baughman, W. A., Maher, M. A., Costanza, D. P., & Supinski, E. P. (1997). Process-based measures of creative problem-solving skills 4: Category combination. *Creativity Research Journal, 10*, 59–71. doi:10.1207/s15326934crj1001_7

Mumford, M. D., Baughman, W. A., Supinski, E. P., & Maher, M. A. (1996). Process-based measures of creative problem-solving skills: 2. Information Encoding. *Creativity Research Journal, 9*, 77–88. doi:10.1207/s15326934crj0901_7

Mumford, M. D., & Hunter, S. T. (2005). Innovation in organizations: A multi-level perspective on creativity. In Yammarino, F. J., & Dansereau, F. (Eds.), *Research in multi-level issues* (*Vol. 4*, pp. 11–74). Oxford, England: Elsevier.

Mumford, M.D., Lonergan, D.C. & Scott, G.M. (2002). Evaluating creative ideas: Processes, standards, and context. *Inquiry: Critical thinking across the disciplines, 22*, 21-30.

Mumford, M. D., Mobley, M. I., Uhlman, C. E., Reiter-Pamon, R., & Doares, L. (1991). Process analytic models of creative capacities. *Creativity Research Journal, 4*, 91–122. doi:10.1080/10400419109534380

Mumford, M. D., Reiter-Palmon, R., & Redmond, M. R. (1994). Problem construction and cognition: Applying problem representations in ill-defined domains. In Runco, M. (Ed.), *Problem finding, problem solving, and creativity* (pp. 3–39). Norwood, NJ: Ablex.

Mumford, M. D., Supinski, E. P., Baughman, W. A., Costanza, D. P., & Threlfall, K. V. (1997). Process-based measures of creative problem-solving skills: 5. Overall prediction. *Creativity Research Journal, 10*, 77–85.

Mumford, M. D., Supinski, E. P., Threlfall, K. V., & Baughman, W. A. (1996). Process-based measures of creative problem-solving skills, 3: Category selection. *Creativity Research Journal, 9*, 395–406. doi:10.1207/s15326934crj0904_11

Reiter-Palmon, R., Herman, A. E., & Yammarino, F. J. (2008). Creativity and cognitive processes: Multi-level linkages between individual and team cognition. In Mumford, M. D., Hunter, S. T., & Bedell-Avers, K. E. (Eds.), *Multi-level issues in creativity and innovation* (pp. 203–267). Oxford: Elsevier.

Robertson, B. F., & Radcliff, D. F. (2009). Impact of CAD tools on creative problem solving in engineering design. *Computer Aided Design, 41*, 136–146. doi:10.1016/j.cad.2008.06.007

Rosenberg, V. (1967). Factors affecting the preferences of industrial personnel for information gathering methods. *Information Storage and Retrieval, 3*, 119–127. doi:10.1016/0020-0271(67)90005-8

Rostan, S. M. (1997). A study of young artists: The development of artistic talent and creativity. *Creativity Research Journal, 10*, 175–192. doi:10.1207/s15326934crj1002&3_6

Runco, M. A., & Okuda, S. M. (1988). Problem-discovery, divergent thinking, and the creative process. *Journal of Youth and Adolescence, 17*, 211–220. doi:10.1007/BF01538162

Schneiderman, B. (1998). Codex, memex, genex: The pursuit of transformational technologies. *International Journal of Human-Computer Interaction, 10*(2), 87–106. doi:10.1207/s15327590ijhc1002_1

Schneiderman, B. (1999) User interfaces for creativity support tools. *Proceedings of the 3rd Conference on Creativity & Cognition,* (pp. 15-22). New York: ACM Press.

Schneiderman, B. (2000). Creating creativity: User interfaces for supporting innovation. *Transactions on Computer-Human Interaction, 7*(1), 114–138. doi:10.1145/344949.345077

Schneiderman, B. (2002). Creativity support tools. *Communications of the ACM, 45*(10), 116–120.

Schneiderman, B., Hewett, T., Fischer, G., Jennings, P., Czerwinski, M., & Kules, B. (2006). Creativity support tools: Report from a U.S. National Science Foundation sponsored workshop. *International Journal of Human-Computer Interaction, 20*(2), 61–77. doi:10.1207/s15327590ijhc2002_1

Scott, G. M., Leritz, L. E., & Mumford, M. D. (2004). The effectiveness of creativity training: A quantitative review. *Creativity Research Journal, 16*, 361–388. doi:10.1080/10400410409534549

Scott, S., & Bruce, R. (1994). The influence of leadership, individual attributes, and climate on innovative behavior: A model of individual innovation in the workplace. *Academy of Management Journal, 37*, 580–607. doi:10.2307/256701

Simon, H. (1973). The structure of ill structured problems. *Artificial Intelligence, 4*(3-4), 181–201. doi:10.1016/0004-3702(73)90011-8

Skovvang, M., Elbaek, M. K., & Hertzum, M. (2006). Personometrics: Mapping and visualizing communication patterns in R&D projects. In F. Crestani and I. Ruthven (Eds.), *COLIS5: Proceedings of the Fifth International Conference on Conceptions of Library and Information Sciences (Glasgow, UK, June 5-8)*, (pp. 141-154).

Smilansky, J. (1984). Problem solving and the quality of invention: An empirical investigation. *Journal of Educational Psychology, 76*, 377–386. doi:10.1037/0022-0663.76.3.377

Sternberg, R. J., O'Hara, L. A., & Lubart, T. I. (1997). Creativity as investment. *California Management Review, 40*(1), 8–21.

Taggar, S. (2001). Group composition, creative synergy, and group performance. *The Journal of Creative Behavior, 35*(4), 261–286.

Van Der Zee, J.T.M. & De Jong, B. (1999). *Alignment is not enough: Integrating business and information technology management with the balanced business scorecard.*

Von Hippel, E. (1994). "Sticky information" and the locus of problem solving: Implications for innovation. *Management Science, 40*(4), 429–439. doi:10.1287/mnsc.40.4.429

Von Seggern, M., & Jourdain, J. M. (1996). Technical communications in engineering and science: The practices within a government defense laboratory. *Special Libraries, 87*, 98–119.

Wallas, G. (1926). *The art of thought.* New York: Harcourt Brace.

Ward, T. B., Smith, S. M., & Finke, R. A. (2009). Creative cognition. In Sternberg, R. J. (Ed.), *Handbook of Creativity* (pp. 189–212). New York: Cambridge University Press.

Warr, A., & O'Neill, E. (2005). Understanding design as a social creative process. *Proceedings of the 5th conference on Creativity & Cognition,* (pp. 118-127). New York: ACM Press.

West, M. A. (1990). The social psychology of innovation in groups. In West, M. A., & Farr, J. L. (Eds.), *Innovation and creativity at work: Psychological and organizational strategies* (pp. 4–36). Chichester, UK: Wiley.

Zaltman, G., Duncan, R., & Holbek, J. (1973). *Innovation & organizations*. New York: John Wiley & Sons.

Zipperer, L. (1993). The creative professional and knowledge. *Special Libraries, 84*, 69–78.

## ADDITIONAL READING

Arias, E., Eden, H., Fischer, G., Gorman, A., & Scharff, E. (2000). Transcending the individual human mind – Creating shared understanding through collaborative design. *ACM Transactions on Human-Computer Interaction, 7*(1), 84–113. doi:10.1145/344949.345015

Basadur, M., Graen, G. B., & Green, S. G. (1982). Training in creative problem solving: Effects on ideation and problem finding and solving in an industrial research organization. *Organizational Behavior and Human Performance, 30*, 41–70. doi:10.1016/0030-5073(82)90233-1

Besemer, S. P., & O'Quin, K. (1999). Confirming the three-factor Creative Products Analysis Matrix model in an American sample. *Creativity Research Journal, 12*(6), 287–296. doi:10.1207/s15326934crj1204_6

Bonnardel, N. (1999). Creativity in design activities: The role of analogies in a constrained cognitive environment. *Proceedings of the 3rd conference on Creativity & cognition,* New York: ACM Press, 158-165.

Bonnardel, N., & Marmèche, E. (2005). Toward supporting evocation processes in creative design: A cognitive approach. *International Journal of Human-Computer Studies, 63*, 422–435. doi:10.1016/j.ijhcs.2005.04.006

Brophy, D. R. (1998). Understanding, measuring, enhancing collective creative problem-solving efforts. *Creativity Research Journal, 11*(3), 199–229. doi:10.1207/s15326934crj1103_2

Candy, L. (1997). Computers and creativity support: Knowledge, visualization, and collaboration. *Knowledge-Based Systems, 10*, 3–13. doi:10.1016/S0950-7051(97)00008-7

Chen, M., & Kaufmann, G. (2008). Employee creativity and R&D: A critical review. *Creativity and Innovation Management, 17*(1), 71–76. doi:10.1111/j.1467-8691.2008.00471.x

Edmonds, E., Candy, L., Jones, R., & Soufi, B. (1994). Support for collaborative design: Agents and emergence. *Communications of the ACM, 37*(7), 41–47. doi:10.1145/176789.176793

Elam, J., & Mead, M. (1990). Can software influence creativity? *Information Systems Research, 1*(1), 1–22. doi:10.1287/isre.1.1.1

Fischer, G., Giaccardi, E., Eden, H., Sugimoto, M., & Ye, Y. (2005). Beyond binary choices: Integrating individual and social creativity. *International Journal of Human-Computer Studies, 63*, 482–512. doi:10.1016/j.ijhcs.2005.04.014

Hewett, T. (2005). Informing the design of computer-based environments to support creativity. *International Journal of Human-Computer Studies, 63*, 383–409. doi:10.1016/j.ijhcs.2005.04.004

Hewett, T. T., & DePaul, J. L. (2000). Toward a human centered scientific problem solving environment. In Houstis, E., Gallopolous, S., Rice, J. R., & Bramley, R. (Eds.), *Enabling Technologies for Computational Science: Frameworks, Middleware, and Environments* (pp. 79–90). Boston: Kluwer Academic Publishers.

Hunter, S. T., Bedell-Avers, K. E., & Mumford, M. D. (2007). Climate for creativity: A quantitative review. *Creativity Research Journal, 19*(1), 69–90.

Hunter, S. T., Friedrich, T. L., Bedell, K. E., & Mumford, M. D. (2006). Creative thought in real world innovation. *Serbian Journal of Management, 1*, 29–39.

Johnson, S. (1997). *Interface culture: How new technology transforms the way we create and communicate*. New York: Basic Books.

Massetti, B. (1996). An empirical examination of the value of creativity support systems on idea generation. *Management Information Systems Quarterly, 20*(1), 83–97. doi:10.2307/249543

Mumford, M. (2000). Managing creative people: strategies and tactics for innovation. *Human Resource Management Review, 10*(3), 313–351. doi:10.1016/S1053-4822(99)00043-1

Mumford, M. D. (2001). Something old, something new: Revisiting Guilford's conception of creative problem solving. *Creativity Research Journal, 13*(3), 267–276. doi:10.1207/S15326934CRJ1334_04

Mumford, M. D., & Baughman, W. A. (1996). Process-based measures of creative problem-solving skills: I. Problem construction. *Creativity Research Journal, 9*, 63–76. doi:10.1207/s15326934crj0901_6

Mumford, M. D., Baughman, W. A., Maher, M. A., Costanza, D. P., & Supinski, E. P. (1997). Process-based measures of creative problem-solving skills, IV: Category combination. *Creativity Research Journal, 10*, 59–71. doi:10.1207/s15326934crj1001_7

Mumford, M. D., Baughman, W. A., Supinski, E. P., & Maher, M. A. (1996). Process-based measures of creative problem-solving skills: II. Information Encoding. *Creativity Research Journal, 9*, 77–88. doi:10.1207/s15326934crj0901_7

Mumford, M. D., & Gustafson, S. B. (1988). Creativity syndrome: Integration, application, and innovation. *Psychological Bulletin, 103*(1), 27–43. doi:10.1037/0033-2909.103.1.27

Mumford, M. D., & Hunter, S. T. (2005). Innovation in organizations: A multi-level perspective on creativity. In Yammarino, F. J., & Dansereau, F. (Eds.), *Research in multi-level issues* (*Vol. IV*, pp. 11–74). Oxford, England: Elsevier.

Mumford, M. D., Hunter, S. T., & Bedell, K. E. (2008). *Research in Multi-level Issues: A Focus on Innovation* (*Vol. VII*). Oxford, England: Elsevier.

Mumford, M.D., Lonergan, D.C., & Scott, G.M. (2002). Evaluating creative ideas: Processes, standards, and context. *Inquiry: Critical thinking across the disciplines, 22*, 21-30.

Mumford, M. D., Mobley, M. I., Uhlman, C. E., Reiter-Pamon, R., & Doares, L. (1991). Process analytic models of creative capacities. *Creativity Research Journal, 4*, 91–122. doi:10.1080/10400419109534380

Mumford, M. D., Reiter-Palmon, R., & Redmond, M. R. (1994). Problem construction and cognition: Applying problem representations in ill-defined domains. In Runco, M. (Ed.), *Problem finding, problem solving, and creativity* (pp. 3–39). Norwood, NJ: Ablex.

Mumford, M. D., Supinski, E. P., Baughman, W. A., Costanza, D. P., & Threlfall, K. V. (1997). Process-based measures of creative problem-solving skills: V. Overall prediction. *Creativity Research Journal*, *10*, 77–85.

Mumford, M. D., Supinski, E. P., Threlfall, K. V., & Baughman, W. A. (1996). Process-based measures of creative problem-solving skills, III: Category selection. *Creativity Research Journal*, *9*, 395–406. doi:10.1207/s15326934crj0904_11

Regli, W. C., & Cicirello, V. A. (2000). Managing digital libraries for computer-aided design. *Computer Aided Design*, *32*, 119–132. doi:10.1016/S0010-4485(99)00095-0

Reiter-Palmon, R., Mumford, M. D., & Threlfall, K. U. (1998). Solving everyday problems creatively: The role of problem construction and personality type. *Creativity Research Journal*, *11*, 187–198. doi:10.1207/s15326934crj1103_1

Robertson, B. F., & Radcliff, D. F. (2009). Impact of CAD tools on creative problem solving in engineering design. *Computer Aided Design*, *41*, 136–146. doi:10.1016/j.cad.2008.06.007

Salter, A., & Gann, D. (2003). Sources of ideas for innovation in engineering design. *Research Policy*, *32*, 1309–1324. doi:10.1016/S0048-7333(02)00119-1

Schneiderman, B. (1998). Codex, memex, genex: The pursuit of transformational technologies. *International Journal of Human-Computer Interaction*, *10*(2), 87–106. doi:10.1207/s15327590ijhc1002_1

Schneiderman, B. (2000). Creating creativity: User interfaces for supporting innovation. *Transactions on Computer-Human Interaction*, *7*(1), 114–138. doi:10.1145/344949.345077

Schneiderman, B. (2001). Supporting creativity with advanced information-abundant user interfaces. In Earnshaw, R., Guedj, R., van Dam, A., & Vince, J. (Eds.), *Frontiers of Human-Centered Computing, Online Communities, and Virtual Environments* (pp. 469–480). London: Springer.

Schneiderman, B. (2003). *Leonardo's laptop: Human needs and the new computing technologies*. Cambridge: MIT Press.

Schneiderman, B. (2007). Creativity support tools: Accelerating discovery and innovation. *Communications of the ACM*, *50*(12), 20–32. doi:10.1145/1323688.1323689

Skovvang, M., Elbaek, M. K., & Hertzum, M. (2006). Personometrics: Mapping and visualizing communication patterns in R&D projects. In: Crestani, F. and Ruthven, I. (Eds.): COLIS5: Proceedings of the Fifth International Conference on Conceptions of Library and Information Sciences (Glasgow, UK, June 5-8), pp, 141-154.

Sternberg, R. J., O'Hara, L. A., & Lubart, T. I. (1997). Creativity as investment. *California Management Review*, *40*(1), 8–21.

Ward, T. B., Smith, S. M., & Finke, R. A. (2009). Creative cognition. In Sternberg, R. J. (Ed.), *Handbook of Creativity* (pp. 189–212). New York, NY: Cambridge University Press.

West, M. A. (1990). The social psychology of innovation in groups. In West, M. A., & Farr, J. L. (Eds.), *Innovation and Creativity at Work: Psychological and Organizational Strategies* (pp. 4–36). Chichester: Wiley.

West, M. A. (2002). Sparkling fountains or stagnant ponds? An integrative model of creativity and innovation implementation in work groups. *Applied Psychology: An International Review*, *51*(3), 355–424. doi:10.1111/1464-0597.00951

Yamamoto, Y., & Nakakoji, K. (2005). Interaction design tools for fostering creativity in the early stages of information design. *International Journal of Human-Computer Studies*, *63*, 513–535. doi:10.1016/j.ijhcs.2005.04.023

Ziegler, R., Diehl, M., & Zijlstra, G. (2000). Idea production in nominal and virtual groups: Does computer-mediated communication improve group brainstorming? *Group Processes & Intergroup Relations*, *3*, 141–158. doi:10.1177/1368430200032003

## KEY TERMS AND DEFINITIONS

**Computed Aided Design (CAD):** Any of a body of computer software that allows users to create and examine design models, as well as share such models among one another.

**Conceptual Combination:** In the context of Mumford's (1991) creative process, the stage at which designers actively combine pieces of information in novel ways, in order to develop creative ideas, which may be later fleshed out into a workable solution.

**Concept Selection:** In the context of Mumford's (1991) creative process, the stage during which designers sort and catalogue information that may be useful in addressing the problem at hand.

**Creative Process:** A conceptualization of creativity that allows for the examination of not just inputs and products, but the detailed cognitive and behavioral elements that influence creative performance, as designers work from identification of a broad problem to a specific creative solution.

**Monitoring:** The final stage of Mumford's (1991) creative process, in which designers systematically seek feedback and information about the conditions and success of an implemented solution.

**Personometric Maps:** A conceptualization of interactions and communication patterns among designers, useful for identifying expertise and how data are shared as part of the design process.

**Rapid Prototyping:** A design technique in which simplified, low-cost versions of a design are quickly and efficiently produced, in order to examine the functions or effectiveness of specific design elements.

# Chapter 3
# Problem-Solving Style, Problem Complexity and Knowledge Generation:
## How Product Development Teams Learn When They Carry on Innovation

**Corrado lo Storto**
*University of Naples Federico II, Italy*

## ABSTRACT

*This chapter presents the findings of a study aimed at investigating how the fit between the problem-solving style of a product development team and the cognitive environment induced by the perceived problem of complexity affects the amount and type of knowledge generated.*

*It is assumed that organizational knowledge is created as a by-product of collective creative technical problem-solving, when people work together to deal with unfamiliar and unexpected situations. Two major outcomes emerge from the analysis of experimental data: (1) different cognitive environment patterns are more conducive than others to organizational learning; (2) there exists some fit between the cognitive environment pattern and the team technical problem-solving style, as some cognitive practices and social behaviours adopted during technical problem-solving are more effective than others in certain cognitive environments. Particularly, practices and behaviours that are more associated to creativity have a not negligible weigh in the generation of knowledge. Ninety-one cases of technical problem-solving occurred during product innovation within 35 small firms studied.*

DOI: 10.4018/978-1-60960-519-3.ch003

## INTRODUCTION

A large body of literature clearly distinguishes between creativity, innovation, and learning and knowledge generation at the organizational level. While creativity is more associated to the generation of new concepts and the exploration of new ideas, innovation is associated to the conversion and implementation of the ideas into tangible products and processes, and learning to the adoption of new knowledge and behaviors within the organization (Cook, 1998; Craft, 2005; Oldham and Cummings, 1996). However, these processes are strongly interconnected and it is usually difficult to separate them (Brennan and Dooley, 2005; McAdam and McClelland, 2002a). Creativity is often considered the front end of the innovation process, while learning is the output of innovation (Amabile, 1983; Majaro, 1988; McAdam and McClelland, 2002b). When people carry on innovation, they manifest their creativity when they are able to find new solutions to old problems or identify new problems to solve (Treffinger and Isaksen, 1992). In the same time, knowledge and learning are critical inputs of the creative process, and vice versa, creativity is a form of knowledge creation. Urabe (1988, p. 3) describes innovation as a "never one-time phenomenon, but a long and cumulative process of a great number of organizational decision-making process, ranging from the phase of generation of a new idea to its implementation phase", and recent models of the innovation process emphasize how it is often a chaotic, iterative and interactive and not easily planned process in which several people, teams, organizations and institutions work together to search, refine, recycle, nurture information and knowledge to solve problems, generate new ideas and develop products and processes (Kline and Rosenberg, 1986; Chesbrough, 2004; Cheng and Van de Ven, 1996; Van de Ven, Polley, Garud and Venkataraman, 1999). According to Argyris and Schon (1978, p. 3) organizational learning is:"[...] a process in which members of an organization detect an error or anomaly and correct it by restructuring organizational theory of action, embedding the results of their inquiry in organizational maps and images". Thus learning is affected by the organization capabilities, practices and behaviors adopted to search, process, interpret and transfer information during technical problem-solving. An important source of learning – particularly in small organizations – is creative technical problem-solving that teams and individuals carry on during the innovative activity. This technical problem-solving usually proceeds randomly, without any planned or deliberated choice, but individuals conceptualize problems only when drawbacks occur and do not make any attempt to anticipate and prevent problems carrying on institutional R&D. Moreover, it is not rare that solving complex technical problems requires that people have to deal with problems of management.

Technical problem-solving is thus of crucial importance for all these processes, both as it is a common piece of the intrinsic nature of them and as it can shape and influence their dynamics. Creative technical problem-solving that individuals and teams implement during innovation affects how learning occurs and its output, e.g. knowledge created in its different shapes and amount. People working together in product development teams adopt different problem-solving styles, combining together a number of social behaviours and cognitive practices, depending on the amount of the perceived problem complexity (i.e. activating relationships to exchange information with the customer, developing either internal or external communication networks, approaching to problem solving by problem framing or problem widening, implementing experimentation and planning when dealing with problems, or divergent thinking) (Amabile, 1983; Andrews, 1975; Bell, 1982; Clark and Fujimoto, 1991; McKee, 1992; Raaheim, 1974; West and Farr, 1989). Problem-solvers perceive problem complexity as a consequence of the perceived state of ambiguity and uncertainty that characterizes the problem-solving cognitive environment (Daft and Lengel, 1986). Both the amount of perceived problem complexity and how technical problem-solving is managed by people (e.g., the problem-solving style) are particularly important to learning and the generation of new organizational knowledge. In particular, (1) the cognitive practices and social behaviours implemented in technical problem-solving influence the nature and amount of knowledge generated during learning, and (2) the perceived cognitive environment has a moderation effect on the generation of knowledge.[1] The investigation of these relationships is the main goal of this chapter.

The organization of the chapter is the following. In section "Background" it is explained how technical problem-solving is a major ingredient of learning and knowledge generation; in section "Modeling the generation of organizational knowledge" a model that explains how new organizational knowledge is generated during technical problem-solving. In particular, the cognitive environment induced by perceived problem complexity, problem-solving collective cognitive practices and social behaviors that have an effect on the generation of new knowledge are discussed. A detailed presentation of the research methodology is provided in section "Methodology", while findings are presented and discussed in sections "Results" and "Discussion and conclusion". Finally, new research trajectories are proposed in the "Future research" section.

## BACKGROUND

Knowledge is at the base of the construction of the capabilities of an organization and, it often allows to raise barriers that prevent competitors to imitate its products. A large part of the value added generated inside an organization can be explained through the generation of new knowledge that according to Grant (1996, p. 375) is: "[...] the most strategically significant source of advantage of the firm".

To acknowledge the role that knowledge has as a means to improve the organization performance stimulates scholars to investigate more in depth how organizations absorb, generate, transfer, save and use it. However, the issue of new knowledge generation inside organizations remains particularly critical. Generating new knowledge means, indeed, not only to accumulate new knowledge, but also to substitute existing knowledge that is obsolete and ineffective, modifying and changing it from one shape to another or refining existing knowledge. The organizations that use intensively knowledge to remain competitive in the market need to create it continuously, and can survive only if their stock of knowledge is re-generated with a certain frequency. Of course, even though less critical, new knowledge generation remains a relevant strategic issue for those organizations that compete in less dynamic and not technologically drive markets.

Even though literature has often emphasized the role and the importance of knowledge as a source of competitive advantage (Rumelt et al., 1992; Teece et al., 1990), there is a relatively small number of studies (theoretical and empirical) that propose models capable to shed light on the way new knowledge is created inside organizations and only in the last years some scholars focused attention on this issue (Nonaka, 1994; Davenport & Prusak, 1997; Nonaka a7nd Konno, 1998; Brown & Duguid, 1991). The proposed models emphasize the interaction within teams, in which individuals having different backgrounds and experiences but convergent interests, informally meet to exchange ideas, opinions, judgments and evaluations. A crucial element of organizational knowledge generation is therefore social interaction, through which individual knowledge is transformed into collective knowledge and becomes explicit from tacit.

However, models based on interaction do not take into account many aspects of the process of generation of organizational knowledge that develops from the individual knowledge, in particular the characteristics of the different learning structures where organizational learning occurs. Another critical ingredient of the process of knowledge generation is creativity, both at the individual and collective level. Individuals manifest their creativity when they deal with unusual and not trivial situations, during problem-solving that occurs whenever innovation is implemented within organizations (Kolb, 1974; Smith, 1989; von Hippel, 1994; von Hippel & Tyre, 1993). Problems can be well-structured or ill-structured, depending on the extent to which individuals involved in problem-solving feel they are familiar with the initial and/or the final state of the problem, and the transformation process to get from the initial state to the final state (Mayer, 1983; Simon, 1979). If individuals and team face a problem in which they are familiar both with the initial and the final states, and the transformation process, then they are dealing with a well-structured problem. In this case individuals will adopt standard operating routines to solve the problem, as the situation is repetitive (Argyris & Schon, 1978; Glaser, 1986).[2] Vice versa, if they are unfamiliar with one or both the states of the problem or the way to pass from the initial to the final state of it they are confronting ill-structured problems to solve which they need innovative responses. In this situation they have to devote great efforts to identify effective strategies to solve them. In highly complex situations problem-solvers will modify their assumptions, search strategies, adopting new logical mental categories for giving meaning to things and interpret reality, and selecting new behaviors (Argyris & Schon, 1978, Jelinek, 1979; Winograd & Flores, 1986). This change in the perspective of analysis of a situation typical of ill-structured problem-solving generates new organizational knowledge. Learning occurs during technical problem-solving because the intermediate and final outcomes of the problem-solving activity are stored in the collective and organizational memory of the organization, in order to be used in other circumstances - similar or not. These outcomes are incorporated in the organization memory in form of new knowledge (either situational or procedural type) (Tulving & Schachter, 1990). Cycles of individual and collective creative problem-solving – planned or random -, as a consequence, become a not minor element of the process of generation of new organizational knowledge as a means for the activation of creativity, social behaviors and cognitive practices necessary to produce knowledge (Amabile, 1983; Corti & lo Storto, 1997; Corti & lo Storto, 1999; Rickards & Freedman, 1978; Zwicky & Wilson, 1965).

It often happens that a great bulk of organizational knowledge (even not technical knowledge) is generated and accumulates as a by-product of technical problem-solving associated to a not systematic search for new knowledge (such as R&D). Indeed, according to Nelson (2004, p. 458): "much of practice in most fields remains only partially understood, and much of engineering design practice involves solutions to problems that professional engineers have learned 'work' without any particularly sophisticated understanding of why". Looking at the creation of knowledge as an activity which is separated from other activities of the general process of knowledge management (i.e., adoption, embodiment, diffusion, etc.) within an organization is indeed another controversial issue in the literature. Many activities of the knowledge management process are interacting and cannot be easily split. For instance, when knowledge is transferred inside the organization, it is usually refined, contextualized, and new knowledge is consequently generated.

Moreover, even though the importance of studying how knowledge is created inside organizations is well acknowledged since time, empirical literature remains still scarce and the small number of empirical studies focused only on some aspects of the subject, such as the transfer of knowledge within or between organizations, the interaction between different shapes of knowledge (i.e., tacit and explicit). It should also be added that small size organizations (i.e., small companies or teams) are fundamentally missing in empirical studies, thus contributing to diffuse a false belief that the creation of knowledge is not a matter of small companies or teams. In particular, small companies, for their nature and characters (lack of resources, ownership and management infrastructure, organizational structure, etc.) need the development of ad hoc interpretive models and not only the adaptation of models developed for large companies as they – it is now well accepted – are not large firms in their infancy.

Locating creative technical problem-solving in a central position of the process of knowledge creation allows to have a tool to build a contingent model of the organizational learning process that may be particularly useful to generate both scientific and practical implications. If creative technical problem-solving both at the individual and collective level is a major determinant of the generation of new knowledge within organizations, how problem-solving occurs and the effective fit between problem-solving practices and behaviors with the perceived problem complexity have an influence on the typology and amount of knowledge generated and learning trajectories. Consequently, as to a certain extent technical problem-solving can be purposely managed, also learning and knowledge creation inside organizations can be more effectively guided.

## MODELING THE GENERATION OF ORGANIZATIONAL KNOWLEDGE

In the following sections a model that explains the generation of organizational knowledge as an output of organizational problem-solving within a cognitive-constructionist perspective (Meindl, Stubbart & Porac, 1996) is presented.

### The Nature of Organizational Knowledge

Several scholars (e.g. Davenport & Prusak, 1998; Nissen et al., 2000; von Krough et al., 2000) have conceptualized a sort of hierarchy of knowledge, distinguishing between knowledge, information and data[3].

Knowledge differs from information. While information is rather objective, knowledge is associated to a particular cognitive structure of individuals that develop and use it. At the individual and the organizational level knowledge derives either from the interpretation of new or existing pieces of information (Daft & Weick, 1984) or, developing – more or less consciously – a new comprehension of surrounding events (Fiol, 1994; Weick, 1995). For instance, Nonaka & Takeuchi (1995, p. 58) describe knowledge as "a meaningful set of information", while Davenport et al. (1998, p. 43) underlines how knowledge is "[…] information combined with experience, environment, interpretation and reflection […]". Knowledge is not independent, but fundamentally situated, being partially a product of the activity, the environment and the culture in which it is generated. The transformation of individual knowledge into organizational knowledge critically depends on organizational culture characters (Schein, 1996). Shared norms, values, interpretive categories, and behaviors are all indicators of organizational knowledge.

Organizational knowledge is based on individual knowledge, but it substantially differs from the second one. According to Hedberg (1981) even though organizational learning occurs through individuals, it would be a mistake to believe that organizational learning – and henceforth knowledge – is only the cumulated learning of the single organizational members. Organizations do not have brains, but have cognitive systems and memories. Members move inside and outside the organization and leadership changes, but the memories of the organization preserve certain behaviors, mental maps, norms and values.

A large amount of organizational knowledge remains in an informal shape, as tacit knowledge that is created and used when formal results leading to formal knowledge are generated.[4] Tacit knowledge is made of facts, ideas, opinions, judgments, assumptions, meanings, questions, decisions, stories that cannot be saved in some physical support. It is rather invisible, idiosyncratic and not easily imitable (Nelson & Winter, 1982), transitory and ephemeral, contrarily to explicit and formal knowledge (Polanyi, 1958; Winter, 1987; Corti & lo Storto, 2000), "disorganized, informal and relatively inaccessible, making it potentially ill-suited for direct instruction" (Wagner & Sternberg, 1985, p. 439). It is almost impossible to grasp it for an individual who is not familiar with the original organizational processes that lead to the accumulation of that stock of knowledge inside the organization, even though this latter can observe and speak with the organization experts. Tacit knowledge remains difficult to communicate or share, but, it is a "rich untapped source of new knowledge" (Nonaka, 1994, p. 16; Nonaka & Takeuchi, 1995, p. 8).

There is a great level of unawareness in tacit knowledge and even the expert finds it difficult to codify it or transfer the underlying decision-making rules to other people. Tacit knowledge cannot be easily extracted as it is embedded in a complex and invisible system of social relationships or individual artisanal practices associated to the specific organization. On the contrary, explicit knowledge includes knowledge contained in books, manuals, documents, instruments and machines. It can be produced and stored by the members of the organization as technical reports, technical literature, plans, electronic spreadsheets, drawings, blueprints, formulas, patents, notes, etc. The organizations do not have great difficulties to capture this kind of knowledge. Nonaka (1991) and Itami (1987) suggest that explicit and formal knowledge is more systematized than tacit knowledge and, as a consequence, it can be easily communicated, transferred and shared (Badaracco, 1991). Some scholars clam that not all tacit knowledge can be made explicit (e.g. Ambrosini & Bowman, 2001; Janik, 1988; Tsoukas, 2003).

Organizational knowledge can be classified in terms of its being or not associated to action, as routines or procedural knowledge and declarative knowledge (Winograd & Flores, 1986). While declarative knowledge only provides the base and the tools to interpret facts and circumstances (i.e., mental categories), it does not help us to manage these circumstances, how to change things and facts to achieve a desired goal, but we need procedural knowledge to use declarative knowledge and modify a situation. Procedural knowledge provides the organization with a capability to use efficiently and effectively declarative knowledge. Thus, procedural knowledge has a strong *behavioral* orientation. Organizational procedural knowledge emerges from the collective behaviors of the members of the organization and it manifests as a collection of routine systems (Gersick & Hackman, 1990; Nelson & Winter, 1982). Routines, rules and standard operating procedures acquire a particular weight in shaping the organizational knowledge (Nelson & Winter, 1982; Chow, 1998), as they make declarative knowledge and organizational memory operative and useful for the organization. According to Levitt & March (1988) they include forms, rules, conventional procedures, strategies and technologies enabling the organization to work and compete. They also include the belief structures, schemata, paradigms, codes, culture that strengthen, change or substitute formal routines. Organizational routines represent a particular way to do that the organization has developed and learned. As the organizations, teams and individuals acquire experience and tend to adopt habitual modes of behavior (e.g. routines), devoting attention and efforts to modify them rarely. Routines can be viewed as a concatenation of actions saved in the memory of the organization (Walsh & Ungson,

1991). Behind these "routinized" behaviors there lie specific shared mental models (Johson-Laird, 1983)[5]. For instance, there exist mental models to carry on problem-solving activity. Organizational routines tend to institutionalize a particular competence of the organization. A successful story to solve a problem or to execute a task can lead the organization to create one or more routines to link individual skills, behaviors, knowledge and the task. By using routines, individuals and organizations adopt a collection of rules accumulated in the organizational memory that allows them to perform actions with a reduced mental effort. They can be more or less equivalent to working standard procedures that are formulated more explicitly and have a normative standing (Cohen & Bacdayan, 1994). Usually, these routines have a strong tacit dimension that make them difficult to imitate or to be modified (Johnson-Laird, 1983)[6]. But, notwithstanding there is a common belief that routines are difficult to be changed so that stability is usually used as an attribute characterizing them, a great bulk of organizational routines can undergo substantial changes as a consequence of organizational learning (Feldman, 2000)[7]. Pentland & Rueter (1994) use the metaphor of grammar to explain how routines can be changed. As a grammar allows some individuals to create a variety of sentences that have a meaning for other individuals who know the grammar by combining the elements of the language according to well defined rules, also routines allow some individuals to select certain elements from a repertory (in the same way of a language) to put together sequences of actions that make sense for the other members belonging to the same organization.

A recent empirical study carried on by Feldman (2000) showed that even the grammar can be changed inside organization, because its members not only use common rules to combine elements of a repertory, but develop a new repertory of interpretive categories and new rules, i.e. new modes to put together the elements. Empirical literature focusing on learning during product innovation in small manufacturing firms support this perspective (Corti & lo Storto, 1997; Corti and lo Storto, 2000). This conceptualization of routines is consistent with the idea proposed in the literature that considers two levels of the organizational learning (Argyris & Schon, 1978). Routines are the more efficient way for the organization to embed knowledge (having a procedural nature) within collective models of behaviors shared by its members and considered successful, and consequently to institutionalize some particular competences (Cohen, 1991; Cohen & Levinthal, 1990; Huber, 1991; Nelson & Winter, 1982).

## Organizational Learning, Creativity, and Technical Problem-Solving

Research carried on in the fields of cognition and behavioral science describes individual learning as a process that implies the acquisition and interpretation of new knowledge to solve problems (Spearman, 1927). As individual learning, organizational learning can be viewed either as a change in the system of beliefs (e.g. the cognitive view), or as a change in the system of routines, rules and behaviors (e.g. the behavioural view) (Dodgson, 1993; Hwang 2003; Polito & Watson, 2002). Many times, organizational learning is induced by a reply to a stimulation from the external environment (i.e., the request from the customer for a new product), as the identification and correction of drawbacks or experimentation, just as individual learning (Cook & Yanow, 1993). In a more comprehensive knowledge-based view, organizational learning is the process of creating, acquiring and transferring knowledge within the organization (Teare & Rayner, 2002).

The relation between individual and organizational learning remains obscure (Duncan & Weiss, 1979; Weick & Westley, 1996; Edmondson & Moingeon, 1998). Rarely there is agreement among scholars relatively to the origin of organizational learning and the way it occurs (Fiol & Lyles, 1985).

Some scholars affirm that only a few number of individuals (key individuals) are learning agents in their organization. According to Simon (1991, p. 125) "[...] All learning takes place inside individual human heads; an organization learns in only two ways: (a) by the learning of its members, or (b) by ingesting new members who have knowledge the organization didn't previously have[...]". Several scholars underlined how organizational learning differ from individual learning and it is not the summation of pieces of learning relative o single individuals belonging to the organization (Argyris & Schon, 1978; Hedberg, 1981; March & Olsen, 1975). Indeed, Argyris and Schon (1978) claim that individual leaning is a necessary condition but not sufficient to organizational learning. According to Huber (1991, pp. 89-90): "[...] An organization learns if any of its units acquires knowledge that it recognizes as potentially useful to the organization [...] more organizational learning occurs when more of the organization's components obtain this knowledge and recognize it as potentially useful, when more and more varied interpretations are developed, and when more organizational units develop uniform comprehensions of the various interpretations [...]". The core of organizational learning is: "[...] the acquiring, sustaining, and changing of inter-subjective meanings[...]" (Cook & Yarrow, 1993, p. 449). Organizational learning is a process that is mediated by the investigative collaboration of single members of the organization (Argyris & Schon, 1978). New organizational knowledge is created through a process of amplification and articulation of individual knowledge (Nonaka, 1994; Nonaka & Takeuchi, 1995). It is complex as it implies not only the generation or acquisition of new knowledge, but rather the existence of mechanisms and practices capable to activate storing and retrieval processes of knowledge from the organization memory (Jelinek, 1979).

Literature identifies different typologies of organizational learning. Some of them are based on the degree of completeness of the learning process (March & Olsen, 1975), and the mechanisms that lead to the solution of the problem that stimulated learning - *learning before doing, learning by doing, learning by trial and error, learning by failing, learning by using* (e.g. Arrow, 1962; Fleck, 1994; Maidique & Zirger, 1984; Rosenberg, 1982; von Hippel & Tyre, 1993). Some other typologies are based on the size of effect that learning has upon organization and/or the level of complexity and radicalism of the process itself (Argyris & Schon, 1978; Hedberg, 1981; Nystrom & Starbuck, 1984). Particularly, two different typologies have been identified: *single-loop* and *double-loop learning* (Argyris & Schon, 1978), *adaptive* and *generative learning* (Senge, 1990; Schein, 1996). These classification are indeed very similar. Generally, adaptive learning levers upon existing stocks of knowledge, comprehension of events and experience that exist in the organization. Adaptive learning occurs when the organization is capable to identify a problem or there exists a gap between the state where it is and the state where in should be in. Finding a solution to the problem makes possible to close this gap. Adaptive learning is based on a frame of past experience, looking at the environment as stable and foreseeable. Organizational learning aims at achieving a level of stability by means of incremental improvements. Generative learning is associated to a process of *re-framing* (Schein, 1996). Indeed, this kind of learning occurs within a reference *frame* that differs from the existing one, and is aimed at activating a discontinuous or transformative change substituting past assumptions with new concepts and ideas, making possible a new comprehension of events by developing new knowledge. Schein (1996) has emphasized how generative learning happens when organizations find that in order to identify a problem or the gap between the desired and the actual state it is necessary to modify the way problems are perceived and dealt.

Technical problem-solving has a critical importance in organizational learning and the development of new knowledge within teams and organizations that carry on innovation (Newell & Simon, 1972). According to Argyris (1982, p. 38) learning is: "[…] a process in which people discover a problem, invent a solution to the problem, produce the solution, and evaluate the outcome, leading to the discovery of new problems [...]". There is a strong association between learning modes and problem-solving activity. von Hippel and Tyre (1993, p. 5) affirm that: "[…] learning by doing is simply a form of problem-solving that involves application of a production process (or product, service or technique) in its intended use environment […]".

Creativity is an essential ingredient of technical problem-solving. It is needed to find effective solutions to novel technical problems. Indeed, Lumsdaine and Lumsdaine (1995, p14) recall that […] creativity is playing with imagination and possibilities, leading to new and meaningful connections and outcomes while interacting with ideas, people, and the environment." Guilford (1970, p. 161) claims that: "[...] problem-solving and creative thinking are closely related. The very definitions of those two activities show logical connections. Creative thinking produces novel outcomes, and problem-solving involves producing a new response to a new situation, which is a novel outcome […]". Moreover, Guilford (1977) considers problem-solving as the activity of dealing with a situation without being prepared to do that. This occurrence stimulates people to move beyond the available amounts of information and knowledge, starting a new and more effective intellectual activity which makes possible to find new answers.

However, problem-solving is in the same time both a creative and a rational activity (Guilford, 1977; Torrance and Myers, 1970). Thus, creative thinking embedded in problem-solving may be either "bounded" or more effectively managed by adopting cognitive practices and individual and collective behaviours suggested by experience and literature.

Problem-solving activity may be extremely complex and develop through a multeplicity of stages, with several tasks and sub-activities (Brightman et al., 1988). In general, scholars identify three stages (Marples, 1961; Wagner, 1993; Mintzberg, Duru and Theoret, 1976): the initial stage of problem formulation when a drawback is perceived as a problem, the problem is identified, described and its implications are rationalized; the stage of diagnosis, when hypotheses are generated, the search for causes is planned, and finally causes are determined; the final stage of identification of the solution, in which strategies for searching problem solution are planned and implemented, a solution is eventually found, the solution is implemented and the outcome is evaluated.

Problem-solving may be more or less difficult depending on the representation and classification of the problem which is done during the stage of problem formulation (Hayes, 1978). For instance, a problem may be represented in a real or abstract way, using a mathematical formula, logical symbols, or words or drawings. Problems may be either "ill structured" or "well structured". Newell, Shaw and Simon (1979) claim that problems that need a creative approach to be solved are, on principle, ill structured and generally ambiguous, and a great amount of the problem-solving effort reduce to the formulation of correct logical frame of the problem. The newness, complexity and ambiguity of problem are all characteristics that influence individuals' capability to find a useful logical frame of the problem (Mintzberg, Duru & Theoret, 1976).

In a cognitive perspective, the more critical stage of problem-solving activity is that of problem formulation. All problem-solving activity begins and is affected by problem formulation, even though the problem is still loosely specified and individuals are far from being fully aware of its implications. A large part of scholars that investigated problem-solving emphasized the importance of this stage, which they refer to using various terms (e.g. problem setting, problem formulation, problem identification, problem framing, etc.) (Lyles & Mitroff, 1980; Schwenk & Thomas, 1983; Rein & Schon, 1977). This stage requires a particular effort of problem-solvers (Mintzberg, Raisinhighani & Theoret, 1976; Polya, 1945; Duncker, 1945; Wertheimer, 1945).

According to Smith (1989a and 1989b) the stage of problem formulation may be identified in terms of a number of mental states of people who are engaged in problem-solving activity. These states are characterized by specific cognitive processes that allow problem-solvers to move from the initial state to the final state of the formulation stage and, henceforth, to complete the stage of problem definition (Pounds, 1969).

Problem-solving begins with the perception of stimuli that are associated to the existence of a problem that induces a state of apprehension of people. In this state the problem identification occurs. This cognitive state finishes when a full awareness and convincement that the problem exists, moving to the next state of problem definition. In this state, the problem-solver specifies the problem with the purpose to construct a comprehensible representation of the problem useful to communicate and make it understandable to others. The cognitive process that lead to the conceptualization or construction of problem uses the stock of information and knowledge available. In the next state, a *frame* is give to the problem (Dery, 1983; Rein & Schon, 1977). Framing the problem is of paramount importance, as it gives the problem-solver the possibility to decide how to deal with the following solution of the problem, and finally, to choose an effective solving strategy (Smith, 1989a). The stage of problem formulation becomes critical as a *preparation* step for the problem-solving activity. The logical sequence of these states is not rigid, but problem-solvers can move back to refine or change the definition, or even, the identification of problems before passing to the framing of it. Getzels and Csikszentmyhalyi (1976) demonstrated that problem formulation is more intimately associated to creativity and interaction among individuals than the other stages of problem-solving activity, and the identification of the solution itself can be more or less difficult depending on the representation and framing of the problem that individuals generate during the formulation stage (Hayes, 1978).

## Problem-Solving Contextual Factors, Cognitive Practices and Behaviors Affecting Organizational Learning

Problem-solving generally takes place through an iterative cycle of search and selection (Marples, 1961). With each cycle, as different technical solutions are identified and tested and a subset among them is chosen, the gap between the initial state and the desired state of the system becomes more and more narrow. In searching for a solution of the problem, the problem-solver needs to receive feedbacks about how each alternative will perform in the new environment defined by the modified operating choices and conditions. This search proceeds iteratively and allows the problem-solver to gain important insights about the performance of each potential solution.

Problems can be well-structured or ill-structured, depending on the extent to which individuals involved in problem-solving feel they are familiar with the initial and/or the final state of the problem, and the transformation process to get from the initial state to the final state (Simon, 1973). If individuals face a problem in which they are familiar both with the initial and the final states, and the transformation process, then they are dealing with a well-structured problem. In this case the initial search for problem causes and solutions seems to be guided by three heuristic rules: search for a model of the problem recalling from the organizational memory similar problems already faced; search for problem causes in the "neighbourhood" of problem symptoms; search for problem solutions in the "neighbourhood" of current already adopted alternatives. Therefore, problem-solvers will make their attempt to solve a problem treating the new problem as an old well known problem. The individuals will adopt standard operating routines to solve the problem, as the situation is repetitive (Schrader, Riggs & Smith, 1991). In an organization, routines for solving problems exist in the head of individuals or are codified in the form of personal memos, operating procedures or organizational manuals.

However, in the stage of problem construction, problem-solvers may feel unfamiliar with one or both the states of the problem or the way to pass from the initial to the final state of it, perceiving the problem as being particularly ill-structured as a consequence of its complexity and, consequently, of the cognitive environment uncertainty and ambiguity. These conditions require the problem-solvers to devote great effort to identify innovative responses to solve the problem (Argyris & Schon, 1978; Glaser, 1986). In highly ambiguous situations problem-solvers will modify their assumptions, search strategies, adopting new logical mental categories for giving meaning to things and interpret reality, and select more effective behaviors. This change in the perspective of analysis of a situation typical of ill-structured problem-solving generates new knowledge that is stored in the organizational memory (Argyris & Schon, 1978, Jelinek, 1979; Winograd & Flores, 1986). Uncertainty is the cognitive state in which a problem-solver falls as a consequence of the lack of information (Daft & Lengel, 1986). This lack of information increases with task complexity, and in particular, with its variety. Indeed, the increase of the number of secondary tasks increases the number of activities not strictly correlated that should be executed. Uncertainty arises both as a consequence of the increased variance of the knowledge domains that managers and technicians have to master in the same time, and of the difficulty of planning. When a great number of tasks have to be executed it is not easy to define accurately all the details before problem-solving is started, and new information has to be acquired during the process as the not expected events produce a gap in the knowledge or competences. Therefore, when uncertainty increases, organizations need to collect and process a larger amount of information.

Effective problem-solving clearly depends upon accurate and efficient exchange of information among individuals participating in problem-solving (Fiol, 1994). Although problem-solvers need information which is both reliable and relevant, the complexity of the problem and the characteristics of the situation may require them to process information at a level that exceeds their cognitive abilities, thus declining efficiency in their abilities to use information. Networking - internal and external - allows to lower uncertainty related to environment as it makes it possible to collect information and transfer knowledge to the locus of problem-solving. When individuals participate in decision-making exercitating influence, interacting and exchanging information, they can provide ideas relatively to new and more efficient ways to work. A decentralized structure allows to take into account different points of views, and with great probability, to produce a greater diversity of ideas. The variety of perspectives relatively to a problem improves the quality of solution. New knowledge to solve technical problems often emerges in the interface between different areas of technological knowledge. von Hippel (1978) found that in many industries the customer has a primary role in stimulating innovation, suggesting technical specifications, innovative solutions, technical targets that go beyond the ordinary targets that the firm is used to deal with, providing information and knowledge.

Henceforth, intense and frequent relationships with the customer firm during product innovation usually provides the supplier firm with valuable information helpful to substantially improve its technical and functional features. Furthermore, the exchange of information during technical problem-solving between the two parties has also a positive effect on the generation of different alternatives, while the transfer of information and explicit knowledge from external sources makes possible the accumulation of new organizational knowledge through mechanisms of "interpretation" and socialization.

Although generating a large number of alternatives can help to reduce uncertainty, it cannot resolve the uncertainty about the effect of these alternatives. A very direct way to test the effect of the alternatives is to apply each alternative to a model of the problem. The most important approach to generate and facilitate these feedbacks is experimentation (Ulrich & Eppinger, 1995). By adopting a trial and error approach, experimentation enables problem-solvers to simulate only rough approximations of what my happen, but it makes it possible (and convenient) to try a large number of alternatives. By conducting experiments on the model of the problem, the consequences can be anticipated, and uncertainties can be more readily managed. Experimentation takes several forms and can be conducted under a variety of conditions (Pisano, 1994).

Planning and formalising tasks can alleviate the stress of individuals by reducing the amount of information they have to process at one time, helping them to search for information, knowledge, and competences when they are needed. Planning and formalising tasks induce problem-solvers to reflection, as a consequence of the separation of idea generation from idea evaluation (Olton, 1980). This separation may positively affect the problem-solving activity enhancing the interpretative capabilities of problem-solvers (Andrews, 1975; Braybrooke & Lindblom, 1963; Rickards & Freedman, 1978).

Unfortunately, the introduction into the organization of new information to be processed often adds further ambiguity to the initial ambiguity of the original formulation of the problem. Ambiguity defines the cognitive state perceived by a problem-solver when several different interpretations, usually contrasting, of a problem are possible. When technical problems generated by the exceptions encountered during technological innovation cannot be correctly framed, the search for a solution cannot be easily planned and has ambiguous aspects. Ambiguity appears when knowledge becomes more tacit or less articulated (Daft & Lengel, 1986; Polanyi, 1958; Weick, 1995; Winter, 1987). If knowledge cannot be articulated, technical problems can be defined only vaguely, and engineers and technicians cannot easily address problem-solving. A decreasing level of articulability of knowledge is associated with a growing difficulty of search, transfer of requested information, and contextualization of knowledge (Perrow, 1967; Winter, 1987). By only increasing the amount of available information and knowledge it is not possible to satisfy all the requirements of information and knowledge within the organization. New data can generate confusion, and even increase uncertainty if they increase the number of possible interpretations in the same time conflictual of a certain situation (Daft & Lengel, 1986). New data could not provide any solution when ambiguity is high. Reframing the problem-solving activity by splitting and decomposing the problem into sub-problems or identifying controllable factors permits the use of specialized capabilities. Further, it helps problem-solvers to increase their information processing capabilities, by grouping information into categories (or chunks) and arranging them by order of importance.

Frequently, the "true problem" of problem-solving activity is in the apparent familiarity of task. Ways to deal with a problem and conventional assumptions that proved to be successful in the past might not work when the characteristics of the problem differ. Changes in the problem representation, of the identification of sequences of search heuristics more efficient can burgeon the domain within which people who are involved in the solution of the problem move. Creative ways to deal with problems that criticize the past experience allow to overcome the inertia enriching the cognitive behavior of individuals. A change in the assumption defining the boundaries of the problem-solving domain, stimulating problem-solvers to relate previously unrelated ideas, can be very effective in generating more appropriate alternatives (Amabile, 1983; Davis & Manski, 1966). Both internal and external networking, and certain human resource practices enhancing individual and collective divergent thinking (i.e., brainstorming) may support problem re-framing by transferring rich information and tacit knowledge to the locus of problem-solving allowing to test several alternatives and widen problem representation. Moreover, a high level of interaction between customer and supplier allows both formal and informal information to be processed, creating the conditions for enhancing trust between parties, resolving conflicts, and reducing ambiguity.

## METHODOLOGY

### Sample and Data Collection

Ninety-one cases of technical problem-solving occurred within small firms were studied in depth. Only ill-structured or partially structured technical problems were included in the sample, while data were collected using structured interviews. Even though technical problems had a various nature, all of them were linked to the ill or no functioning of a new machine or equipment under development. Furthermore, they were all mechanical-type problems. Firms belong to the food-equipment manufacturing industry and are located in Southern Italy. They manufacture products for an industry where processes and products are neither based on sophisticated and complex technologies, nor carry on any formalized R&D activity. They are low-technology firms and their markets are local and of niche.

Any change in the amount of declarative and procedural knowledge accumulated by the organization as a by-product of the specific problem-solving activity were considered as a performance variable of the learning process. Declarative knowledge denotes the ability of describing and understanding situations and facts, while procedural knowledge the capability to execute a set of activities to modify situations efficiently. To resolve the conceptual/operative difficulty of assessing the emergence of new procedural knowledge, routinized behavior was associated to a set of condition-action rules prescribing the actions to be made for every condition. The change in procedural knowledge is thus identified by the change in the corresponding set of rules of actions and decision-making criteria. To measure constructs reflecting the nature of variables, Likert-type scales were developed following a logical-deductive reasoning, and using both results of an in depth case-study investigation and the literature (Churchill, 1979). Each scale contained different items proposed as questions during the in-field data collection. Measures were obtained as summation of the scores given to the single items of each scale after standardizing the value distribution for every item. The construct validity of measures developed for the study was established by factor analysis of item sets designed to measure the constructs. Measures reliability and internal consistency was assessed by using the alpha cronbach index (Nunnally, 1967). Values for alpha ranged from 0.64 to 0.92. Table 1 shows variables affecting the creation of declarative and procedural knowledge during technical problem-solving.

### Data Analysis and the Measurement of Fit

A set of organizational cognitive environment typologies was derived as patterns of the combination of states of ambiguity and uncertainty perceived by problem-solvers. Cluster analysis was used to uncover homogeneous patterns of environment (Arabie & Hubert, 1994). Variables were normalized to make them more commensurable by standardizing them obtaining variables with “0” mean and “1” standard deviation (Mil-

*Table 1. The description of variables used in the study*

| Variables | Description |
|---|---|
| uncertainty (UNCERTAINTY) | Problem-solving environments are perceived uncertain when there is a lack of information or knowledge relative to the state, preference, goals, and present and future tasks (Daft & Lengel, 1986; Duncan, 1972; Galbraith, 1973; Garner, 1962). (*Amount of environment uncertainty perceived by problem-solvers*) |
| ambiguity (AMBIGUITY) | Problem-solving environments are perceived ambiguous when (Weick, 1995; Machlup & Mansfield, 1983; Daft & Macintosh, 1981; Daft & Weick, 1984; March & Olsen, 1975; Martin & Meyerson, 1988, McCall, 1977): (a) individuals are unable to interpret the meaning of facts and actions, as they do not have enough information and knowledge that can help them to get useful hints and thus perceive events as not familiar and are unable to give a judgment; (b) there are multiple and redundant interpretations of the same thing (*Amount of ambiguity environment perceived by problem-solvers*) |
| interaction with customer (CUSTINT) | This category includes a specific network typology, i.e. the dyadic relationship between the customer and the supplier firm. An intense relationship is particularly useful to alleviate ambiguity, for instance in the stage of definition of functional and technical specification of a product (*Amount and quality of formal and informal interaction with the customer technicians and engineers during the specific problem-solving activity*) |
| internal networking (INTNETWORK) | It includes inter-personal relationships between individuals involved in problem-solving, and/or between individuals involved in problem-solving and experts within the organization. This kind of network allows the exchange of rich information, fostering the connection and association of different facts, concepts, and ideas that stimulate the generation of new knowledge. This network typology supports creative processes through the articulation of problems, inspiring new ideas, exploring new concepts, exciting imagination and suggesting novel alternatives (*Amount and quality of internal networks developed through interpersonal contacts involving in the problem-solving activity experts of the organization*) |
| external networking (EXTNETWORK) | This type of network allows the organization to come in touch with several external sources of information and knowledge. By means of these networks individuals exchange explicit knowledge (*Amount and quality of external networks developed both through informal and formal relationships with experts outside the organization to collect useful information and knowledge*) |
| problem-solving framing (FRAMING) | Individuals who participate in problem-solving split up the problem into sub-problems, focus on the pieces of it that can be analyzed, and identify the boundaries of the problem. In such a way they build a frame of the problem. The construction of the problem frame includes therefore a set of assumptions, insights and attitudes that address problem-solving which depends on the problem-solvers mental model (*Amount and quality of problem-solving framig practices adopted*) |
| problem-solving widening (WIDENING) | Individuals implement problem-solving by changing problem representation, modifying or substituting mental models used to deal with the problem, identifying sequences of heuristics that question past experience, thus overcoming organization inertia (*Amount and quality of problem-solving widening practices adopted*) |
| experimentation during problem-solving (EXPERIMENTATION) | By means of experimentation, often adopting a "trial-and-error", problem-solvers are able to simulate rough approximations of what may happen, with the aim to analyze a great number of alternatives, even when they do not have a good knowledge of the phenomenon (*Amount and quality of problem-solving experimentation practices adopted*) |
| problem-solving planning (PLANNING) | Individuals preliminarily identify a set of hypothesis to test, resources and task sequences to execute during problem-solving (*Amount and quality of problem-solving planning practices adopted*) |
| Divergent thinking in problem-solving (DIVERGENT THINKING) | Divergent thinking activation occurs when people adopt cognitive behaviors and practices that allow the free thinking with the generation of ideas, without being in dread of receiving negative judgment (*Amount and quality of problem-solving divergent thinking practices adopted*) |
| declarative knowledge (KNOWLEDGE) | Change in the declarative knowledge asset of the firm consequently to the specific problem-solving activity |
| procedural knowledge (ROUTINE) | Change in the routine asset of the firm consequently to the specific problem-solving activity |

ligan & Cooper, 1988; Morrison, 1967). In order to identify the appropriate number of clusters and the optimal clustering strategy, the VRC (Variance Ratio Criterion) index was calculated (Milligan, 1981; Milligan & Cooper, 1985; Baker & Hubert, 1976). Appendix reports details of the VRC analysis. The results of cluster analysis were validated by splitting the sample into two groups, randomly reducing it twice by randomly eliminating 35% of its cases. Every time the largest sub-sample was cluster re-analyzed separately, and the results were then compared for the three situations. Cluster centroids in the three cases remained almost stable.

A hierarchical moderated regression analysis was performed to investigate fit between knowledge amount and type and technical problem-solving cognitive practices and behaviors.

The existence of fit between the patterns of cognitive environment and problem-solving style was investigated adopting a standard procedure suggested by literature. In general, in order to explore the existence of fit between cognitive environment and structural or process variables scholars adopt different approaches and methodologies, influenced by the aim of the study (Drazin & Van de Ven, 1985; Gresov, 1989; Venkatraman, 1985). Indeed, in the strategic and organizational literature many different conceptualizations of fit have been proposed (Fry, & Slocum, 1984; Mohr, 1982). Drazin and Van de Ven (1985) classified these conceptualizations into three main approaches: fit as interaction, fit as selection, and fit as a system. Here a bivariate interactive conceptualization of fit is adopted. In the interaction approach it is assumed that the interaction between couples of factors of environment and structure affects the performance of the unit analyzed (Schoonhoven, 1981; Venkatraman, 1989).

This perspective of fit is more reflective of the logic behind the study. The profile of problem-solving style is indeed a reaction that problem-solvers have when they perceive to be upset with a situation. Furthermore, the lack of a comprehensive theoretical framework providing conceptual models of the way new knowledge is generated in small firms and the relationship between cognitive environment and learning process does not make it possible to test the fit of empirical data to an ideal type a priori theoretically identified (Schoonhoven, 1981).

In order to test the interaction effects of fit, the hierarchical moderated regression analysis technique was performed. As literature suggests, the moderated regression analysis represents a technique to test hypothesized contingency relations since it permits to consider and examine the effects of the interaction between two variables (Covin & Slevin, 1988; Schoonhoven, 1981; Drazin & Van de Ven, 1985). Following a procedure proposed by Sharma, Durand, and Gur-Arie (1981), given a dependent variable $Y$ (i.e. a variable which indicate a performance), an independent variable $X$, and an independent variable $Z$ (the variable hypothesized as moderator of the effect of $X$ on the dependent variable $Y$), it is possible to assess the magnitude of the effect of the interaction of $Z$ with $X$ computing the regressions:

$$Y=a+b_1X \quad (1)$$

$$Y=a+b_1X+b_2Z \quad (2)$$

$$Y=a+b_1X+b_2Z+b_3XZ \quad (3)$$

If the addition of the interaction term significantly increases "the power" of the regression equation to explain the variance of the dependent variable (i.e., the value of $R$-squared), then it can be affirmed that an interaction effect exists. Particularly, the coefficient $b_3$ must be significantly different from zero so that it is possible to affirm that $X$ and $Z$ interact, and consequently there is an influence on $Y$. Additionally, a positive coefficient $b_3$ which significantly differs from zero implies that the positive influence of $X$ on $Y$ is larger for the higher values of $Z$. A negative value

*Figure 1. The relations between the generation of organizational knowledge, problem-solving cognitive environment, practices and behaviors investigated in the study*

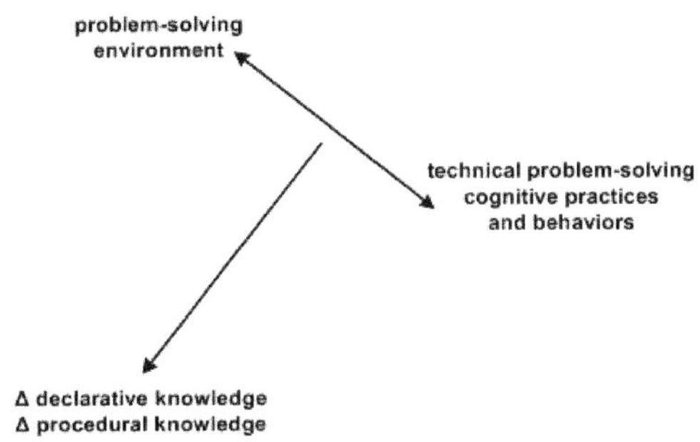


of $b_3$ and significantly different from zero implies the contrary.

Independent variables to be used in the hierarchical moderated regression analysis were obtained running a stepwise forward regression. This analysis provided an ordered sequence of the factors defining the problem-solving style that mostly affect the generation of new knowledge. That was done to limit the number of independent factors considered in the regression analysis and to obtain a ranking of them for relevance of effect on knowledge (Figure 1).

## RESULTS

Table 2 shows the statistics and comparison across clustering identifying different environment patterns. The VRC index procedure supports a 4-clusters solution (see Appendix for details). Indeed, there clearly emerge 4 clusters that describe four different environment patterns. The first cluster identifies a cognitive environment pattern in which centroid means for both ambiguity and uncertainty achieve average values. This cluster includes 36 technical problem solving processes. The second cluster identifies an environment pattern in which the mean value of the perceived uncertainty amount is much lower than average, but – vice versa - the perceived ambiguity amount is higher than the average. This cluster contains 7 technical problem-solving processes.

The third cluster refers to a cognitive environment pattern where both ambiguity and uncertainty achieve very high values (higher than average). This cluster includes 23 technical problem-solving processes. Finally, the fourth cluster (25 problem-solving processes) is associated to an environment pattern in which both the uncertainty and ambiguity amounts are lower than the respective average values in the sample. Figure 2 plots all 4 clusters in a two dimension (ambiguity-uncertainty) quadrant.

Table 3 reports the ANOVA outcome relative to variables describing the technical problem-solving style across the four cognitive environment patterns. The last column shows the results of the Tukey test for differences across means of variables. Couples of patterns whose problem solving style variable means significantly differ at 90% are indicated in brackets. From data in Table 3, it emerges that several profiles of problem-solving style are associated to the four patterns. Furthermore, there exist differences in the amount and type of knowledge (KNOWLEDGE and ROUTINE) acquired with problem-solving activity. In

*Table 2. Statistics and comparison across clusters for environment (clustering) variables*

| variable | pattern 1 (n=36) | | pattern 2 (n=7) | | pattern 3 (n=23) | | pattern 4 (n=25) | | F | prob. |
|---|---|---|---|---|---|---|---|---|---|---|
| | mean | st.dev. | mean | st.dev. | mean | st.dev. | mean | st.dev. | | |
| uncertainty | 0.247 | 0.043 | 0.080 | 0.055 | 0.391 | 0.064 | 0.108 | 0.052 | 136.1 | 0.000 |
| ambiguity | 0.275 | 0.044 | 0.357 | 0.035 | 0.401 | 0.071 | 0.128 | 0.057 | 103.5 | 0.000 |

*Figure 2. Plot of points belonging to identified clusters. Cluster boundaries are only indicative*

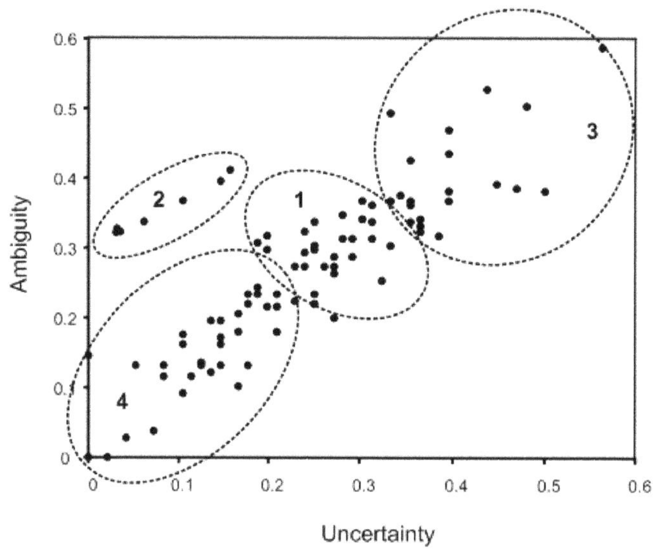


particular, pattern 2 in which there are higher values of ambiguity but lower values of uncertainty, the amount of knowledge generated during technical problem-solving is higher than in the other patterns. Moreover, when all patterns are examined, it clearly appears that the lower amount of knowledge is generated in pattern 4 with lower values for ambiguity and uncertainty.

Thus, an association between the typology of pattern and amount of knowledge generated during problem-solving is evident. Pattern 1 is characterized for having values of the problem-solving style variables in the average. Pattern 2 has high values for EXPERIMENTATION and DIVERGENT THINKING, but lower values for variables that are associated to social behaviors. Pattern 3 is characterized for having high values

*Table 3. Statistics and comparison across clusters for problem-solving style variables*

| variable | pattern 1 | pattern 2 | pattern 3 | pattern 4 | F | Tukey test |
|---|---|---|---|---|---|---|
| INTNETWORK | -0.078 | -0.831 | 0.596 | -0.182 | 5.220 | (1,3) (2,3) (3,4) |
| EXTNETWORK | 0.010 | 0.175 | 0.653 | -0.638 | 8.213 | (1,3) (1,4) (3,4) |
| CUSTINT | 0.273 | -0.800 | -0.024 | -0.148 | 2.722 | (1,2) |
| FRAMING | -0.118 | 0.011 | 0.708 | -0.484 | 7.200 | (1,3) (3,4) |
| WIDENING | 0.020 | 0.112 | 0.644 | -0.653 | 8.435 | (1,3) (1,4) (3,4) |
| PLANNING | 0.170 | 0.257 | 0.066 | -0.378 | 1.768 | |
| EXPERIMENTATION | -0.007 | 0.391 | -0.197 | 0.082 | 0.703 | |
| DIVERGENT THINKING | 0.106 | 1.128 | -0.399 | -0.102 | 4.999 | (1,2) (2,3) (2,4) |
| KNOWLEDGE | -0.023 | 0.988 | 0.583 | -0.780 | 14.412 | (1,2) (1,3) (1,4) (2,4) (2,4) |
| ROUTINE | 0.042 | 0.607 | 0.436 | -0.631 | 6.743 | (1,4) (2,4) (3,4) |

Legend: Tukey HSD multiple comparison was implemented to test for significance of difference between cluster means. In brackets clusters whose means differ significantly at least with prob. <.10 are indicated. Variables were standardized.

for WIDENING, FRAMING, EXTNETWORK, and INTNETWORK. In pattern 4, values for most variables are substantially below the average – particularly WIDENING, FRAMING, EXTNETWORK - except for DIVERGENT THINKING, and EXPERIMENTATION.

Table 4 shows the output of the stepwise forward regression analysis. This analysis was performed twice, adopting each time KNOWLEDGE and ROUTINE as dependent variables. In both cases, four variables entered the regression equation. Some of the variables are common to both equations (FRAMING, and EXTNETWORK), others are specific (EXPERIMENTATION, DIVERGENT THINKING, WIDENING, and CUSTINT). These variables were used as independent variables in the moderated regression analysis.

Table 5 and Table 6 respectively show the output of the hierarchical regression analyses performed using KNOWLEDGE and ROUTINE as dependent variables. A 4-state categorical variable CLU was introduced to take into account the moderation effect of the cognitive environment pattern. The introduction of this variable allowed to assess both the existence and amount of a moderation effect of the cognitive pattern on technical problem-solving style. Information contained in the original 4-state categorical variable was codified in a set of 3 dummy binary variables ($CLU_i$, for $i$=1, 2, 3). Using these dummies, all cases that belong to a particular category were assigned a code of 1; cases not in that category received a code of 0. Following this coding convention, a set of dummy variables for the categorization was constructed so that any particular environment was coded 1 on one and only one dummy variable in the set. Pattern 4 was chosen as a reference category for the categorical variable CLU, both as it contains a sufficient number of cases to allow a reasonably precise estimate of the pattern centroid means, and as it denotes a low-complexity cognitive environment in which problem solving occurred. Henceforth, the bivariate regression coefficient for the dummy variable expresses the average increase (or reduction) in KNOWLEDGE (or ROUTINE) of Pattern "*i*" respect to the reference Pattern 4. In other words, the regression coefficient expresses the difference between the average contribution of pattern *i* to KNOWLEDGE (or ROUTINE) and the average of reference pattern.

Three regression models were explored. In the simplest model (Model 1) the dependent variable is a linear function of the set of variables obtained from the stepwise regression analysis. The second model (Model 2) differs from the first model as it includes three dummy variables to account for environmental pattern differences. This model makes it possible to determine whether the expected value of the dependent variable differs in a systematic way for given values of environment perception. To test for interaction effects between the environment pattern and the problem-solving style, a specification to estimate differential effects

*Table 4. Stepwise regression analysis*

| number of variable in: | dependent variable: KNOWLEDGE | dependent variable: ROUTINE |
|---|---|---|
| 1 | FRAMING | WIDENING |
| 2 | EXTNETWORK | FRAMING |
| 3 | EXPERIMENTATION | CUSTINT |
| 4 | DIVERGENT THINKING | EXTNETWORK |
| F to enter | 1.76 | 2.24 |
| minimum tolerance | 0.796 | 0.692 |
| R | 0.674 | 0.786 |

*Table 5. Moderated regression analysis with KNOWLEDGE as dependent variable*

| variable | Model 1 | | Model 2 | | Model 3 | |
|---|---|---|---|---|---|---|
| | std. coeff. | prob | std. coeff. | prob | std. coeff. | prob |
| CONSTANT | 0.000 | 0.921 | 0.000 | 0.003 | 0.000 | 0.007 |
| FRAMING | 0.468 | 0.000 | 0.396 | 0.000 | 0.089 | 0.558 |
| EXTNETWORK | 0.304 | 0.001 | 0.196 | 0.029 | 0.430 | 0.011 |
| EXPERIMENTATION | 0.114 | 0.181 | 0.146 | 0.066 | 0.536 | 0.000 |
| DIVERGENT THINKING | 0.114 | 0.187 | 0.021 | 0.804 | -0.189 | 0.268 |
| CLU1 | | | 0.245 | 0.011 | 0.275 | 0.011 |
| CLU2 | | | 0.358 | 0.000 | -1.061 | 0.190 |
| CLU3 | | | 0.299 | 0.006 | 0.296 | 0.019 |
| CLU1_FRAMING | | | | | 0.269 | 0.040 |
| CLU2_FRAMING | | | | | -0.198 | 0.037 |
| CLU3_FRAMING | | | | | 0.175 | 0.160 |
| CLU1_EXTNETWORK | | | | | -0.192 | 0.120 |
| CLU2_EXTNETWORK | | | | | -0.227 | 0.047 |
| CLU3_EXTNETWORK | | | | | -0.134 | 0.264 |
| CLU1_EXPERIMENTATION | | | | | -0.204 | 0.047 |
| CLU2_EXPERIMENTATION | | | | | 0.785 | 0.087 |
| CLU3_EXPERIMENTATION | | | | | -0.365 | 0.007 |
| CLU1_DIVERGENT THINKING | | | | | 0.189 | 0.119 |
| CLU2_DIVERGENT THINKING | | | | | 1.162 | 0.066 |
| CLU3_DIVERGENT THINKING | | | | | 0.122 | 0.357 |
| R-squared | 0.454 | | 0.559 | | 0.698 | |
| F | 17.480 | 0.000 | 14.672 | 0.000 | 8.396 | 0.000 |
| ΔR-squared (Model i+1 - Model i) | | | 0.105 | | 0.139 | |
| ΔF | | | 5.192 | | 1.811 | |

and ascertain their significance was introduced in the regression equation. For every quantitative variable three product terms were constructed and added in the specification, thus obtaining Model 3. This last model allows to test for the existence of fit between the environment perceived and technical problem-solving style for knowledge generation. The new variables, $CLU_i$_VARIABLE, were computed by multiplying the three dummies $CLU_i$ ($i$=1, 2, 3) for each one of the four variables describing the problem-solving style. Twelve new variables were included in the regression equation.

In Model 1 of Table 5, EXTNETWORK and FRAMING significantly affects KNOWLEDGE. Shifting to Model 2 and controlling for the environment pattern effect, these two variables continue to affect KNOWLEDGE, even though, their effect diminishes. Furthermore, EXPERIMENTATION acquires a weight. The significativity of all dummies indicates a strong effect of the cognitive environment pattern. In particular, in all cases there is a positive increase of KNOWLEDGE, even though in the environment pattern 2 this increase is greater. Model 3 is still statistically significant as the $R$-squared and $\Delta F$ show val-

*Table 6. Moderated regression analysis with ROUTINE as dependent variable*

| variable | Model 1 | | Model 2 | | Model 3 | |
|---|---|---|---|---|---|---|
| | std. coeff. | prob | std. coeff. | prob | std. coeff. | prob |
| CONSTANT | 0.000 | 0.837 | 0.000 | 0.754 | 0.000 | 0.953 |
| WIDENING | 0.503 | 0.000 | 0.511 | 0.000 | 0.644 | 0.000 |
| FRAMING | 0.292 | 0.001 | 0.311 | 0.000 | 0.142 | 0.325 |
| CUSTINT | 0.144 | 0.041 | 0.189 | 0.009 | 0.015 | 0.936 |
| EXTNETWORK | 0.113 | 0.139 | 0.116 | 0.131 | 0.200 | 0.198 |
| CLU1 | | | 0.019 | 0.820 | 0.022 | 0.831 |
| CLU2 | | | 0.199 | 0.009 | 0.226 | 0.090 |
| CLU3 | | | -0.081 | 0.399 | -0.010 | 0.937 |
| CLU1_WIDENING | | | | | -0.088 | 0.379 |
| CLU2_WIDENING | | | | | -0.060 | 0.840 |
| CLU3_WIDENING | | | | | -0.170 | 0.100 |
| CLU1_FRAMING | | | | | 0.236 | 0.069 |
| CLU2_FRAMING | | | | | -0.011 | 0.960 |
| CLU3_FRAMING | | | | | 0.013 | 0.906 |
| CLU1_CUSTINT | | | | | 0.109 | 0.479 |
| CLU2_CUSTINT | | | | | 0.087 | 0.486 |
| CLU3_CUSTINT | | | | | 0.106 | 0.323 |
| CLU1_EXTNETWORK | | | | | -0.084 | 0.467 |
| CLU2_EXTNETWORK | | | | | -0.087 | 0.764 |
| CLU3_EXTNETWORK | | | | | -0.022 | 0.842 |
| R-squared | 0.618 | | 0.667 | | 0.714 | |
| F | 34.019 | 0.000 | 23.153 | 0.000 | 9.077 | 0.000 |
| ΔR-squared (Model i+1 - Model i) | | | 0.049 | | 0.047 | |
| ΔF | | | 3.456 | | 0.811 | |

ues. Further, some of the product variables have statistically significant coefficients, indicating the existence of interaction effects. The effect of the environment pattern controlling for all other variable remains significant and positive for pattern 1 and pattern 3. Indeed, pattern 2 is more sensitive to interaction effects. Negative signs for the regression coefficients of product terms indicate differentiated effects of the environment pattern on the specific variables shaping the profile of technical problem-solving style. FRAMING and EXTNETWORK have a negative effect on KNOWLEDGE in environment pattern 2. Vice versa, EXPERIMENTATION and DIVERGENT THINKING have a strong and positive effect in pattern 2, but negative in the other two environment patterns.

In Table 6, the moderated regression analysis performed with ROUTINE as dependent variable shows different findings as the introduction of interaction terms in Model 3 is not significant. WIDENING, FRAMING, and CUSTINT have a positive effect on ROUTINE not controlling for the effect of environment pattern and cross-effects (Model 1). When control for the environment pattern is introduced, the effect of the previous variables remains positive and strong. Moreover, an environment effect is evident as in pattern 2 there is an increase in the value of ROUTINE.

## DISCUSSION AND CONCLUSION

The analysis of empirical data shows that the perceived cognitive problem-solving environment has a statistically significant effect on the type and amount of knowledge generated. Particularly, the environment cognitive patterns where ambiguity and uncertainty achieve higher values seem to be more conducive to new organizational knowledge. Data reported in Table 3, Table 5 and Table 6 show this findings. The outcome of new organizational routines of pattern 2 is relevant.

Data in Table 5 and Table 6 reveal that certain cognitive practices and behaviors modeling the problem-solving style are more effective in certain problem-solving cognitive environment patterns when the performance of organizational learning is assessed in terms of new declarative knowledge. Vice versa that is not true when procedural knowledge is the outcome of learning. Focusing our analysis on declarative knowledge, two issues merit further consideration. The great positive weight that experimentation and divergent thinking have in environment pattern 2 (the interaction effects amplify the linear effect of both factors) and the negative effect of problem framing and external networking in pattern 2. These findings put emphasis on a uniqueness of cognitive environment pattern 2. Problem-solvers who tackled with the technical problems considered in the study tended to use divergent thinking as a tool when they felt that they were fully aware that the problem could not be tackled in the traditional way, but it required a considerable mental effort to find new approaches to solve it. Appealing to divergent thinking during problem-solving (for instance, fostering individual or group brainstorming) introduces new alternatives that could be further sources of ambiguity. But, this increase in ambiguity is more manageable than before, as people generate alternatives having clear in their mind how to refine and select them. Divergent thinking works differently from problem-solving widening. In problem-solving widening problem solvers modify the problem representation when they understand that the original representation does not enable them to proceed. To modify the representation of the problem means to modify the knowledge domain of the problem in terms of variables, relationships, constraints, solving strategies. Changing problem representation implies to alter the environment pattern in which problem-solvers have to move. People find a solution to the problem identifying a new problem, and solving this one:

*"[...] we have understood that it is necessary to take into account the effective conditions of use of the machine. Particularly for the grated-ice drink machines. Indeed, we suggest our customer to use a certain percentage of sugar (15%), but in most cases we know that this indication is rejected to save money. Therefore, we decided to build machines able to work satisfactorily in conditions slightly different from the conditions we suggest" [source: field interviews]*

Ambiguity decreases because the problem changes. Problem-solving widening is henceforth a strategy of "ambiguity avoidance". In order to reduce this increase in ambiguity problem solvers do experiments, that is to say, by trial and error they simulate possible solutions of the problem. Experimentation is indeed a strategy of ambiguity reduction. Whenever people do not have information and need to generate useful data, or, more frequently, whenever they do not have enough information to choose among alternatives (when ambiguity is high) and there is a lack of objectivity, experimentation remains the most effective way to proceed. Problem solving framing is a typical strategy of ambiguity reduction. As it reduces problem complexity by splitting the original problem into a set of linked sub-problems easy to deal with, it is able to alleviate ambiguity because it reduces the number of variables to consider in the same time. A problem representation with too many variables means to have a great number of combinations of variables, a great number of connections among them, too many interpretations to search. The fragmentation of the problem when it is framed allow problem solvers to re-organize the information set they have into new categories. In this way they are able to give an orientation to the problem solving activity, stimulating the collection of new information, thus reducing uncertainty.

The study shows that cognitive practices and collective behaviors modeling the problem-solving style that are more intimately associated to creativity are more critical in the process of organizational knowledge creation and learning when teams carry on innovation.

As it usually occurs with empirical data analysis, the study presents both theoretical and empirical limitations. Most of the results confirm the idea that the problem-solving style influences the amount and type of knowledge created during learning associated to innovation and that this influence may be more o less important depending on the amount of complexity perceived by problem-solvers. Of course, the problem-solving process was simplified and the large part of its tasks and sub-processes are with no doubt more complex than assumed in the study.

The assumption was done that the variables included in the statistical analysis have reached a stable state and the relationships among them were explored since that time. Vice versa, learning is fundamentally dynamic. Furthermore, as it was underlined, the model includes only a limited subset of factors that may affect the generation of new knowledge. Indeed, as a choice, the focus of this study is on those factors (cognitive practices and social behaviors) which can be purposely modeled to positively affect learning.

Problems of construct and statistical conclusion validity, common in all empirical research, are also present in this study. That was primarily due to the substantial lack in the literature of pre-tested scales to measure the variables of the model and to the absence of a theory of knowledge creation which made it hard to identify factors influencing the process of knowledge generation.

The contingent approach without any doubt has a number of limitations. Contingency theories postulate a direct linkage between environment and organization characteristics as a materialistic ontology. The perspective of fit as interaction presents some weakness points (Drazin & Van de Ven, 1985; Venkatraman, 1989). First, the meaning of fit is strictly tied to the performance variable. The amount and type of knowledge is just one of the set of possible measures of performance of the learning process. A further limitation is the intrinsic reductionism. The theory is unable to capture the gestalt of the overall learning process (Drazin & Van de Ven, 1985; Venkatraman, 1989). The focus on a simplified environment attribute measured on a categorical scale does not capture the complexity of the way processes develop. Some scholars suggest to develop theories of multiple contingencies (Gresov, 1989) or follow a systemic/configurational approach (Venkatraman, 1989). Of course, the interpretation of the effect of multiple contingencies is not always immediate and easy.

However, even though limits are present, the findings have relevant implications in the practitioner perspective as they give insights for an active intervention to guide the process of learning during technical problem-solving in order to enhance the knowledge asset of the organization.

## FUTURE RESEARCH

The study has explored how new organizational knowledge is generated during technical problem-solving when teams are engaged in product development. But, the need to simplify the model, and particularly the empirical analysis, required to focus the study only a specific aspect of the knowledge creation process, e.g. the effect of the problem-solving style on the learning output. Consequently, the cognitive and social mechanisms that - both at the individual and organization level – govern the dismantling of obsolete knowledge and the retrieval and storing of new knowledge in the organizational memory were not investigated. As it was emphasized in the introductive discussion, knowledge creation and knowledge accumulation are processes that cannot be separated. Henceforth, the organizational memory process should be included in the model.

The model does not distinguish between individual and team problem-solving styles, but focuses on collective cognitive practices and social behaviors assuming the team as the unique actor of the problem-solving process and associated learning. This view is extremely reductive and the dualism "individual-team" merits a more in depth analysis. Both problem-solving and learning are largely influenced by the interaction between individuals and groups. Thus, the individual and the organizational dimensions and the interaction between them has to be investigated.

Finally, even though the choice of the research methodology adopted (a survey) was induced by the goal to have generalizable relations that were not rigidly linked to the specific case study, both future research trajectories outlined require a major reflection about its real effectiveness and practicability.

## REFERENCES

Amabile, T. M. (1983). *The Social Psychology of Creativity*. New York: Springer-Verlag.

Ambrosini, V., & Bowman, C. (2001). Tacit knowledge: Some suggestions for operationalization. *Journal of Management Studies, 38*(6), 811–829. doi:10.1111/1467-6486.00260

Andrews, F. M. (1975). Social and psychological factors that influence the creative process. In Taylor, I. A., & Getzels, W. (Eds.), *Perspectives in creativity* (pp. 117–145). Chicago: Aldine.

Arabie, P., & Hubert, L. J. (1994). Cluster analysis in marketing research. In R.P. Bagozzi (Ed.), *Advanced methods of marketing research,* (pp.160-189). Basil: Blackwell.

Argyris, C. (1982). *Reasoning, learning and action: Individual and organizational*. San Francisco: Jossey-Bass.

Argyris, C., & Schon, D. A. (1978). *Organizational learning: A theory of action perspective*. Reading, MA: Addison-Wesley.

Arrow, K. (1962). The economic implications of learning by doing. *The Review of Economic Studies, 29*, 155–173. doi:10.2307/2295952

Badaracco, J. L. (1991). *The knowledge link: How firms compete through strategic alliances*. Boston: Harvard Business School Press.

Baker, F. B., & Hubert, L. J. (1976). A graph-theoretic approach to goodness-of-fit in complete-link hierarchical clustering. *Journal of the American Statistical Association, 71*, 870–878. doi:10.2307/2286853

Bell, M. A. (1982). Phases in group problem-solving. *Small Group Behavior, 13*, 475–495. doi:10.1177/104649648201300405

Bouthillier, F. & Shearer, K. (2002). Understanding knowledge management and information management: The need for an empirical perspective. *Information Research, 8*(1).

Braybrooke, D., & Lindblom, C. E. (1963). *A strategy of decision*. New York: The Free Press.

Brennan, A., & Dooley, L. (2005). Networked creativity: A structured management framework for stimulating innovation. *Technovation, 25*, 1388–1399. doi:10.1016/j.technovation.2004.08.001

Brief, A. P., & Downey, H. K. (1983). Cognitive and organizational structures: A conceptual analysis of implicit organizing theories. *Human Relations, 36*, 1065–1090. doi:10.1177/001872678303601201

Brigthman, H. J., Elrod, R., & Ramakrishna, H. (1988). Matching problem diagnostic tools to managers' decision styles: A contingency approach. *Omega, 16*, 1–9. doi:10.1016/0305-0483(88)90002-3

Brown, J. S., & Duguid, P. (1991). Organizational learning and communities-of-practice: Toward a unified view of working, learning, and innovation. *Organization Science, 2*, 40–57. doi:10.1287/orsc.2.1.40

Calinski, R. B., & Harabasz, J. (1974). A dendrite method for cluster analysis. *Communication Statistics, 3*, 1–27. doi:10.1080/03610927408827101

Cheng, Y. T., & Van de Ven, A. H. (1996). Learning the innovation journey: Order out of chaos? *Organization Science, 7*(6), 593–614. doi:10.1287/orsc.7.6.593

Chesbrough, H. (2003). *Open innovation: The new imperative for creating and profiting from technology*. HBS Press.

Choo, C. W. (1998). *The knowing organization: How organizations use information to construct meaning, create knowledge, and make decisions*. New York: Oxford University Press.

Churchill, G. A. (1979). A paradigm for developing better measures of marketing constructs. *JMR, Journal of Marketing Research, 15*, 64–73. doi:10.2307/3150876

Clark, K. B., & Fujimoto, T. (1991). *Product development performance: Strategy, organization, and management in the world auto industry*. Boston: Harvard University Press.

Clement, J. (1989). Learning via model construction and criticism: Protocol evidence on sources of creativity in science. In Glover, J., Ronning, R., & Reynolds, C. (Eds.), *Handbook of creativity: Assessment, theory and research* (pp. 341–381). NY: Plenum.

Cohen, M. D., & Bacdayan, P. (1994). Organizational routines are stored as procedural memory: Evidence from a laboratory study. *Organization Science, 5*, 554–568. doi:10.1287/orsc.5.4.554

Cohen, W. M. (1991). Individual learning and organizational routine: Emerging connections. *Organization Science, 2*, 135–139. doi:10.1287/orsc.2.1.135

Cohen, W. M., & Levinthal, D. A. (1990). Absorptive capacity: A new perspective on learning and innovation. *Administrative Science Quarterly, 35*, 128–152. doi:10.2307/2393553

Cook, P. (1998). *Best practice creativity*. Aldershot, UK: Gower Publishing.

Cook, S. D. N., & Yanow, D. (1993). Culture and organizational learning. *Journal of Management Inquiry, 2*(4), 373–390. doi:10.1177/105649269324010

Corti, E., & lo Storto, C. (1997). Technical problem-solving and learning processes during product innovation: Some empirical evidence for the development of a theoretical framework of knowledge creation in small firms. *Proceedings of the PICMET'97*, Portland (Oregon), July.

Corti, E. & lo Storto, C. (1999). Factors to sustain the creation of knowledge, competences and routines in small firms during product development. *Piccola Impresa/Small Business, 2*, 33-57.

Corti, E., & lo Storto, C. (2000). Knowledge creation in small firms during product innovation: An empirical analysis of cause-effect relationships among its determinants. *Enterprise & Innovation Management Studies, 1*(3), 245–263. doi:10.1080/14632440010023190

Covin, J. G., & Slevin, D. P. (1988). The influence of organization structure on the utility of an entrepreneurial top management style. *Journal of Management Studies*, *25*, 217–234. doi:10.1111/j.1467-6486.1988.tb00033.x

Craft, A. (2005). *Creativity in schools: Tensions and dilemmas, Oxford: Routledge Falmer*. London: QCA. doi:10.4324/9780203357965

Cyert, R. M., & March, J. G. (1963). *A behavioral theory of the firm*. Englewood Cliff, NJ: Prentice Hall.

Daft, R. L., & Lengel, R. H. (1986). Organizational information requirements, media richness and structural design. *Management Science*, *32*, 554–571. doi:10.1287/mnsc.32.5.554

Daft, R. L., & Macintosh, N. B. (1981). A tentative explanation into the amount and equivocality of information processing in organizational work units. *Administrative Science Quarterly*, *26*, 207–224. doi:10.2307/2392469

Daft, R. L., & Weick, K. E. (1984). Toward a model of organizations as interpretation systems. *Academy of Management Review*, *9*, 284–295. doi:10.2307/258441

Davenport, T. H., Long, D. W. D., & Beers, M. C. (1998). Building successful Knowledge Management projects. *Sloan Management Review*, *39*(2), 45–57.

Davenport, T. H., & Prusak, L. (1997). *Information ecology*. Oxford University Press.

Davis, G. A., & Manski, M. E. (1966). An instructional method for increasing originality. *Psychonomic Science*, *6*, 73–74.

Dery, D. (1983). Decision-making, problem-solving, and organizational learning. *Omega*, *11*, 321–328. doi:10.1016/0305-0483(83)90025-7

Dodgson, M. (1993). Organizational learning: A review of some literatures. *Organization Studies*, *14*, 375–394. doi:10.1177/017084069301400303

Drazin, R., & Van de Ven, A. H. (1985). Alternate form of fit in contingency theory. *Administrative Science Quarterly*, *30*, 514–539. doi:10.2307/2392695

Duncan, R., & Weiss, A. (1979). Organizational learning: Implications for organizational design. *Research in Organizational Behavior*, *1*, 75–123.

Duncan, R. B. (1972). Characteristics of organizational environments and perceived environmental uncertainty. *Administrative Science Quarterly*, *17*, 313–327. doi:10.2307/2392145

Duncker, K. (1945). *On problem solving. Psychological Monographs Series, 58(5)*. California: American Psychological Association.

Edmondson, A., & Moingeon, B. (1998). From organizational learning to the learning organization. *Management Learning*, *29*(1), 5–20. doi:10.1177/1350507698291001

Feldman, M. S. (2000). Organizational routines as a source of continuous change. *Organization Science*, *6*, 611–629. doi:10.1287/orsc.11.6.611.12529

Fiol, C. (1994). Consensus, diversity, and learning in organizations. *Organization Science*, *5*, 403–420. doi:10.1287/orsc.5.3.403

Fiol, C., & Lyles, L. (1985). Organizational learning. *Academy of Management Review*, *10*, 803–813. doi:10.2307/258048

Fleck, J. (1994). Learning by trying: The implementation of configurational technology. *Research Policy*, *23*, 637–652. doi:10.1016/0048-7333(94)90014-0

Fry, L. W., & Slocum, J. W. (1984). Technology, structure, and work group effectiveness: A test of a contingency model. *Academy of Management Journal, 27*, 221–246. doi:10.2307/255923

Galbraith, J. R. (1973). *Designing complex organizations*. Reading, MA: Addison-Wesley.

Garner, W. R. (1962). *Uncertainty and structure as psychological concepts*. New York: John Wiley & Sons.

Gersick, C. J. G., & Hackman, J. R. (1990). Habitual routines in task-performing groups. *Organizational Behavior and Human Decision Processes, 47*, 65–97. doi:10.1016/0749-5978(90)90047-D

Getzels, F., & Csikszentmyhalyi, M. (1976). *The creative vision: A longitudinal study of problem solving in art*. New York: Wiley.

Glaser, R. (1986). On the nature of expertise. In Klix, F., & Hagendorf, H. (Eds.), *Human memory and cognitive capabilities: Mechanisms and performances* (pp. 915–928). Amsterdam: Elsevier Science Publishing.

Grant, R. M. (1996). Prospering in dynamically-competitive environments: Organizational capability as knowledge integration. *Organization Science, 7*, 375–387. doi:10.1287/orsc.7.4.375

Gresov, C. (1989). Exploring fit and misfit with multiple contingencies. *Administrative Science Quarterly, 34*, 431–453. doi:10.2307/2393152

Guilford, J. P. (1977). *Way beyond the IQ*. Buffalo, NY: Bearly Limited.

Hayes, J. R. (1978). *Cognitive psychology. Thinking and creating*. Homewood, IL: Dorsey Press.

Hedberg, B. (1981). How organizations learn and unlearn. In Nystrom, P. C., & Starbuck, W. H. (Eds.), *Handbook of organizational design* (pp. 3–27). New York: Oxford University Press.

Hedberg, B., Nystrom, P. C., & Starbuck, W. H. (1976). Camping on seesaws: Prescriptions for a self-designing organization. *Administrative Science Quarterly, 21*, 41–65. doi:10.2307/2391877

Huber, G. P. (1991). Organizational learning: The contributing processes and a review of the literature. *Organization Science, 2*, 88–115. doi:10.1287/orsc.2.1.88

Hwang, G. J., Hsiao, C. L., & Tseng, J. C. R. (2003). A computer-assisted approach to diagnosing student learning problems in science courses. *Journal of Information Science and Engineering, 19*(2), 229–248.

Itami, H. (1987). *Mobilizing invisible assets*. Cambridge, MA: Harvard University Press.

Janik, A. (1988). Tacit knowledge, working life and scientific method. In Göranzon, B., & Josefson, I. (Eds.), *Knowledge, skill and artificial intelligence* (pp. 53–63). London, Berlin: Springer & Verlag.

Jelinek, M. (1979). *Institutionalizing innovation*. New York: Praeger.

Johnson-Laird, P. N. (1983). *Mental modes*. Cambridge, MA: Harvard University Press.

Kline, S. J., & Rosenberg, N. (1986). An overview of innovation. In Landau, R., & Rosenberg, N. (Eds.), *The positive sum game* (pp. 275–305). Washington, DC: National Academy Press.

Kolb, D. A. (1974). Learning and problem solving on management and the learning process. In Kolb, D. A., Rubin, I. M., & McIntyre, J. M. (Eds.), *Organizational psychology: A book of readings* (pp. 27–42). Englewood Cliffs, NJ: Prentice-Hall.

Levitt, B., & March, J. G. (1988). Organizational learning. *Annual Review of Sociology, 14*, 319–340. doi:10.1146/annurev.so.14.080188.001535

Lumsdaine, E., & Lumsdaine, M. (1995). *Creative problem solving: Thinking skills for a changing world*. McGraw-Hill.

Lyles, M. A., & Mitroff, I. I. (1980). Organisational problem formulation: An empirical study. *Administrative Science Quarterly, 25*, 102–119. doi:10.2307/2392229

Machlup, F., & Mansfield, U. (1983). *The study of information: Interdisciplinary messages*. New York: John Wiley & Sons.

Maidique, M. A., & Zirger, B. J. (1984). The new product learning cycle. *Research Policy, 14*, 299–313. doi:10.1016/0048-7333(85)90001-0

Majaro, S. (1988). *The creative gap–managing ideas for profit*. Longman.

March, J. C., & Olsen, J. P. (1975). The uncertainty of the past: Organizational learning under ambiguity. *European Journal of Political Research, 3*, 147–171. doi:10.1111/j.1475-6765.1975.tb00521.x

March, J. G., & Simon, H. A. (1958). *Organizations*. New York: John Wiley and Sons.

Marples, D. L. (1961). The decisions of engineering design. *IEE Transactions on Engineering Management, 20*, 55–71. doi:10.1109/IRET-EM.1961.5007593

Martin, J., & Meyerson, D. (1988). Organizational cultures and the denial, channeling and acknowledgement of ambiguity. In Pondy, L. R., Boland, R. J., & Howard, T. (Eds.), *Managing ambiguity and change* (pp. 93–125). New York: Wiley.

Mayer, R. E. (1983). *Thinking, problem solving, and cognition*. San Francisco: Freeman.

McAdam, R., & McClelland, J. (2002a). Individual and team-based idea generation within innovation management: Organizational and research agendas. *European Journal of Innovation Management, 5*(2), 86–97. doi:10.1108/14601060210428186

McAdam, R., & McClelland, J. (2002b). Sources of new product ideas and creativity practices in the UK textile industry. *Technovation, 22*, 113–121. doi:10.1016/S0166-4972(01)00002-5

McCall, M. W. (1977). Making sense with nonsense: Helping frames of reference clash. In Nystrom, P. C., & Starbuck, W. H. (Eds.), *Prescriptive model of organizations* (pp. 111–123). Amsterdam: North-Holland.

McKee, D. (1992). An organizational learning approach to product innovation. *Journal of Product Innovation Management, 9*, 232–245. doi:10.1016/0737-6782(92)90033-9

Meindl, J. M., Stubbart, C., & Porac, J. F. (1996). Introduction. In Meindl, J. M., Stubbart, C., & Porac, J. F. (Eds.), *Cognition within and between organizations* (pp. ix–xxiii). Thousand Oaks, CA: Sage Publications.

Milligan, G. W. (1981). A Monte Carlo study of thirty internal criterion measures for cluster analysis. *Psychometrika, 46*, 187–199. doi:10.1007/BF02293899

Milligan, G. W., & Cooper, M. C. (1985). An examination of procedures for determination of clusters in a data set. *Psychometrika, 50*, 159–179. doi:10.1007/BF02294245

Milligan, G. W., & Cooper, M. C. (1988). A study of standardization of variables in cluster analysis. *Journal of Classification, 5*, 181–204. doi:10.1007/BF01897163

Mintzberg, H. (1976). Planning on the left side and managing on the right side. *Harvard Business Review, 54*, 49–58.

Mintzberg, H., Duru, R., & Theoret, A. (1976). The structure of unstructured decision process. *Administrative Science Quarterly, 21*, 246–275. doi:10.2307/2392045

Mohr, L. B. (1982). *Explaining organization behavior*. San Francisco: Jossey-Bass.

Morrison, D. G. (1967). Measurement problems in cluster analysis. *Management Science, 13*, B774–B780. doi:10.1287/mnsc.13.12.B775

Nelson, R. R. (2004). The market economy and the scientific commons. *Research Policy, 33*, 455–471. doi:10.1016/j.respol.2003.09.008

Nelson, R. R., & Winter, S. G. (1982). *An evolutionary theory of economic change*. Cambridge, MA: Harvard University Press.

Newell, A., Shaw, J. C., & Simon, H. A. (1979). The processes of creative thinking. In Simon, H. A. (Ed.), *Models of thought* (pp. 144–174). New Haven: Yale University Press.

Newell, A., & Simon, H. A. (1972). *Human problem solving*. Englewood Cliffs, NJ: Prentice Hall.

Nissen, M., Kamel, M., & Sengupta, K. (2000). Integrated analysis and design of knowledge systems and processes. *Information Resources Management Journal, 13*(1), 24–43.

Nonaka, I. (1991). The knowledge creating company. *Harvard Business Review*, 96–104.

Nonaka, I. (1994). A dynamic theory of organizational knowledge creation. *Organization Science, 5*, 14–36. doi:10.1287/orsc.5.1.14

Nonaka, I., & Konno, N. (1998). The concept of 'ba': Building a foundation for knowledge creation. *California Management Review, 40*, 40–54.

Nonaka, I., & Takeuchi, H. (1995). *The knowledge-creating company*. New York: Oxford University Press.

Nunnally, J. C. (1967). *Psychometric theory* (1st ed.). New York: McGraw Hill.

Nystrom, P. C., & Starbuck, W. H. (1984). To avoid organizational crises. *Organizational Dynamics*, (Spring): 53–65. doi:10.1016/0090-2616(84)90011-1

Oldham, G. R., & Cummings, A. (1996). Employee creativity: Personal and contextual factors at work. *Academy of Management Journal, 39*(3), 607–655. doi:10.2307/256657

Olton, R. M. (1980). Experimental studies of incubation: Searching for the elusive. *The Journal of Creative Behavior, 13*, 9–13.

Pentland, B. T., & Rueter, H. H. (1994). Organizational routines as grammar of actions. *Administrative Science Quarterly, 39*, 484–510. doi:10.2307/2393300

Perrow, C. (1967). *Organizational analysis: A sociological view*. Belmont, CA: Wadsworth.

Pisano, G. P. (1994). Knowledge, integration, and the locus of learning: An rmpirical snalysis of process development. *Strategic Management Journal, 15*, 85–100. doi:10.1002/smj.4250150907

Polanyi, M. (1958). *Personal knowledge: Toward a post-critical philosophy*. Chicago: University of Chicago Press.

Polito, T., & Watson, K. (2002). Toward an interdisciplinary organizational learning framework. *The Journal of American Academy of Business, 2*(1), 162–166.

Polya, G. (1945). *How to solve it: A new aspect of mathematical method*. Princeton University Press. reprinted in 2004

Pounds, W. F. (1969). The process of problem finding. *Industrial Management Review, 11*, 1–19.

Raaheim, K. (1974). *Problem solving and intelligence*. Oslo: Norwegian Universities Press.

Rein, M., & Schön, D. (1977). Problem setting in policy research. In Weiss, C. (Ed.), *Using social research in public policy making* (pp. 235–251). Lexington, MA: D. C. Heath.

Rickards, T., & Freedman, B. (1978). Procedures for management in idea deficient situations: An examination of brainstorming approaches. *Journal of Management Studies*, *15*, 43–55. doi:10.1111/j.1467-6486.1978.tb00908.x

Rosenberg, N. (1982). Learning by using. In Rosenberg, N. (Ed.), *Inside the black box: Technology and economics* (pp. 120–140). Cambridge, UK: Cambridge University Press.

Rumelt, R., Schendel, D. & Teece, D.J. (1992). Strategic management and economics. *Strategic Management Journal, Winter Special Issue, 12*, 5-29.

Schein, E. H. (1996). Organizational learning as a cognitive re-definition: Coercive persuasion revisited. Retrieved from http:/learning.mit.edu/res/wp/10010.html

Schoonhoven, C. B. (1981). Problems with contingency theory: Testing assumptions hidden within the language of contingency theory. *Administrative Science Quarterly*, *26*, 349–377. doi:10.2307/2392512

Schrader, S., Riggs, W. M., & Smith, R. P. (1991). Problem solving in the management of technology and innovation: Choosing the uncertainty-ambiguity boundary. MIT Working Paper # 3345-91-BPS.

Schwenk, C., & Thomas, H. (1983). Formulating the mess: The role of decision aids in problem formulation. *Omega*, *11*, 239–252. doi:10.1016/0305-0483(83)90049-X

Senge, P. M. (1990). *The fifth discipline*. New York: Doubleday.

Sharma, S., Durand, R. M., & Gur-Arie, O. (1981). Identification and analysis of moderator variables. *JMR, Journal of Marketing Research*, *18*, 291–300. doi:10.2307/3150970

Simon, H. A. (1973). The structure of ill-structured problems. *Artificial Intelligence*, *4*, 181–201. doi:10.1016/0004-3702(73)90011-8

Simon, H. A. (1979). Rational decision-making in business organizations. *The American Economic Review*, *69*, 493–513.

Simon, H. A. (1987). Making management systems: The role of intuition and emotion. *The Academy of Management Executive*, *1*, 57–64.

Smith, G. F. (1989a). Defining managerial problems: A framework for prescriptive theorizing. *Management Science*, *35*, 963–981. doi:10.1287/mnsc.35.8.963

Smith, G. F. (1989b). Managerial problems identification. *OMEGA International Journal of Management Sciences*, *17*, 27–36. doi:10.1016/0305-0483(89)90017-0

Spearman, C. (1927). *The abilities of man*. New York: Macmillan.

Teare, R., & Rayner, C. (2002). Capturing organizational learning. *International Journal of Contemporary Hospitality Management*, *14*(7), 354–360. doi:10.1108/09596110210440657

Teece, D., Pisano, G., & Shuen, A. (1990). *Firm capabilities, resources, and the concept of strategy*. Center for Research on Management, University of California, Berkeley, CCC Working Paper 90-8.

Torrance, E. P., & Myers, R. E. (1970). *Creative learning and teaching*. New York: Dodd.

Treffinger, D.J. & Isaksen, S.G. (1992). *Creative problem solving: An introduction*. Center for Creative Learning: Sarasota.

Tsoukas, H. (2003). Do we really understand tacit knowledge? In Easterby-Smith, M., & Lyles, M. A. (Eds.), *The Blackwell handbook of organizational learning and knowledge management* (pp. 411–427). Cambridge, MA: Blackwell Publishing.

Tulving, E., & Schachter, D. L. (1990). Priming and human memory systems. *Science*, *247*, 301–306. doi:10.1126/science.2296719

Ulrich, K. T., & Eppinger, S. (1995). *Product design and development*. New York: McGraw-Hill.

Urabe, K. (1988). Innovation and the Japanese management system. In Urabe, K., Child, J., & Kagono, T. (Eds.), *Innovation and management international comparisons* (pp. 3–26). Berlin: Walter de Gruyter.

Van de Ven, A. H., Polley, D. E., Garud, R., & Venkataraman, S. (1999). *The innovation journey*. New York: Oxford University Press.

Venkatraman, N. (1989). The concept of fit in strategy research: Toward verbal and statistical correspondence. *Academy of Management Review*, *14*, 423–444. doi:10.2307/258177

von Hippel, E. (1978). A customer-active paradigm for industrial product idea generation. *Research Policy*, *3*, 240–266. doi:10.1016/0048-7333(78)90019-7

von Hippel, E. (1994). Sticky information and the locus of problem solving: Implications for innovation. *Management Science*, *40*, 429–439. doi:10.1287/mnsc.40.4.429

von Hippel, E., & Tyre, M. (1993). How learning by doing is done: Problem identification in novel process equipment. Massachusetts Institute of Technology, SSM Working Paper#BPS 3521-93.

Von Krogh, G., Ichijo, K., & Nonaka, I. (2000). *Enabling knowledge creation: How to unlock the mystery of tacit knowledge and release the power of innovation*. New York: Oxford University Press.

Wagner, C. (1993). Problem solving and diagnosis. *OMEGA International Journal of Management Sciences*, *21*, 645–656. doi:10.1016/0305-0483(93)90006-7

Wagner, R. K., & Sternberg, R. J. (1985). Practical intelligence in real world pursuits: The role of tacit knowledge. *Journal of Personality and Social Psychology*, *49*, 436–458. doi:10.1037/0022-3514.49.2.436

Walsh, J. P., & Ungson, G. R. (1991). Organizational memory. *Academy of Management Review*, *16*, 57–91. doi:10.2307/258607

Weick, K. E. (1995). *Sensemaking in organizations*. Thousand Oaks, CA: SAGE Publications.

Weick, K. E., & Westley, F. (1996). Organizational learning: Affirming an oxymoron. In Clegg, S. R., Hardy, C., & Nord, W. R. (Eds.), *Handbook of organization studies* (pp. 440–458). London: Sage.

Wertheimer, M. (1945). *Productive thinking*. New York: Harper.

West, M. A. (2002). Sparkling fountains or stagnant ponds: An integrative model of creativity and innovation implementation in work groups. *Applied Psychology: An International Review*, *51*(3), 355–424. doi:10.1111/1464-0597.00951

West, M. A., & Farr, J. L. (1989). Innovation at work: Psychological perspectives. *Social Behaviour*, *4*, 15–30.

Winograd, T., & Flores, F. (1986). *Understanding computers and cognition: A new foundation for design*. Norwood, NJ: Ablex.

Winter, S. G. (1987). Knowledge and competence as strategic assets. In Teece, D. J. (Ed.), *The competitive challenge: Strategies for industrial innovation and renewal* (pp. 159–184). Cambridge, MA: Ballinger.

Zwicky, F., & Wilson, A. (1965). *New methods of thought and procedure*. New York: Springer-Verlag.

## ENDNOTES

1 Even though it is here fully acknowledged that problem-solving cognitive practices and social behaviors implemented are not the only factors that influence both the problem-solving success and the learning

process, but some other factors may be even more critical, such as (see West, 2002) people experience and knowledge, people personal attributes (i.e., the problem-solver leadership and influence, task motivation), organization characters (i.e., the communication and decision-making structure, the organization climate), the main focus is on those factors – cognitive practices and collective behaviors – that are more easily purposely modeled to guide the learning process.

2 In an organization, routines for solving problems exist in the head of individuals or are codified in the form of personal memos, operating procedures, or organizational manuals.

3 Some scholars and practitioners use interchangeably "information" and "knowledge" (Bouthillier and Shearer, 2002), thus generating great confusion.

4 Nonaka (1994) and Nonaka and Takeuchi (1995, p. 8) also distinguish technical tacit knowledge (know-how and skills) from cognitive tacit knowledge (mental models).

5 Mental models guide individuals' behaviors in the organization, determining what is relevant to understand a certain phenomenon or to solve a particular problem (Mintzberg, 1976; Brief & Downey, 1983; Simon, 1987; Clement, 1989), and a well defined mental model pre-determines the relevant domain to identify the solution to the problem (Clement, 1989). According to Mintzberg (1976, p. 54): "[…] in effect, managers (like everyone else) use their information to build mental models of their world, which are implicit synthesized apprehensions of how their organizations and environments function. Then, whenever an action is contemplated, the manager can simulate the outcome using his implicit model".

6 More rarely routines are in the form of procedures or formalised rules inside the organization (Cyert & March, 1963).

7 For instance, some *switching rules* that indicates what organizational tasks should be executed when determined conditions occur can give a certain degree of flexibility to routines (March & Simon, 1958). Further, routines can be modified – even though not immediately and easily – after new behaviours inside the organization have been experienced (Hedberg, 1981; Hedberg et al., 1976).

## APPENDIX

The Calinski and Harabasz (1974) index (*VRC, variance ratio criterion*) is calculated as follows:

$$VRC=\frac{(n\text{-}k)}{(k\text{-}1)}\frac{BGSS}{WGSS},$$

where

n is the sample size and k is the number of clusters

BGSS is the between-group sum of squares

WGSS is the within-group sum of squares

According to the Calinski and Harabasz criterion, the right number of cluster is the number k for which the VRC has an absolute or first local maximum.

It was assumed k=2, 3,..., 10, as the maximum number of 10 cluster solutions was set, and for every solution the VRC index was computed. Figure 3 shows the results of this analysis. Values reported are limited to 5 clusters. The maximum VRC was obtained for k=4.

*Figure 3. The variance ratio criterion index*

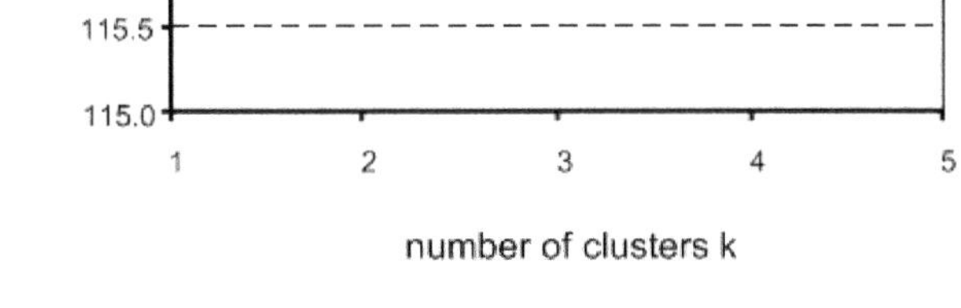

# Chapter 4
# Towards a Theoretical Framework for Creative Participation:
## How Personal Characteristics Influence Employees' Willingness to Contribute Ideas

**Natalya Sergeeva**
*University of Reading, UK*

**Milan Radosavljevic**
*University of Reading, UK*

## ABSTRACT

*To retain competitiveness, succeed and flourish, organizations are forced to continuously innovate. This drive for innovation is not solely limited to product/process innovation but more profoundly relates to a continuous process of improving how organizations work internally, requiring a constant stream of ideas and suggestions from motivated employees. This chapter investigates some recent developments and proposes a conceptual framework for creative participation as a personality driven interface between creativity and innovation. Under the assumption that employees' intrinsic willingness to contribute novel ideas and solutions requires a set of personal characteristics and necessary skills that might well be unique to each organizational unit, the chapter then explores personal characteristics associated with creativity, innovation and innovative behavior. Various studies on the correlation between creativity and personality types are also reviewed. The chapter provides a discussion of solutions and future development together with recommendations for the future research.*

DOI: 10.4018/978-1-60960-519-3.ch004

## INTRODUCTION

Changing and continuously evolving conditions force the vast majority of organizations to proactively seek new and improved solutions in order to retain their positions. The proportion of top innovative companies among hundred biggest companies in the world has almost tripled (from 11 to 30) and their part of capitalization in total capitalization has almost doubled (from 20 to 39%) in the period between 2005 and 2009[1]. Whether we are willing to admit this or not, continuous improvement is an indispensable process for any such proactive organization requiring highly motivated and participative employees. Most such advanced organizations are finding it challenging to effectively encourage employees to participate in the ever more innovative continuous improvement process. This demands fundamental structural shifts in organizations that would lead employees to be more willing to accept change, embrace new ideas or problem solutions, and develop necessary skills and capabilities. It is therefore not an issue whether employees are creative or not and even innovation process *per se* may already be well established. The main problem is that although individual employees may generate plenty of creative ideas they may not be willing to contribute some of these ideas due to not yet well understood circumstances on the interface between idea generation and implementation (defined here as *creative participation*). These circumstances may well be influenced by individual personalities and/or organizational and social settings. According to our concept, creative participation is influenced by personal characteristics including knowledge, motivation, self-confidence, personality, etc. and organizational factors like rewards, encouragement, organizational culture, etc. Although arising from separate sources organizational factors and personal characteristics are interdependent (Shalley, Zhou, & Oldham, 2004; Woodman, Sawyer, & Griffin, 1993), human creativity is based on intrinsic personal characteristics and organization can only help to recognize, develop and implement ideas but it is an individual who generates ideas. It is therefore of scientific interest to isolate the influence of personal characteristics from organizational factors before studying their interdependencies and as a result we have focused on personal characteristics in order to investigate their influence on creative participation.

The chapter is structured in several sections. The first section presents a review of past work on creativity examining the influence of personal characteristics including studies on the correlation between creativity and personality types. In the second section we examine the relevant studies on innovation with a particular interest in the influence of personal characteristics on the innovative process. This is then followed by a critical review of studies on the correlation between creativity and innovation identifying additional stages between the two. From this information a new theoretical framework for creative participation as an overlapping interface between creativity and innovation is derived. Some applications of this concept in organizational and management practices are given. This chapter also provides further discussion of possible solutions, recommendations and insights for future research. Finally, all thoughts and discussions are summarized in the conclusion.

## BACKGROUND

On its way to the viable innovation, creative idea must be transferred from the concept to the reality (Roffe, 1999). Individuals generate ideas through various different ways and once generated, these ideas could potentially be translated into practical actions, processes or products. The process of innovation can be considered as a two-stage process including ideas generation and their implementation (Zaltman, Duncan, & Holbek, 1973; West, 2001). Recently, some studies have identified additional stages in this process (Dorenbosch, Van Engen, & Verhagen, 2005; Janssen,

2003; Jong & Kemp, 2003; Mostert, 2007; Rank, Pace, & Frese, 2004; Scott & Bruce, 1994; Van der Meer, 2007). They examined the connection between creativity and innovation and found that these two stages do not fully describe innovation process since it also includes ideas development stage through involvement of employees and organizational support. However, they do not provide any further explanations or theoretical perspectives on the role and context of the ideas development process. Furthermore, they focus primarily on the two processes themselves but lack any rigorous insight into an influence of various factors on willingness of individuals to participate (*i.e.* interface between creativity and innovation). Therefore, the knowledge about this interface is still very limited and purely narrative in nature.

While creativity refers to ideas generation process at the individual level, innovation refers to the implementation of these ideas at the organizational level (Amabile, 1996), creative participation refers to the contribution of these ideas at the individual level before they are even considered for implementation. Innovation is thus much more dependent on relations, team dynamics, environmental and organizational factors (Van de Ven, Polley, Garud, & Venkataraman, 2008). However, whilst individuals may think of something new or improved, for various reasons they may not be willing to contribute these ideas. With other words, they do not necessarily creatively participate, which represents potentially significant and unaccounted for waste of human capital. There might be environmental factors at the group and organizational levels, including organizational culture and climate, that influence their willingness to make creative contributions, but there are also their individual characteristics. Previous research found that certain personal characteristics and abilities describe creative people, including knowledge, motivation, curiosity, intelligence, self-confidence, flexibility, among others (Amabile, 1996; Csikszentimihaly, 1992; Oldham & Cummings, 1996; Quinn, 1985). The correlation between personality types and creativity has been also studied extensively (Furnham, Crump, Batey, & Chamorro-Premuzik, 2009).

## CREATIVE PARTICIPATION: AN INTERFACE BETWEEN CREATIVITY AND INNOVATION

### Creativity

The literature on creativity does not provide its exact definition. Three main approaches to describe this term can be considered:

1. Creativity has been defined as the process of production of novel, useful and appropriate ideas or problem solutions (Amabile, 1996; Sternberg, 1999; Weisberg, 1993).
2. In contrast, Boden (2004) and Heilman (2005) have described creativity as the human ability to generate novel and useful ideas and thoughts.
3. Another approach to define creativity belongs to Glover, Ronning and Reynolds (1989), who have connected creativity with:
    - A parameter of intelligence.
    - An unconscious process.
    - A parameter of problem solving.
    - A social process.

Combining the three approaches it is possible to define creativity as a social process constituted by a specific individual mental state and social influence that leads to the generation of novel, useful and appropriate ideas or problem solutions.

Creativity has been studied for over fifty years (Barron & Harrington, 1981; Stein & Heinze, 1960). However, recently a strong demand for fundamental research of factors that influence creativity has appeared due to the inspiration of some commercial businesses (Kanter, 1981). Two types of factors that influence creativity have been distinguished:

- Organizational factors are those which influence creativity through organization (e.g. organizational culture, support, rewards, resources, strategy etc.).
- Individual factors are those which influence, on the other hand, creativity through personal characteristics (e.g. character, personality traits, emotions etc.).

Studies on creativity (Amabile, 1996; Ahmed, 1998; Barron & Harrington, 1981; Csikszentmihalyi, 1992; Glover et al., 1989; Oldham & Cummings, 1996; Shalley et al., 2004) have found that there is a reasonable consensus that certain personal characteristics are associated with relatively high level of creativity. They include knowledge, intrinsic motivation, curiosity, self-confidence, intelligence, personality types, among others. Although more factors were found by these studies, this chapter will focus on the ones stated above since they have been found as the most significant factors.

## Personal Characteristics that Influence Creativity

### Knowledge

The importance of knowledge for creativity is based on development and recognition of creativity that can be divided into two parts:

- Importance of domain (knowledge about specific subject area).
- Importance of field (the context in which the knowledge is elaborated) (Sternberg, 1999).

Knowledge is essential in order to generate novel ideas or solve a problem. Although creative ideas can be generated regardless of the level of knowledge, it can be tremendously difficult if not possible to independently judge their value in a particular problem-solving situation (Glover et al., 1989). This is supported by Sternberg, O'Hara, and Lubart (1997) who found positive correlation between knowledge and creativity. In their experimental study, people were asked to produce two creative products in each of the following four areas: writing, art, advertising and science. Each area included three to ten topics. The products that people produced were judged by how creative they are. The findings reported that people who did well in advertising, for instance, did not necessarily did well in art, writing or science and vice versa. One of the reasons for that is the considerable amount of time needed to acquire sufficient knowledge in an unknown area. Similarly, Amabile (1996) has defined creativity as a function of three components: expertise, creative-thinking skills and motivation. Expertise has been described as technical, procedural and intellectual knowledge. People with greater expertise have also been found to be more creative in a particular area of their expertise. Knowledge has also been emphasized as necessary skill for effective group interactions and exchange of ideas (Paulus, 1999).

### Motivation

Motivation is viewed as one of the drivers of creativity in organizations (Amabile, 1996; Csikszentmihalyi, 1992; Sternberg, 1999). Individual's motivation can be intrinsic or extrinsic in nature. Intrinsic rewards are those which arise from individuals through performing an interesting and pleasurable task, whereas extrinsic rewards come out from the organization. Amabile, Barsade, Mueller, and Staw (2005) have discovered "motivation effect" which means employees perform better when they are intrinsically motivated (e.g. by their own satisfaction of work, challenge, enjoyment and interest of performing job) compared to those who are extrinsically motivated (e.g. by money, managers' support, increased salary). Describing creative people, Csikzentmihalyi (1992) has emphasized their enjoyment in the process of creation for their own sake. Feist and Baron (2003) have characterized creative students as being driven by intrinsic motivation and interested in wide range of different areas. Individual's intrinsic motivation has been considered as a key for making creative contributions (Sternberg et al., 1997). However, there are some studies that believe in more profound influence of extrinsic rewards on creativity than intrinsic ones (Schwartz & Waterman, 2006) and more or less equal influence of both rewards on creativity (Glover et al., 1989).

**Curiosity**

Curiosity has been defined as a driver, desire and tendency to seek new information and experience (Litman, 2008). Creative people consistently describe themselves as having a wide range of interests and ambitious, and as curious and enthusiastic individuals (Barron & Harrington, 1981; Glover et al., 1989). According to Torrance (1963) curiosity is a key factor towards enhancing creativity. He found that curious students are more creative than non-curious ones. In order to make creative contributions people have to be unusually curious and interested in the subject (Csikszentmihalyi, 1992). There are two possible reasons where this interest may come from. The first reason can be a result of interaction with people who are interested in the same subject; the second reason can be a result of being born with "sensitivity" to some domain of expertise. There are some studies focused on children's curiosity and its effect on creativity. For instance, Siegle (2008) postulated that encouraging curiosity is important since it causes better understanding of the surroundings, expands knowledge through various explorations and leads to better problem-solving process. Thus, parents should encourage and motivate their children to be more curious and help them in translating their curiosity into creative and productive life. Success, engagement in the work and community are the result of children's intellectual curiosity. However, curiosity is often lost before children reach adulthood (Torrance, 1963). Social influence, unmet expectations are the main reasons for loosing curiosity during the life.

**Self-Confidence**

One of the major personal traits that influence creativity is self-confidence. Creative people are expected to possess high level of self-confidence (Barron & Harrington, 1981; Glover et al., 1989). Without the perception of confidence, it would be difficult or, perhaps, impossible for employees to put forward their novel ideas and inventions. Although self-confidence has been emphasized as a driver of creativity, there is a lack of empirical investigations towards support of this statement. There are still some arising questions whether self-confidence predicts creative achievements or creative achievements enhance creativity.

**Intelligence**

There are contradictory reports in the case of intelligence. In general, intelligence has been considered as individual's capacity, skill, or talent with relation to three mental functions, including information process, problem-solving, and abstract reasoning (Feist & Barron, 2003). Five possible correlations between creativity and intelligence have been identified:

- Creativity is a subset of intelligence. Guilford (1968) was one of the first who studied creativity as a subset of intelligence. As a result of this study several psychological tests (including test of divergent production of semantic unit, test of production) that can measure creativity were developed. He found correlation between the California Test of Mental Maturity (CCTM, an IQ test measure) and Divergent Production Test (DPT). In his study, students with low IQ were also low in DPT, whereas students with high IQ were scattered over much in the DPT.
- Intelligence is a subset of creativity. Sternberg et al. (1997) have considered six elements which are related to creativity, including intelligence, knowledge, thinking styles, personality, motivation and en-

vironment. Thus, intelligence is one of the six drivers of creativity.

- Creativity and intelligence are overlapping sets. Barron and Harrington (1981) have found similarities between creativity and intelligence as both represent human ability to solve problems. Roe (1976), in contrast, has considered the differences between them. Creative process is not necessary always logic and does not necessarily have a clear goal in comparison to intelligence.
- Creativity and intelligence are coincident sets. Weisberg (1993) has viewed creativity and intelligence as based on the same underlying mechanisms of the problem-solving process. To support this statement, people were asked to connect nine dots which were put in the shape of the square with three rows of three dots each, using only four straight lines never arriving at a given dot twice and never lifting their pencils from the sheet of paper. The solution of this problem is allowing their line segments to go outside the periphery of the dots. By this task Weisberg (1993) has demonstrated that although people were given the insight, they were still struggling with solving this problem.
- Creativity and intelligence are only moderately correlated. Torrance (1963), in contrast, has emphasized only moderate correlation between creativity and intelligence. Similarly, although MacKinnon (1965) has identified intelligence as one of the characteristics of a creative person, he has found that intelligence is not significantly correlated to creativity. According to this being more or less intelligent does not identify the level of creativity. Similarly, intelligence has been emphasized to be an important but not sufficient prerequisite for creativity (Barron & Harrington, 1981; Glover et al., 1989).

**Personality Types and Creativity**

In general researchers agree that personality is related to creativity (Barron & Harrington, 1981; Gough, 1957; Sternberg, 1999). There are many developed personality inventories to measure personality types. Myers-Briggs Type Indicator (MBTI) is one such well known Jungian based inventory that applies a self-report format. It constitutes four scales: Introversion-Extraversion, Sensing-Intuition, Thinking-Feeling and Judging-Perceiving (Myers, 1962).

Respondents can be categorized into one of the sixteen personality types. On average, people are more creative when they are:

- Intuitive rather than Sensing (N>S).
- Perceiving rather than Judging (P>F).
- Extraverted rather than Introverted (E>I).
- Feeling rather than Thinking (F>T) (Furnham et al., 2009).

Kirton Adapter-Innovator (KAI) scale is a psychological approach assessing two dimensions of creativity is another well known inventory to identify individual's attitude toward creative achievements (Kirton, 2003). It differentiates "adaptive" and "innovative" problem-solving styles:

- Adapters are associated with production of ideas based on existing knowledge and definitions of a problem. These ideas help to solve problems by improving existing products/processes/services.
- Innovators are characterized by reconstruction the problem and come up with unexpected solutions. Innovators are more likely to produce different and novel ideas or problem solutions rather than improvements of existing objects.

Studies using the KAI suggest that it can be used as a predictor of employees' creativity at work. People with an adaptive style produce less creative outcomes than people with an innovative style since they are more inclined towards risk-taking producing more novel and useful outcomes for organizations (Puccio, Treffinger, & Talbot, 1995; Sadler-Smith & Badger, 1998).

NEO-PI-R (Neuroticism (N), Extraversion (E), Openness to Experience (O) Personality Inventory (PI) – Revised (R)) - personality inventory based on the "Big Five" model is another attempt to examine the relationship between personality traits and divergent thinking (Costa & McCrae, 1993). It measures Neuroticism (N), Extraversion (E), Openness to Experience (O), Agreeableness (A) and Conscientiousness (C). Each factor consists of six points which can be summarized to form a total score. Most studies on the "Big Five" and divergent thinking based creativity have emphasized positive correlation with Extraversion, Openness to Experience and negative with Agreeableness (Furnham et al., 2009). Recently, however, Batey, Chamorro-Premuzik, and Furnham (2009) have emphasized Openness to Experience to be the only positively correlated and Neuroticism to be negatively correlated to divergent thinking based creativity.

Although, the above personality inventories have been widely used to measure personality types and their correlation with creativity, it is necessary to consider their limitations. Personality inventories are based on individual preferences at certain time. Thus, personality type to some extent depends on situations and could be changed during the life. Special care should therefore be made in order to correlate personality types with creativity (Ahmed, 1998; Gentry, Mondore, & Cox, 2007; Stricker & Ross, 1964).

## Innovation

The ability of organizations to compete, thrive, improve processes and successfully participate in the market derives from innovation. There is an agreement in the researcher community that appropriate innovation improves competitive advantage and drives organizations to succeed. Kanter (1983) has considered innovation as the process of generation, acceptance and implementation of ideas into the product, process or service where the last two processes are central in this definition. Consequently, Hauser, Tellis, and Griffin (2006) have defined innovation as the process of implementing new products, processes or services into the market. Amabile (1996) has defined innovation as successful implementation of creative ideas into novel products, processes or services within organizations. Similarly, Van de Ven et al. (2008) have described innovation as the process of ideas development and their implementation. Although, there are other definitions of innovation, they have more or less the same meaning. Once ideas are generated, developed (communicated, contributed) and implemented, they become innovations making implementation the main defining characteristic. Based on all these definitions we can summarize that innovation is a successful implementation of already generated and contributed ideas or solutions.

In today's uncertain and competitive environment, organizational success depends on using skills and internal capabilities to put forward novel ideas and solve everyday problems. Rigorous consideration of various employees' personality types has become a significant aspect and an important part of organizational strategy towards effective development. The literature on innovative behavior has emphasized the importance of this issue. Innovative behavior has been considered as employees' involvement in the processes of generation, development and realization of new ideas within organizations in order to improve overall performance (Scott & Bruce, 1994). Like for creativity there are two groups of factors that influence innovation:

- Organizational factors are those that influence innovation through organization (job challenge, organizational strategy, support-

ive environment, external contacts, autonomy, freedom etc.).
- Individual factors are those that influence innovation through personal characteristics (e.g. self-confidence, intrinsic motivation, curiosity).

Adair (1990), Glynn (1996), Quinn (1985) and West (2001) have found that knowledge, intrinsic motivation, curiosity, intelligence, self-confidence and flexibility are positively correlated with innovation. Although they have identified other personal characteristics that influence innovation, this chapter will focus only on those above as they are recognized by the literature to be most significant.

## Personal Characteristics that Influence Innovation

### Knowledge

Taking into account the value of knowledge for creativity, it should be extremely important and useful for innovation as well. Knowledge supports creativity, motivation and ability to improve intellectual development of organizations. Since creativity is a starting point of innovation, knowledge could be considered as an analogy to raw materials (Yusuf, 2009). Knowledgeable employees have been found to make most creative contributions and by using employees' brain capacity organizations improve their innovativeness (Carneiro, 2000).

### Motivation

In addition to structural attributes of organizations, the motivational states play an important role in predicting innovative capacity in an organization. Many agree that motivation is associated with innovation and leads to growing employees' self-esteem and greater participation in the innovation process (Pierce & Delbecq, 1977). It is also recognized that organizations should provide motivational conditions (e.g. rewards and other supports) to foster employees' motivation to innovate.

### Curiosity

Curiosity has been associated with innovation as being the original driving force followed by global market force (Humble & Jones, 1989). Although the order of significance may be questionable, curiosity is one of the most prominent characteristics of innovative people leading to greater employees' involvement and improved commitment. The leaders' ability to encourage innovation is based on intellectual curiosity and interest to seek new approaches to solve emerging problems (Oke, Munshi, & Walumbwa, 2009).

### Self-Confidence

Self-confidence has also been found to be positively correlated to innovation. For example, Parker (1998) reveals that employees, who feel confident at performing their work tasks, are usually more successful and perform better than those who do not. Employees who are confident and proud of their abilities are found to be more motivated and open to change. Building self-confidence helps to promote employees' commitment to organizational and personal goals (Locke, 1999).

### Flexibility

Flexibility has been defined as the human capacity to adapt to a changing environment and bring a stream of new ideas quickly and efficiently (Georgsdottir & Getz, 2004). Flexibility has been also defined as cognitive ability (Runco, 2004; Torrance, 1963; West, 2001). Since change and adaptation are essential for innovation, flexibility is one of the personal characteristics that describe innovative individuals. Flexible employees can quickly redefine the problem into a simpler one in order to find optimal solution. They tend to be complex, open to change and have an attraction to novelty.

*Figure 1. Additional stages between creativity and innovation*

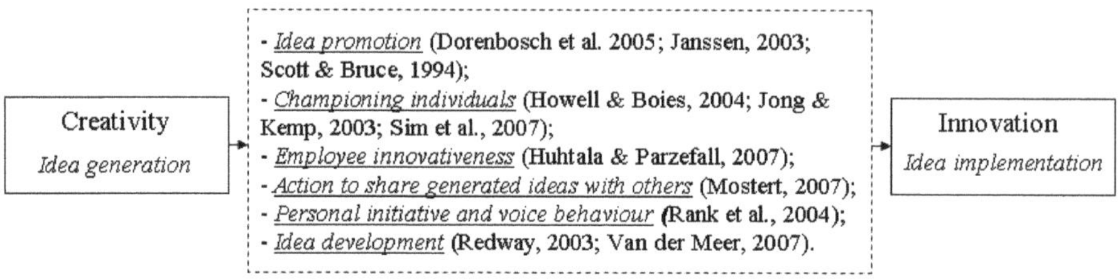


## Creative Participation

As it was mentioned above most of the earlier studies distinguished two stages in the innovation process: ideas generation and their implementation. Creativity relates to the first stage where ideas are generated, whereas innovation relates to the last stage where contributed ideas are on a path to implementation. However, recently some studies suggest that there is a connection between creativity and innovation criticizing the two-stage approach since innovation process also includes idea development process (Figure 1). These studies will be considered in more detail below.

### Additional Stages between Creativity and Innovation

#### Idea Promotion

The literature on employees' innovative behavior in the workplace (Dorenbosch et al., 2005; Janssen, 2003; Scott & Bruce, 1994) has distinguished "idea promotion" stage between idea generation and its realization. Once an employee has generated an idea, he/she has to engage in social activities to find a support from co-workers or sponsors necessary for the implementation of this idea into the market. In effect this stage comprises promotion of novel ideas or suggestions to superiors and colleagues followed by realization through prototyping or modeling. While "idea promotion" stage relates to employees' engagement in certain social activities, creative participation refers to day-to-day problems requiring contribution of ideas/suggestions to work colleagues responsible for their development and implementation.

#### Championing Individuals

Howell and Boies (2004), Jong and Kemp (2003), Sim, Griffin, Price, and Vojak (2007) have studied a so-called "championing individuals" phase with emphasis of the role of "champions" in the organizations who are informally involved in the job and enthusiastically promote new ideas. For example, Howell and Boies (2004) describe champions as being committed to the organization, who promote ideas or suggestions for solving problems through active engagement and willingness to take risks of their positions or reputation on its way to realization. Similarly, Jong and Kemp (2003) characterize champions as employees who strongly feel committed to the organization appointed by employers to put effort into their creative ideas in order to bring them into the market. Willing to be innovative, firms are trying to find, keep and encourage champions who put forward novel ideas and problem solutions. Further Sim et al. (2007) discuss the role of champions cross initiation gap between ideas generation and recognition opportunities as individuals who understand the technology and connect it to the market need by recognizing its potential. Usually champions do not create innovative ideas but find them in an organization. The "championing individuals" phase is characterized by interaction between co-workers, including negotiations, collaborations and pushing new ideas. However, Howell and Boies (2004), Jong and Kemp (2003) attribute this phase to innovation process only, whereas the process of putting forward creative ideas and share them with colleagues often occurs before innovation even starts to be recognized. Moreover, all these studies emphasize the role of "champions", i.e. solving particular problems occurring in certain parts of the organization. Creative participation, on the other hand, relates to everyday problems requiring quick solutions, involvement of all employees in all parts of organizations, and is not limited to only a handful of employees.

**Employee Innovativeness**

Huhtala and Parzefall (2007) provide another view and recognize "employee innovativeness" forming the basis of the transformation from an idea into a full innovation. It is characterized as a complex behavior consisting of idea generation, idea promotion and realization. Nevertheless, employees have different personality traits which impact their behavior and correspondingly shape their innovativeness. There is a question of varying willingness to participate that employee innovativeness does not address explicitly.

**Action to Share Generated Ideas with Others**

Mostert (2007) distinguishes "action to share generated ideas with others" between creativity and innovation. After problems or potentials for improvements are identified, employees think and come up with possible solutions through communication and frequent interaction with their superiors or other co-workers. However, employees may or may not wish to participate and explore their thoughts and ideas. Thus, one of the organizational strategies is motivation of employees to participate and share their ideas or solutions. Although Mostert (2007) attributes this phase to creativity, against commonly accepted definitions, the process of sharing new ideas with others is actually the process of employees' participation. Before ideas are realized into innovation, they need to be fist generated and most importantly shared or communicated with other employees.

**Personal Initiative and Voice Behavior**

Rank et al. (2004) state that "personal initiative and voice behavior" sits between ideas generation and their implementation. It is characterized by communicative and innovative behavior that emphasizes expression of constructive challenge for improvements. In essence employees speak up with suggestions for change or improvements. However, there is again a question of willingness as employees may not be willing to promote their ideas or suggestions and personality is among the reasons behind their willingness. Employees' personal characteristics play a crucial role in their willingness to make creative contributions but this still remains an unexplored area of research.

**Idea Development**

Another attempt to better understand the interaction between creativity and innovation is made by Redway (2003) and Van der Meer (2007). They call this an "idea development" phase. For example, Redway (2003) has divided this phase into following five sub-stages:

1. **Enthusiasm:** when employees are highly confident and believe in successful implementation of their ideas.
2. **Struggle:** when employees are still willing to participate, however it becomes more difficult to put forward generated ideas.
3. **Disaster:** at this sub-stage something unexpected might happen. In that case employees' confidence in success is lower than at the beginning. In order to solve this situation manger should help employees to manage the situation and further develop their ideas.
4. **Recovery:** this sub-stage started when customers and experts have been consulted and their advice has been received. The progress is slow.
5. **Negotiation:** struggle and disaster might change the original created idea. The employees' expectations might be met or unmet with the outcome.

Within these five sub-stages creative participation could be attributed to the fist two – when employees believe in the progress and success, might struggle about putting forward generated ideas and still willing to participate in idea contribution process. However, creative participation is a generic process that applies to everyday work, any ideas, at any level and not merely to new product development. Furthermore, negative experience in the past may cause significantly reduced enthusiasms in the future. Different personal characteristics can be manifested by potentially different levels of enthusiasm and willingness to pursue ideas or suggestions and what might be a struggle for some employees might be a normal process for others. The last three sub-stages are attributed to implementation process of these generated and contributed ideas.

Van der Meer (2007) also identifies this as an "idea development" phase but with a different meaning, as transforming of employees' ideas into projects. In this context this phase is more likely to be attributed to a business environment (i.e. idea implementation process), whereas identification of ideas is more likely to be related to creativity (i.e. idea generation process). However, creative participation is positioned between ideas identification and their development when employees are willing to contribute identified and generated ideas.

Bragg and Bragg (2005) have divided development of ideas into four sub-stages:

1. Recognition of opportunities.
2. Generation of new ideas.
3. Evaluation and selection of ideas.
4. Business plan for implementation.

Within these four sub-stages, creative participation is a state of mind between generation and evaluation of ideas. Before ideas are evaluated and selected, they need to be shared. Furthermore, employees' intrinsic willingness to speak up their suggestions is influenced by their personal characteristics and environmental factors. For instance, the same individual can thrive in one organization but can be completely isolated and passive in another one. This may be a result of organizational differences and personality.

## A Theoretical Framework for Creative Participation

Many of the above scholars call for a rethink of current understanding of innovation with attempts at identifying additional stages in the innovation process. However, these attempts focus primarily at breaking down existing stages into ever finer detail but fail to address the influence of a complex blend of individual personality and organizational factors. The intention here is to address these by proposing a theoretical framework of creative participation as in interface between creativity and innovation (Figure 2).

While innovativeness is clearly a desirable trait and innovation processes are rigorously in-

*Figure 2. Creative participation as the crucial interface between creativity and innovation*

vestigated, the literature offers very little to enable appropriate investigation of the impact of personality on the willingness to participate with creative ideas. These are indispensable for any kind of development and not merely for innovation. Even at most basic levels in everyday work environment there is constant need for new ideas and improvements resulting from ever more pervasive and frequent challenges. But how could one say with any degree of certainty that a particular individual with a specific set of personal characteristics will be freely willing to creatively participate? The literature on creativity and innovation do not provide any reliable answer to this question either so there seems to be a substantial gap in generic understanding of conditions that lead an individual to engage and willingly contribute new ideas. Team dynamics is an area of research that promises such answer (Donnellon, 1996; Tarricone & Luca, 2002) but unfortunately the investigations are more concerned with team effectiveness paying little regard to creative participation. For example, these studies stated that groups more often make creative contributions than individuals working alone since teams have the resilience, range of skills, abilities, and experience to ensure that creative ideas are put into the innovation. However, team is a group of individuals and in order to understand what drives team effectiveness, there is a need to investigate what drives each team member to make these creative contributions.

## Solutions and Recommendations

Before idea is implemented, it needs to be generated but even more importantly it needs to be contributed (i.e. shared with people who are directly responsible for its implementation). Many of such ideas are simple and often neglected quick suggestions shared with colleagues. It is a continuous stream of such ideas from every corner of an organization that very often characterizes most successful companies and their unique culture. Creating an environment that will engage employees in the continuous stream of creative ideas is now one of the key organizational strategies. However, the solution is not as simple as it might seem at first sight. Because of engrained mistrust, personal differences and other organizationally induced problems even most creative employees may not wish to participate. This kind of a 'brain' loss is difficult to measure and is often unaccounted for inevitably resulting in reduced competitiveness against the most successful organizations that manage to align themselves more closely with their employees' individual characteristics. Companies should thus pay more attention to idea contribution process since not all employees are naturally self-driven or self-motivated. Needless to say, even highly proactive individuals may not be willing to participate to their full potentials if the conditions are not appropriate. On the other hand, managers can not force employees to creatively participate but they can change the organizational settings creating right conditions for employees to become intrinsically willing to contribute ideas/suggestions. When managers communicate with employees and encourage them, they are more creatively participative and highly engaged; the turnover rate is low. Managers should be always open to new ideas or suggestions from all employees avoiding an immediate selective response. Many ideas may not be immediately applicable or may not be applicable at all but an immediate negative response may lead to lower creative participation in the future. All ideas should be accepted but only useful and appropriate ideas should be supported, contributors recognized and appropriately rewarded. Fully participative employees bring in a full range of their skills, abilities significantly boosting creative potential of an organization. Nevertheless, even though one may assume that individuals possess relevant abilities and skills they may not be willing to fully share their ideas or solutions. There are many reasons behind the peculiarities of individual willingness to creatively participate,

*Table 1. Some applications of the theoretical framework in management practices*

| Personal characteristics | The influence of personal characteristics on employee creative participation | Managerial practices |
|---|---|---|
| Knowledge, expertise and understanding of the work | The more knowledgeable employees (i.e. experts with deep understanding of the work) – the more they are creatively participative and the higher the levels of their idea contribution in comparison to less knowledgeable employees, non-experts with lower understanding of the work | Invest in employees' education, organize special trainings with possibility to discuss, communicate or share thoughts and suggestions, opportunity to interact with work colleagues and professionals in the field |
| Self-confidence | The higher the level of employees' self-confidence – the higher the levels of their idea contribution in comparison to employees with the lower level of self-confidence | Be open to change, accept useful and appropriate ideas, avoiding an immediate selective response since rejection of employees' ideas in the past could lead to lower self-confidence and therefore lower creative participation in the future |
| Human curiosity | The more profound employees' curiosity – the more likely it is that they are willing to contribute ideas in comparison to less curious employees | Provide the opportunity for employees to learn and develop (focus on multi-tasks instead of routine job), continuously improving organizational settings |
| Personality types | People with some personality types (e.g. ENFP) contribute more creative ideas than other people characterized by other personality types | Taking into account employee personality type for arranging individual to the appropriate job position |

but one of the more important influences appears to be individual's personal characteristics. Based on the literature review examining the correlation between creativity and personal characteristics (Amabile, 1996; Ahmed, 1998; Barron & Harrington, 1981; Csikszentmihalyi, 1992; Furnham et al., 2009; Oldham & Cummings, 1996; Shalley et al., 2004), innovation and personal characteristics (Adair, 1990; Glynn, 1996; Quinn, 1985, West, 2001) we assume that knowledge and understanding, curiosity, self-confidence and personality types are the main individual characteristics that profoundly influence employees' creative participation (Table 1).

The contribution of novel ideas or solutions to problems requires a certain degree of knowledge and expertise, the willingness to communicate or share them with manager and people who are directly responsible for their implementation. Employees' intrinsic willingness to contribute ideas/suggestions requires curiosity and interest in the process. It would also be natural to expect creatively participative individuals to be highly satisfied with their work and with their self-confidence rising as a result of acceptance from organization. Better understanding of creative participation and its improvement would thus inevitably lead to overall organizational improvements.

## FUTURE RESEARCH DIRECTIONS

During the past few decades researchers have devoted a growing level of attention to the subject of creativity, innovation and employee innovative behavior examining the influence of personal and organizational factors on these processes. Another stream of research that developed recently involves the connection between creativity and innovation, distinguishing additional stages between the two. However, these studies fail to provide any further theories and do not explain how personal characteristics and organizational settings impact idea development process. They have also not attempted to examine the influence of personal characteristics and organizational settings on this process. New research directions will likely involve following aspects:

- Studying creative participation as a crucial interface between creativity and innovation;
- Focusing on the influence of personal characteristics and organizational factors on employees' intrinsic willingness to contribute novel ideas or problem solutions;
- Expanding the range of personal characteristics examining their influence on employee creative participation would be beneficial for a better understanding of idea contribution process;
- Identifying how other factors at different levels such as the team context, organization, and larger society influence employee creative participation; For example, one individual might possess the necessary characteristics and relevant knowledge to contribute new ways of solving problems and might feel intrinsically willing to creatively participate. But if the work environment is poor and inappropriate, these ideas will be suppressed and will not be contributed. Therefore, it is crucial to examine the influence of organizational parameters on employee creative participation as well;
- Investigating interdependencies between individual characteristics and organizational factors that influence employees' creative participation by including a variety of measures. Most of the studies on creativity, innovation and innovative behavior are based on methods using questionnaires and interviews. These methods are often overly subjective snapshots in time, depend on self-reports of behavior and provide less control over the situation. Research is now needed that based on both quantitative and qualitative approaches, including psychological experiments, observations and established measures of personality.

## CONCLUSION

In this chapter we have reviewed past work on creativity, innovation and innovative behavior with emphasis on how they are influenced by personal characteristics. The innovation process has been analyzed and theoretical framework of creative participation as a state-of-mind interface between creativity and innovation has been proposed. Based on the literature on creativity, innovation and various personality-related studies we have assumed that knowledge and understanding, curiosity, self-confidence and certain personality types are the main personal characteristics that should profoundly influence employees' intrinsic willingness to contribute novel ideas and/or problem solutions. In order to prosper and succeed organizations should much more focus on employees' creative participation in order to attain the culture in which they can contribute more useful and appropriate ideas and/or suggestions for improvements. To maintain competitive edge top companies need a continuous flow of ideas from motivated and willingly participating individuals. Market and other forces demand regular innovations and constant improvement of their existing products, processes and services.

Better understanding of creative participation, including personal characteristics and wider organizational factors, would provide organizations an opportunity to potentially rethink employment and training practices and improve internal processes.

## REFERENCES

Adair, J. (1990). *The challenges of innovation*. England: The Talbot Adair Press.

Ahmed, P. K. (1998). Culture and climate for innovation. *European Journal of Innovation Management*, *1*(1), 30–43. doi:10.1108/14601069810199131

Amabile, T. M. (1996). *Creativity in context*. Boulder, CO: Westview Press.

Amabile, T. M., Barsade, S. G., Mueller, J. S., & Staw, B. M. (2005). Affect and creativity at work. *Administrative Science Quarterly*, *50*(3), 367–403. doi:10.2189/asqu.2005.50.3.367

Barron, F., & Harrington, D. (1981). Creativity, intelligence, and personality. *Annual Review of Psychology*, *32*, 439–476. doi:10.1146/annurev.ps.32.020181.002255

Batey, M., Chamorro-Premuzik, Y., & Furnham, A. (2009). Intelligence and personality as predictors of divergent thinking: The role of general, fluid and crystallized intelligence. *Thinking Skills and Creativity*, *4*(1), 60–69. doi:10.1016/j.tsc.2009.01.002

Boden, M. A. (Ed.). (2004). *The creative mind*. London: Routledge.

Bragg, A. & Bragg, M. (2005). *Developing new business idea: The fast-track to creating viable new businesses for executives and entrepreneurs*. Edinburg: Pearson Education Limited.

Carneiro, A. (2000). How does knowledge management influence innovation and competitiveness? *Journal of Knowledge Management*, *4*(2), 87–98. doi:10.1108/13673270010372242

Costa, P. T., & McCrae, R. R. (1993). *Revised NEO personality inventory and NEO five-factor inventory, professional manual*. Odessa, FL: Psychological Assessment Resources.

Csikszentmihalyi, M. (1992). Motivation and creativity. In Albert, R. S. (Ed.), *Genius and eminence* (pp. 19–33). Oxford: Pergamon Press.

Donnellon, A. (1996). *Team talk: The power of language in team dynamics*. Harvard Business School Press.

Dorenbosch, L., Van Engen, M. L., & Verhagen, M. (2005). On-the-job innovation: The impact of job design and human management through production ownership. *Creativity and Innovation Management*, *14*(2), 129–141. doi:10.1111/j.1476-8691.2005.00333.x

Feist, G. J., & Barron, F. X. (2003). Predicting creativity from early to late adulthood: Intellect, potential, and personality. *Journal of Research in Personality*, *37*(2), 62–88. doi:10.1016/S0092-6566(02)00536-6

Furnham, A., Crump, J., Batey, M., & Chamorro-Premuzik, T. (2009). Personality and ability predictors of the "consequences" test of divergent thinking in a large non-student sample. *Personality and Individual Differences*, *46*(4), 536–540. doi:10.1016/j.paid.2008.12.007

Gentry, W. A., Mondore, S. P., & Cox, B. D. (2007). A study of managerial derailment characteristics and personality references. *Journal of Management Development*, *26*(9), 857–873. doi:10.1108/02621710710819348

Georgsdottir, A. S., & Getz, I. (2004). How flexibility facilitates innovation and ways to manage it in organizations. *Creativity and Innovation Management*, *13*(3), 166–175. doi:10.1111/j.0963-1690.2004.00306.x

Glover, J. A., Ronning, R. R., & Reynolds, C. R. (1989). *Handbook of creativity*. New York: Plenum Press.

Glynn, M. A. (1996). Innovative genius: A framework for relating individual and organizational intelligences to innovation. *Academy of Management Review*, *21*(4), 1081–1111. doi:10.2307/259165

Gough, H. G. (1957). *California psychological inventory manual*. Consulting Psychologists Press.

Guilford, J. P. (1968). *Intelligence, creativity and their educational implications*. San Diego: Knapp.

Hauser, J., Tellis, G. J., & Griffin, A. (2006). Research on innovation: A review and agenda for Marketing Science. *Marketing Science*, *25*(6), 687–717. doi:10.1287/mksc.1050.0144

Heilman, K. M. (2005). *Creativity and the brain*. New York: Taylor & Francis group.

Howell, J. M., & Boies, K. (2004). Champions of technological innovation: The influence of contextual knowledge, role orientation, idea generation, and idea promotion on champion emergence. *The Leadership Quarterly*, *15*(1), 123–143. doi:10.1016/j.leaqua.2003.12.008

Huhtala, H., & Parzefall, R. (2007). A review of employee well-being and innovativeness: An opportunity for a mutual benefit. *Creativity and Innovation Management*, *16*(3), 299–306. doi:10.1111/j.1467-8691.2007.00442.x

Humble, J., & Jones, G. (1989). Creating a climate for innovation. *Long Range Planning*, *22*(4), 46–51. doi:10.1016/0024-6301(89)90081-2

Janssen, O. (2003). Innovative behavior and job involvement at the price of conflict and less satisfactory relations with co-workers. *Journal of Occupational and Organizational Psychology*, *76*(3), 347–364. doi:10.1348/096317903769647210

Jong, J. P. J., & Kemp, R. (2003). Determinants of co-workers' innovative behaviour: An investigation into knowledge intensive services. *International Journal of Innovation Management*, *7*(2), 189–212. doi:10.1142/S1363919603000787

Kanter, R. M. (1983). *The change masters*. Cox & Wyman Ltd.

Kirton, M. J. (2003). *Adaption-Innovation: in the context of diversity and change*. London: Routledge.

Litman, J. A. (2008). Interest and deprivation factors of epistemic curiosity. *Personality and Individual Differences*, *44*(7), 1585–1595. doi:10.1016/j.paid.2008.01.014

Locke, E. A. (1999). *The essence of leadership: the four keys to leading successfully*. Boston: Lexington Books.

MacKinnon, D. W. (1965). Personality and the realization of creative potential. *The American Psychologist*, *20*(4), 273–281. doi:10.1037/h0022403

Mostert, N. M. (2007). Diversity of the mind as the key to successful creativity at Unilever. *Creativity and Innovation Management*, *16*(1), 93–100. doi:10.1111/j.1467-8691.2007.00422.x

Myers, I. B. (1962). *The Myers-Briggs type indicator*. Palo Alto, CA: Consulting Psychologists Press.

Oke, A., Munshi, N., & Walumbwa, F. O. (2009). The influence of leadership on innovation process and activities. *Organizational Dynamics*, *38*(1), 64–72. doi:10.1016/j.orgdyn.2008.10.005

Oldham, G. R., & Cummings, A. (1996). Employee creativity: Personal and contextual factors at work. *Academy of Management Journal*, *39*(3), 607–634. doi:10.2307/256657

Parker, S. K. (1998). Enhancing role breadth self-efficacy: The role of job enrichment and other organizational interventions. *The Journal of Applied Psychology*, *83*(6), 835–852. doi:10.1037/0021-9010.83.6.835

Paulus, P. B. (1999). Group creativity. In Runco, M. A., & Pritzker, S. R. (Eds.), *Encyclopedia of creativity* (pp. 779–795). CA: Academic Press.

Pierce, J. L., & Delbecq, A. L. (1977). Organization structure, individual attitudes and innovation. *Academy of Management Review*, *2*(1), 27–37. doi:10.2307/257602

Puccio, G. J., Treffinger, D. J., & Talbot, R. J. (1995). Exploratory examination of relationships between creativity styles and creative products. *Creativity Research Journal*, *8*(2), 157–172. doi:10.1207/s15326934crj0802_4

Quinn, J. B. (1985). Managing innovation: Controlled chaos. *Harvard Business Review*, *63*(3), 73–84.

Rank, J., Pace, V. L., & Frese, M. (2004). Three avenues for future research on creativity, innovation, and initiative. *Applied Psychology: An International Review*, *53*(4), 518–528. doi:10.1111/j.1464-0597.2004.00185.x

Redway, K. (2003). *Make it happen! A step-by-step guide from creativity and innovation*. UK: Judy Pitkus.

Roe, A. (1976). Psychological approaches to creativity in science. In Rothenberg, A., & Hausman, C. R. (Eds.), *The creativity question* (pp. 165–175). Durham, NC: Duke University Press.

Roffe, I. (1999). Innovation and creativity in organizations: a review of the implications for training and development. *Journal of European Industrial Training*, *23*(4), 224–237. doi:10.1108/03090599910272103

Runco, M. A. (2004). Creativity. *Annual Review of Psychology*, *55*, 657–687. doi:10.1146/annurev.psych.55.090902.141502

Sadler-Smith, E., & Badger, B. (1998). Cognitive style, learning and innovation. *Technology Analysis and Strategic Management*, *10*(2), 247–265. doi:10.1080/09537329808524314

Schwartz, S. J., & Waterman, A. S. (2006). Changing interests: A longitudinal study of intrinsic motivation for personality salient activities. *Journal of Research in Personality*, *40*(6), 1119–1136. doi:10.1016/j.jrp.2005.12.003

Scott, S. G., & Bruce, R. A. (1994). Determinants of innovative behavior: A path model of individual innovation in the workplace. *Academy of Management Journal*, *37*(3), 580–607. doi:10.2307/256701

Shalley, C. E., Zhou, J., & Oldham, G. R. (2004). The effects of personal and contextual characteristics on creativity: Where should we go from here? *Journal of Management*, *30*(6), 933–958. doi:10.1016/j.jm.2004.06.007

Siegle, D. (2008). Promoting creativity through curiosity and engagement: Wonder is not enough. *Parenting for High Potential,* 3.

Sim, E. W., Griffin, A., Price, R. L., & Vojak, B. A. (2007). Exploring differences between inventors, champions, implementers and innovators in creating and developing new products in large, mature firms. *Creativity and Innovation Management*, *16*(4), 422–436. doi:10.1111/j.1467-8691.2007.00457.x

Stein, M. I., & Heinze, S. J. (1960). *Creativity and the individual*. Illinois, USA: Free Press of Glencoe.

Sternberg, R. J. (Ed.). (1999). *Handbook of creativity*. New York: Cambridge University Press.

Sternberg, R. J., O'Hara, L. A., & Lubart, T. I. (1997). Creativity as investment. *California Management Review*, *40*(1), 8–21.

Stricker, L. J., & Ross, J. (1964). An assessment of some structural properties of the Jungian personality typology. *Journal of Abnormal and Social Psychology*, *68*(1), 62–71. doi:10.1037/h0043580

Tarricone, P., & Luca, J. (2002). Employees, teamwork and social interdependence–a formula for successful business? *Team Performance Management*, *8*(3/4), 54–59. doi:10.1108/13527590210433348

Torrance, E. P. (1963). *Creativity*. USA: American Educational Research Association of the National Education Association.

Van de Ven, A. H., Polley, D. E., Garud, R., & Venkataraman, S. (2008). *The innovation journey*. New York: Oxford University Press.

Van der Meer, H. (2007). Open innovation–the Dutch treat: Challenges in thinking in business models. *Creativity and Innovation Management*, *16*(2), 192–202. doi:10.1111/j.1467-8691.2007.00433.x

Weisberg, R. W. (1993). *Creativity: Beyond the myth of genius*. New York: W. H. Freeman & Co.

West, M. A. (2001). Management of creativity and innovation in organizations. In Smelser, N., & Baltes, P. (Eds.), *International encyclopedia of the social and behavioral sciences* (pp. 2895–2900). Oxford: Pergamon Press.

Woodman, R. W., Sawyer, J. D., & Griffin, R. W. (1993). Toward a theory of organizational creativity. *Academy of Management Review, 18*(2), 293–321. doi:10.2307/258761

Yusuf, S. (2009). From creativity to innovation. *Technology in Society, 31*(1), 1–8. doi:10.1016/j.techsoc.2008.10.007

Zaltman, G., Duncan, R., & Holbek, J. (1973). *Innovations and organizations*. New York: Wiley.

## ADDITIONAL READING

Amabile, T. M., Conti, R., Coon, H., Lazenby, J., & Herron, M. (1996). Assessing the work environment for creativity. *Academy of Management Journal, 39*(5), 1154–1184. doi:10.2307/256995

Angle, H. L. (1989). Psychology and organizational innovation. In Van de Ven, A. H., Angle, H. L., & Poole, M. S. (Eds.), *Research on the management of innovation: The Minnesota studies* (pp. 135–170). New York: Harper & Row.

Axtell, C. M., Holman, D. J., Unsworth, K. L., Wall, T. D., Waterson, P. E., & Harrington, E. (2000). Shopfloor innovation: Facilitating the suggestion and implementation of ideas. *Journal of Occupational and Organizational Psychology, 73*(3), 265–285. doi:10.1348/096317900167029

Bessant, J., & Tidd, J. (2007). *Innovation and entrepreneurship*. West Sussex, England: John Wiley & Sons Ltd.

Bharadwaj, S., & Menon, A. (2000). Making innovation happen in organizations: individual creativity mechanisms, organizational mechanisms or both? *Journal of Product Innovation Management, 17*(6), 424–234. doi:10.1016/S0737-6782(00)00057-6

Binnewies, C., Ohly, S., & Sonnentag, S. (2007). Taking personal initiative and communicating about ideas: What is important for the creative process and for idea creativity? *European Journal of Work and Organizational Psychology, 16*(4), 432–455. doi:10.1080/13594320701514728

Burns, D. J. (2007). Toward an explanatory model of innovative behavior. *Journal of Business and Psychology, 21*(4), 461–488. doi:10.1007/s10869-007-9037-x

Dodgson, M., Gann, D., & Salter, A. (2005). *Think, play, do: technology, innovation, and organization*. New York, USA: Oxford University Press.

Dubrin, A. J. (Ed.). (2007). *Leadership: research findings, practice, and skills*. Boston, USA: Houghton Mifflin Company.

Fairbank, J. F., & Williams, S. D. (2001). Motivating creativity and enhancing innovation through employee suggestion system technology. *Creativity and Innovation Management, 10*(2), 68–73. doi:10.1111/1467-8691.00204

Flood, P., Turner, T., Ramamoorthy, N., & Pearson, J. (2001). Causes and consequences of psychological contract among knowledge workers in the high technology and financial services industries. *International Journal of Human Resource Management, 12*(7), 1152–1165. doi:10.1080/09585190110068368

Garfield, M. J., Taylor, N. J., Dennis, A. R., & Satzinger, J. W. (2001). Research report: modifying paradigms – individual differences, creativity techniques, and exposure to ideas in group idea generation. *Information Systems Research, 12*(3), 322–333. doi:10.1287/isre.12.3.322.9710

Gumusluoglu, L., & Ilsev, A. (2009). Transformational leadership, creativity, and organizational innovation. *Journal of Business Research*, *62*(4), 461–473. doi:10.1016/j.jbusres.2007.07.032

Hoegl, M., & Parboteeah, K. P. (2007). Creativity in innovative projects: How teamwork matters. *Journal of Engineering and Technology Management*, *24*(1-2), 148–166. doi:10.1016/j.jengtecman.2007.01.008

Jones, B., & Miller, B. (2007). *Innovation diffusion in the new economy: the tacit component*. Abingdon, UK: Routledge.

Jones, O., & Tilley, F. (Eds.). (2003). *Competitive advantage in SMEs: Organising for innovation and change*. West Sussex, England: John Wiley & Sons Ltd.

Knights, D., & Willmott, H. (2007). *Introducing organizational behaviour & management*. Australia: Thomson.

Leonard, D., & Sensiper, S. (1998). The role of tacit knowledge in group innovation. *California Management Review*, *40*(3), 112–132.

Mainemelis, C., & Ronson, S. (2006). Ideas are born in fields of play: Towards a theory of play and creativity in organizational settings. *Research in Organizational Behavior*, *27*, 81–131. doi:10.1016/S0191-3085(06)27003-5

Martins, E. C., & Terblanche, F. (2003). Building organizational culture that stimulates creativity and innovation. *European Journal of Innovation Management*, *6*(1), 64–74. doi:10.1108/14601060310456337

Mathisen, G. E., & Einarsen, S. (2004). A review of instruments assessing creative and innovative environments within organizations. *Creativity Research Journal*, *16*(1), 119–140. doi:10.1207/s15326934crj1601_12

Miettinen, R. (2006). The sources of novelty: a cultural and systematic view of distributed creativity. *Creativity and Innovation Management*, *15*(2), 173–181. doi:10.1111/j.1467-8691.2006.00381.x

Mumford, M. D., & Licuanan, B. (2004). Leading for innovation: Conclusions, issues, and directions. *The Leadership Quarterly*, *15*(1), 163–171. doi:10.1016/j.leaqua.2003.12.010

Nguyen, L., & Shanks, G. (2009). A framework for understanding creativity in requirements engineering. *Information and Software Technology*, *51*(3), 655–662. doi:10.1016/j.infsof.2008.09.002

Porter, L. W., Bigley, G. A., & Steers, R. M. (Eds.). (2003). *Motivation and work behavior*. USA: The McGraw-Hill Companies, Inc. Prajogo, D. L., & Ahmed, P. K. Relationships between innovation stimulus, innovation capacity, and innovation performance. *R & D Management*, *36*(5), 499–515.

Shalley, S. E., & Gilson, L. L. (2004). What leaders need to know: A review of social and contextual factors that can foster or hinder creativity. *The Leadership Quarterly*, *15*(1), 33–53. doi:10.1016/j.leaqua.2003.12.004

Sternberg, R. J. (2003). *Wisdom, intelligence and creativity synthesized*. Cambridge, UK: Cambridge University Press. doi:10.1017/CBO9780511509612

Sundström, P., & Zika-Viktorsson, A. (2009). Organizing for innovation in a product development project: Combining innovative and result oriented ways of working – A case study. *International Journal of Project Management*, *27*(8), 745–753. doi:10.1016/j.ijproman.2009.02.007

Tesluk, P. E., Farr, J. L., & Klein, S. A. (1997). Influences of organizational culture and climate on individual creativity. *The Journal of Creative Behavior*, *31*(1), 27–41.

Utterbback, J. M. (1994). *Mastering the dynamics of innovation: how companies can seize opportunities in the face of technological change*. Boston, Mass: Harvard Business School Press.

Van der Vegt, G. S., & Janssen, O. (2003). Joint impact of interdependence and group diversity on innovation. *Journal of Management*, *29*(5), 729–751. doi:10.1016/S0149-2063_03_00033-3

## KEY TERMS AND DEFINITIONS

**Creative Participation:** Employees' intrinsic willingness to contribute generated ideas or problem solutions; an interface between creativity and innovation.

**Creativity:** A social process constituted by a specific mental state and social influence that leads to the generation of novel, useful and appropriate ideas or problem solutions.

**Curiosity:** A human desire to learn, know and seek new information and experience.

**Individual Factors that Influence Creative Participation:** Personality-based factors (e.g. knowledge and understanding, curiosity and self-confidence, personality types).

**Innovation:** A successful implementation of already generated and contributed ideas or problem solutions.

**Knowledge:** An expertise acquired by a human through learning and experience.

**Organizational Factors that Influence Creative Participation:** Organizationally induced factors (e.g. organizational culture, support, rewards, structure and strategies).

**Personality:** A sum of personal characteristics, cognitive abilities and behavioral events.

**Self-Confidence:** A belief in oneself, in own abilities and skills.

**Understanding:** A human ability to use and interconnect existing knowledge, and apply it into various contexts.

## ENDNOTE

1 A list of the top 50 innovative companies was taken from "The world's leading companies", Forbes, 2000-2005 and "The top 100 most innovative companies ranking", Business Week, 2006-2009 and compared with a list of 100 biggest companies that was taken from "Global 500", Financial Times, 2005-2009. Those companies which were presented in both lists were taken into account.

# Chapter 5
# Cultivating Innovation through Social Relationships:
## A Qualitative Study of Outstanding Australian Innovators in Science and Technology and the Creative Industries

**Ruth Bridgstock**
*Queensland University of Technology, Australia*

**Shane Dawson**
*University of British Columbia, Canada*

**Greg Hearn**
*Queensland University of Technology, Australia*

## ABSTRACT

*In this chapter, social relationship patterns associated with outstanding innovation are described and explored. In doing so, the chapter draws upon the findings of 16 in-depth interviews with award-winning Australian innovators from science & technology and the creative industries. The interviews covered topics relating to various influences on individual innovation capacity and career development. For all of the participants, innovation was a highly social process. Although each had been recognised individually for their innovative success, none worked in isolation. The ability to generate innovative outcomes was grounded in certain types of interaction and collaboration. The chapter outlines the distinctive features of the social relationships which seem to be important to innovation, and ask which 'social network capabilities' might underlie the ability to create an optimal pattern of interpersonal relationships. The implications of these findings for universities play a key role in the development of nascent innovators.*

DOI: 10.4018/978-1-60960-519-3.ch005

## INTRODUCTION

Modern economies are driven by innovation, which we suggest is comprised of new knowledge, combined with the capacity to turn this new knowledge into valued commodities. It is widely accepted that activity in the science and technology sectors drives innovation systems, and recently it has also been argued that innovation in the creative industries may have a similar effect on the entire economy, driving economic growth beyond sectoral effects (Potts & Cunningham, 2008). Thus, the study of innovation facilitators and ways to maximise innovation potential within these sectors is a valuable and much pursued scholarly topic.

Innovation and creativity research are both now a mature, multifaceted fields. The most common distinction between the two is that creativity is often thought of as an individual capacity (Amabile, 1983), whilst innovation is a systemic output of organisations. It can therefore be argued that creativity is a necessary but insufficient condition for innovation to occur. Put simply, innovation within an organisational context is affected by individual decisions about whether or not to choose "creative" options over more "traditional" options. Pink (2005) for example, argued that emerging business opportunities will depend increasingly on the ability to sense, predict and creatively capitalise on new market opportunities in consumer markets. Hearn (2006) proposed that new product innovation comprises technical, cultural and business model innovation which are all undergirded by creative skills sets. However, the creativity of individuals does not guarantee innovation. Assink (2006) suggested that other individual and organisational factors, including a successful business strategy and risk-reducing culture, are need to translate creativity into innovation.

Early studies of the determinants of innovation tended to emphasise individual creative or entrepreneurial performance, and the characteristics of individuals (e.g., Barron & Harrington, 1981; Simonton, 1988). However, most theorists now agree that while individual skills and knowledge, and traits like personality and intelligence, are important foundations for innovation, in actuality innovation thrives on social interaction and collaborative efforts. It involves the active combination of people, knowledge and resources. An important recent addition to the field of innovation studies is the idea of open innovation. Chesborough (2003) argued that open innovation needs a different mindset and company culture from traditional or closed innovation (Chesborough, 2003). The emphasis is on building external and internal networks of knowledge resources and finding ways to link them. Moreover, this knowledge structure is facilitated by the advent of the internet with continually evolving tools for collaboration at a distance and between companies. These internet tools enable processes that correspond to information search; brainstorming; structured problem solving; and feedback. All kinds of information are being produced in this way – from open source software to artistic creativity – via the Creative Commons movement.

In line with the 'open innovation' concept, recent studies of the determinants of innovation have focussed on team innovation performance (Perry-Smith, 2006) and innovation in social networks within and outside organisations (Ahuja, 2000; Burt, 2000). This body of literature has contributed much in terms of documenting the advantages and disadvantages of various types of social ties and tie configurations in networks (e.g., Burt, 2000). However, much less is known about the nature of these social ties at the level of the individual. In addition, despite a plethora of popular literature in the area, we know a surprisingly modest amount about what people need to know and do in order to create and maintain 'optimal' social networks for success in innovative ventures (c.f., Seibert, Kraimer, & Liden, 2001). In the present chapter, we attempt to shed light on these issues by documenting the findings of a qualitative study into the social factors which appear influential to the success of sixteen outstanding Australian innovators. We begin with a discussion of the nature of innovation.

## INNOVATION AS A SOCIAL PROCESS

Innovation has become a buzzword discussed enthusiastically in boardrooms and by policymakers. Beneath the rhetoric, however, the process of innovation is pivotal to modern economies. The value of an investment in primary resources such as in mining or fertile land run down over time as the resource is exhausted, whereas the value of investments in new knowledge from innovation (for example, Viagra or Google) increases as the new knowledge is turned into products, and consumers are captured. Increasing returns result because the cost of product development has already been made, and so the unit cost of a product decreases as sales increase (Arthur, 1996). Not only do unit costs decrease, but new products often become early market leaders and enjoy competitive advantages (Arthur, 1996). Thus, waves of whole new sectors of economic activity — film, pharmaceuticals, computers, aircraft, and telecommunications, software, digital content and so on, have undergirded economic growth for the last 100 years.

It is the ability to generate wholly new products and services, rather than deriving greater efficiency and economies of scale from existing production processes, that has been a defining factor in the transition from an industrial to a creative economy. However, the difference between new products and new efficiencies is somewhat illusory. Many increases in efficiency are actually developed by new companies supplying service innovation to a different industry's supply chain. Innovation processes are therefore also involved in renovating basic modes of production or grafting new features onto existing products or services. Innovation occurs right across the supply chain from production to consumption. In each case, the heart of innovation is new knowledge, and thus our focus should be on the source of this new knowledge: the people who create it.

There is also strong economic evidence that human capital is central to success in modern economies where knowledge, creativity, and innovation are of particular importance (e.g., De la Fuente & Ciccone, 2002; Florida, 2003; Machin & Vignoles, 2005). As Florida, suggests, "studies of national growth find a clear connection between the economic success of nations and their human capital, as measured by the level of education" (Florida, 2003, p.222). However, human capital cannot be understood simply as individuals with skills and competencies. Social relationships are carriers of knowledge, facilitators of learning, problem solving and creativity (cf. Wenger, 1999). We diffuse knowledge by communicating: put simply, by exchanging ideas, arguing and asking questions. The nature of these interactions and their innovative capacity are influenced by the structure, quality and quantity of those relations (Rooney, Hearn, Mandeville, & Joseph, 2003). Networks of ideas, people, and technologies are therefore central to innovation (c.f., Chia, 1998; Rooney & Schneider, 2005; Stacey, 2001).

Innovation has traditionally been defined as a 'process' involving the production, growth, dissemination and execution of new ideas among socially and institutionally contextualised actors in networks (e.g., Van de Ven, 1986) From an information science perspective, networks are ideal mechanisms of information resource allocation and flow. Structurally, they put people in direct contact via the provision of horizontal links across institutional boundaries, thus facilitating rapid information transfer. In addition to transmitting information, networks also help create it. New ideas may develop as each person in the network receives and synthesises information; information easily builds on information. Thus, new ideas are both shared and created via networks. In short, innovation is best viewed as a social process, and the theory of social networks can be expected to be a valuable lens with which to view this process.

Based on this premise, in this study we sought to understand the social dimension of innovation at an individual level, and by extension, the individual's network. We aimed to identify and describe distinctive relationship patterns and relationship behaviour amongst outstanding innovators, as a means to enhance understanding of the factors and processes which underlie innovation success.

## METHODOLOGY

We used in-depth, semi-structured interviews and Grounded Theory analytic strategies (Glaser & Strauss, 1967; Strauss & Corbin, 1998) to explore the career-related experiences of the sample of sixteen outstanding innovators. The aim of Grounded Theory is to construct an emergent theory from data collected from participants relating to their experiences of the phenomena under study. The theory is constructed through an iterative process of data collection, data coding, and conceptualisation / theorisation. Emergent theoretical constructions are repeatedly verified, modified and enhanced through the addition of new data, until theoretical saturation is reached; that is, further data collection does not result in further changes or additions to the theory.

As is customary in qualitative research, a theoretically drawn sample (Huberman & Miles, 2002) was used in this research study. Sixteen innovators were sourced progressively through media searches, awards scheme archival information, and by word of mouth. They were selected for the study according to eligibility criteria, and also in accordance with the constant comparative method for maximum variation within the sampling frame, which helps ensure maximal generalisability of the generated theory (Glaser & Strauss, 1967). Eligibility criteria included: (1) Australian nationality (by birth or naturalisation), (2) working in science / technology or the creative industries (sectors which are argued to drive national innovation systems, e.g., Potts & Cunningham (2008), (3) national or international level award/s or other peer-based recognition (4) innovative work (work involving generation new knowledge of some kind, combined with capacity to turn this new knowledge into valued commodities, as agreed by a panel of three researchers), (5) less than 10 years since completion of formal education (to limit the participant pool to mid-career, and ensure relatively good recall of formal educational experiences). The fields represented by the science / technology innovators were: aeronautical engineering, physics, epidemiology, artificial intelligence, polymer science, aerospace avionics, and chemical engineering, and for the creative industries innovators were: games programming, visual art, interior design / architecture, communication design, film production, animation, theatre making, music technology, and festival / media production. Twelve of the innovators were living in urban areas of Australia, and the remaining four were living in cities abroad. Three of the creative and three of the science / technologist innovators were female.

An initial recruitment email contained information about the nature of the study and eligibility criteria, and invited the innovator to participate in an initial semi-structured interview by phone or in person with a researcher. Follow-up recruitment telephone calls yielded a participation rate of 88% - one potential participant declined because she was no longer working and was thus ineligible, and one other was too busy to be interviewed. A researcher then either travelled to conduct face-to-face interviews, or where innovators were located overseas or were otherwise not available for a face-to-face meeting, conducted telephone interviews. The 60-90 minute interviews were recorded and fully transcribed for analysis, and in addition the researcher took notes and "memos" (Glaser & Strauss, 1967; Strauss & Corbin, 1998) during the interviewing process.

In accordance with the Grounded Theory approach (Glaser & Strauss, 1967; Strauss & Corbin, 1998), the present study did not begin with any specific hypotheses relating to the determinants of innovation, or the career development of outstanding innovators. Rather, we asked exploratory questions based on existing literature relating to the conditions under which innovation occurs, influences on individual career development, and determinants of career success. An emergent theoretical model was constructed from successive iterations of data collection and coding, including components relating to intrapersonal factors such as motivations, personal background and developmental factors, and socio-environmental conditions. This chapter concentrates on one such distinct component of the theoretical model: the current social context of the innovator.

In discussing our theorisation of the social relationships described and discussed by the innovators in this chapter, we use some terminology and concepts from social network analysis (SNA) which seems to lend itself particularly well to the interpretation of the interview findings. Social network analysis emphasises the patterns of interactions and information flows between people, and uses mathematical and visual methods to investigate these relationships (Wasserman & Faust, 1994). Social network analysis evolved from a need to examine and quantify the *centrality* or *positional properties* an individual occupies within the context of the broader network. In this context, the properties of an individual located within a social network can be expressed mathematically. Various centrality measures such as *degrees*, *closeness* and *betweenness* provide insight into the types of relationships an individual establishes and their impact upon the overall network in terms of information flow and access to resources both internal and external (Otte & Rousseau, 2002).

While centrality measures provide a quantifiable assessment of an individual's position in a network, they do not provide an indication of the overall "strength" of the relationship formed between 2 or more actors. The relationship ties between individuals are commonly expressed as either strong or weak. This binary was initially proposed by Mark Granovetter (1973) in his influential work, 'The strength of weak ties'. In essence, weak ties represent those relationships where individuals are infrequently in contact, and unlikely to readily share resources. In contrast, strong ties involve frequent exchanges of information and exchange of resources. These types of ties are commonly observed between family members and close friends.

## FINDINGS

The participants were selected for this study because they had received a significant degree of individual recognition for their innovative work. However, irrespective of whether they were scientist / technologists (ST) or creatives (C), participants indicated consistently that they had in no way "innovated in isolation". Rather, the participants relied heavily on input of various kinds from other people. Put another way, their social networks provided a significant proportion of the 'ingredients' required for innovation success. Indeed, several participants refused to discuss their achievements in individual terms, indicating that being 'too much of an individualist' was counterproductive:

*If you are in a team and you involve yourself in many projects and you give your ideas into the big pot of ideas and others do the same, then you won't have that output just for yourself but you will have more output overall, and in the long term it will benefit you... because two brains or three brains do more than one. In collaborations it's very important to say, you had an idea, okay, but now it is the idea of the team...you are going to get paid back multi-fold for that and you will have a very good network, will have made friends... I don't think that any one of us would have had these sorts of careers if we didn't work in a team. (ST2)*

Three distinctive patterns of productive social relationships emerged from the innovators' interview material: strong ties, bridging ties, and weak/indirect ties. These patterns are discussed in this section. In addition, we noted that participants appeared to embody certain types of attitudes, abilities and behaviours in relating to others; what we have termed 'social networking capability' (Hearn & Bridgstock, 2010), which enabled them to build the kinds of relationships which were particularly conducive to innovation. In our discussion of findings, we use a system similar to Rhodes, Hill, Thompson & Elliot (1994) as follows: "most participants" indicates the characteristic response of 12 or more participants, "some participants" indicates responses from 5 to 11 participants, and "a few participants" indicates responses from 4 or fewer.

## A Small Number of Strong, Highly Productive Ties

Most of the participants maintained a small number of extremely productive close working relationships. These strong ties were essential in ensuring the effectiveness of the participants' ventures. The participants were highly selective in their choice of ties. Strong ties were chosen according to criteria such as: shared values, a high level of intrinsic motivation and commitment to the field, and possession of important skills and abilities. The participants were generous in terms of time, information and resource sharing with their strong ties, and were prepared to shape projects and broader work agendas according to input from them. In this study, participants referred to their strong tie relationships using terms such as, 'my team', 'my people', 'the core group', or simply as, 'we'.

*I guess with my team I select very heavily on drive and enthusiasm and how they'll fit within the team, not so you don't get the odd problem but really trying to understand people's drivers and what excites and interests them – and to the best of my ability trying to create roles that are flexible that way. (ST5)*

One key type of strong tie discussed by the participants was the mentor-mentee relationship. Most of the participants had built ties with one or two mentors early in their careers, often through proactive development of informal contacts rather than formal mentoring schemes. These mentors possessed advanced industry, occupational or discipline-based skills and knowledge. Participants reported that their mentors had spent time 'showing them the ropes', providing skills and career-related guidance on an ongoing basis.

*Our Head of School was very firm on giving people advice on what they should do if they wanted to get their career going. It was very much 'you must apply for a (national competitive) grant every year, you need to get this many papers and be on committees'. He just had a formula that seemed to work, and he really pushed me to make sure that I kept up to the mark on that. (ST 1)*

*You know, it was a bit of a struggle. She made demands and she made more demands about where the work was heading and it was for the best. She said to me it needed another three or four weeks' creative development and she was absolutely right, it did. It was a bit hard for me to absorb at first but I said, yes I'll do it. She helped me understand what a massive project it is to get your work out there and to articulate your work. In some ways it was good because it made me fight harder for what I wanted to get on and what I wanted to say as an artist. (C7)*

Many of the participants indicated that they currently had a number of mentees for whom they provided similar help and guidance. Participants collaborated with their mentees, and saw mentoring activities as essential to growing not only their own work, but for the development of their field/s more broadly. Mentees were rigorously selected, and were expected to perform to a high standard once inducted into the team. In return, participants provided unstinting support to their mentees.

*I'm mentoring a senior scientist at the moment who wants to become a professor and he wanted to join my group and I said, 'you can do this but hard work and 150% dedication is a pre-requisite. Otherwise we are not even talking any further. If you shy away from the 60 hour week and more than two weeks vacation a year you are in the wrong place'. So you must be able to take up challenges. Go to the right place, select a good mentor and then be generous. Don't be too egocentric or too vain. (ST2)*

*I work with a very good bunch of people and I couldn't do what I do without them. I run a very large research group at the moment. I have 19 PhD students which is a few more than normal. They always have automatic first rights of entry to my office – they are the most important part of my professional life. I treat my group as my family to a certain extent. (ST7)*

Another important type of strong tie was based on complimentary skill sets. For instance, some of the participants possessed well-developed business and entrepreneurship skills themselves. However, several of the interviewees who lacked this entrepreneurial nous made greater efforts to forge strong ties with like-minded people who also possessed business and entrepreneurship expertise. In essence, these participants recognised that commodification was an essential part of innovation, and thus they built and maintained close relationships with people who were able to provide this skill set when and where necessary.

*Coming up with ideas, yeah, there's a bit of fun in that, but you know everybody's got a hundred ideas and if you're working with fifty people that's five thousand ideas that get developed during the time you've produced one game. You've got to back them up with what you call the execution. An idea that other people can't interact with has no real value. Successful implementation has been a big part of what I've been about. The first game sold two million copies and the second game sold one million copies. (C1)*

*I think that if you've got a good product, you need to find the best of way of getting it out there, or it's pointless. We knew we had something fantastic going on, but we really had no idea at all when it came to business models or marketing or any of that. To be honest, it just isn't my thing. Thankfully, I knew this guy. (C6)*

The formation of strong ties between the innovators and business / entrepreneurship experts is also a good example of the second type of relationship pattern for innovation discussed by the participants: trans-disciplinary ties, in which the individual 'forms a bridge' across different network clusters.

## Forming a Bridge between Disciplines or Sub-Disciplines

In line with current thinking about the trans-disciplinary nature of innovation (Hearn & Bridgstock, 2010) many of the participants reported that their work involved the intersection of several knowledge regimes. First, most of the participants reported having diverse skill sets and eclectic educational backgrounds (e.g., maths and visual art, astrophysics and avionic engineering, dance and science, business and animation), which afforded them a unique, and perhaps 'more creative' perspective on problem solving and problem identification / conceptualisation at work:

*I have a technology background. I graduated from a technology high school in Shanghai which is a selective high school focusing on maths and physics, which requires a good development of the left brain. But my nature is more interested in visual art and literature, which means naturally I'm more of a right brain person. So I switch between left and right, the balance between these two [brains]. (C2)*

Further, in some cases participants' varied backgrounds may have assisted them to build strategic relationships, by giving them 'disciplinary agility': the ability to traverse different disciplinary perspectives and terminologies, thus assisting them to communicate with people both within and outside their core specialisms. Several innovators noted that they often adopted a 'brokerage role' between networks of people from different disciplines. This finding is in keeping with Burt's (2000, 2004) suggestions about the optimal conditions for innovation involving structural holes in social networks, where one person forms a relational 'bridge' between previously disconnected network clusters.

*The current vision I've had that has led to this new Institute, has really been about trying to break down some of the boundaries between research areas and research disciplines – and to bring together physics, chemistry, biology and other areas to solve some, I guess, what you call really big problems that you can't tackle with any one discipline alone. (ST5)*

*There was a project that wasn't traditional, delivering a new engineering box and making it work. It was an interdisciplinary communication. But it was one where I knew I would learn a lot because it required me to learn about the engineering and the science and all the areas in between. And of course meet and work with a lot of the people with much, much more experience in each of these particular fields. (ST6)*

Some participants talked about needing a particular set of skills and attitudes when performing a trans-disciplinary network brokerage role, including an openness to take on new learning and revise old ideas, respect for others of diverse backgrounds, and the ability to stimulate co-operation in others towards common goals.

*It (trans-disciplinary work) has to be because you are genuinely interested. You have to have to have enough understanding and it is just through your own research and your own reading and in many cases collaboration with the people who are the experts to get your skills up to a reasonable level of understanding of that field. If you don't understand that field you can't answer the problem properly. (ST 4)*

*You need an understanding of another person's priorities and you also have to connect with that person. The group has a dynamic – just working well together from a social point of view, whether one individual has better emotional intelligence than the others, or whatever it is. There has to be a personality compatibility and a – I suppose a work ethic compatibility and it has to be more than just academic. It has to be a personal compatibility. (ST 4)*

Rather than remaining the only tie in the region of a trans-disciplinary structural hole, participants worked actively to build additional collaborative ties among their colleagues of different disciplines, a concept analogous to the '*tertius iungens* orientation' suggested by Obstfeld (2005). By contrast, in using a '*tertius gaudens*' strategy (Simmel, 1950), an individual acts selfishly and benefits from structural holes by acting as a gatekeeper between networks. A '*tertius iungens*' strategy is concerned with cooperation and group benefit, and works by introducing previously disconnected individuals, or facilitating new coordination between individuals who are already connected. One senior manager described one way that she ensured her team collaborated with others for the benefit of all:

*There's a lot of cross-fertilisation of ideas amongst our various disciplines and studios. So at any one time I make sure I have people from our discipline travelling throughout Australia and Asia. One of my senior designers has been in Bangkok for almost a year working on a couple of major workplace projects. It makes sure that we're at the forefront of what the discipline is doing, and it builds productive relationships outside the studio. (C4)*

The participants' bridging ties with other disciplines tended to complement their highly productive strong intra-disciplinary ties. Their bridging ties seemed to function primarily to provide access to diverse ideas, skills and resources. Their strong ties were also a source of new ideas, but their crucial role seemed to be in forming a centralised powerhouse for idea integration and implementation. The strong bridging ties some participants had formed with business / entrepreneurship experts seemed to contain both elements: market-driven information input & idea generation, and innovation implementation / commodification.

One participant summarised his position on intra and extra-network social ties:

*You have to keep both (types of relationships) up. You need a strong team of creative people with a shared vision, and you get together and get the work done and then get it out there, but you also need to not be afraid to go outside the team for new ideas and inspiration. Then you draw those new outside people in as well, see what happens, mix it up a bit. (C6)*

## A Large Number of Weak and/or Indirect Ties

Another emergent theme in the interviews concerned the roles of weak and indirect ties. All of the participants were time poor, and many of them indicated that that their strong ties involved significant upkeep. Therefore, they also tended to retain a large number of weak ties, and in addition rely on the secondary networks of those they had established strong tie relationships with. These weak or indirect ties could then be transmuted into future strong ties, should circumstances necessitate. The participants used a variety of strategies to promote themselves and their work, and thus acquire strategic weak ties. Common examples of this self-promotion included: entry into industry-based awards schemes, media exposure, and strategic committee participation.

*Getting awards was a targeted strategy. There's no shortage of awards around. In fact, there are hundreds. You can get an award for anything. So I kind of use it as a way to build profile quickly. (ST7)*

The science / technologists and the creatives exhibited much equivalence in the strong-tie and bridging tie relationship patterns we described previously. Participants from both sectors similarly exploited the opportunities afforded by weak and indirect ties. However, there was some variation in the nature of weak / indirect ties described by the two types of participants. Perhaps this is in part due to the science / technologists being employed by large organisations, whereas the creatives tended to be self-employed or engaged in freelance opportunities - often working with an agent, curator, or producer. Some creatives were particularly good at exploiting indirect ties via the networks of these key people, or via contacts within professional industry bodies. Other creatives saw the potential in the internet and social networking sites. None of the scientist / technologists mentioned online social networking as a means to acquire weak ties. One creative described her experiences with social networking media:

*I think networking is definitely a really great way of finding the new project or finding your next interest. If you think about (my) illustration being adopted by Twitter, it's because of iStockphoto. It's not only a great showcase, but also it's a good way to network with a lot of people. You never know who is going to look at your work and you never know who is maybe interested in getting in touch with you and work with you or use your work. That's why I am on Facebook and I am on Twitter and all the other social network media. I think definitely social network media is a great tool for opening new possibilities. (C2)*

## DISCUSSION

Our overall aim in this study was to explore social influences on individual innovation success, particularly those relating to relationships and social networks. Social relationships were found to be critically important to the outstanding Australian innovators we interviewed. Their relationship patterns shared a number of features, including a balance between strong intra-disciplinary ties and bridging trans-disciplinary ties, and a large number of weak or indirect ties. The innovators were selective in their choices of strong-ties, but once a relationship was formed, they shared resources and ideas unstintingly. Rather than keeping bridging ties to themselves, innovators encouraged additional trans-disciplinary collaborative relationships by introducing their ties to one another. The innovators' relationships were geared towards mutual benefit.

Comparing the findings with the existing social network literature, we find strong arguments that trans-disciplinary ties are important to innovation. For instance, Granovetter (1985, 2005) argued that networks were beneficial to innovation because they were an ideal mechanism for transmitting unique and non-redundant (i.e., potentially innovative) information between clusters of actors. "Innovation means breaking away from established routines. Development of resources outside their usual spheres may often be a source of profit and new institutional forms can facilitate such deployment" (Granovetter, 2005, p.46). Conversely, too many intra-disciplinary strong ties can lead to redundancy of knowledge and ideas, and not provide enough fodder for innovation. In a study of innovation within the Australian music industry, Ninan (2005) found that innovative entrepreneurs used trans-disciplinary networks to ensure a flow of non-redundant information. He also found that too much embeddedness in a disciplinary network exposed the innovative entrepreneur to group-think through conformity and long term enculturation.

However, there is evidence that strong ties can be beneficial to innovation. Our finding that clusters of strong-ties seemed to function as idea integration and implementation 'powerhouses' was borne out by recent literature. Strong ties are characterised by trust and sharing, reciprocal assistance, and open communication (Dyer & Singh, 1998), and these are ideal conditions for the integration of knowledge, resources, and ideas. Without strong ties, Obstfeld (2005) observed that innovative projects can suffer from the 'action problem' and stall at the level of concept integration and implementation.

The finding in our study that the innovators engaged in *both* bridging and strong-tie relationships runs somewhat counter to traditional social network theory. Strong tie and bridging tie strategies have often been assumed to be exclusive of one another, and in fact operate in opposition and in competition with one other (e.g., Coleman, 1988). However, some recent empirical work has suggested that the 'alliance ambidexterity' just described is indeed possible, and is highly advantageous for established firms (Tiwana, 2008) and new ventures (Stam & Elfring, 2008). There is some support for this contention at the level of the individual as well. For instance, Mehra, Dixon, Brass and Robertson (2006) conducted a study of sales team leaders' social ties, and found that the most successful seemed to not only maintain a balance of internal and external friendship networks, but were somehow able to play a central role in both types of networks. This alliance ambidexterity may well be a key to innovation success.

To take the implications of our findings further into the organisational realm, it may be that traditional firm structures, which often emphasise cohesiveness and uni-disciplinarity, have the potential to stifle the trans-disciplinary backdrop of creative idea generation. Conversely, too little social cohesion within an organisation can mean that a new idea does not reach fruition (Ray, Barney, & Muhanna, 2004). Leaders in organisations need to be aware of the complementary and balanced power of bridging and strong ties, and then develop each as appropriate. Balkundi and Kilduff (2005) noted that it is important for leaders to recognise that important bridging and strong ties can be formal or informal, and intra- or extra-organisational – that beneath most formal alliances between organisations, "lie a sea of informal ties" (Powell, Koput, & Smith-Doerr, 1996). The innovators studied in the present research were no exception to this: interpersonal friendships can lead to business alliances, just as business alliances can lead to warmth and trust between two organisations, two organisational units, or two individuals.

In addition, organisations may struggle with the '*tertius iungens*' strategy, which emphasises bringing others together to join common effort ('win-win'). Unlike terius gaudens ('dog-eat-dog') strategies, tertius iungens benefits can be advantageous for the entire network of an organisation, and long-term benefits can flow to the brokering organisation, because of delayed reciprocity (Obstfeld, 2005). Uniting behavior creates new ties and fills structural holes in networks (Obstfeld, 2005) which opens information channels and creates opportunities for knowledge exchange, innovation, and other determinants of competitive advantage. For instance, Paquin (2007) demonstrated that through a process of 'industrial symbiosis' facilitated by tertius iungens strategies, co-located firms in the Danish city of Kalundberg were able to create an innovative industrial ecosystem involving an optimal exchange of wastes, by-products, and energy.

## Individual Social Network Capability: Reflective and Proactive Development of Relationships for Innovation

Despite significant diversity in their fields of work, career experiences and backgrounds, the participants in this study were distinguished collectively by their understanding of how to build, maintain and navigate social networks to maximise the probability of, and the scale of, innovation success. None of the participants had received formal instruction in social network theory, but their interviews revealed significant consistency and distinctiveness in their relationship patterns and dealings with others. All were highly 'social network capable' (Hearn & Bridgstock, 2010).

We have proposed elsewhere (Hearn & Bridgstock, 2010) that individual social networking capability: (1) involves the capacity to effectively build and maintain personal and professional relationships with others for mutual benefit in work or career, and (2) that it is through these relationships that information and resource transmission for generation and commodification of new knowledge, i.e., innovation, occurs. Being social network capable also involves what McWilliam and Dawson (2008) refer to as 'network agility'—the ability to develop and navigate social networks in a strategic and enterprising manner.

Although all of the innovators were highly skilled and knowledgeable in their own fields (and often in other fields as well), they knew that the right kinds of relationships and interactions enhanced their chances of venture success. Many of the innovators stated in their interviews that they did not need to possess all of the necessary 'ingredients' for innovation individually; indeed, several said outright that they did not possess all of the necessary skills and knowledge themselves. Participants were, however, able to identify what (or who) was needed to propel innovation forward at a given point in time, and were then able to make the most of relationships with others to meet those needs in a mutually beneficial way.

## Implications for Education

The findings of this study have important implications for university learning and teaching practice in terms of cultivating and promoting the necessary graduate skills associated with innovation. A recent discourse in higher education literature holds that conventional pedagogies will not provide adequate skills and knowledge for productive participation in the 21st century creative economy (Bentley, 2008; Bridgstock, 2009; Pink, 2005), in which innovation plays such a vital role. We suggest that social networking capabilities are one such category of neglected skill.

While current pedagogical practice does emphasise the need for social interaction in order to promote learning, this remains largely welded to the provision of content and discipline-specific knowledge and skills. In short, any emphasis towards developing student social network capability is commonly devolved to group work activities associated with the delivery of discipline content, and it is often not present in any more sophisticated form than implied by the ubiquitous generic skill labels 'working with others', 'interpersonal skills', or 'team work'.

The present findings suggest that students need a far more nuanced understanding of social relationships and networks in order to thrive in a working world where innovation is prized. We suggest that students should not be expected to tackle all learning scenarios solo, relying upon individual skills, knowledge and other resources exclusively. Students need to learn how to make the most of the fact that they are operating in a social environment. Programs should therefore include elements of relationship building (particularly with those of other disciplines), collaboration and resource sharing, and social network theory, to assist in the development of social networking capabilities.

The innovators' interview findings also give rise to a possibly more controversial suggestion with respect to higher education provision. We have noted already that several of the innovators found that they had skill or knowledge deficiencies in important areas, but they were able to make the most of relationships with others to redress those deficiencies. One striking example of the innovators relying on their social networks to provide needed skills or knowledge was the formation of relationships with business / entrepreneurship experts. The question then becomes: if students are developing skills at university which mean they can source needed expertise and skill sets via relationship building, are there then domains of knowledge which are no longer essential to include in undergraduate programs? For example, if social network capability becomes a key skill acquired through core curricula, is entrepreneurship education (incorporating such topics as new venture creation, intellectual property, and small business management) for creative and science / technology students still necessary? Or is it sufficient to ensure that these students work with business students on shared projects involving commodification of creative work? In this way, it may be that social network capability can provide at least a partial solution to the increasing problem of university curriculum overcrowding.

Our call for the development of social networking capabilities in universities complements wider commentaries on pedagogies for the 21$^{st}$ century. For instance, Robinson (2000) calls for new forms of teaching and learning practice that draw upon social, communicative, and creative learning activities. This emphasis on the necessity for a change in teaching approaches from the didactic and transmissive to engaged and networked also resonates with findings of this current study. In short, contemporary pedagogical practice must be better equipped for fostering *hands on*, *minds on* and *linked in* learners.

It is important to note in this context, that engaged and networked does not merely imply the implementation of a greater number of group work opportunities. Rather, it refers to the pedagogical practices associated with what McWilliam and Dawson (2008) have termed as "flocking". The authors discuss the parallels between innovative and creative teams and the 'flocking behaviour' of social organisms. As McWilliam further elaborates, "flocking together allows birds to fly higher and exhibit greater scheduling and routing capacities than each bird can do alone" (McWilliam, 2008, p. 144). Learning and teaching activities involving team work, play, interchange of leadership, networking, and trans-disciplinarity are more closely aligned with the development and cultivation of skills associated with innovation than didactic or mandated group activities.

Recent advances in web-based technologies have provided educators and students with a multitude of tools that actively support teaching and learning strategies that are more focused on the principles of social learning and developing student networking capacity. Through the use of social media such as twitter, Flickr, Elgg, youtube and others, students and teachers alike have greater and more flexible opportunities to engage in team work, trans-disciplinary practices, thereby promoting diversity, leadership and followship skills and importantly in the context of this chapter, an understanding of the role and benefits associated with effective social networking. Importantly, engagement with these forms of information and communication technologies (ICTs) also provides an opportunity for educators to observe individual and group networking behaviour as a result of the captured student interaction data with the technology. For instance, Dawson (2008, 2009) has advocated for the greater integration of academic analytics (data-mining student interaction data) in teaching practice to develop a picture of the student social network as a means for (a) self-evaluating teaching practices designed to promote social interactivity, and (b) to identify students disengaged from the learning network. Through the implementation of these pedagogical principles and practices, these necessary student skills and attributes underpinning social network capability can be actively developed, evaluated and demonstrated.

## CONCLUSION

Our aim in this chapter has been to draw attention to the individual within their social context, as a key element of what drives innovation in modern economies. While the human capital approach emphasises individual skills and attributes needed for innovation, and the social network approach is concerned with identifying the group or organisation-level characteristics which are necessary to innovation, surprisingly few have attempted to combine the two via an exploration of the relationships of the innovative individual.

The present findings find innovative value in both the individual and their social network, with which the individual is inextricably linked. A key challenge for education, which has hitherto maintained an unrelenting focus on the development and assessment of individual competence, must now accept the idea that individuals operate within networks of people, technologies, knowledge, and ideas. It is these networks, and the behaviour of individuals within these networks, that have a meaningful influence on the ability to innovate. Further, if it is to maintain currency in the creative economy, education must also move beyond traditional disciplinary silos in order to accommodate the inherently trans-disciplinary nature of innovative work. Finally, it must also recognise that a crucial underpinning of innovation is the creation of new ideas, and that creating new ideas involves a different kind of thinking to that usually promoted via formal education.

In closing, we suggest that education systems themselves need to embody the practice of innovation in a process of self-reinvention for the creative economy. Higher education has much strength in disciplinary content, and in order to capitalise on this strength in the innovation age, it will certainly be necessary to renovate not only curriculum and pedagogy, but also policy, resourcing, digital infrastructure, and teacher professional development.

## REFERENCES


Ahuja, G. (2000). Collaboration networks, structural holes, and innovation: A longitudinal study. *Administrative Science Quarterly*, *45*(3), 425–455. doi:10.2307/2667105

Amabile, T. M. (1983). The social psychology of creativity: A componential conceptualization. *Journal of Personality and Social Psychology*, *45*(2), 357–376. doi:10.1037/0022-3514.45.2.357

Arthur, W. (1996). Increasing returns and the new world of business. *Harvard Business Review*, *74*, 100–109.

Assink, M. (2006). Inhibitors of disruptive innovation capability: A conceptual model. *European Journal of Innovation Management*, *9*(2), 215. doi:10.1108/14601060610663587

Balkundi, P., & Kilduff, M. (2005). The ties that lead: A social network approach to leadership. *The Leadership Quarterly*, *16*(6), 941–961. doi:10.1016/j.leaqua.2005.09.004

Barron, F. B., & Harrington, D. M. (1981). Creativity, intelligence, and personality. *Annual Review of Psychology*, *32*, 439–476. doi:10.1146/annurev.ps.32.020181.002255

Bentley, T. (2008). Learning beyond the classroom. *Educational Management Administration & Leadership*, *28*(3), 353–364. doi:10.1177/0263211X000283008

Bridgstock, R. (2009). The graduate attributes we've overlooked: Enhancing graduate employability through career management skills. *Higher Education Research & Development*, *28*(1), 27–39. doi:10.1080/07294360802444347

Burt, R. S. (2000). The network structure of social capital. *Research in Organizational Behavior*, *22*, 345–423. doi:10.1016/S0191-3085(00)22009-1

Burt, R. S. (2004). Structural holes and good ideas. *American Journal of Sociology*, *110*(2), 349–399. doi:10.1086/421787

Chesbrough, H. W. (2003). *Open innovation: The new imperative for creating and profiting from technology*. Harvard Business Press.

Chia, R. (1998). From complexity science to complex thinking: Organization as simple location. *Organization*, *5*(3), 341–369. doi:10.1177/135050849853003

Coleman, J.S. (1988). Social capital in the creation of human capital. *American Journal of Sociology*, *94*(Supplement: Organizations and Institutions: Sociological and Economic Approaches to the Analysis of Social Structure), 95-120.

Dawson, S. (2008). A study of the relationship between student social networks and sense of community. *Journal of Educational Technology & Society*, *11*(3), 224–238.

Dawson, S. (2009). 'Seeing' the learning community: An exploration of the development of a resource for monitoring online student networking. *British Journal of Educational Technology*.

De la Fuente, A., & Ciccone, A. (2002). *Human capital in a global and knowledge-based economy*. Brussels: European Commission Directorate-General for Employment and Social Affairs.

Dyer, J. H., & Singh, H. (1998). The relational view: Cooperative strategy and sources of interorganizational competitive advantage. *Academy of Management Review*, *23*(4), 660–679. doi:10.2307/259056

Florida, R. (2003). *The rise of the creative class: And how it's transforming work, leisure, community and everyday life. North Melbourne*. Pluto Press.

Glaser, B. G., & Strauss, A. (1967). *The discovery of Grounded Theory: Strategies for qualitative research*. New York: Aldine.

Granovetter, M. (1973). The strength of weak ties. *American Journal of Sociology*, *78*(6), 1360–1380. doi:10.1086/225469

Granovetter, M. (1985). Economic action and social structure: The problem of embeddedness. *American Journal of Sociology*, *91*(3), 481–510. doi:10.1086/228311

Granovetter, M. (2005). The impact of social structure on economic outcomes. *The Journal of Economic Perspectives*, *19*(1), 33–50. doi:10.1257/0895330053147958

Hearn, G. (2006). Value creating ecologies: Understanding next generation business systems. *Foresight*, *8*(1), 55–65. doi:10.1108/14636680610647147

Hearn, G., & Bridgstock, R. S. (2010). Education for the creative economy: Innovation, transdisciplinarity and networks. In Peters, M., & Araya, D. (Eds.), *Education in the creative economy*. New York: Peter Lang.

Huberman, A. M., & Miles, M. B. (2002). *The qualitative researcher's companion*. Thousand Oaks, CA: Sage.

Machin, S., & Vignoles, A. (2005). *What's the good of education? The economics of education in the UK*. Princeton: Princeton University Press.

McWilliam, E. (2008). *The creative workforce: How to launch young people into high flying futures*. Sydney: UNSW Press.

McWilliam, E., & Dawson, S. (2008). Teaching for creativity: Towards sustainable and replicable pedagogical practice. *Higher Education*, *56*(6), 633–643. doi:10.1007/s10734-008-9115-7

Mehra, A., Dixon, A. L., Brass, D. J., & Robertson, B. (2006). The social network ties of group leaders: Implications for group performance and leader reputation. *Organization Science*, *17*(1), 64–79. doi:10.1287/orsc.1050.0158

Ninan, A. (2005). *What are the roles of networks and clusters in the operation of an industry? The case of Queensland music*. Brisbane: Queensland University of Technology.

Obstfeld, D. (2005). Social networks, the tertius iungens orientation, and involvement in innovation. *Administrative Science Quarterly*, *50*(1), 100–130.

Otte, E., & Rousseau, R. (2002). Social network analysis: A powerful strategy, also for the information sciences. *Journal of Information Science*, *28*(6), 441–453. doi:10.1177/016555150202800601

Paquin, R. (2007). *Brokering industrial ecosystems: Creating regional institutional forms*. Paper presented at the Conference on the Institutional Foundations for Industry Self-Regulation, Boston, MA.

Perry-Smith, J. E. (2006). Social yet creative: The role of social relationships in facilitating individual creativity. *Academy of Management Journal*, *49*(1), 85–101.

Pink, D. H. (2005). *A whole new mind*. New York: Penguin.

Potts, J., & Cunningham, S. (2008). Four models of the creative industries. *International Journal of Cultural Policy*, *14*(3), 233–247. doi:10.1080/10286630802281780

Powell, W., Koput, K., & Smith-Doerr, L. (1996). Interorganisational collaboration and locus of innovation: Networks of learing in biotechnology. *Administrative Science Quarterly*, *41*, 116–145. doi:10.2307/2393988

Ray, G., Barney, J. B., & Muhanna, W. A. (2004). Capabilities, business processes, and competitive advantage: choosing the dependent variable in empirical tests of the resource-based view. *Strategic Management Journal*, *25*(1), 23–37. doi:10.1002/smj.366

Rhodes, R. H., Hill, C. E., Thompson, B. J., & Elliott, R. (1994). Client retrospective recall of resolved and unresolved misunderstanding events. *Journal of Counseling Psychology*, *41*(4), 473–483. doi:10.1037/0022-0167.41.4.473

Robinson, K. (2000). *Out of our minds: Learning to be creative*. Oxford: Capstone.

Rooney, D., Hearn, G., Mandeville, T., & Joseph, R. (2003). *Public policy in knowledge-based economies: Foundations and frameworks*. Cheltenham, UK: Edward Elgar.

Rooney, D., & Schneider, U. (2005). A model of the material, mental, historical and social character of knowledge. In Rooney, D., Hearn, G., & Ninan, A. (Eds.), *Handbook on the knowledge economy* (pp. 19–36). Cheltenham, UK: Edward Elgar.

Seibert, S. E., Kraimer, M. L., & Liden, R. C. (2001). A social capital theory of career success. *Academy of Management Journal*, *44*(2), 219–237. doi:10.2307/3069452

Simmel, G. (1950). *The sociology of Georg Simmel* (Wolff, K., Trans.). Glencoe, IL: Free Press.

Simonton, D. K. (1988). *Scientific genius: A psychology of science*. New York: Cambridge University Press.

Stacey, R. D. (2001). *Complex responsive processes in organizations: Learning and knowledge creation*. London: Routledge.

Stam, W., & Elfring, T. (2008). Entrepreneurial orientation and new venture performance: The moderating role of intra-and extra-industry social capital. *Academy of Management Journal*, *51*(1), 97–111.

Strauss, A. L., & Corbin, J. M. (1998). *Basics of qualitative research: Techniques and procedures for developing Grounded Theory* (2nd ed.). Thousand Oaks, CA: Sage Publications.

Tiwana, A. (2008). Do bridging ties complement strong ties? An empirical examination of alliance ambidexterity. *Strategic Management Journal*, *29*(3), 251. doi:10.1002/smj.666

Van de Ven, A. H. (1986). Central problems in the management of innovation. *Management Science*, *32*(5), 590–607. doi:10.1287/mnsc.32.5.590

Wasserman, S., & Faust, K. (1994). *Social network analysis: Methods and applications*. Cambridge: Cambridge University Press.

Wenger, E. (1999). *Communities of practice: Learning, meaning and identity*. Cambridge: Cambridge University Press.

## KEY TERMS AND DEFINITIONS

**Alliance Ambidexterity:** Involves the establishment and maintenance of both bridging and strong-tie relationships.

**Bridging Ties:** Are relationships between individuals which span structural holes – i.e., networks which are otherwise unconnected with one another. Their main role in innovation seems to be to provide access to diverse ideas, skills and resources.

**Innovation:** Comprised of new knowledge, combined with the capacity to turn this new knowledge into valued commodities. Innovation drives the growth of modern economies.

**Social Network Capability:** Involves the capacity to effectively build and maintain personal and professional relationships with others for mutual benefit in work or career. It is through these relationships that innovation occurs.

**Social Network Theory (SNA):** Emphasises the patterns of interactions and information flows between people, and uses mathematical and visual methods to investigate these relationships. social network analysis evolved from a need to examine and quantify the *centrality* or *positional properties* an individual occupies within the context of the broader network.

**Strong Ties:** Are relationships which involve frequent exchanges of information and exchange of resources. These types of ties are commonly observed between family members, and close friends/ colleagues, their crucial role in innovation seems to be in forming centralised powerhouses for idea integration and implementation.

**Tertius Iungens (c.f. tertius gaudens):** A social networking strategy emphasising cooperation and group benefit, and works by introducing previously disconnected individuals, or facilitating new coordination between individuals who are already connected.

**Weak Ties:** Are relationships where individuals are infrequently in contact, and unlikely to readily share resources. These relationships take relatively little time and energy to maintain.

# Chapter 6
# Methods against Methods

**Marc Stierand**
*NHTV Breda University of Applied Sciences, The Netherlands*

**Viktor Dörfler**
*University of Strathclyde, UK*

## ABSTRACT

*This chapter intends to clarify some issues about the often misunderstood terminology of creativity and innovation methods. Following the train of thought outlined in this chapter, it is argued that neither creativity nor innovation is guided by a method. There are only methods against methods that can help the extraordinary individual to step faster and easier into a state of mind that is conducive to creativity, but which has no effect on whether the creative output becomes an innovation. In order to support this claim, three major reasons that seem to be responsible for making people believe that such methods for creativity and innovation exist are outlined here. Next, the chapter addresses the phenomenon of creativity and continues with a discussion on the systemic character of creativity and innovation. Finally it shows that there are no methods for creativity, but methods against non-creativity by explaining in particular how one of these methods against non-creativity works. What this chapter outlines here is a necessarily one-sided and partial view, aiming not to convince the readers of the correctness of the view, but rather to make them think by presenting one possible consistent approach.*

## CREATIVITY AND INNOVATION

Are there steps that, when followed, lead not only to a new idea (i.e. creativity), but also to a successful idea (i.e. innovation)? We do not think that there are, but we think that there are methods against methods that can help the creative individual to step faster and easier into a state of mind that is conducive to creativity. But there is still no connection to innovation, because whether an idea becomes an innovation is not decided by

DOI: 10.4018/978-1-60960-519-3.ch006

the creative individual, but, as we shall see later, by the perceptive value of this idea.

Outside the academic world innovation is often seen as enigma, as the work of a creative genius or as serendipity. However, this view is mostly dismissed by academics as unscientific, because its validity is very difficult to test. Therefore, academia has created a preference for seeing innovation as a continuous, rational, and purposive process (Nelson & Winter, 1982; Rogers, 1962/2003). Researchers interested in innovation often apply micro perspectives and frequently seem to be lost in thought trying to answer questions of explanation, prediction, and correspondence. This has led to a lack of understanding about the phenomenology of innovation, which we see as the foundation on which any explanation, prediction and correspondence should be built. Moreover, there is still the problem that we have to deal with creativity when we engage with the topic of innovation. Although many researchers acknowledge the link between creativity and innovation, they just close creativity into a black box or go around in circles by trying to blueprint the creative process.

The problem about creativity is that we know it exists, but we do not know how it works. In other words, we do not know what the creative process looks like. We have never seen it. We have only seen the creative product that is the result of a person's creativity. Another widespread issue is linked to the creative individual. Many people think that everyone can be creative. But if this would be the case, why are there so few Mozart's and Shakespeare's? Our Western and postmodern society tends to equate equality with sameness and thereby neglects the creative extraordinariness of a few. Our society does not welcome outliers. In fact, we even cut off these tall poppies, because their talents naturally distinguish them from the rest of us. People can be equal, but they are not the same. Not everyone can paint like Matisse and not everyone can write like Goethe. Even if one could prove that all people start with the same genetic makeup for creativity from birth, not everyone will be able or will have the chance to develop a creative ability that can produce creations of the quality and influence as those of the aforementioned masters.

In order to support our arguments, we are going to outline our understanding of creativity and innovation and show how both phenomena are linked. Based on this outline we focus on the creative individual. We particularly draw on Gardner's conception of extraordinariness in order to illustrate that creating something new and valuable requires an individual with substantial knowledge and not just the ability to enter a state of mind that is conducive to creativity. Then we expand our discussion to the phenomenon of innovation by showing that innovations are not produced solely by an individual but require the socio-cultural world for validation and co-creation of the new value. At the end we look at de Bono's work on creativity methods.

What we want to present in this essay is our approach to creativity and innovation. This is a necessarily partial and one-sided approach, just as anyone else's. Conversely, we do not offer a comprehensive literature review; we are only covering the literature necessary to outline our position. Specifically, we do not engage with the issues of team dynamics, thus we disregard the sources aimed at encouraging creativity in organization and facilitating teams regardless whether those are supposed to be creative teams or not. We also limit our study of creativity to extraordinary performance; we do not engage with the creativity of people at below-expert knowledge level. Our argument is not directed towards a particular method or groups of methods, but rather we are questioning the idea of having methods for creativity/innovation. Thus this paper does not include description or comparison of various methods. There is one exception to this, a group of methods baring specific relevance to our argument; this group is outlined briefly to support understanding. We do not aim at convincing the readers that this is the only right way to address creativity and innovation, only that this is one possible way. By doing so, we would like to make the readers think more deeply about creativity and innovation. However, before doing so it is necessary to explain what we mean by creativity.

## UNDERSTANDING CREATIVITY

Let us start explaining creativity through the notion of problem solving. Simon (1973) distinguished between ill-structured and well-structured problems and regards the first as a residual concept, which means that a problem is ill-structured when it is not well-structured. An ill-structured problem is one that is deficiently defined, whereas as a well-structured problem possesses some or all of the following characteristics (Simon, 1973, p. 183):

1. There exist definite criteria to test the solution;
2. The initial problem state, the goal state and all intermediate states may be represented;
3. The transitions between the previous states can be represented;
4. The acquired knowledge can be represented;
5. The effects of the environment can be represented;
6. And a feasible amount of search and computing is required;

Simon, however, admitted that his proposed criteria for well-structured problems are vague and not completely definite, because there are many shades of definiteness along the way between ill-structured and well-structured problems. People were always fascinated by the idea of discovering nature's book of explanations in which they believe they can find the answer to how the human mind works. In the 17th century, for example, the great French mathematician and philosopher René Descartes tried to find a universal method to solve problems. But, as Pólya (1957/1990) remarks, Descartes failed to understand how the human mind solves problems in the first place. Over 300 years later Simon took on Descartes challenge and tried for most of his career to build what he called a General Problem Solver (GPS). The working principle of the GPS is very sensible: find what is identical in all problems, the common part, and use this as the base and then you only need to sort out the details that make the difference (e.g. Newell & Simon, 1972). This is sensible *in principle*. But can you tell what the common part is in finding a cure for cancer and making great paella? Nothing has changed since Descartes, and we still do not know how the mind solves problems and especially ill-structured ones. The only thing we know is that the same in all ill-structured problem-solving is that the problem-solver sees things differently from how they are usually seen. In other words, the problem is solved through creativity. The reasons why we still do not know how the mind uses creativity, is simply because the part of thinking, which is responsible for seeing things differently, is non-algorithmic. This means that this part of thinking cannot be put into a finite sequence of instructions for solving a problem. De Bono (1971, 1994) calls this non-algorithmic part of thinking lateral thinking or sometimes parallel thinking. The notion of lateral thinking can be easily grasped through explaining how good jokes work (Baracskai, 1998). In good jokes the joke teller takes us on a vertical way of thinking. Something like:

*A man and a friend are playing golf one day at their local golf course. One of the guys is about to chip onto the green when he sees a long funeral procession on the road next to the course. He stops in mid-swing, takes off his golf cap, closes his eyes, and bows down in prayer.*

This vertical way sets us on a way towards the obvious conclusion: probably the man stops playing golf, because he wants to be respectful towards the mourners. This is the essence of vertical thinking; there is a single outcome and the thinking converges towards it. But in a joke there is a jump out from the vertical storyline into a lateral direction. The obvious is rearranging to form a new order, to make new sense when we get to the punch line. In our example the friend says:

*"Wow, that is the most thoughtful and touching thing I have ever seen. You truly are a kind man." The man then replies: "Yeah, well we were married 35 years."*

If there were to be no new order it would not be a joke. Nobody would laugh. The reason why we laugh is because we understand that there is another way of thinking according to which the punch line is perfectly logical. It makes sense, but we would not have thought of it. We have now seen that creativity is highly anti-methodological in Feyerabend's (1993) sense and the new, the thing that astonishes us, that makes us laugh, or that makes us say "why did I not think of that before?", is a lateral detour. In other words, it is a discontinuity, a sudden change that is not usual in the given situation and that shows no obvious connection between the factors under consideration (de Bono, 1971b). But with hindsight, and that might be the reason why people think that there are steps to produce new ideas, the lateral thinking is logical, which means that the new solution is obvious. However, there is again bad news for those who think that we should just do a bit of reverse engineering and follow back the lateral route to blueprint the steps the creator or joke teller took to see how creativity works. But unfortunately we have no evidence that the later explanation reflects the way how the creator or joke teller got to the novelty (Gladwell, 2005). Of course, saying that creativity is about seeing things differently, does not mean seeing in any different way. Only in ways that make sense, just no one has seen it before. Hadamard (1954), for example, investigated how new results are born in the field of mathematics, which is usually thought of as being completely logical. His investigation showed, among other things, that the creator requires deep knowledge and that the novelty is born in a flash of intuition. According to Hadamard, the first phase is the conscious hard work of trying to solve a problem followed by a kind of forgetting phase, which is a sort of unconscious continuation of the work. After that phase comes the sudden insight accompanied by a sense of certainty. At the end the mathematician proves the result on paper in full consciousness. We now have a reasonably solid explanation why there can be no methods for being creative: First, the creative jump cannot be seen in advance, but only retrospectively. And second, an algorithm cannot go into a place that cannot be seen. This is only possible through imagination and intuition. We now continue to outline the systemic character of creativity and innovation by referring to the link between creativity and innovation, the creatives, and idea creation and value creation. These aspects, as we will show, are reasons why we cannot apply steps for creativity and innovation.

## THE SYSTEMIC CHARACTER OF CREATIVITY AND INNOVATION

### The Link between Creativity and Innovation

Csíkszentmihályi (1997, 2006) speaks of two types of creativity: Creativity (with a capital C) and creativity (with a lower case c) as can be seen in the following figure. Creativity (with a capital C) is a system of three inter-related parts: the domain, the field, and the individual. The domain is the area in which the individual has chosen to work. Each domain has its specific rules, knowledge, tools, practices and values. The field, on the other hand, consists of the persons and institutions that judge the individual creator's quality of work. In other words, the field consists of the gatekeepers to the domain. The individual creator, in contrast,

*Figure 1. A systems model of creativity. Source:Csíkszentmihályi (2006, p. 4)*

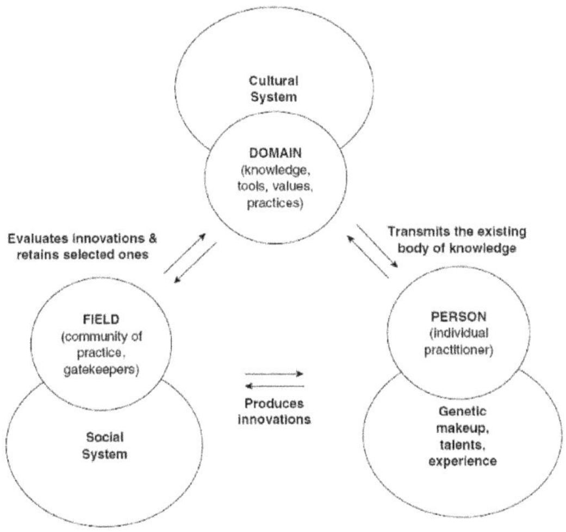


is influenced by her/his personal creativity (that is creativity with a lower case c) and by her/his specific genetic makeup, talent and experience (Figure 1).

Hence, Creativity is concerned with changing or transforming an existing domain, whereas creativity is concerned with the actions or thoughts of a creative person that have the power to change or transform an existing domain. In other words, creativity is concerned with the creation of a new idea and Creativity is concerned with realizing a new value, that is, the successful innovation from the idea. This link between creativity and innovation can be expressed as a two stage heuristic process. The first stage is a creative process of solving an ill-structured problem in which the problem solver rearranges her/his existing knowledge in order to obtain a solution for the problem (Dörfler, 2004). The validation of the idea happens then in the network of gatekeepers. The second stage is concerned with how the idea is converted into a value. The validation of the new value is then executed by the field, which actually co-creates the value by promoting it to the domain (Figure 2).

Going back about one century to the beginnings of innovation research tells us that the interest in innovation actually emerged from diverse branches within the social sciences. While these branches applied different background knowledge, they had the common intention to describe and give reasons for social changes. Anthropologists like Alfred Kroeber (1876-1960) and Ralph Linton (1883-1953), for example, explored cross-cultural diffusions of technical and social practices, which they called borrowing inventions. And sociologists like William Ogburn (1886-1959) explained social change as continuous cultural lag. The economist Joseph Schumpeter (1911/1934, 1939/1961, 1947/1976), however, is certainly acknowledged by many as the initiator of innovation research. He derived his ideas about innovation from his analysis of economic and social systems and many believe that his thoughts gave the impetus to recognize innovation as the crucial factor for economic progress and change. In his early work, he assumed that a company's size affects its ability to innovate and therefore smaller companies, while being more flexible, seem to be better positioned to innovate than larger and more bureaucratic companies. Later, though, he suspected that larger companies, in particular those with a monopolistic market power, might be better positioned to innovate than smaller firms, because of their power and because

*Figure 2. Heuristic process of innovation. Source: own figure*

they often have better resources. Fundamentally, he saw innovations as waves of creative destruction that can revolutionize a whole market. This view challenged the status quo of capitalism at the time, because it rejected the view that capitalism is about managing its existing structures and proposed that it actually is about destroying its existing structures in order to create new ones. This also meant that firms that can respond fast enough and can take hold of discontinuities are better positioned that firms that have difficulties to do so. Hence, innovation is therefore not the response to a well-structured problem for which we are able to apply exact criteria that we can test and neither can we blueprint each single phase of the innovation process. Approaches that try to make us believe that this is possible just create an artificially validated process under a condition that Bessant and Caffyn (1997) termed continuous innovation. Under this condition, innovation is nothing more than a process of improvement that takes place in a framework of existing and known rules. Simply said, continuous innovation means doing things as usual, only better. This does not exclude significant changes but implies that changes occur within an established framework. Of course, continuous innovation is already a complex phenomenon, but research cannot ignore that innovation can also be influenced by discontinuous and chaotic conditions. Discontinuity can be a scary notion, because it is not an everyday event. Innovators are forced to experiment in order to accumulate new knowledge that can help them to keep track in an unpredictable world. During times of experimentation a so-called dominant design emerges that in some way predicts the most popular but not necessarily the most sophisticated trajectory of the future. The old trajectories, however, are still in place and normally undergo rapid improvements, which in turn sharpens the conditions for all actors. But many still see innovation as a linear and continuous process and argue that social (i.e. non-technological) and technological innovations are opposite ends of a continuum. Yet, social innovation just means that it is the result of accepting something new that changes the domain or creates a new domain. Consequently, every innovation, whether technological or non-technological, is also a social innovation, because the creative output has to pass social validation before it can become an innovation. This becomes obvious, if we look at how some great technical innovations that affected our lives. For instance, the first airplane was designed in 1903 but this had no effect on our lives whatsoever. The real impact of flying happened only in the 1960's, when we actually started to use airplanes to fly all over the world.

We have now seen that innovation lies outside the influence of the creator, because it is a social validation of the creator's idea. It is obvious that there cannot be a method for controlling this social validation. This raises interesting questions regarding the responsibility of the innovator and also of various social institutions, but we do not engage with these in this chapter.

The term "social innovation" has been used throughout the literature to express a range of different ideas (Pol & Ville, 2009). Taylor and Gabor were probably the first who mentioned the term explicitly. As a behavioural scientist, Taylor (1970, p. 70) argued that social innovation is a response to social needs by introducing, for example, an *"innovative new school, a new way of dealing with poverty, a new procedure for resocializing delinquents, a new technique for rehabilitating the schizophrenic."* Gabor (1970), on the other hand, saw social innovation as a tool to stimulate new arrangements in society such as new technologies or laws. Since 1970 it seems that a headless chicken run started to add yet another definition of social innovation to the literature and very often without improving the understanding of the concept. We are great believers in going back to the origins and, indeed, Taylor teaches us that his understanding of social innovation really explains a specific type of innovation; one that is expected to disrupt valued roles, identities, and skills of high complexity. In other words, social innovations may enrage and put whole communities to the test. Taylor offers here the amusing contrast with the invention of a new and better mousetrap. A new mousetrap will most likely be accepted quite quickly given sufficient advertising and right distribution. The reason is that accepting a new mousetrap involves no revolution in a person's identity or life style, whereas a social innovation requires a radical and revolutionary change of values and beliefs. And we all know how difficult that is.

After this detour into the social aspect of innovations, let us now look at the creatives themselves before we continue to discuss the topics of idea validation and value creation in detail.

## The Creatives

Social psychologists have long argued for a link between extraordinary individuals and creativity (see e.g. Gardner, 1993; Guildford, 1950; Koestler, 1964; Osborn, 1953). More recently, however, they have begun suggesting that creativity is not exclusive to extraordinary individuals, but is a primary component of every human life. The problem is that situational factors, such as childhood and education, are very influential and finally decide whether a person is able to actualize on her/his creative potential. In other words, we face the same problem as earlier mentioned that, even if we assume that at birth all people start with the same genetic makeup for creativity, not everybody has the chance to exploit it. Gardner (1998, pp. 11-12) distinguishes between four main types of extraordinary people: the master, the maker, the introspector, and the influencer. For example, he describes Wolfgang Amadeus Mozart as a master who typically *"gains complete mastery over one or more domains of accomplishment; his or her innovation occurs within established practice."* Sigmund Freud represents the maker, because a *"maker may have mastered existing domains, but he or she devotes energies to the creation of a new domain."* The introspector is represented by Virginia Woolf, because the *"primary concern to this individual is an exploration of his or her inner life."* And finally Mahatma Gandhi embodies the influencer, who *"has as a primary goal the influencing of other individuals."* Further, it is important to note, argues Gardner, that individuals may comprise more than one type of extraordinariness. What most extraordinary people have in common is that they have often failed dramatically, but they have the ability to reflect on their failures and learn from them in such a way that they were able to clearly identify their strengths, exploit them, and thus turn defeats into opportunities. These lessons can be nicely linked to Schumpeter's description of the entrepreneur and the creative destroyer. For the early Schumpeter (1911/1934) entrepreneurs were a small dynamic minority of agents that, in comparison to the rest of us, actively respond to changing environments and are able to create something new through overcoming internal and external resistance. This special breed of people, which Schumpeter (1947/1976) later calls the creative destroyer, has the ability to successfully combine already available economic possibilities in completely new ways. Positive economic development is therefore achieved through innovations that have their source in the entrepreneurial spirit. Innovators are thus a kind of temporary monopolists, because they can exploit their position until imitators copy their ideas (Dörfler, 2010). Schumpeter (1911/1934), however, strictly differentiated between the inventor and the innovator:

*"Economic leadership in particular must be hence distinguished from 'invention'. As long as they are not carried into practice, inventions are economically irrelevant. And to carry any improvement into effect is a task entirely different from inventing of it, and a task, moreover, requiring entirely different kind of aptitudes. Although entrepreneurs of course may be inventors just as they may be capitalists, they are inventors not by nature of their function but by coincidence and*

*vice versa. Besides, the innovations which it is the function of the entrepreneur to carry out need not necessarily be any inventions at all" (Schumpeter, 1911/1934, pp. 88-89).*

Schumpeter's distinction between the inventor and the innovator can be nicely linked to Csíkszentmihályi's dual flow between the field and the individual mentioned earlier, because both are needed to produce innovations. In other words, the field catches the idea of the inventor and co-produces the innovation by converting the idea into a new value by promoting it to the domain. Another way of arguing for a link between the extraordinary and creativity is the relation between creativity and domain knowledge. Here, Csíkszentmihályi draws two arrows between the individual and the domain. One goes from the domain to the individual signifying that the domain transfers knowledge to the individual. The other one goes from the individual to the domain signifying that the domain has accepted that the individual has created new knowledge that is valuable (i.e. that it is an innovation) and worth adopting. With regard to the link between domain and individual, Einstein (1956/1984) stressed that even if the individual inherits some knowledge from the domain, the actual knowledge always originates in the mind of the individual. Prietula and Simon (1998) also talk about the extraordinary, but use the term expert. For them expertise goes far beyond just knowing a multitude of facts. In other words, they see a difference between people who know a myriad of facts and those few extraordinary individuals who can use their knowledge beyond the borders of reasoning to create new creative solutions. Johann Wolfgang von Goethe, for example, said that you can often hear amateur painters saying that their work is not finished yet. Goethe argued that they will never be finished, because they paint without awareness. The extraordinary or the master, on the other hand, knows from the first stroke what the final painting will look like (Goethe cited in Senge, et al., 1999, p. 157). Goethe's example shows that the greatest distinction between amateur and extraordinary is the degree of awareness. By referring to Csíkszentmihályi's model, this means that the extraordinary creator has a deep awareness about the knowledge, values, tools and practices of the domain. Without this awareness a creation would have only a small chance to get accepted by the domain. Based on this train of thought we can now summarize that in order to create an innovation it needs an extraordinary person who is able to cope with the balancing act of using their knowledge beyond the borders of reasoning to create new creative solutions and being aware of the rules, knowledge, tools, practices and values of the domain. Now we are going to outline how a new idea is validated and how a new value is created.

## Idea Validation and Value Creation

There is much disagreement about what qualifies a creation to be creative. This disagreement is certainly linked with the problem that the creative process itself (i.e. the process of coming up with a new idea) is unknown. One could argue that 'being creative' could also mean to come up with ideas that are widely considered 'useless' as long as the inventor went through a recognizable creative process. The Japanese Kenji Kawakami created the art of inventing objects that are practically useless by virtue of their disproportionate usefulness. He called his art *chindogu.*Zizek (2008) explains that chindogu objects must meet two criteria: they must be feasible to build, but they should be impractical in the sense that they cannot be marketed, because they would never receive the consent of the gatekeepers of the field to enter the domain. Consequently, the only possibility to judge whether an idea is creative or not, is to judge the value of the outcome of creativity. Amabile (1988; 1996) and George (2007) define creativity as the creation of ideas that are both new and useful. In other words, there is a clear distinction between the creation of ideas and the creation of values. First, an idea is created, which, abstractly seen is always new knowledge, which is needed to solve the ill-structured problem. Only then can the new idea be pitched and transferred into a new value. If there is only a new idea and no value creation, the idea remains unknown, which means that the idea needs to be passed on by the value. The management literature speaks in this respect of the perceived newness or degree of innovativeness of the idea. Some authors formulate that an idea is creative when it is new and valuable and it is innovative when the idea has become realized. Others propose that the newness has to be validated in relation to the firm or market. Many years ago Zaltman, Duncan and Holbeck (1973) have introduced a broader concept of perceived newness, which they called relevant units of adoption. They argued that newness is context specific and evaluated along continua that describe the quality of newness. Of essence is that both the individual creator and the field require knowledge in order to produce an innovation and to validate its value. Hence, a person that aims at altering or changing a domain by his/her idea must convince the field of the value of the idea. The tricky thing with new ideas is that it is difficult to judge beforehand how new and valuable they will be perceived. Rogers (1962/2003) introduced the concept of diffusion of innovations, which is the planned, but also spontaneous spread of ideas. Diffusion is a type of communication of new ideas, which, because the idea is new, involves uncertainty. Rogers (1962/2003, p. 6) describes this uncertainty as *"the degree to which a number of alternatives are perceived with respect to the occurrence of an event and the relative probability of these alternatives."* Hence, information is used to overcome the lack of structure and predictability implied in uncertainty, because it is important that innovations are understood by the members of the field and domain. Understanding helps to limit the perceived risk and uncertainty. If a new idea is diffused and gets either adopted or rejected, this leads to a change in the social system. Consequently, whether a new idea becomes an innovation depends on the one hand how rule-breaking the idea is and on the other how compatible it is with the value system of the relevant unit of adoption. Innovations may disturb the sense-making of the domain and this is why innovation is often seen as dangerous, because it requires space and freedom from direction and control. In other words, the culture of the domain tends to remunerate individuals for their conformity and tends to punish those who challenge its culture. Let us discuss this two way influence between the domain and its gatekeepers (i.e. field) and the individual by means of the concept of value systems. According to von Bertalanffy (1981, p. 13) *"values are things or acts which are chosen by and are desirable to an individual or to society within a certain frame of reference."* This means people within a value system are concerned with what is good and what is bad. Our individual value system can be described as muddled, because in the course of our lives, due to many different events, we change or reject values or adopt new ones. On the other hand, the value system of a domain is born from complex interactions of the value systems of its individual members and from the influence of other domains. The interesting thing is, however, that after the domain's value system is formed, it becomes independent of its members. This phenomenon can be explained by Hamel and Prahalad's (1994, pp. 55-56) story about a group of monkeys. A number of monkeys were put in a cage. In the middle of this cage was a pole and at the top of this pole were bananas. As soon as the monkeys tried to get up to the bananas they received a cold shower. Now, the interesting part is that the monkeys very quickly learned not to go up the pole to get the bananas, because they would get a cold shower, which they obviously did not like. Later, all the monkeys were re-placed one by one and still no new monkey touched the bananas, because it became part of their group value system and each new monkey was "told" not to get the bananas. Again, we have shown that there is a difference between the perception of an individual and the perception of others. Therefore, it became now even clearer that there can be no method to make other people positively validate an idea and transform it into a new value. Now we are going back to creativity itself, which we have shown is a prerequisite for innovation, to explain its anti-methodological character.

## NO METHODS ALLOWED

As a professional anarchist, Feyerabend (1987, 1993) rejects the use of any methods, as he observes that there is absolutely nothing that is present in all creativity but is absent in all other enterprises. Thus he declares that *"Anything goes!"* In his various books de Bono describes creativity as unexpected, non-linear, non-algorithmic, anti-methodical mode of thinking. This is what he calls lateral thinking to contrast it to convergent thinking, or parallel thinking to contrast it with vertical thinking. Yet, he offers a series of tools (used when needed) and habits (always present) for lateral thinking. Let us look at what these methods really do: de Bono (1973) offers a word, "PO", that goes beyond the true-false dichotomy; the "Six Thinking Hats" (de Bono, 1990) represent roles that can be used in brainstorming; the "Focus and Purpose" keeps one from forgetting the purpose; in the "Forward and Parallel" the forward thinking is a step-by-step inference along the path and the parallel thinking makes one stop and look around; the "Perception and Logic" emphasises that the logic is usually all right, but the perception should be improved for better outcomes; the "Values" serve as the basis for rejecting a logically correct solution, thus one should pay attention to what values are involved and who are affected by them; the "Outcome and Conclusions" are about not forgetting to consider the implications; the "Aims, Goals and Objectives" (AGO) similarly aim to prevent forgetting any of them; the "Consider All Factors" (CAF) suggests thoroughness instead of quick recipes; the "Other People's Views" (OPV) emphasises open-mindedness through considering who is affected (it can be considered a version of the stakeholder analysis); the "Alternatives, Possibilities and Choices" (APC) encourages both exploring and constructing alternatives; the "First Important Priorities" (FIP) focuses on selecting what really matters; the "Consequence and Sequel" (C&S) urges exploring the consequences of the alternative actions; and finally the "Plus, Minus and Interesting" (PMI) is an assessment tool (de Bono, 1993, pp. 63-150). Is this not infuriating? Why would someone who obviously understands the essence of creativity offer methods for it? Using Feyerabend we could say that anything goes but what you would normally do. These methods do not aim at making the creative process happen but rather at preventing us using the methods that we learned so well. De Bono's point is that we are so badly brainwashed that we would use the learned methods even subconsciously. Therefore his methods should not be seen as methods *for creativity* but rather as *methods against non-creativity*. Methods against methods.

Let us look closer at the example of PO. As mentioned above, de Bono (1973) offers PO as a word that goes beyond the true-false dichotomy. It is so to say the opposite of NO and NOT, without meaning YES. The essence of vertical or logical thinking is to make a selection by either accepting or rejecting something. As soon as we learn when to say no, we have learned how to use logical thinking. Or, as de Bono (1973, p. 196) says: *"Logic could be said to be the management of NO."* In contrast, lateral thinking can be explained as a restructuring and rearrangement of information that helps us to break out from rigid patterns we have learned from experience. To break out from these patterns de Bono offers the use of the word PO and therefore describes lateral thinking as the management of PO. Of course, both NO and PO are language tools, but they have totally different functions. By saying NO we give a judgment, whereas by saying PO we are anti-judgmental. By saying NO we think within our learned framework of reasoning, whereas by saying PO we think outside our framework. But how does PO work? As earlier said, PO is, like NO, a language tool, but one that we are not used to. Therefore, it can help us to break out from the safe and familiar environment of our language and way of thinking. We just do not use the word PO on a daily basis and thus we are not used to it. Because we are not used to PO, it can help our mind put together information in new ways even if these new arrangements of information are unjustified. In other words, PO can help liberate us from our mind by disrupting our established patterns of thinking. This means that we have in addition to the selective options of YES and NO used in logical thinking also the option of PO, which allows us to select, or better accept, an option that might seem totally illogical and even absurd. Let us refer here to an example given by de Bono (1971, pp. 205-206). He provides the following statement to explain the use of PO: *"PO water flows uphill if it is colored green."* Most of us would probably say: "NO, water never flows uphill if it is colored green. In fact, water will never flow uphill!" But water can flow uphill. By adding a tiny amount of a particular sort of plastic, the water solidifies slightly with the effect that when we pour out this plastic-water mixture from a jug and then again hold the jug upright the water continues to flow out by climbing up the sidewalls of the jug. Is this not incredible? De Bono further suggests understanding the response to PO before we actually use PO. Why is that? In our common form of communication, when we use vertical thinking, we always give selective judgments about another person's statement. In other words, we either agree or disagree with what the other person has said. The same will happen when beginners try to use the word PO. We are so indoctrinated by the selective mechanisms of our language that we think the PO-statement is also a judgment. But it is not. This is the very essence of PO. Maybe it is helpful to picture PO as a sticky tape with which we can tape those possibilities that we would normally reject as being completely irrelevant, useless, illogical or impossible to realize. However, this does not imply that the PO-statement is a better alternative or even an alternative at all. In this sense PO is never judgmental and therefore it is a protection that allows following new ways and that can help us to use information in a provocative way. Moreover, PO can bring relaxation in situations where we stuck in the rigidity of thinking. It can make us smile, because we can formulate seemingly alien possibilities that seem so absurd that they can help us break the rigidity to see the problem from a different angle from which we might easier grasp a solution.

## CONCLUSION

At the beginning we asked whether there are steps that, when followed, lead not only to a new idea (i.e. creativity), but also to a successful idea (i.e. innovation). We argued that there are no such steps, but that there are methods against methods that can help the creative individual to step faster and easier into a state of mind that is conducive to creativity. Then we identified a number of reasons why some people think that such steps exist. The first reason is that academia has created a preference for seeing innovation as a continuous, rational, and purposive process, because the fuzzy aspects of creativity are difficult to validate and cannot result in testable explanations, predictions, and correspondence. The second reason is that many academics close creativity into a black box or run around in circles trying to blueprint the creative process. We see this is as a problem, because on the one hand we cannot ignore that creativity is a prerequisite for innovation, but on the other hand we have to accept that nobody knows how creativity works. The only possibility that remains is therefore to come as close as possible to the phenomenon of creativity. The third reason is connected to the creative individual. Academia tends to assume that everyone can be creative and often rejects that a truly creative idea requires an extraordinary mind. This has probably created the strongest belief that steps exist that, when followed, bring us to the novum. In order to support these earlier mentioned reasons, we discussed the systemic character of creativity and innovation by particularly focusing on the link between creativity and innovation, the creatives, and how ideas are validated and how new values are created. Based on this discussion we concluded that creativity, and for that matter innovation, does not allow for methods. In other words, there are no methods for creativity, but methods against non-creativity. This claim was further supported by particularly focusing on the method "PO" as proposed by de Bono.

In this paper we outlined our particular view of creativity and innovation, more specifically one aspect of this view – our view on the impossibility of methods for creativity and innovation. This is a necessarily one-sided and limited view; but rather than providing a shallow and more inclusive overview of a topic we chose to immerse more deeply in this narrow interpretation. Our aim was not to convince the readers that our view is correct (let alone the only possible correct one) but to make them think more deeply about their own view of creativity and innovation. This is why we did not address areas that would be necessary for a more comprehensive discussion, such as the levels or types of creativity and especially the particular techniques/methods aimed at creativity and innovation.

There are also topics in this chapter that we marginally discussed, only to the extent needed for our present theme, some of which may be candidates for future research. Particularly we plan to engage more deeply into the social aspects of creativity and innovation including the questions of responsibility. We have further plans to research various aspects of creativity. In a project that we completed, Marc interviewed 19 of the greatest chefs in the world. In another, presently ongoing project, Viktor interviewed so far 15 grandmasters (people at the highest knowledge level), including 12 Nobel Laureates. We have further plans to carry out similar research projects in various creative industries. The rich qualitative data that we are collecting will be analyzed in the framework outlined in this chapter.

## REFERENCES


Amabile, T. (1988). A model of creativity and innovation in organizations. In Staw, B., & Cummings, L. (Eds.), *Research in organizational behaviour* (*Vol. 10*, pp. 123–167). Greenwich, CT: JAI Press.

Amabile, T. (1996). *Creativity in context*. Boulder, CO: Westview Press.

Baracskai, Z. (1998). *Profi problémamegoldó (Master of Problem Solving)*. Nyíregyháza: Szabolcs-Szatmár-Bereg megyei Könyvtárak Egyesülés.

Bessant, J., & Caffyn, S. (1997). High-involvement innovation through continuous improvement. *International Journal of Technology Management*, *14*(1), 7–28. doi:10.1504/IJTM.1997.001705

Csíkszentmihályi, M. (1997). *Creativity: Flow and the psychology of discovery and invention*. New York: HarperCollins.

Csíkszentmihályi, M. (2006). A systems perspective on creativity. In Henry, J. (Ed.), *Creative management and development* (3rd ed., pp. 3–17). London: Sage.

de Bono, E. (1971). *Lateral thinking for management: A handbook*. London: McGraw-Hill.

de Bono, E. (1973). *Po: Beyond yes and no*. London: Penguin Books.

de Bono, E. (1990). *Six thinking hats*. London: Penguin Books.

de Bono, E. (1993). *Teach your child how to think*. London: Penguin Books.

de Bono, E. (1994). *Parallel thinking: From socratic to De Bono thinking*. London: Viking, Penguin Group.

Dörfler, V. (2004). *Descriptive model of learning capability.* Paper presented at the 3rd Annual Conference on Information Science Technology and Management, Alexandria, Egypt.

Dörfler, V. (2010). The fitness of innovation in an evolutionary framework. In Jolly, A. (Ed.), *The innovation handbook: How to develop, manage and protect your most profitable ideas*. London: Kogan Page.

Einstein, A. (1956/1984). *The world as I see it*. New York: Kensington Publishing.

Feyerabend, P. K. (1987). *Farewell to reason*. New York: Verso.

Feyerabend, P. K. (1993). *Against method* (3rd ed.). London: Verso.

Gabor, D. (1970). *Innovations: Scientific, technological and social*. Oxford: Oxford University Press.

Gardner, H. (1993). *Creating minds*. New York: Basic Books.

Gardner, H. (1998). *Extraordinary minds*. London: Phoenix.

George, J. (2007). Creativity in organizations. *The Academy of Management Annals, 1*(1), 439–477. doi:10.1080/078559814

Gladwell, M. (2005). *Blink: The power of thinking without thinking*. London: Penguin Books.

Guildford, J. (1950). Creativity. *The American Psychologist, 5*, 444–454. doi:10.1037/h0063487

Hadamard, J. (1954). *The Psychology of invention in the mathematical field.* New York: Dover Publications.

Hamel, G., & Prahalad, C. (1994). *Competing for the future*. Boston: Harvard Business School Press.

Koestler, A. (1964). *The act of creation*. New York: Macmillan.

Nelson, R., & Winter, G. (1982). *An evolutionary theory of economic change*. Cambridge: Belknap Press.

Newell, A., & Simon, H. A. (1972). *Human problem solving*. Englewood Cliffs, NJ: Prentice Hall.

Osborn, A. (1953). *Applied imagination*. New York: Charles Scribner's Sons.

Pol, E., & Ville, S. (2009). Social innovation: Buzz word or enduring term? *Journal of Socio-Economics, 38*, 878–885. doi:10.1016/j.socec.2009.02.011

Pólya, G. (1957/1990). *How to solve it: A new aspect of mathematical method.* London: Penguin Books.

Prietula, M., & Simon, H. (1998). The experts in your midst. *Harvard Business Review*, (January-February): 120–124.

Rogers, E. (1962/2003). *Diffusion of innovations* (5th ed.). New York: Free Press.

Schumpeter, J. (1911/1934). *The theory of economic development*. Cambridge: Harvard University Press.

Schumpeter, J. (1939/1961). *Business cycles*. New York: MacGraw Hill.

Schumpeter, J. (1947/1976). *Capitalism, socialism and democracy* (5th ed.). London: George Allen & Unwin.

Senge, P., Kleiner, A., Roberts, C., Ross, R., Roth, G., & Smith, B. (1999). *The dance of change: The challenges to sustaining momentum in learning organizations*. New York: Doubleday Currency.

Simon, H. (1973). The structure of ill structured problems. *Artificial Intelligence, 4*(3-4), 181–201. doi:10.1016/0004-3702(73)90011-8

Taylor, J. (1970). Introducing social innovation. *The Journal of Applied Behavioral Science*, *6*(1), 69–77. doi:10.1177/002188637000600104

von Bertalanffy, L. (1981). *A systems view of man*. Boulder, CO: Westview Press.

Zaltman, G., Duncan, R., & Holbek, J. (1973). *Innovations and organisations*. New York: John Wiley & Sons.

Zizek, S. (2008). The prospects of radical politics today. *International Journal of Baudrillard Studies*, *5*(1).

# Section 2
# Techniques for Creativity

# Chapter 7
# Methods to Improve Creativity and Innovation:
## The Effectiveness of Creative Problem Solving

**Fernando Sousa**
*INUAF and Apgico, Portugal*

**Ileana Monteiro**
*University of Algarve and Apgico, Portugal*

**René Pellissier**
*UNISA, South Africa*

## ABSTRACT

*This chapter focuses on the development of organizational creativity, using the CPS methodology, aiming at demonstrating its effectiveness in using the individual and team divergent thinking improvement in identifying organizational problems. A study was undertaken using problem solving teams in seven companies, in which each individual was submitted to a pre-post test in attitudes towards divergent thinking and asked to express the evaluation of the method. All the information reported in the sessions was recorded.*

*The results indicate a change in attitude favourable to divergent thinking, the provision of a professional, efficient method of organizing knowledge in such a way that can help individuals to find original solutions to problems, and an important way to lead teams to creativity and innovation, according to companies' different orientations.*

DOI: 10.4018/978-1-60960-519-3.ch007

## INTRODUCTION

As the world becomes more complex and changes at an accelerated rhythm, the organizations find themselves working in ways poorly adapted to this new and discontinuous environment. The traditional solutions, which granted success for a long time, are no longer suitable, yelling for new and innovative ways of doing business. Lately, the importance of creativity and innovation has gathered full recognition in western societies, emphasizing its contribution to social well being and to organizational effectiveness (Mumford & Gustafson, 1988), and underestimating the possible nuisances. In the global environment, organizations need flexibility and adaptation to face the changing market conditions as well as efficiency to maintain successful routines (Basadur, 1997). Efficiency refers to the daily routine operation, fulfilling and improving the organizational quality standards; when facing unexpected market changes, the organization must be flexible enough to react appropriately and, finally, the organization must analyze and reflect upon its routines, in order to anticipate environmental changes and therefore adapt by creating new products, services or processes. The tension between these paradoxical forces (routine and change) calls for an organizational system capable of managing innovation without losing quality.

This chapter presents the first results of an action research on an organizational innovation model (Sousa, Monteiro & Pellissier, 2008), and concentrates only in the team problem solving step of the model, trying to demonstrate its effectiveness in developing team creativity. The approach suggested in this chapter illustrates the intervention in seven companies, using a creative problem solving methodology and a divergent thinking questionnaire adapted from Basadur (1997, 1999, 2000), showing that the co-workers of different organizational levels are willing and capable of defining accurate problems, produce useful solutions, and involve themselves in projects to improve the organizational sustainability.

In the first section the main concepts of innovation and creativity will be analyzed and the general theoretical framework explained. Then the principles and the method of creative problem solving (CPS) will be described, providing a global understanding of the organizational interventions. The empirical study, involving seven organizations, seven teams and sixty nine persons of diverse educational backgrounds and hierarchical levels, will be presented and discussed. Also, evidence will be presented, regarding the CPS's capability to change the participants' attitudes, help people and organizations to define important problems and find interesting solutions, and organize the team members in the implementation of a project, without interfering with their daily work.

## INNOVATION, CULTURE AND KNOWLEDGE CREATION

Innovation, within the framework of a knowledge-based economy goes far beyond the linear or chain linkage models that have long been used in innovation theory to explain innovation processes in high-tech knowledge industries. Here, innovation is seen as a social, spatially embedded, interactive learning process that cannot be understood independently of its institutional and cultural context (Cooke, Heidenreich, & Braczyk, 2004; Lundvall, 1992). Innovation results from the involvement of a diversity of stakeholders, each one contributing with its own body of knowledge in order to build a new common and shared perspective of reality.

Strambach (2002) suggests that the interdisciplinary view of innovation systems is concerned with understanding the general context of the generation, diffusion, adaptation and evaluation of new knowledge, which determines innovativeness. It follows that the focus is on non-technical forms of innovation as defined above. Common characteristics of the different approaches to innovation, identified by Edquist (1997), include (1) innovation and learning at the centre, (2) a holistic and evolutionary perspective, and (3) an emphasis on the role of institutions, meaning that innovation do not happen in isolation: organizations are shaped and influenced by institutions and, simultaneously, influence and shape the institutions, setting a two way relationship, thereby influencing the innovation processes and systems.

The increasing interdependence of technological and organizational change is a significant feature of systems of innovation, which means that technological innovation and organizational innovation have become increasingly important. These are combined with more diverse knowledge requirements which include not only technical know-how, but also economic, organizational, and sociological knowledge and competencies. The second reason for the increased interest in non-technical innovations is associated with the connection between the organizational innovation and the corresponding learning capacity. The acceleration of change that is part of the globalisation process means that organizational learning processes are more and more important for creating and maintaining competitiveness.

Ultimately, whether innovation is successfully diffused, it requires some absorptive capacity on the part of the target audience. Cohen & Levinthal (1990: 128) define absorptive capacity as (…) *the ability of a firm to recognise the value of new, external information, assimilate it and apply it to commercial ends*. The diffusion of the innovation is normally dependent upon the specific innovation typology, the innovation champions, the time element to successful diffusion and the absorptive capacity of the adopters.

In organizational innovation, the unit for innovation is the organization itself (Wolfe, 1994). Although the outcome of the innovation may be process, product or service, the innovation needs to be undertaken through the creative inputs of the individuals and/or the management. As to measures of innovation, Dalal (2008) mentions the qualitative measure of emotional and psychological impact the innovation produces on the users (the "Aaa Aaa" moments); the quantitative measures of the total population of end users using the new innovation (and even helping co-create it), and the net revenue generated by the company that can be attributed to the new innovation.

The importance of leadership and organizational or group culture must be acknowledged as a primary condition for innovation to occur, namely with organizational innovation. In fact, innovation requires the agreement, support and commitment of managers towards an innovative project. The participative or collaborative leadership is largely recognised as a style fostering innovation (Kanter, 1983; West, 1990, for example); as a leader's role, which is not to have new ideas, but to build and help the team to engage in problem solving activities and project implementation (Basadur, 2004); time management will help reflexivity, which is critical to individual and collective learning (West, 2002).

Connected with this research, West (1990) described the four psychosocial features that, in interaction, will foster innovation: vision, psychological safety, climate and norms favouring innovation. Through vision, the team may reach common goals and create an image of a desired and valued future – clear and valued goals also encourage team innovation (De Dreu & West, 2001; West, 1990). Psychological safety is achieved through valuing individual contributions in team discussion and decision making allowing failure and the deferral of judgement, and promoting team reflexivity. A climate promoting team cohesion and trust, allowing minority dissent (Nemeth, Personnaz & Goncalo, 2004), has repeatedly been associated to innovation. Finally, the definition of norms, explicitly or implicitly formulated, foster new work practices or improved processes, namely if there is coherence between espoused theories and theories in use (Argyris, 1999), i.e., between what people say about how processes are run and how they really act in those processes.

This coherence can be enhanced through an innovation oriented culture; organizational culture being viewed as a shared, taken for granted, implicit assumptions held by a group that determines how it perceives, thinks and reacts to the environment (Schein, 1992). Organizational culture arises from interactions between the diverse actors (or stakeholders) as they try to adapt to external environments and solve integration problems, in an ongoing action and conversation making sense of their experience and common ideology (Stacey & Griffin, 2005). As Collins & Porras (1994) pointed out in their research, successful companies have a clear and communicated ideology, translated in a strategy and in consistent goals, well articulated with practices and actions. A collaborative leadership promotes employee commitment and involvement in the improvement of organizational processes, creating a future oriented system, and valuing knowledge, experimentation and innovation.

## CREATIVITY AND INNOVATION

The construct of innovation has largely been studied in organizations, referring to "...the intentional introduction and application within a role, group or organization of ideas, processes, products or procedures, new to the relevant unit of adoption, designed to significantly benefit the individual, the group, organization or wider society" (West & Farr, 1990: 9). This definition is wide enough to consider radical and incremental forms of innovation, technological and non technological innovation, industrial and service innovation, as well as organizational innovation, and shows innovation as a social process or, as Spence (1994) mentions, "the processes of implementation of ideas, relying mainly on organizational communication and power". But, if definitions of innovation appear to be relatively simple in the literature, things become more complex when we try to define creativity and to establish its relationship with innovation.

One of the first difficulties is that not everyone interprets and values creativity the same way. In fact, as Woodman & Schoenfeld (1990) recall, the term "creativity" can be seen either as a social concept, expressed by people's implicit theories, or as a theoretical construct, developed by researchers in the field.

Considering the theoretical definitions, and after carefully analysing the propositions evidenced by Kasof (1995), it is possible to conclude that the construct of creativity was (and still is) used in the scientific literature to designate something perceived by others. Amabile (1983) states that *a product or response is creative to the extent that appropriate observers independently agree it is creative. (...) and it can also be regarded as the process by which something so judged is produced.* Stein (1953; 1984; 1994) maintains that *creativity is a process that results in novelty which is accepted as useful, tenable, or satisfying by a significant group of others at some point in time.* These examples illustrate what may be designated as *hetero-attributed creativity*, something pertaining to the *communication process*.

Envisaged as a sort of "persuasive communication", in which the creator is the source, the original product the message, and the judge or audience is the recipient (Kasof, 1995; Csikszentmihalyi, 1994), creativity enters the broad domain of exceptional personal influence (Sawyer, 1998; Simonton, 1991; 1995); and the social processes of the making of a reputation (Ludwig, 1995; Jones, 1997; Mace, 1997); or just the processes underlying the capacity to shift roles, in which the creator develops a dialogue with his or her work, anticipating the audience's reaction (Stein, 1993).

As the product of that communication process, creativity appears connected to what is perceived as new by someone other than its originator, or as the "putting to use of an idea" (Kanter, 1983; West & Farr, 1990), in the domains of production, adoption, implementation, diffusion, or commercialisation of creations (Kaufmann, 1993; Rogers,

2003; Spence, 1994). In these cases, creativity is seen as *innovation*.

Runco (1998), and Kokot & Colman (1997), also see creativity as a *self-attributed* construct. And Baer (1997) considers creativity to be *anything that someone does in a way that is original to the creator and that is appropriate to the purpose or goal of the creator*.

Recognising creativity as a self-attributed concept, used by people to describe their acts at any moment is, in a sense, using implicit theories of creativity. It lies in how each individual organises and incorporates the perception of reality in his or her own self. Striving for mastery and perfection, the expression of one's own individuality and sharing with others, become essential parts of the core construct of creativity, which may, then, encompass a wider array of activities, products, processes and performances.

Creativity seems then to acquire its full meaning as a process of communication between the creator (or the product) and the judges or audience (hetero-attributed), or between the creator and the product (self-attributed). Innovation seems to be more appropriate to designate the resulting attribution made by the audience apropos the product, as depicted in Figure 1.

As a consequence, hetero-attributed creativity can only be measured through socio-cultural judgements, and is therefore context-dependent. Quoting Csikszentmihalyi (1994), *creativity is located in neither the creator nor the creative product but rather in the interaction between the creator and the field's gatekeeper who selectively retains or rejects original products*.

*Figure 1. The construct of creativity*

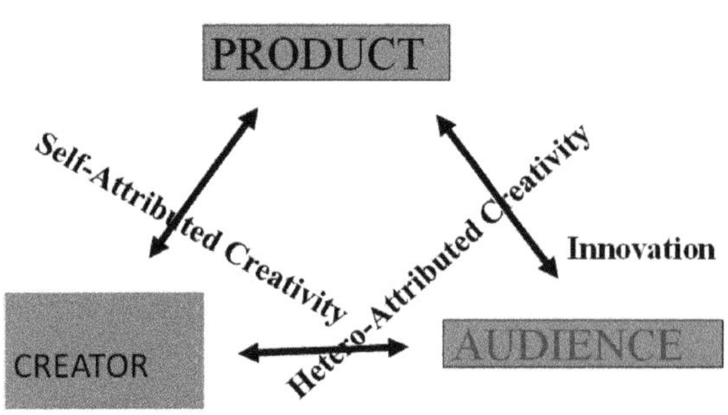


This way, the theoretical construct of creativity relies on people's implicit theories of creativity, i.e. in the ways they consider a specific product, person or process as representative of their conceptions of creativity.

Thus, while innovation concerns the processes of implementation, relying mainly on organizational communication and power, in the domains of production, adoption, implementation, diffusion, or commercialisation of creations (Spence, 1994), creativity remains exclusive to the relation established between the creator and the product, where nor even originality and usefulness are important, but only the "trying to do better", connected to cognitive and emotional processes taking place at the individual level (Sousa, 2008).

If we relate creativity to problem definition, and innovation to decision implementation, this last step requires a series of problem definitions, in order to carry out a decision or an idea, thereby making it difficult to separate these concepts at an organizational level. In fact, when we move from the individual level to the team and organizational levels, creativity and innovation become more and more difficult to separate, so that we must agree with Basadur (1997), when he says there is no difference between organizational creativity and innovation. Therefore, the moment we move to other levels besides the individual, we will use these terms (creativity and innovation) as synonyms, and we refer to organizational creativity as a *system devoted to enhance creativity in organizations,* thus using the definition proposed by Basadur (1997).

As to the several approaches to identify types of innovation, either by separating the adoption of products and processes from its development (Cebon, Newton & Noble, 1999) or, in a more classical way, product and process innovation (Adams, 2006), most authors agree that innovativeness, or organizational innovation, is a third important type of innovation, which represents the potential of the workforce to promote changes to benefit of the organization.

## CREATIVE PEOPLE

As Huhtala & Parzefall (2007: 299) mention, *(...) to remain competitive in the global market, organizations must continuously develop innovative and high quality products and services, and renew their way of operating*, and they also maintain that companies increasingly rely on the employees continuous ability to innovate. Also, even though innovation may take place through the adoption or development an existing product or service, through investments in R&D or in technology acquisition, it is only through developing and sustaining a creative workforce that the organization will succeed in maintaining the necessary potential to overcome difficult problems and situations that cannot be solved through investments only (Cebon, Newton & Noble, 1999).

This creative workforce potential is both the ability to retain creative managers and employees (McAdam & McClelland, 2002) and to provide an environment where each one will feels free and willing to contribute to organizational success. Aspects like raising job complexity, employee empowerment and time demands, together with low organizational controls (decision making, information flow and reward systems), are said to raise employee creativity (Adams, 2006). However, more elements are necessary in order to make people willing and able to contribute to organizational effectiveness. For instance, supportive leadership, knowledge acquisition, and team work procedures favouring creativity (Unsworth, 2005) can add to success. Creative people, either managers or employees, are committed to their work and organization, and so they may bring in important issues, provided that top management values their work and ideas. In fact, according to a Gallup Management Journal (GMJ) survey (Hartel, Schmidt & Keyes, 2003), engaged employees are more likely to "think outside of the box" and produce creative ideas than disengaged people; they also are more receptive to new ideas. The research concludes that engaged people tend to find and suggest new ways to improve their work and business processes, which may lead to the assumption that creative people have a deeper understanding of the organizational processes, by being in a privileged position to identify, define and find organizational problems.

To a certain extent, all of this can be achieved by elevating the importance of creativity in the organization and providing a system through which individual potential may be channelled into profitable innovation. What are required are freedom to create, content and process skills to be able to create, and a supportive human environment (peers and team leader). The issues surrounding the potential of an organization to innovate are still in its beginnings, although Mclean (2005) and Puccio, Firestien, Coyle & Masucci (2006) and especially, Basadur (1997, 1999, 2000), did some empirical research. The major challenges are to define criteria to evaluate the impact of organizational innovation on process and product innovation (Wolfe, 1994).

## THE CREATIVE PROCESS

Graham Wallas, in his book *The art of thought*, published in 1926, presented one of the first models of the creative process. The simplicity of its four steps (preparation, incubation, illumination and verification) made it resist time and maintain until recently:

1. **Preparation:** Includes the whole education process, the individual acquisition of the necessary skills in a specific domain and the development of flexible thinking;
2. **Incubation:** An unconscious or partially conscious process. In this stage, the flex-

ible knowledge gathered previously begins to be organized and generates new mental structures. This is still an unknown process.

3. **Illumination or insight:** The creative idea bursts from the preconscious into the conscious awareness. The idea will in most cases need to be worked and reorganized, for it is seldom entirely operational.
4. **Verification:** Reviews, reorganizes and works on the idea. The dialogue between the creator and his or her creation occurs at this point.

The creative process studies developed significantly after the Great World War, within the psychometric perspective which promoted divergent thinking process and a whole set of techniques to foster creativity, as the pioneer work of Paul Torrance and Wallach & Kogan (1965). Only in the eighties, the concept of convergence (until then seen as the opposite of creative thinking) was integrated in the creative process as a skill of creative problem solving.

Research on the creative process has reached the conclusion that creative people do not reveal unique or distinct thinking processes (Weisberg, 1999), or present stages of *inspiration* or *illumination*, or unconscious processes (Fryer, 1996). Instead, as Langley & Jones (1991) stated, the memory system just waits for a clue to trigger an analogy, so creating a false feeling of sudden insight. Actually we believe with Dean K. Simonton (1991) that "luck only favours the prepared mind"

## Creative Problem Solving (CPS)

Several systems in creative team work are available since Alex Osborn (Osborn, 1953) introduced the brainstorming method to produce ideas. Sidney Parnes and Ruth Noller (Parnes & Noller, 1972), for example, worked on Creative Problem Solving (CPS) - a method that has been subjected to investigation by researchers like Isaksen, Dorval, & Treffinger (2000) and, especially, Min Basadur. Other methods, like *brainwriting* or *635 Method*, *Six Sigma*, *Synectics*, *TRIZ*, *Soft Systems*, De Bono's *Six Thinking Hats,* do not possess the scientific research background that CPS does, and so they were not considered in this project.

From the Creative Problem Solving (CPS) approach, Basadur (1997, 1999, 2000) proposed a new model, the Simplex model. Basadur's Simplex is a cyclic process in three distinct phases and eight steps. In each step there is a moment for active divergence, when individuals or groups produce as many ideas or options they can find, in a supporting climate in which judgment is deferred to allow the perception of new relationships between facts. During the divergence moments everyone must make extended efforts to avoid stopping too early, before all possible options have been produced. During active convergence, the participants will select one or more options to carry on to the next step. One last skill will allow the process to go on systematically through its eight steps and three phases: it's called vertical deferral of judgment. This skill helps the participants to distinguish between unclear situations and well defined problems, and between defining a problem and solving a problem.

### First Phase: Problem Definition

The following steps are involved.

#### Problem Finding

This step consists in identifying problems and opportunities for change or improvement within or outside the organization. In the first moment of active divergence, judgment deferral is required and sustained until the participants feel they cannot collect more relevant problems or changes opportunities. It is then time for active convergence, selecting the problems that will deserve further exploration.

**Fact Finding**

Begins with a divergence moment, when the group defers judgment in order to gather as many information as possible on the selected problem, always accepting all the data that is produced. When there is a perception that all useful or possible facts have been collected, the group can converge and select a few facts that are considered to deserve further expansion.

**Problem Definition**

In this step the group will reformulate the facts selected into creative opportunities or challenges. Then the more promising problem will be selected to carry on to the next step. For Basadur (1997) this is a crucial step and skilled participants will really help the process by asking the right questions that will be answered further on. In this step they elaborate maps reframing the problems using the question *How might we...*, considered the most important question in the Simplex process. Another question will help to deepen the problem: *What is blocking...*, *What is stopping...*, or *Why*. The challenge mapping process helps to see the hierarchy or problems and the relations between them, clarifying the big picture.

## Second Phase: Problem Solving

The following steps are involved.

**Generating Potential Solutions**

This step requires the participants to actively create as many potential solutions as possible to solve the selected problems or challenges. Divergence moment allows creating the most radical and apparently impossible solutions. In the convergence moment, some of them will be selected for evaluation.

**Evaluating Potential Solutions**

Here it is required to generate as many criteria as possible to help evaluating the potential of each solution that has been developed in the previous step. Having established the criteria, participants will evaluate the potential solutions against each criterion and decide which should be implemented.

## Third Phase: Solution Implementation

The following steps are involved.

**Action Planning**

Divergence skills are required to generate a number of specific actions that may help the implementation of solutions generated previously. Then convergence skills will allow selecting the most adequate actions.

**Gaining Acceptance**

This step aims at overcoming resistance to change and involve people needed in the process to assure its feasibility. This is directed essentially to people who did not participate in the earlier steps, but whose commitment is indispensable to bring the project to success.

**Taking Action**

Taking action is not the final step of the model, assumed as a circular process. As Basadur (2000) mentions, the organizational level is a continuous flow of products, services and processes that foster a better interaction with the environment. In this step, participants may find reasons not to fully implement the project, as a result of fear of failure and of resistance to change. To undermine these problems the author adopts Lakein (1973) techniques that advise to start with simple, specific and realistic actions, to address the fear of unknown by analyzing what could happen and then generating ideas to cope with fear of failure, trying to turn it into advantages.

After a series of trials, Basadur's model was reduced to five steps (Figure 2), in order to adapt it to the three 4-hour session design. In the model we considered that the session's objective, defined by management during an interview, was not part of the cycle. The same happened with taking ac-

*Figure 2. Simplified Basadur's creative problem solving method*

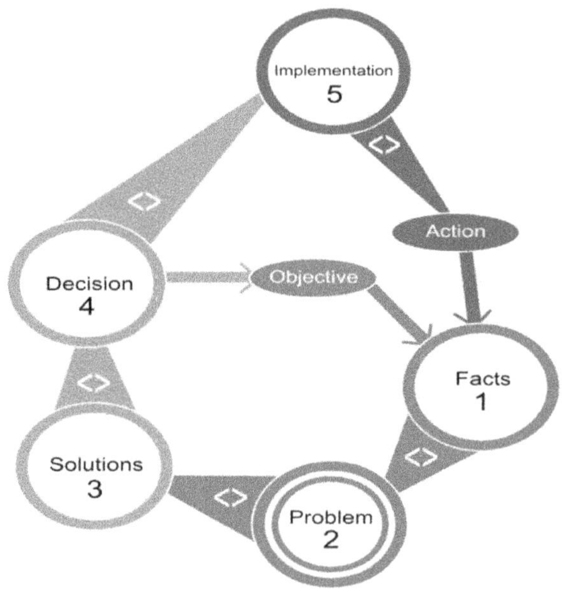


tion, where the innovation project is implemented. The intention is that the implementation process may give rise to other CPS teams.

According to Puccio et al. (2006) research, the impact of CPS in the workplace can take place in three areas: the individual's attitudes; the individual's behaviour and; its effects on groups. For example, in the study run by Basadur & Hausdorf (1996), they concluded that CPS procedures produced changes in behaviour when attitudes towards divergent thinking had been changed into a positive way; also, CPS training improved the fluency in producing solutions to problems. As to groups, CPS training improved work group climate, communication, interpersonal relations and problem solving outcomes. Finally, Puccio et al. (2006) reported several studies, concerned with CPS impact on organizational effectiveness, which revealed aspects like cost reduction, high revenue solutions, or a culture that inspired innovative design concepts.

If successful, the model will allow for the creation of a culture of innovation within the organization, committing more and more of its constituents, as more development projects become profitable innovations (Basadur & Paton, 1993; Isaksen et al., 2000). Also, the outcomes of the change process will establish a different culture in the organization, allowing for a shared thinking process that will facilitate knowledge management and the fit between the organization and its changing environment (Basadur & Gelade, 2006).

Therefore, it is necessary to understand how the success of the system in individuals and teams can help to attain profitable innovations at organizational level. One possibility is by identifying organizational problems which, once solved, may contribute to organizational internal efficiency and to the necessary market adaptation (effectiveness). And so, this research will focus on the development of organizational creativity, using the CPS methodology, aiming at demonstrating its effectiveness in using the individual and team divergent thinking improvement in identifying organizational problems.

## METHOD

The CPS process was conducted in seven small and medium national enterprises of different economic sectors, where one team of five to ten people was designated by the company administrator, involving 48 people, aged 24 to 59 (average age was 42,5).

One group of companies in the agriculture sector, produces sustainable irrigation systems, grows and commercializes vegetables and fruits. The team integrated the directors of each company and workers in the administrative and engineering unit, dealing with integration problems. The seven members presented a diverse educational level from university degree to the basic level.

A second company in the construction field, specialized in the installation of electrical systems, counting 80 workers, concerned with budget calculation, designated a seven elements' team, whose education level varied from basic to university degree.

A construction firm nominated seven co-workers to "anticipate the market fluctuations". The majority had a university degree, one completed college and one had the basic level of education.

Six team members joined in a team from a beverage retail and distributor, counting about 100 workers, to solve the following problem: "how might we improve the salesmen and distributor's competencies". This team presented the same diversity of education level as in the previous ones.

One five star hotel, integrated in a national group of companies, counting an average of 300 employees, was concerned with cost reduction. Most hotel departments and hierarchical levels were represented by six team members whose level of education and responsibility was also diverse.

In a computer company, the team was composed by eight members, all of them having completed a university degree. The problem suggested by the administrator had to do with the company billing process.

The last team worked for a group of service enterprises in mediation. eight members with university degrees, joined to work on "how to make all employees recommend the group's services".

In each company, the administration identified a problem, normally a complex problem they had been struggling with for some time. This was taken as the objective (or a fuzzy situation) and presented to the team engaging in the process. The management was kept informed with the process, intervening at the problem definition and decision stages, and whilst building the action plan.

All the team members were designated by the administration as organizational talents having the necessary knowledge in the field.

The team members participated in three CPS sessions (four hours each): the first session was an introductory one, to explain the CPS process; the second session was dedicated to define the problem and the third to solve it and build an action plan.

As stated above, the CPS process begins with the management's objective, engaging the team in active divergence to find the more relevant facts that will help to define the problem. The average number of facts each team produced was 84. This was an important contribution to help bringing the team members' tacit knowledge into explicit knowledge and magnify the company's understanding of its organizational concerns. All this knowledge was registered so that the problems could be fully analysed.

Also, a 14-item questionnaire, adapted from Basadur, Pringle, Speranzini, & Bacot (2000), addressing the participant's preference for avoiding premature closure and deferring judgement, was administered twice, before and after the three 4-hour creative problem solving sessions. Each item had a 5 point scale (1 *totally disagree* to 5 *totally agree*) and the closer to 5, the closer to divergent think preference (Table 2). The effects of the method (*X*) were tested comparing the gains from $O_1$ (observation before) to $O_2$ (observation after). Both questionnaires were submitted to statistical analysis with SPSS software (version 17), enabling to assess the respondents' attitude evolution.

At the end of the third session, the participants were asked to evaluate the process and write their opinion about it. These responses, together with all facts registered by external observers, were submitted to content analysis, in order to reduce its complexity and aggregate them into a reduced number of categories, thus allowing for a deeper comprehension and to draw perceptual maps, using DTMc40 software. This statistical technique, as Hair, Anderson, Tatham, & Black (1987) stated, allows the dimensional reduction and conducts perceptual mapping by associating sets of attributes.

*Table 1. Mean and test significance for mean differences, in avoiding concentrating in idea quality and deferring judgement, before and after the creative problem solving sessions*

| Moment | *N* | Factor | | | |
|---|---|---|---|---|---|
| | | **Deferring judgement** | | **Avoiding concentrating in idea quality** | |
| | | *M* | *SD* | *M* | *SD* |
| Before Sessions | 48 | 2.74 | .73 | 2.93 | .67 |
| After Sessions | 47 | 3.48 | 1.04 | 3.61 | .75 |
| Sig. | | .00 | | .00 | |

## RESULTS

In presenting the research results the focus was twofold: the first related to the assessing of the CPS effectiveness, at the individual and team levels, by comparing the responses to the questionnaire before and after the process and analyzing their evaluations; and the second aimed at giving an insight of the organizational problems which, once solved, might lead the company to profitable innovations.

The questionnaires filled in were submitted to factor analysis, which extracted two factors, i.e. *deferring judgement* (explaining 31% of the variance), with five items such as *I should do some prejudgement of my ideas before telling them to other (R); we should cut off ideas when they get ridiculous and focus on the other (R)* (Cronbach's Alpha.66) and *avoiding concentrating in idea quality* (explaining 28% of the variance) with another 5 items like, for example: *a group must be focused to produce worthwhile ideas (R); I like to listen to other people's crazy ideas because even the wackiest often lead to the best solutions; I wish people would think about whether or not an idea is practical before they open their mouths (R)* (Cronbach's Alpha.73). Because they did not fit this 2-factor structure, 4 items were left out, thus reducing the 14-item questionnaire to ten items.

A paired sample *t* Test showed significant differences in each factor in both moments of application (Table 1).

As can be seen from the table, both variables show a significant improvement after the creative problem solving sessions, meaning the team members think they should defer judgement and avoid concentrating in idea quality, more than before the sessions. In this case, the difference was enough to bring the average opinion from slight disagreement (below 3) to slight agreement.

The questionnaire submitted to the participants, at the end of the sessions, included an open question asking them to express their opinion about the three session process. Each participant wrote an evaluation of the creative problem solving session and their comments were submitted to a content analysis and categorization, in order to reduce the corpus, and a correspondence analysis was carried out. The perceptual map may be analysed in Figure 3.

As can be seen in the figure, the first two axes organize the participant's perceptions in four quadrants: the horizontal axe opposes *efficiency to innovation* and the vertical axe *individual* versus *collective perspective*. The participants thought the method was necessary and useful at four levels of analysis, i.e. personal, professional, organizational and at the team level, but in different ways. The method was seen as useful at the *organizational* level by fostering *efficiency newness*, and *knowledge*; at the *team* level by promoting *openness to ideas, creativity* and *innovation*; at the *personal* level the participants thought the process could *help* them to *find different solutions*; at a *professional* level the par-

*Figure 3. Perceptual map of the evaluations produced during the CPS sessions*

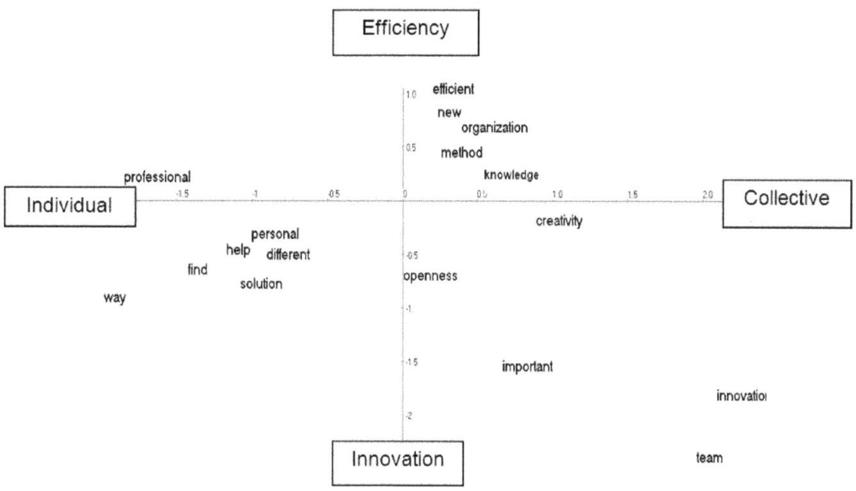


ticipant's perceptions were quite similar to the individual level.

It seems the method was seen as changing the individual, in his personal and professional sphere; as a new method, useful to bring creativity and innovation at team level, and as a means to foster organizational knowledge and efficiency.

## Problem Identification

Each administration stated the objective that was meant to start the team CPS process, All were related to market requirements or client needs, which implied a change in internal procedures in the company. The seven companies produced 586 facts (short 3-6 word sentences), which were submitted to content analysis and categorization in order to reduce the corpus. Correspondence analysis was carried out and the perceptual map is presented in Figure 4.

As we can see, the first two axes organize the facts in four quadrants: the horizontal axis opposes *Flexibility* to *Effectiveness* and the vertical axis the *Internal* versus *External* focus. So, in a more external focus and in a balanced position, between flexibility and effectiveness, a five star hotel (01) closer to *objectives, marketing, image* and *employment* and a construction company (07) closer to *market_demands* and *partnership.* In the quadrant defined by effectiveness and internal focus, two companies: one in the distribution business (02) closer to *motivation, billing* and *clients_problems,* and one in construction (04) closer to *time_problem, administration, client_communication.* Finally, in the quadrant of flexibility and internal focus, three organizations: a computer technology firm (03), one (05) in the agriculture field, and another (06) in construction and real estate mediation, being the former two closer to *trust, communication, incentives, change, structure* and *process,* and the latter, to *identity, quality* and *competition.*

## DISCUSSION

The creative problem solving method has proved to be able to provide effectiveness in changing the individual's attitude towards divergent thinking, namely by avoiding premature closure, acceptance of others ideas and less self-censorship. These data were obtained using Basadur's questionnaire and the results are totally consistent with his findings (Basadur & Hausdorf, 1996; Basadur et al., 2002),

*Figure 4. Perceptual map of the facts produced during the CPS sessions*

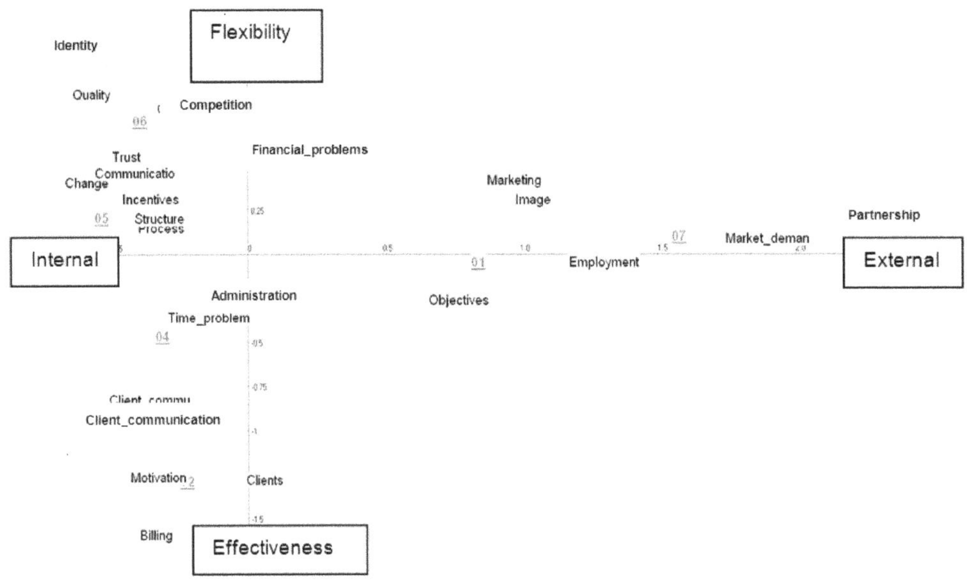


even though we cannot make cultural generalizations, as it was not proven that the same change in attitudes could be achieved with people from other cultures and nationalities. The training sessions proved to be sufficient, at least in the short term, for the assimilation of CPS active divergence and active convergence skills. Further research is needed to understand if this attitude change will last, after project implementation.

The subjects also agreed as to the method's capability in providing a professional, efficient way of organizing knowledge in such a way that can help individuals to find original solutions to problems, and an important instrument to lead teams to creativity and innovation. CPS seems to be an important source of learning and attitude change, as participants also state its efficiency both at work (at the team level and organizational) and in their personal life. As teams involve in project implementation it will be easy to follow up not only their attitudes but their behaviour as well.

The analysis of the knowledge produced by the teams provided rich and useful data to understand the main problems the companies are facing and how they deal with the actual turbulent environment. By providing the identification of facts pertaining to organizational problems, the method allowed for a diagnosis of the main areas of concern in companies and it was possible to build a typology of these areas, matching each company and the main categories of concern and grouping them in the space provided by the axes extracted from correspondence analysis. Even though most problem solving sessions started with external objectives, defined by the administration, the teams provided data revealing different orientations. The major concentration was around internal flexibility, grouping three firms that are fragmented in small companies, thus requiring several improvements in communication. Another quadrant – internal effectiveness – includes two low technology companies that need to organize themselves in order to develop a more effective relationship with the clients. Finally, in the external orientation, two companies that are under serious market threats and need to adapt to new requirements in order to survive.

Even though further evidence is needed, it became clear that innovation mechanisms and priorities differ according with the company or, at least, with specific variables, internal and external, which imply that innovation theoretical models cannot be used as if every company would have to work the same way. In fact, as evidence shows, teams tend to define areas of improvement around internal company efficiency, in areas like management control, communication, motivation, supervision and relationships with the clients. This way, organizational problems, seen from the point of view of the employees, tend to contradict the necessity of innovation or, put in another way, to stress the need for the organization to adapt to market, technological and management changes by providing some internal stability. Constant time and management pressures over the employees do not provide conditions for them to see innovation as a benefit in itself, but their attitude may change if they are given enough credit to balance constraints and initiatives, and to make innovation a collective internal effort. That is why the existence of organizational innovation systems is fundamental.

The companies that took part in this research represent a reality which cannot be ignored and that must be understood in the light of a specific culture, economy and management skills. Even if we try to do our best to bring companies to the standards reflected in our management study books, which are based on the best company examples in the world, the gap between reality and perfection is too deep to be easily reduced. And so, it is by making an extra effort to understand the kind of problems raised by the personnel, and not by trying to make ideal solutions work, regardless of company reality, that researchers and consultants may bring help to improve company innovation, even if this means making the best to prevent external and internal changes from disrupting the ability of the company to innovate from within.

This problem solving model has proved to give useful contributions to organizational innovation, thus contributing to the success of the whole model (Sousa, Monteiro & Pellissier, 2008). As the creative problem solving tools have already demonstrated their usefulness in finding solutions and helping organizations to improve, what remains to be proved is the value of selecting and organizing talented people in an organization, by giving them time, space, knowledge and the opportunity to team up and direct their individual creativity to the organizational problems. The process of developing organizational innovation and creativity is complex and non-linear, with ups and downs, which can only give rise to a culture of innovation with the management's total commitment. Future research will allow for testing of the model, in its wide complexity, and will provide new insights into the process of organizational creativity and innovation.

The use of management control measures, as described by Adams (2006), in order to evaluate the impacts of the innovation projects into the final results of the organization, will provide the necessary frames of reference to evaluate the progress of other organizational variables. First, as Hartel et al. (2003) explain, an increase in employee commitment, as more and more people become trained in CPS procedures and involved in innovation projects. Then, a systematization of explicit knowledge (Borghini, 2005), derived from the team work necessary to carry out CPS decisions. Also, the improvement of formal and informal communication channels (Moss & Ritossa, 2007), due to the involvement of the whole organization in carrying the projects through; and, finally, the movement towards a culture of innovation, through creative leadership level improvement, described by Xu & Rickards (2007), as the practice of project implementation values aspects like delegation, employee empowerment, trust and support to creative work.

The present study should expand and determine the relevant measures of behaviour in each organization, in order to understand long term effects of training and project involvement on team learning. As the methodology will involve more companies, it will be possible to draw control measures that can prove the method's capacity to improve organizational efficiency and sustainability. The projects' follow up will provide useful insights into organizational development. And so further research is needed by bringing in more companies and about what follows solution planning, i.e. project development, in order to analyse what can be done to improve the method's effectiveness in developing innovations.

## CONCLUSION

Research in organizational creativity and innovation is still in its beginning, as people realize the usual approach to innovation is insufficient to acknowledge the complexity of such a process. There is a growing number of researchers who realize the importance of employee involvement in innovation projects in order to foster organizational change and company sustainability. This chapter resumes a first attempt to prove the effectiveness of the creative problem solving approach in changing the employees' attitudes towards creativity skills, and revealed itself to be a good instrument to solve complex problems at individual, group and organizational levels. If three sessions of training were enough to change the participant's views, it is legitimate to anticipate that the involvement in project implementation will contribute to raise people's commitment to organizational goals, so that they may be willing to participate in continuous improvement projects, therefore contributing to build a culture favourable to change and to organizational success. It is also important to remind that the participants in the CPS sessions were very diverse, in age, educational level and background, job experience and department expertise, and that this diversity resulted in better solutions, as it helped to acknowledge different points of view representative of the various organizational sectors and levels. Simultaneously, cohesion may improve as the different colleagues have the opportunity to discuss openly, in a non threatening environment, the different perspectives on organizational reality. We believe the progressive involvement of more and more teams within a company will foster cultural change and increase the well-being among employees.

## REFERENCES

Adams, R. (2006). Innovation measurement: A review. *International Journal of Management Reviews*, *8*(1), 21–47. doi:10.1111/j.1468-2370.2006.00119.x

Amabile, T. M. (1983). *The social psychology of creativity*. New York: Springer-Verlag.

Argyris, C. (1999). *On organizational learning*. Oxford: Blackwell Publishers.

Baer, J. (1997). Gender differences in the effects of anticipated evaluation on creativity. *Creativity Research Journal*, *10*, 25–31. doi:10.1207/s15326934crj1001_3

Basadur, M. (2004). Leading others to think innovatively together: *Creative* leadership. *The Leadership Quarterly*, *15*(1), 103–121. doi:10.1016/j.leaqua.2003.12.007

Basadur, M., & Gelade, G. (2006). The role of knowledge management in the innovation process. *Creativity and Innovation Management*, *15*(1), 45–61. doi:10.1111/j.1467-8691.2006.00368.x

Basadur, M. & Paton, B. (1993). Using creativity to boost profits in recessionary times. *Business Strategies*, Jan-Feb., 14-19.

Basadur, M. S. (1997). Organizational development interventions for enhancing creativity in the workplace. *The Journal of Creative Behavior*, *31*(1), 59–73.

Basadur, M. S. (1999). *Simplex: A flight to creativity*. Buffalo, NY: The Creative Education Foundation.

Basadur, M. S. (2000). *The economic, social and psychological outcomes of implementing a deliberate process of organizational creativity.* Working paper n° 100. McMaster University, Management of Innovation and New Technology Research Center.

Basadur, M. S., & Hausdorf, P. A. (1996). Measuring divergent thinking attitudes related to creative problem solving and innovation management. *Creativity Research Journal, 9*(1), 21–32. doi:10.1207/s15326934crj0901_3

Basadur, M. S., Pringle, P. F., & Kirkland, D. (2002). Crossing cultures: Effects on the divergent thinking of Spanish-speaking South American Managers. *Creativity Research Journal, 14*(3-4), 395–408. doi:10.1207/S15326934CRJ1434_10

Basadur, M. S., Pringle, P. F., Speranzini, G., & Bacot, M. (2000). Collaborative problem solving through creativity in problem definition: Expanding the pie. *Creativity and Innovation Management, 9*(1), 54–76. doi:10.1111/1467-8691.00157

Borghini, S. (2005). Organizational creativity: Breaking equilibrium and order to innovate. *Journal of Knowledge Management, 9*(4), 19–33. doi:10.1108/13673270510610305

Cebon, P., Newton, P., & Noble, P. (1999). *Innovation in firms: Towards a framework for indicator development.* Melbourne Business School Working Paper #99-9, September.

Cohen, W. M., & Levinthal, D. A. (1990). Absorptive capacity: A new perspective on learning and innovation. *Administrative Science Quarterly, 35*, 128–152. doi:10.2307/2393553

Collins, J. C., & Porras, J. I. (1994). *Built to last: Successful habits of visionary companies.* New York: HarperBusiness.

Cooke, P., Heidenreich, M., & Braczyk, H. J. (2004). *Regional innovation systems.* London: Routledge.

Csikszentmihalyi, M. (1994). The domain of creativity. In Feldman, D. H., Csikszentmihalyi, M., & Gardner, H. (Eds.), *Changing the world: A framework for the study of creativity.* London: Praeger.

Dalal, S. (2008). *The innovation boot camp.* Orange, CA: The Institute for Effective Innovation.

De Dreu, C. K. W., & West, M. A. (2001). Minority dissent and team innovation: The importance of participation in decision making. *The Journal of Applied Psychology, 86*(6), 1191–1201. doi:10.1037/0021-9010.86.6.1191

Dowd, E. T. (1989). The self and creativity. Several constructs in search of a theory. In Glover, J. A., Ronning, R. R., & Reynolds, C. R. (Eds.), *Handbook of creativity* (pp. 233–241). New York: Plenum Press.

Edquist, C. (1997). *Systems of innovation: Technologies, institutions and organizations.* London: Printer.

Fryer, M. (1966). Creative teaching and learning. London: Paul Chapman.

Hair, J. F., Anderson, R. E., Tatham, R. L., & Black, W. C. (1987). *Multivariate data analysis* (4th ed.). Englewood Cliffs, NJ: Prentice Hall.

Hartel, J., Schmidt, F., & Keyes, L. (2003). *Well-being in the workplace and its relationship with business outcomes: A review of the Gallup studies* (pp. 205–224). Washington, D.C.: American Psychological Association.

Huhtala, H., & Parzefall, M.-R. (2007). A review of employee well-being and innovativeness: An opportunity for a mutual benefit. *Creativity and Innovation Management, 16*(3), 299–306. doi:10.1111/j.1467-8691.2007.00442.x

Isaksen, S., Dorval, K., & Treffinger, D. (2000). *Creative approaches to problem solving: A framework for change.* Buffalo, NY: The Creative Problem Solving Group.

Jones, E. (1997). The case against objectifying art. *Creativity Research Journal, 10*(2/3), 207–214. doi:10.1207/s15326934crj1002&3_9

Kanter, R. (1983). *The change masters*. New York: Simon & Schuster.

Kasof, J. (1995). Social determinants of creativity: Status expectations and the evaluation of original products. *Advances in Group Processes, 12*, 167–202.

Kaufman, G. (1993). The logical structure of creativity concepts: A conceptual argument for creativity as a coherent discipline. In Isaksen, S. G., Murdock, M., Firestein, R., & Treffinger, D. (Eds.), *Understanding and recognizing creativity: The emergence of a discipline, (141 - 158).* Norwood, NJ: Ablex Publishing Corporation.

Kokot, S. J., & Colman, J. (1997). The creative mode of being. *The Journal of Creative Behavior, 31*(3), 212–226.

Lakein, A. (1973). *How to get control of your time and your life*. New York: Peter W. Wyden, Inc.

Langley, P. & Jones R. (1991). Computational model of scientific insight. In R.S. Sternberg (Ed.), The nature of creativity. Contemporary psychological perspectives. Cambridge, UK: Cambridge University Press.

Ludwig, A. M. (1995). What explaining creativity doesn't explain. *Creativity Research Journal, 8*(4), 413–416. doi:10.1207/s15326934crj0804_8

Lundvall, B.A. (1992). National systems of innovation: Towards a theory of innovation and interactive learning. London: Printer.

Mace, M. (1997). Toward an understanding of creativity through a qualitative appraisal of contemporary art making. *Creativity Research Journal, 10*(2/3), 265–278. doi:10.1207/s15326934crj1002&3_15

McAdam, R., & McClelland, J. (2002). Sources of new product ideas and creativity practices in the UK textile industry. *Technovation, 22*, 113–121. doi:10.1016/S0166-4972(01)00002-5

McLean, L.D. (2005). Organizational culture's influence on creativity and innovation: A review of the literature and implications for human resource development. Advances in Developing Human Resources, 7, 226-246.

Moss, S. & Ritossa, D. (2007). The impact of goal orientation on the association between leadership style and follower performance, creativity and work attitudes. Leadership, 3(4), 433-456.

Mumford, M. D., & Gustafson, S. B. (1988). Creativity syndrome: Integration, application, and innovation. *Psychological Bulletin, 103*, 27–43. doi:10.1037/0033-2909.103.1.27

Nemeth, C. J., Personnaz, B., Personnaz, M., & Goncalo, J. A. (2004). The liberating role of conflict in group creativity: A study in two countries. *European Journal of Social Psychology, 34*, 365–374. doi:10.1002/ejsp.210

Osborn, A. F. (1953). *Applied imagination: Principles and procedures of creative problem-solving*. New York: Scribner's Sons.

Parnes, S. J., & Noller, R. B. (1972). Applied creativity: The creative studies project: Part 1–the development. *The Journal of Creative Behavior, 6*, 11–22.

Puccio, G.J., Firestien, R.L., Coyle, C. & Masucci, C. (2006). A review of the effectiveness of CPS training: A focus on workplace issues. Creativity and Innovation Management, 15(1), 19-33.

Stein, M.I. (1984). Making the point. Amagansett, NY: Mews Press.

Rogers, E. M. (2003). *Diffusion of innovations* (5th ed.). New York: Free Press.

Sawyer, R. K. (1998). The interdisciplinary study of creativity in performance. *Creativity Research Journal*, *11*(1), 11–19. doi:10.1207/s15326934crj1101_2

Schein, E. (1992). *Organizational culture and leadership*. San Francisco: Jossey-Bass.

Simonton, D. (1995). Exceptional personal influence: An integrated paradigm. *Creativity Research Journal*, *8*(4), 371–376. doi:10.1207/s15326934crj0804_3

Simonton, D. K. (1991). Creativity, leadership and chance. In Sterneberg, R. S. (Ed.), *The nature of creativity. Contemporary psychological perspectives*. Cambridge, NY: Cambridge University Press.

Sousa, F. (2008). Still the elusive definition of creativity. *International Journal of Psychology: A Biopsychosocial Approach, 2*, 55-82

Sousa, F., Monteiro, I & Pellissier, R. (2008). Creativity and problem solving in the development of organizational innovation. *ERIMA 08 Proceedings*, 5-11.

Spence, W. R. (1994). *Innovation: The communication of change in ideas, practices and products*. London: Chapman & Hall.

Stacey, R., & Griffin, D. (2005). Introduction, leading in a complex world. In Griffin, D., & Stacey, R. (Eds.), *Complexity and the experience of leading organizations* (pp. 1–16). London: Routledge.

Stein, M. I. (1953). *Stimulating creativity: Group procedures* (*Vol. I*). Amagansett, NY: Mews Press.

Stein, M. I. (1993*). Moral issues facing intermediaries between creators and the public*. Unpublished paper.

Stein, M. I. (1994). *Stimulating creativity* (*Vol. II*). New York: The Mews Press, Ltd.

Strambach, S. (2002). Change in the innovation process: New knowledge production and competitive cities-the case of Stuttgart. *European Planning Studies*, *10*(2), 215–231. doi:10.1080/09654310120114508

Unsworth, K. L. (2005). Creative requirement: A neglected construct in the study of employee creativity? *Group & Organization Management*, *30*, 541–560. doi:10.1177/1059601104267607

Wallach, M. A., & Kogan, M. (1965). *Modes of thinking in young children*. New York: Holt, Rinehart & Wiston.

Weisberg, R. W. (1999). Creativity and knowledge: A challenge to theories. In Sternberg, R. J. (Ed.), *Handbook of creativity* (pp. 227–250). Cambridge: Cambridge University Press.

West, M. A. (1990). The social psychology of innovation in groups. In West, M. A., & Farr, J. L. (Eds.), *Innovation and creativity at work, psychological and organizational strategies* (pp. 309–333). Chichester, UK: John Wiley & Sons.

West, M. A., & Farr, J. L. (1990). Innovation at work. In M.A. West and J.L. Farr (Eds.), *Innovation and creativity at work: Psychological and organizational strategies,* (3-13). Chichester, UK: John Wiley.

Wolfe, R. A. (1994). Organizational innovation: Review, critique and suggested research directions. *Journal of Management Studies*, *31*, 405–431. doi:10.1111/j.1467-6486.1994.tb00624.x

Woodman, R. W., & Schoenfeldt, T. (1999). Individual differences in creativity: An interactionist perspective. In Glover, J. A., Ronning, R. R., & Reynolds, C. R. (Eds.), *Handbook of creativity* (pp. 77–93). New York: Plenum Press.

Xu, F., & Rickards, T. (2007). Creative management: A predicted development from research into creativity and management. *Creativity and Innovation Management*, *16*(3), 216–228. doi:10.1111/j.1467-8691.2007.00445.x

## ADDITIONAL READING

Burns, T., & Stalker, G. M. (1996). *The management of innovation*. New York: Oxford University Press.

Csikszentmihalyi, M. (1996). *Creativity: Flow and the psychology of discovery and invention*. New York: HarperCollins.

Glover, J. A., Ronning, R. R., & Reynolds, C. R. (Eds.). (1989). *Handbook of Creativity* (pp. 3–33). New York: Plenum Press.

Kahn, A., Castellieon, K. B., & Griffin, A. (Eds.). (2005). *The PDMA handbook of new product development* (2nd ed.). Hoboken, NJ: John Wiley & Sons.

Puccio, G. J., Murdock, M. C., & Mance, M. (2007). *Creative leadership: Skills that drive change*. New York: Sage.

Robinson, A. (1998). *Corporate creativity: How innovation & improvement actually happen*. San Francisco: Berret Koehler Publishers.

Sawyer, K. (2007). *Group genius: The creative power of collaboration*. New York: Basic Books.

Shapiro, S. M. (2001). *24/7 Innovation: A blueprint for surviving and thriving in an age of change*. New York: McGraw-Hill.

Thompson, L., & Choi, H. (Eds.). (2006). *Creativity and innovation in organizational teams*. New York: Lawrence Erlbaum Associates.

VanGundy, A. B. (1987). *Creative problem solving*. New York: Quorum Books.

West, M. A., & Farr, J. L. (Eds.). (1990). *Innovation and creativity at work: Psychological and organizational strategies*. Chichester: Wiley.

Zhou, J., & Shalley, C. (Eds.). (2008). *Handbook of organizational creativity*. New York: Lawrence Erlbaum Associates.

## KEY TERMS AND DEFINITIONS

**Corporate Innovation:** A system devoted to enhance creativity in organizations.

**Creativity:** Individual cognitive and emotional processes that take place between the individual and the product that is created.

**Creative Workforce Potential:** Is both the ability to retain creative managers and employees, and to provide an environment where each one will feels free and willing to contribute to organizational success.

**Creative Problem Solving:** Structured method for defining problems, generating solutions and developing an action plan.

**Creative Workforce Potential:** Is both the ability to retain creative managers and employees and to provide an environment where each one will feels free and willing to contribute to organizational success.

**Divergent Thinking:** A thought-analogy process or method used to generate creative ideas by exploring many possible solutions.

**Innovation:** Processes of implementation of creations, in the domains of production, adoption, implementation, diffusion, or commercialisation.

**Innovativeness:** The potential of the workforce to promote changes to benefit of the organization.

**Organizational Creativity:** A system devoted to enhance creativity in organizations.

**Organizational Innovation:** A system devoted to enhance creativity in organizations.

## APPENDIX

*Table 2. The Basadur's questionnaire administrated before and after the eight-hour CPS sessions*

| QUESTIONAIRE | | | | | |
|---|---|---|---|---|---|
| **Name** | | | | | |
| **Company** | | | | | |
| Read each statement and indicate the extent to which you agree or disagree with the statements listed below, using the following scale: Strongly Disagree (SD); Disagree (D); Neither Disagree, Nor Agree (NDNA); Agree (A); Strongly Agree (SA). | | | | | |
| | **SD** | **D** | **NDNA** | **A** | **SA** |
| I should do some prejudgment of my ideas before telling them to others | | | | | |
| We should cut off ideas when they get ridiculous and get on with it | | | | | |
| I think that people at work ought to be encouraged to share all their ideas, because you never know when a crazy-sounding one might turn out to be the best | | | | | |
| One new idea is worth 10 old ones | | | | | |
| Quality is a lot more important than quantity in generating ideas | | | | | |
| A group must be focused and on track to produce worthwhile ideas | | | | | |
| Lots of time can be wasted on wild ideas | | | | | |
| I think everyone should say whatever pops into their head whenever possible | | | | | |
| I like to listen to other people's crazy ideas because even the wackiest leads to the best solution | | | | | |
| Judgment is necessary during idea generation to ensure that only quality ideas are developed | | | | | |
| You need to be able to recognize and eliminate wild ideas during idea generation | | | | | |
| I think all ideas should be given equal time and listened to with an open mind, regardless of how zany they seem to be | | | | | |
| The best way to generate new ideas is to listen to others, then tailgate or add on | | | | | |
| I wish people would think about whether or not an idea is practical before they open their mouths | | | | | |

# Chapter 8
# An Interdisciplinary Workshop for Business-Idea Generation

**Astrid Lange**
*Brandenburg University of Technology Cottbus, Germany*

## ABSTRACT

*This chapter describes a workshop concept for small groups that aims at the qualification for creative business-idea generation in interdisciplinary teams of academics. The chapter's aim is to provide a theory-based and application-oriented description of the workshop, including a first-hand report on implementation. The chapter will start with a description of theoretical contributions and underlying research results to illustrate the workshop's framework. Then a description of the aims, methods and target group follows, as well as the organizational settings and training methods. Then the chapter will focus on the realization of the workshop. Finally, the workshop concept's scope of application is discussed and summarized.*

## INTRODUCTION

The workshop was developed in the context of the EXIST-III project at the Brandenburg University of Technology Cottbus (BTU): "Development of a Workshop for Entrepreneurial and Team Competence for Diverse Disciplines and Start-up Teams". This project[1], bringing together economically and technically educated academics to improve business creation processes, is publicly funded by the German Federal Ministry of Economics and Technology as well as the European Social Fund (ESF). Generally, the EXIST-III initiatives pursue objectives as the durable establishment of entrepreneurial culture and the purposive advancement of the potentials of entrepreneurial ideas and personalities at universities (Kulicke, Stahlecker, Lo, & Wolf, 2006). At the Chair of

DOI: 10.4018/978-1-60960-519-3.ch008

Organization, Human Resource Management and General Management, three sub-projects are managed: The so-called "Gründervilla" (counselling, providing infrastructure to entrepreneurs), the so-called "Ideensauger" (detecting, collecting and developing ideas within the BTU), and extra-curricular entrepreneurship education. The workshop presented here is part of the last sub-project.

At the beginning a redefinition of the task of extra-curricular entrepreneurship education and the selection of course-topics were necessary. Then the workshop had to be designed, aiming at a particular target group to practice a selective scope of techniques. The chapter will follow this logic. The main section will illustrate the workshop itself and some of the experiences, providing information for further applications of the workshop.

## BACKGROUND AND CONTEXT OF THE WORKSHOP

The following paragraphs start with some comprehensive deliberations regarding the understanding of entrepreneurship and entrepreneurship education as well as the empirically based derivation of creativity as a topic for extra-curricular entrepreneurship education. Then, we will present more specific deliberations regarding creativity and creative idea generation.

### Defining Entrepreneurship and Entrepreneurship Education

In general, there are various definitions of entrepreneurship without global consensus (e.g., Greene, Katz, & Johannisson, 2004). The European Commission (2004) states that, "[i]n a broad sense, entrepreneurship should be considered as a general attitude that can be usefully applied in all working activities and in everyday life" (p. 10, emphasis omitted). Jones and English (2004), arguing along similar lines, emphasize the distinction between entrepreneurship and business management, with the first containing innovation and creativity (p. 417). The understanding of entrepreneurship in the context of our project work is rather broad too, and refers to Bygrave and Hofer (1991) as well as to Shane and Venkataraman (2000), who sketch the field of entrepreneurship focusing on the detection, assessment, development, and utilization of opportunities by individuals (p. 218).

Entrepreneurship education refers to the education of aspects related to these definitions. The European Commission (2004) directly refers to creativity when it points out that entrepreneurship education has to promote "the development of personal qualities that are relevant to entrepreneurship, such as creativity, spirit of initiative, risk-taking and responsibility" (p. 12, emphasis omitted). In general, "[e]ntrepreneurship education can be viewed broadly in terms of the skills that can be taught and the characteristics that can be engendered in individuals that will enable them to develop new and innovative plans" (Jones & English, 2004, p. 417).

While entrepreneurship becomes increasingly important in particular for universities (e.g., increased competition between universities, increased necessity for knowledge transfer), it still represents a very diverse and non-standardized field (see e.g., Henry, Hill, & Leitch, 2005b; Jones & English, 2004). With reference to entrepreneurship education at German universities, a Kienbaum study (Federal Ministry of Education and Science, 2005, p. 34 ff.) suggests that the offer of broad- ranging entrepreneurship education represents rather the exception than the rule. By structuring the field of entrepreneurship education along the two dimensions "content focus" (i.e., "motivation and awareness rising", "impartment of basic knowledge", and "domain-specific entrepreneurship education") and "method focus" (i.e., "teaching" and "training"), they found a clear dominance of teaching the basic knowledge. They criticize that most offers within universities address only those participants who are already intent on starting their own business and fail to integrate offers aimed at the education of more transferable competencies, which may be important for employment too (p. 37; cf. also Henry et al., 2005a; Jones & English, 2004). This critique addresses the idea that entrepreneurship represents a generally valued cluster of competencies. To quote Henry et al. (2005a): "it is apparent that [...] there will be a greater need for people to have entrepreneurial skills and abilities to enable them to deal with life's current challenges and an uncertain future" (p. 101). With this broad understanding in mind, we went on to redefine the task of extra-curricular entrepreneurship education by determining the contents to be offered.

## Defining the Scope of Extra-Curricular Entrepreneurship Education

The decision to select particular topics for extra-curricular courses depended mainly on our own research results as well as on some theoretical considerations. We apply an approach of specialization, based on deliberations about entrepreneurship as a field of expertise, which indicates the preference for specialized sub-areas of training as opposed to global entrepreneurship courses (Lange, 2009a). This perspective broadens the scope of target groups of entrepreneurship-related courses, as it implies that not everybody involved in the entrepreneurial process has to establish an own business.

Such an approach may result in a kind of typology of entrepreneurial people, depending on fields of entrepreneurship-related expertise. One approach of typing derives from Sim, Griffin, Price, and Vojak (2007) who deduce their categorization from the stages of new product development. They differentiate the inventor, who is settled in the stage of research and development, from the so-called champion, who is tied to the stage of opportunity recognition, and the implementer, who is tied to the stage of project execution. As a fourth type they propose a generalist, who incorporates the skills, attributes and motives of all types figured out. Lentzy and Weineck (2009) underlined the existence of these four entrepreneurial types within a sample of professors and scientific staff at the BTU Cottbus. In concordance with this specialization approach Faltin (2001) argues to "apply the principle of division of labor" (p. 127) to entrepreneurship and thus differentiate innovative entrepreneurs from business administrators. In the context of creative idea generation, Mumford (2000) argues in the same vein, stating that "some people may make a greater contribution to earlier phases of creative problem solving efforts where new ideas must be generated, while other people may be more effective in later phases, where new ideas [...] must be implemented" (p. 317, see also Cooper, 2005).

These considerations justify offering extra-curricular courses for specialized, entrepreneurship-related skills rather than courses for the enhancement of allrounders. Furthermore, the specification of the particular topics to be offered is based on the results of another entrepreneurship-related project[2] at the BTU Cottbus. Applying the Theory of Planned Behaviour (Ajzen, 1991), Attribution Theory (e.g., Hewstone, 1989), and Social Cognitive Theory of Bandura (e.g., 2001), the subjective beliefs that students associate with entrepreneurship were investigated. To examine the particular beliefs students in Germany associate with entrepreneurship, an elicitation study was conducted (Lange, 2009b) and subsequent quantitative studies examined the value of the resulting beliefs within samples of students with differing entrepreneurial intentions (Lange, 2010). Without presenting the full range of results, the elicitation study indicates that creativity and creative ideas are part of control beliefs in terms of Ajzens' theory. The categories "creativity, idea variety, and unconventional thinking" and "having convincing, innovative-creative entrepreneurial ideas" were mentioned as entrepreneurship-enabling factors; the category "lack of entrepreneurial ideas and deficient innovation" was mentioned as an entrepreneurship-impeding factor. The quantitative study revealed that students with low entrepreneurial intentions are, amongst others, characterized by beliefs about a high value of being creative and having ideas but, at the same time, they expect themselves to be less creative and have less innovative ideas than high intentional students.

As a result, course topics were selected, with creative idea generation being one of them. A sound conceptualization of this workshop was necessary, because "[s]ystematic idea development and refinement are rarely ever found in the syllabus of entrepreneurship education" (Faltin, 2001, p. 135). The next section presents some creativity related considerations, followed by a description of the workshop itself.

## Defining Creativity and its Role for Entrepreneurship

Various definitions of creativity exist (e.g., Amabile, 1983; Garfield, Taylor, Dennis, & Satzinger, 2001; Runco, 2004; Sternberg, 2006a; Urban, 2002). According to Barron and Harrington (1981, p. 441), definitions of creativity vary in what they try to capture: creative achievements, creative abilities, or creative attitudes. These approaches represent different perspectives and induce different implications.

The general definition of creativity "as the goal-directed production of novelty" (Weisberg, 2006, p. 761; cf. also Sternberg, 2006a) is related to creative achievement. To categorize an achievement as creative, several criteria were mentioned. These included novelty, originality, usefulness, and social acceptance (e.g., Preiser, 2006; Runco, 2004; Urban, 2002). There is neither unity about the appropriateness of criteria nor about the concrete meaning of several criteria. For instance, Weisberg (2006) differentiates between the creative but nonvalued innovation, and creative, valued innovation, whereby innovation refers to the creative product (p. 763). Dasgupta (1996) differentiates between psychological creativity and historical creativity, with the former relating to the intra-individual newness and the latter relating to historical novelty. Somewhat between considerations about creative products and creative abilities is the statement by Ericsson (1996), that creative products are the manifestation of the highest level of expertise.

Understandings related to creative abilities can be found within the structure-of-intellect model of Guilford (e.g., 1988) or within Sternberg's Theory of Successful Intelligence (1998; 2006b; Sternberg & Kaufman, 1998). Creative ability is one of the talent factors within the Munich Model of Giftedness (Heller, 2004). In contrast, creative attitudes refer to creativity as a style of thinking in general, "as an attitude toward life, a way to come up to grips with the problems of existence" (Smith & Carlsson, 2006, p. 202). These explanations emphasize that creativity is a multi-faceted construct (e.g., Isaksen, Puccio, & Treffinger, 1993).

The importance of creativity for society in general and specifically for entrepreneurship as well as entrepreneurship education was emphasized by a broad range of researchers (e.g., Cooper, 2005; European Commission, 2004; Jones & English, 2004; Kirby, 2004; Runco, 2004; Zampetakis & Moustakis, 2006). As we focus on creative idea generation and, thus, centre creative achievement, the question arises if and how creativity or the generation of creative ideas can be developed and increased. Isaksen et al. (1993, p. 158) and Preiser (2006, p. 168) clearly affirm this question.

One has to differentiate between developing creativity within people and enhancing creative idea production. The first is rather person-centred and long-term oriented and is, for instance, related to expertise research (e.g., Weisberg, 2006). The second focuses on the act of creation itself. As the workshop is about idea generation, the second approach, including creativity methods (e.g., Higgins, 1996) and the process of idea generation (e.g., Griffiths-Hemans & Grover, 2006), becomes more important. Various authors have described the creative process of idea generation. A prominent exemplary stage model comes from Wallace (1926; cit. in Barrett, 1978; Dasgupta, 1996; Mouchiroud & Lubart, 2006). As the stages of preparation, incubation, illumination, and verification are not globally accepted (e.g., Dasgupta, 1996; Heilman, Nadeau, & Beversdorf, 2003), other conceptualizations exist (e.g., the six-stage model of Baer & Kaufman, 2006; and the two-tier model of Runco & Chand, 1995). We decided to use a process model of Scherer (2007) because of its practical applicability for the workshop. Scherer as well as Runco and Chand (1995) remark on the importance of bearing in mind that stage models are useful but the distinctiveness of stages and static frequency are rather unrealistic, as, for instance, feedback loops occur. We will return to some of the topics later in this chapter.

To summarize the previous paragraphs, theoretical and empirical considerations substantiated the decision to invest in a workshop on creative idea generation, addressing a broader range of participants than usually integrated in entrepreneurship education.

## THE WORKSHOP APPROACH

After deciding the overall topic for the workshop, a specification of the detailed objectives and target group became necessary to deduce the particular contents and appropriate teaching methods.

### Aims and Target Group

The workshop pursued multiple objectives. Some general objectives resulted from the task description within the EXIST-III project at BTU Cottbus. Other specific objectives were deduced from the empirical description of the pertaining state at BTU Cottbus as well as from deliberations of several authors regarding the increase of creative idea generation. Thus, the aims of the workshop were specified as follows:

- A first aim is developing competence that goes beyond business creation. Participants should be generally strengthening idea generation competencies.
- This implies the metacognitive reflection about creativity as one important part of the workshop to enhance self-regulation of participants for the case of creative idea generation (cf. Fasko, 2000-2001; Preiser, 2006). For this purpose the workshop should address an understanding of creativity in terms of Dasgupta's psychological creativity (1996) and in terms of a learnable competence. Furthermore, the workshop should illustrate the full process of creative idea generation and provide opportunities to train some particular creativity techniques.
- The contextualization of the idea generation process in entrepreneurship should foster increased awareness regarding entrepreneurship. It was not an aim to convince the participants to become entrepreneurs. Instead, a conscious examination of the theme of entrepreneurship should be induced.
- Encouraging supportive beliefs about the role and appearance of creativity. According to the above-mentioned research results regarding subjective beliefs (e.g., Lange, 2010), building potential entrepreneurial competencies would start with enhancing appropriate feasibility beliefs that seem to be based, amongst others, on beliefs about the value of being creative and having ideas as well as beliefs regarding the appearance of own ideas and own creativity. Thus, both kinds of beliefs should be addressed.

The target group was partly predefined, resulting from the project description. The subtask was defined as the preparation and execution of extracurricular workshops for interdisciplinary students and academic staff. The advantage of workshops for interdisciplinary participants may be a broader range of ideas because multiple perspectives are available (Faltin, 2001; Mumford, 2000; Woodman, Sawyer, & Griffin, 1993). The challenge may be the differences in language and thinking.

Based on considerations in the field of instructional design (e.g., Reigeluth, 1996) and cognitive training approaches (e.g., Silber, 1998), before starting the content building of training, a sound knowledge of potential participants, their needs and attitudes, as well as potential obstructions should be reached (see also Salas & Cannon-Bowers, 2001). Previous experiences with the workshop approach suggest that the target groups value soft-skills training, but suffer from time restrictions. As one of the objectives was increased awareness regarding entrepreneurship, the target group was not limited to those who were willing to start an own business. Students with low entrepreneurial intentions show specific profiles of beliefs regarding entrepreneurship, including beliefs about the importance and appearance of creative skills and idea generation competencies (Lange, 2010). An as yet unpublished qualitative examination of academic staff resulted in comparable results regarding the perceived role of creativity. Depending on the objectives and existing state of the target group, the main contents for the workshop were derived.

## Content Modules of the Workshop

Three broad modules comprise the workshop, which is oriented along the process of creative idea generation and exemplified for the case of entrepreneurial ideas. The three main-modules were called (1) Defining the starting point, (2) Idea generation process and techniques, and (3) From ideas to business conceptions. As mentioned above, the basic model used for the creative idea generation process is the stage model of Scherer (2007). This model incorporates some of the basic concepts of other process models in an application-oriented manner. Scherer suggests four stages (definition, opening, identification and transformation) and illustrates stage-related creativity techniques in the context of new product development within organizations. Applied models, terms and considerations should be easily understandable for participants from any field of study. This is true for the model of Scherer. Because of its vividness, this model seemed to be appropriate for this practical oriented workshop. The comprehensive deliberations of various other authors supplement the basis of this workshop.

### Module (1): Defining the Starting Point

The first main module is equivalent to Scherer's (2007) first stage of definition as well as to the stage of problem finding according to the model of Runco and Chand (1995). A sound definition of the problem to solve cannot be overestimated, as "[c]reative problems tend to be ill-defined or poorly structured" (Mumford, 2000, p. 315). The definition has the purpose of understanding the topic and should result in a detailed formulation of the objectives of creative processes (Scherer, 2007). A reflective approach to the specifications of this module seemed elementary, as decisions made here are likely to affect the success of the whole workshop. The main point is that the definition has to relate to the participants. For instance, Runco and Chand (1995) emphasize that self-discovered problems may be advantageous in terms of intrinsic motivation and in terms of the results of creative thinking (see also Amabile, 1983; Runco & Okuda, 1988). Problem definition or construction seems to thrive more when a connection to broader life experiences exists (Mumford, 2000). Faltin (2001), who argues in the same manner, remarks that this view "challenge[s] the common view [of business administration] that first of all we have to study market needs" (p. 130). Expertise research indicates advantages of broad and elaborated problem-related knowledge for creative idea production (e.g., Heilman et al., 2003; Weisberg, 2006).

Thus, problems have to be meaningful for participants and should relate to their individual backgrounds. As the workshop should address interdisciplinary participants, the discipline-related experiences seemed to be an adequate starting point. Because of time constraints, a rough pre-formulation of the problem was provided for participants, but this pre-formulation enabled the specification of individually meaningful problem formulations. The pre-formulated problem statement was that participants should generate an entrepreneurial idea that would require the particular mix of domain knowledge of workshop participants. Thus, participants were asked to refine the problem statement, referring to their specific working experiences, fields of expertise or interests. For this purpose, some techniques are utilized to evoke team-specific connections to the rough topic. Techniques applied here should foster both divergent and convergent thinking (Baer & Kaufman, 2006). To enhance meaningfulness, the topic "regional future branches" and its connection to the domains of participants was included. This, as well as the authentic context of generating ideas for start-ups should provide individual meaningfulness (cf., Silber, 1998). At the end of this module a problem statement should be fixed which refers to participants' particular experiences and interests.

## Module (2): Idea Generation Process and Techniques

The second main module consists of two submodules, "metacognitive reflection" and "application of creativity techniques". Fasko (2000-2001), referring to Davis (1991), emphasizes that an "increased understanding of creativity would increase creativity consciousness, demystify creativity, and increase creative ideas and products" (p. 318). Runco and Chand (1995) also reflect on the significant role of metacognition and remark that procedural knowledge, i.e., know-how, can enhance creative idea production via explicit instruction. This sub-module contains reflections of what participants know and believe about creativity, as, on the one hand, beliefs often hinder creative thinking (Anderson, 1993) and, on the other hand, these beliefs represent anchors with which workshop contents can be connected. Furthermore, considerations about the definition of creativity, the advantages of and barriers for being creative, remarks about the creative process of idea generation, creative products, and creativity techniques were included. Creativity was illustrated as a competence that can be trained and enhanced with the utilization of creativity techniques. Without referring to alternative classifications (see e.g., Fasko, 2000-2001; McFadzean, 1998), a rough classification of creativity techniques in terms of intuitive and systematic techniques (Garfield et al., 2001) was presented within the workshop. This classification was used because of its simplicity and usefulness when working with interdisciplinary participants, who probably prefer heterogeneous styles of thinking (see McFadzean, 1998). It was assumed that the presentation of creativity techniques as systematic as well as intuitive would increase subjective expectations of one's own abilities to produce something creative. Within the workshop, an illustration as well as a self-reflection regarding the entrepreneurial types mentioned above were presented (see Sim et al., 2007; Lentzy & Weineck, 2009), highlighting that it is not necessarily one person who realizes the full process.

The second sub-module contains the application of diverse creativity techniques, based on the idea that creative thinking involves both divergent as well as convergent thinking (e.g., Baer & Kaufman, 2006; Barrett, 1978; Fasko, 2000-2001; Howard-Jones, 2002; McFadzean, 1998; Woodman et al., 1993) and that "creative ideas are most likely to arise through the use of diverse concepts, multiple features, and multiple strategies" (Mumford, 2000, p. 316). Introducing a broad range of techniques may be advantageous because everybody can find individual fitting techniques. But as workshop time is restricted, we mentioned a broad range of techniques, but practiced only a small range of them comprehensively. Practiced techniques are, amongst others, Mind Mapping, Morphological Box, Combination matrices, Brainstorming, and Brainwriting 635. Metaplan-card techniques were continuously applied. In terms of McFadzean (1998), the workshop preferred to use paradigm preserving techniques rather than paradigm breaking techniques. The reason for this selection is because of the ease with which they can be learned and practiced (cf. McFadzean, 1998, p. 135). The challenge was to adaptively employ the techniques, depending on the progress of the workshop participants.

### Module (3): From Ideas to Business Concepts

The third module consists of the stages of identification and comments on transformation. When the pool of ideas seems adequate, evaluation becomes important. Scherer (2007) differentiates the steps of selection, revision, assessment and documentation. Techniques utilized for selection were idea reviews, idea clustering and prioritization, depending on team interests. Exercises were used to revise business ideas, focusing on market and branch investigation, product/service specification, the unique selling proposition, customer needs, and customer benefits. When adequate, some techniques practiced previously should be applied again, emphasizing their diverse applicability. To assess the business ideas, applied criteria should relate to real-world states and opportunities of transformation of business ideas. Within the workshop, references to the knowledge, skills and expertise fields of the participants were mentioned. This led to the idea of entrepreneurial teams and competence allocation and is related to the type approach presented previously. Because the workshop aimed at the practice of being creative and not to the sound generation of own entrepreneurial ideas, the stage of transformation was addressed only through one particular exercise (idea presentation and persuasion speech).

The third sub-module also contains information about the entrepreneurial process as realized at BTU Cottbus. The process of creative idea generation is put into proper order within this process. The description of entrepreneurial process is aligned to practical considerations and illustrates adequate, i.e. phase-specific, supportive institutions too. Not mentioned yet, the workshop of course consists of some supplementary sub-modules, which contain formal workshop contents such as the get-to-know-each-other and evaluation procedure. Description of the evaluation procedure and the trainer team follow in the next section.

## Organization of the Workshop

Organization depends on the objectives and the target group mentioned above, as well as on available resources. One critical resource is time. Planning the timetable for this kind of extra-curricular workshop seems somewhat ambivalent, because, on the one hand, students and staff have significant time restrictions but, on the other hand, behavioural training that goes beyond pure knowledge arrangement requires some time to practice and reflect procedures (cf. e.g., Edmonds, Branch, & Mukherjee, 1994; Gagné & Dick, 1983; Goldberg, 1980; Koper & Bennett, 2008). The idea generation process itself is fairly time consuming (e.g., McFadzean, 1998; Mumford, 2000; Runco & Chand, 1995). To ensure the participation without opportunity costs that were too great, the duration was fixed at three days. With eight hours a day, the participants worked together for a total of 24 hours. To procure some more time to think and rethink the issues, the three days were not consecutive; there was a gap of two days between the second and third day. The two days were used for rethinking the ideas. For this purpose, participants received some homework. The modules described above were done over three days, but a flexible approach to time allocation was necessary, as this workshop contained some uncertainties in terms of results and processes. Five hours of the first day were assigned to module (1). The last hour of the second day and the whole third day, as well as the homework, were assigned to module (3). The remaining time was assigned to module (2). Also, for the purpose of minimizing opportunity costs, the workshop was offered in the 2009 semester break.

The number of participants for the workshop was rather small, with a maximum of twelve. This is substantiated with learner centralization within the workshop approach. The workshop took place in one of the university's conference rooms. This location was ideal for the workshops' demands, because it is well equipped with techniques and workshop materials (flipchart, Metaplan-board, and beamer). We brought materials such as Metaplan cards, pins, paper, and a laptop with us. Programs used were Microsoft® Power Point®, Microsoft® Excel®, and Mindjet® MindManager® Pro 7.

Multiple information channels were used to search for participants. Registration stopped two weeks before the workshop started. In front of the workshop, participants received a document with information about the agenda, organizational issues, and some short questions. They were asked to send the completed document back up to three days before the workshop started. Because of practical restrictions only basic questions that were important for the workshop curriculum itself were included, as participants' expectations, their preferred career path (entrepreneurship, science, employment), the presence of entrepreneurial ideas (yes/ no), and experiences starting an own business (yes/ no). Some questions were necessary for the application of the typological approach in module (2). Thus, we utilized the pre-questionnaire for the discovery of participants' individual needs and expectations. The workshop was held twice, both times in German, with 12 participants in total. A description of the participants follows later. In terms of the workshop's organization, two issues are of particular importance: Evaluation procedure and instructional considerations.

## Considerations about Evaluation

A sound evaluation has to fulfill multiple targets, and not all targets have been realized yet. In general, there are no universally accepted criteria to evaluate entrepreneurship education courses (Henry et al., 2005b). Evaluation aims at judging the specific and global attainment of goals. Specific goals are those objectives mentioned above that are specific to the workshop. Global goals refer to the comprehensive purpose of offering extra-curricular workshops on entrepreneurship-related topics. To evaluate the attainment of global goals, general evaluation questionnaires, developed at the beginning of the EXIST-III project at BTU Cottbus, were employed. These questionnaires were constructed to assess the attainment of phase-related goals within the entrepreneurial process. The questionnaire for the first phase of increasing awareness contains a global evaluation of extra-curricular courses. The questionnaires, always completed anonymously, require the judgment on a five-point scale, ranging from excellent (1) to dissatisfying (5). A supplementing scale point (0) enables the expression of "I do not know". Some global questions (e.g., judging the quality of the guidance through the course) and questions related to the global aim of increasing awareness regarding various aspects of entrepreneurship (e.g., the occupational option of entrepreneurship, multiple ways of entrepreneurship, and the process of entrepreneurship) were included. Furthermore, the change of intention is questioned, by stating "no", "yes, it increased", and "yes, it decreased" as answer categories. Open questions ask about previous expectations, what participants rate as "particularly positive", and their "suggestions for improvement". One question asks how well the course met the participants' expectations. An improved procedure will be illustrated in the concluding paragraphs of this chapter.

## Considerations about Instruction and Training Method

As factors within the workshop situation will likely have an effect on the generation of creative ideas (cf. Amabile, 1983; Preiser, 2006), they have to be carefully considered in terms of general deliberations about adult learning and training design as well as deliberations specific to entrepreneurial training and creativity training.

Adult learners are centred on the concept of "andragogy" (see Merriam, 2001), pointing out the specifications of adults as learners. For instance, adult learners bring a broad range of life experiences into the classroom, are "interested in immediate application of knowledge" (Merriam, 2001, p. 5), and have learning needs that relate to their social life. Reigeluth (1996) demands a focus on customization rather than standardization, with customization including active learning and authentic tasks. Silber (1998), referring to cognitive psychology, points out the importance of guiding the attention of learners, supporting the discovery of relationships, presenting input in multiple modes, and integrating subjectively meaningful elements. In order to guide the attention of participants it is helpful to know their expectations. Metacognitive reflections could be used to support the discovery of relationships between contents, and multiple media are useful to vary input modes. To integrate meaningful elements, authentic subtasks within the entrepreneurial process are applied.

Coming to terms with training as a specific learning context, Salas and Cannon-Bowers (2001, p. 481) point out some principles that seem to be related to the effectiveness of trainings. One principle states that the relevant knowledge, skills, and abilities should be displayed. Other principles state the importance of practice opportunities and individual feedback (cf. also Silber, 1998). Koper and Bennett (2008) emphasize the substantial role of the executed activities of participants. Such action-oriented, experiential learning has to build upon the participants' existing knowledge (e.g., Silber, 1998). And Jones and English (2004) conclude that action-oriented training styles encourage creativity more than passive-learning approaches (p. 416). This kind of training challenges trainers, who should consist of a multidisciplinary team (Luczkiw, 2008) and who have to be skilled in multiple ways, as the following exemplary list illustrates (see generally Amabile, 1983).

- Trainers have to act like coaches (e.g., Reigeluth, 1996) and encourage the involvement of each participant (Fasko, 2000-2001).
- Trainers have to be aware of group and social processes (Garfield et al., 2001; McFadzean, 1998).
- Trainers should value creativity, originality and unorthodox views (Fasko, 2000-2001) and have to be competent at choosing appropriate creativity techniques (McFadzean, 1998).
- Trainers should be able to model intuitive and analytical creativity techniques; they must be able to change between divergent and convergent thinking (Fasko, 2000-2001; McFadzean, 1998).

In general, trainers have to adaptively regulate the progress of the course, as the direction the generation of ideas will take is uncertain. To meet these demands, a multi-person trainer team was deployed. Three trainers, of which the author was one, attended the workshop, two at a time. One trainer attended the full workshop, while the two others alternated. The trainers are skilled in psychology (including creativity), sociology (including entrepreneurial consulting), and economic engineering (including technological innovation).

Fasko (2000-2001), referring to Sternberg and Williams (1996), presents some teaching strategies that foster creativity. Important for the workshop presented here are the following strategies: Modelling creativity, building self-efficacy (illustration of creativity as a method, providing positive feedback), allowing time for creative thinking (adaptive management of agenda), explicitly challenging creative ideas, tolerating ambiguity, illustrating the value of "mistakes", promoting self-regulation (active involvement of participants in deciding about next steps in the workshop), encouraging creative collaboration, and imagining other viewpoints (interdisciplinarity). Applied instructional strategies were also short lectures, including external referents, guided as well as open group work and discussions, single working sessions, and team presentations. Explicit instructions (e.g., "be creative") were applied too (cf. Runco & Chand, 1995). Media used included laptop presentations with a beamer, paper handouts and working sheets, verbal talks, writings on a flipchart and Metaplan cards. The workshop was designed to encourage the active involvement of all participants, either in the whole group or in smaller teams or individually. The next paragraphs shortly illustrate the particular execution of the workshop and provide some exemplary results.

## Accomplishments of the Workshop

The workshop for Interdisciplinary Idea Generation Processes was completed twice by a total of 12 participants. Four (33%) of the 12 participants were female. Seven participants were academic staff at the BTU Cottbus from different disciplines, including humanities, engineering, electrical engineering, environmental studies, history, and business mathematics. Three participants were students from different fields of study too (engineering, physics, urban management). The entrepreneurial intentions of participants ranged broadly from no intention at all to high intention. About half of the participants said they had entrepreneurial ideas. They revealed various expectations, some of which were entrepreneurship related (e.g., information about starting own businesses, entrepreneurial risks, and financial support structures). Other expectations were creativity related (e.g., sector-related entrepreneurial ideas, getting to know creativity techniques, getting to know tools for the generation and refinement of conceptions); and some others were socially or knowledge oriented (e.g., interdisciplinary collaboration, applicable knowledge, getting to know the regional sectors with high economic potential). Not all expectations met the objectives, and introductory remarks at the beginning cleared this point. One of the male participants did not finish the workshops, for reasons unknown. One workshop was executed with four participants, the other with eight (seven). Both workshops are presented together, and we will comment on significant differences. The following illustrations are geared to summarize the actual execution of the workshop.

### (A) Introduction

The 45-minute introductory session contained an illustration of our motivation to offer this workshop, an accentuation of the three main modules, and a review of the concept "workshop", including the formulation of demands as reflected by the individual activities of each participant, and an open-minded group atmosphere. Further components were the following.

- Connection of the expectations of participants with the objectives of the workshop, including an emphasis on excluded objectives and an accentuation of the objective to qualify for general idea-generation skills.
- Exercise "Getting to know each other": Each person, including the trainers, got a

prepared worksheet with four cue questions relating to the topic of the workshop (e.g., Describe ideas you ever had that you would assess as creative; what do you think are necessary conditions for people who want to start their own businesses?). The task was to interview a randomly assigned partner and then change the roles. After 10 minutes each person was introduced in the plenum by their interviewers. The result was a collection of beliefs regarding entrepreneurship and ideas.

## (B) Defining the Starting Point: Working Areas & Future Sectors (Module 1)

The first main-module started with an exercise. Participants were challenged to answer the question: "Where in everyday life do you experience your own scientific domain?" After a 10-minutes individual brainstorming the answers were collectively put together on the Metaplan board, sorted in terms of similarities. To supplement the map, participants were questioned about their current working areas, explicitly challenging a broad range of answers by asking for private and former fields of "expertise" (e.g., hobbies, side jobs). Participants were asked to complete the answers until everybody felt comfortable to be well represented. This took 60 to 75 minutes, depending on the number of participants. The result was a broad collection of working fields and areas for the whole group of participants.

To enhance the meaningfulness of the contents, the topic "regional future sectors" was included and connected with the working areas of the participants. After a 30 minutes presentation of two external referents the trainers (a) provided a broad definition of sectors as economic fields where comparable services or products are provided, and (b) introduced a regional classification of future sectors, based on information from the Brandenburg Ministry of Economic Affairs and Brandenburg Economic Development Board[3]. Two exercises followed:

- Creation of a collective, multidisciplinary sector cluster. For this purpose, a framework map was prepared, including various sectors. The group constructed an overview of sectors where all group members were represented, as they sorted the cards on the Metaplan board around larger cards that represented the sector names. After about 30 minutes, the multidisciplinary sector collection was completed.
- First preliminary fixing: To select those sectors, on which further tasks and exercises would be related to, each participant got three dots, one dot for each answer to the following questions:
    - The most important sector for my job is...
    - Resulting from my everyday life, the most familiar sector for me is...
    - I see most in the sector...

The resulting selection of sectors included "energy, tourism and cultural and regional management" for Workshop 1 and "energy, environment and general service and consulting" for Workshop 2. The sector specifications provided the framework for upcoming hours, as they defined the problem statement: The task was defined as the generation of entrepreneurial ideas within the specified sectors, which relate to the particular experiences of all participants.

## (C) Where Do Ideas Come From? (Module 2)

After the objective of idea generation was defined, the question emerged where and how to search for ideas. Trainers remarked that a systematic search for business ideas requires the determination of the search field, and this determination makes the search field broad or narrow (cf. e.g., Dasgupta, 1996). To provide an overview of the multiplicity of possible systematizations of the search field, a 30- to 40-minute input presentation was provided. All illustrated examples included a specified generic topic (i.e., sector) and a definition of related sub-topics. As Mind Mapping seemed suited for the search along such relations, all examples were put into Mind Maps. It was emphasized that the formulation of subtopics (second-level entries in Mind Maps) determines the kind, diversity, and concreteness of generated ideas. Some worksheets are exemplarily illustrated in Figures 1 to 3.

Within the workshop, the map in Figure 3, based on the counseling practices in the field of business creation support, was selected for exercise, as it is particularly beneficial in interdisciplinary groups. Each participant can add some inputs to particular fields of the map, as the second-level subtopics are very broad. This map produces comprehensive contents. This is advantageous when a group wants to scan a broad spectrum of information, but may be inappropriate when the search field should rather be narrowly defined. The trainers stressed that an application of various map approaches is helpful to change the perspective. After various maps were presented and discussed, an exercise followed with the aim of individual completion of the worksheet presented in Figure 3.

*Figure 1. Exemplary Mind Map: Fields of application of services*

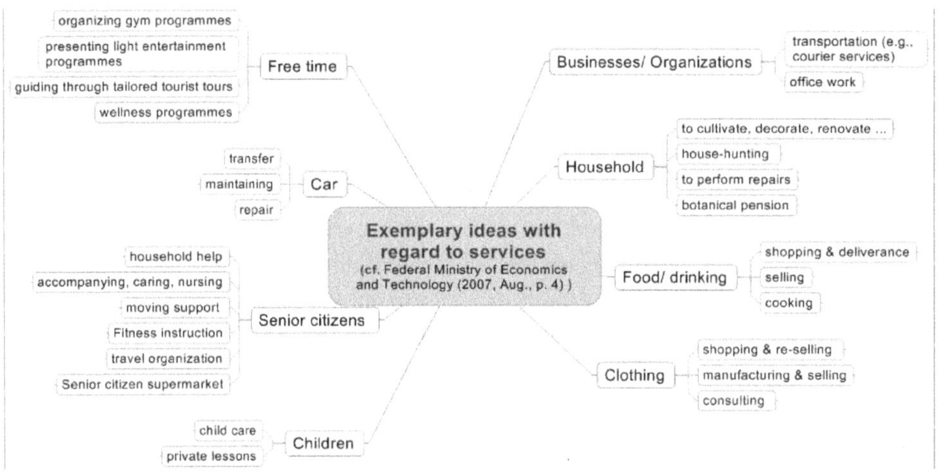


*Figure 2. Exemplary Mind Map: Target groups*

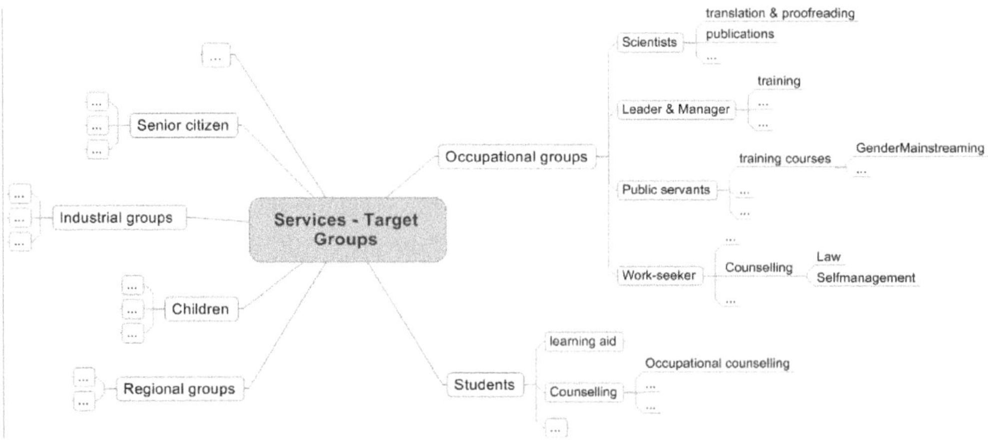

*Figure 3. Basic Mind Map for comprehensive search*

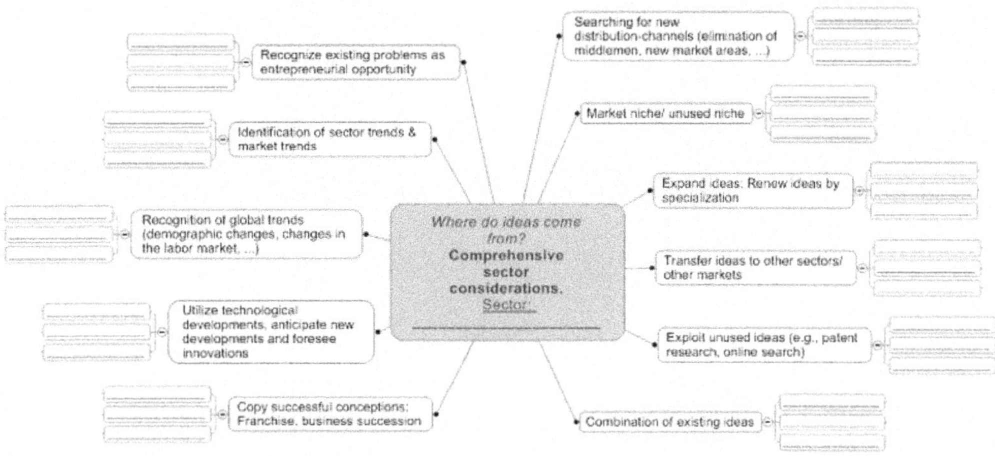


## (D) Sector-Specific Search Fields (Module 2)

Participants were encouraged to fill the map on the worksheet. They were explicitly instructed to overcome their restrictions and write down everything that came to their minds in 30 minutes. They were encouraged to enhance the second-level subtopics when they missed a category. After the individual brainstorming a group session followed to gather all inputs the participants collected. Participants were challenged to gather inspiration from the inputs of others and to add new thoughts. The trainer collected all entries on the laptop and the collective map was beamed on the wall, visible to everybody. This group session went on for 30 to 60 minutes and resulted in a broad collection of entries. For instance, the Workshop 1 group, focusing on the sectors "energy, tourism, and cultural management", collected 132 branches in the map. Exemplary entries were "Lusatia[4] as one region, rather than a separation of Lusatia in Brandenburg, Saxony, and Poland" and "mobile bicycle-repair services in analogy to mobile car-repair services" for the second-level subtopic "market/unused niche."

Participants got some feedback from the trainers, emphasizing that each one contributed to the broad results as they added their own perspective to the sector search and inspired some more ideas in others. After highlighting the advantages of such group-based collection in interdisciplinary teams, the trainers mentioned that for the purposes of exemplary idea generation processes the workshop would take a short-cut immediately and take the next step within the process, whereas real-world applications would require more time to elaborate the search field.

## (E) Combinations and Associations (Module 2)

For the purpose of illustrating how to make use of the resulting collection of first ideas, the approach of combination and association was introduced as one possibility, using the technique "Brainwriting 635" (e.g., Holzbaur, 2007; Preiser, 2006). Participants first had to choose two of the second-level subtopics of the resulting map from the last exercise. For instance, the Workshop 2 group elected to work with the second-level branches "existing problems and market/ unused niche". A worksheet for the application of Brainwriting 635 was handed over to each participant, as illustrated in Table 1. One trainer briefly introduced the technique of Brainwriting 635. Up to six group

*Table 1. Brainwriting 635 – master copy*

| | Idea 1 | Idea 2 | Idea 3 |
|---|---|---|---|
| 1 | | | |
| 2 | | | |
| 3 | | | |
| 4 | | | |
| 5 | | | |
| 6 | | | |

members (represented by six lines in the table) collected up to three ideas (represented by three columns in the table) for a given question within five minutes. Then, each group member gave the worksheet to the right neighbor and advanced the three ideas of the predecessor by picking one particular part of an idea or picking the full idea, and supplementing or combining it with a new element. With six participants, each participant would have up to 18 ideas written on the worksheet at the end of this exercise, which totals 108 ideas for a six-member group.

The trainers encouraged the participants to write down what their associations led to in their minds, without judging their ideas. If a participant signalled problems in generating ideas, the trainers asked cue questions such as, "What comes to your mind when you think of the wellness boom in tourism?" The following example of an idea chain from Workshop 2 illustrates the process of Brainwriting.

- Task: Generate potential working areas by combining an existing problem ("workaholism") with an unused niche ("seniors"). Resulting idea-chain: (1) "procurement of senior mentors who support workaholics", (2) "mentoring partnerships between seniors and employees, particularly for cultural arrangements", (3) "seniors and childcare", (4) "seniors as teachers of special contents (use their experiences)", (5) "training of seniors by pedagogic specialists", and (6) "private senior academy to qualify seniors for further application of their experiences".

After 30 minutes, each participant selected at least two of the written ideas that seemed particularly interesting or promising. Those ideas were collected on the flipchart. As the first workshop day came to its end, the trainers gave a short preview for the next day. The trainers prepared the materials for the following day by adapting some materials to the directions the workshop took and by illustrating the generated ideas of this first day on flipchart papers.

## (F) The Run-Up Questionnaire (Module 2)

As mentioned above, an illustration as well as self-reflection regarding entrepreneurial types was included in the module (2). The aim was to underline the entrepreneurial process and to highlight that it is not necessarily one person who realizes the full process. The trainers began the second day with a welcome back and the agenda for the day. Then they mentioned that the previously completed questionnaire had different objectives, among which was to record the entrepreneurial typology of the participants. One of the trainers introduced the typology, which is illustrated in Table 2.

Participants received individual evaluation of their results. As the process of creative innovativeness involves execution, an innovative group would be defined as being marked in all three areas, which is also true for entrepreneurship. All types were represented within both workshops. Depending on participants' need to discuss the results, the session was finished after 30 to 60 minutes.

*Table 2. Illustration of entrepreneurial types, based on the entrepreneurial process and the New Product Development Process (Lentzy & Weineck, 2009; cf. alsoSim et al., 2007)*

| | | |
|---|---|---|
| Inventor type<br>• develops ideas, creative<br>• yet without reference to application | Marketing type (Champion)<br>• perceives application opportunities<br>• sells ideas<br>• success oriented | Implementing type<br>• actor; makes things, realizes plans<br>• organizes things<br>• applies rules |
| Innovative allrounder<br>• something from each type, consumer-oriented and creative | | |

## (G) Creativity Techniques (Module 2)

The talk about the types gave good reason to ask participants who assessed themselves as creative, in concordance with the typing or not. The trainers asked the participants what criteria they used to base their answers. Furthermore, the understanding of creativity was brainstormed. Answers guided the trainers about what depth of input information was necessary, as they illustrated the group's knowledge and belief structures. The answers as well as the experiences that emerged on the first day of the workshop established the anchors for the trainers to connect the following contents to the knowledge network of the participants.

- Advantages of creative idea-generation skills were stressed, in particular for entrepreneurial contexts.
- Remarks on the definitions suggested that creativity is not uniformly defined and that criteria used to judge a result as creative are diverse and controversial.
- The definition of creativity "as the goal-directed production of novelty" (Weisberg, 2006, p. 761) was introduced. It was discussed that this directedness implies conscious self-regulation and that techniques may enhance this process.
- Different forms of creative acts (see Anderson, 1993; Zampetakis & Moustakis, 2006) and creative contributions (see Sternberg, 2006b) were briefly mentioned to stress the variability of what it means to "be creative".
- The distinction between creativity as ability and creativity as technique was introduced, highlighting that the workshop focuses on creativity as technique. This differentiation seems to be important to enhance creativity-related self-efficacy beliefs.
- Edison's statement that *success is one per cent inspiration and ninety-nine per cent perspiration*, was reviewed, and led to the deduction of creativity-inhibiting factors such as, for instance, fear of failure, negative beliefs, and, particularly important in the context of entrepreneurship, an inappropriate focus on realization while producing ideas (e.g., Amabile, 1983; Griffith-Hemans & Grover, 2006; Higgins, 1996; Mumford, 2000; Runco, 2004; Runco & Chand, 1995; Scherer, 2007; Woodman et al., 1993).
- An overview introduced the creative process in terms of the model of Scherer (2007). The four stages were explained and it was shown that the participants had already gone through the first stage of definition. Objectives and rules of the other stages were explained (e.g., the evaluation of ideas has to be strictly separated from the opening stage). The trainers stated that the activities sometimes lead back to previous stages. Some references to the entrepreneurial types mentioned earlier were

made, as they can be situated in the creative process.

- Creativity techniques were defined as "preparing procedures that are supposed to increase, but not guarantee, the probability of occurring creative ideas" (Häcker & Stapf, 1998, p. 468, translation by author). Intuitive and systematic creativity techniques (e.g., Garfield et al., 2001) were introduced. The first group was pointed out as adequate when aiming towards the generation of manifold, versatile and original ideas. In contrast to this, the second group aimed towards generating solutions for rather complex problems and encouraging structured, logical thoughts as well as rather incremental advancement (e.g., Garfield et al., 2001; McFadzean, 1998).

It was mentioned that the group would have to apply some particular techniques for the specification of more detailed entrepreneurial ideas. For this purpose, certain techniques were presented, with highlighting their variable arrangement. Brainstorming (e.g., Amabile, 1983, p. 190; Higgins, 1996; McFadzean, 1998; Preiser, 2006; Summers & White, 1976) and Brainwriting 635 (e.g., Holzbaur, 2007; Preiser, 2006) as examples of intuitive creativity techniques were illustrated more deeply, including their appropriateness for special kinds of problems, their basic rules, but also their weaknesses. Other intuitive creativity techniques as Synectics (e.g., Amabile, 1983; Barrett, 1978; Hockey, 2004; Summers & White, 1976), Guided Fantasy and imaginary excursion (e.g., Garfield et al., 2001; Higgins, 1996; McFadzean, 1998) were shortly introduced. Morphological Box (e.g., Barrett, 1978; Geschka, 2007; Hockey, 2004) and Combination/Morphological matrix (e.g., Higgins, 1996; Holzbaur, 2007) as examples of systematic creativity techniques were illustrated more deeply too, including a specification of appropriate problems, basic rules, and some exemplary variations. Other systematic creativity techniques as Lotus Blossom (e.g., Higgins, 1996) and Force Field Analysis Technique (Garfield et al., 2001) were shortly mentioned.

As the participants were still searching for entrepreneurial ideas in a defined sector, trainers indicated the upcoming task as specifying an idea path already roughly worked out. For this purpose some techniques were subsequently applied, accentuating a more systematic scanning of ideas.

## (H) Application of Selective Creativity Techniques (Module 2)

A break in problem focus evoked not the generation of as many ideas as possible but a more systematic search for entrepreneurial opportunities. Systematic techniques were employed now, starting with the Morphological Box. The task here was much more ill-defined than those problems typically illustrated for morphological techniques, but for the purposes of the workshop the technique seemed adequate.

Some worksheets for the application of Morphological Boxes were distributed. Participants needed some time to come to terms with the specification of dimensions. These specifications were not exhausted but represented a group-based selection. The selection of attributes for each dimension seemed to be easier for participants. The specified dimensions proved to be very rich, with diverse parameters. The tables grew large, and Microsoft® Excel® was utilized rather than the worksheets because of the breadth of parameters. Participants were very engaged and did not want to shorten their collection. For both workshops, this exercise needed around 60 minutes, which was more time than expected. As a consequence, trainers decreased the number of practiced techniques so as not to disturb the work of participants.

As an example, the participants in Workshop 1, concentrating on the sectors energy and tourism, defined the dimensions "excursion topics" (e.g., history, get to know like-minded people, new technologies), "excursion forms" (e.g., cycling, spaceship, balloon), "overnight-stay location" (e.g., bed & breakfast, houseboat, museum), "provision of energy" (e.g., self-generated, biomass, geothermal heating), and some others. Because of this variety, it was not easy to provide an overview, as the trainers wanted to lead participants to a specification rather than broadening the ideas. Combination matrices were applied next, focusing on only two dimensions, and a maximum of four to six attributes for each dimension. To stay with the example, participants of Workshop 1 chose forms of excursion (bicycle, hiking, water excursions and public transport) and excursion topics (old technologies/industrial memorials, energy, sport, new technologies, landscapes and architecture). Participants were asked to write down their associations for the combined attributes. Time was limited to 20 minutes. Results were discussed. An exemplary result for the combination of bicycle and new technologies was "bicycles built with innovative materials and a new design that produce energy for the heating of a tea bottle".

To enhance insights into the applicability of techniques, the possible use of Brainwriting 635 for the same task was mentioned. While Combination Matrix allowed for the scanning of various combinations, Brainwriting would elaborate just one combination but would produce more ideas for this combination. After 45 minutes, this exercise was concluded.

## (I) Idea Review (Module 2) and Decision (Module 3)

The trainers and participants supplemented the ideas collected on the first day with some of the ideas generated on the second day. Then the trainers recalled the creativity process in terms of Scherer's model (2007) and pointed out that the course now had to leave the stage of opening and move to the stage of identification.

The first task of identification was to specify those ideas that should be evaluated and refined next. The participants had to decide this as a group, using sticking dots to signal preferences. The experiences of both workshops suggest that several participants experienced some problems with this turn in the workshop. This may be substantiated in participants' assimilation of the open way of thinking. To avoid defensive reactions, it was important to refer to the workshop's objectives and the process model, which goes further than idea collection. Here, the diversity of participants in terms of the types presented earlier was an advantage, as some of them (champions, implementers) were curious about exploring the applicability of the ideas too.

The Workshop 1 group selected the idea "Energy CulTourism" for further exploration. This idea was about personally and electronically guided tours through the so-called "Energy region" of Cottbus and Lusatia and the provision of overnight accommodation in former energy-producing buildings. Workshop 2 selected two ideas for two groups. The first idea contained event-management, specializing in social themes. The second idea was called "multi-generation learning". The groups' decisions required between 20 to 30 minutes.

## (J) The Entrepreneurial Process (Module 3)

To introduce the entrepreneurial process, an exercise was undertaken. The trainers prepared small cut plastic cards with magnets for an empty whiteboard. Each card contained a phrase, which represent activities that are implemented within the entrepreneurial process. The trainers sorted the cards in a phase-specific manner and the task for the participants was to assign them to an adequate sequence. The phrases were based on the counseling practices of the "Gründervilla". Five separate phases were distinguished, with one parallel phase for technological innovations only. Table 3 illustrates the five basic phases.

The sorting session led to some discussions and one trainer explained the assignment and

*Table 3. Entrepreneurial phases and exemplary related activities*

| *Phases* | *Basic question to be answered* | *Exemplary activities* |
|---|---|---|
| (1) Awareness growing and Activation | Do I want to start my own business? | Visit idea-generation workshops, perceive entrepreneurship as a career opportunity |
| (2) Motivation and Orientation | With what kind of idea and with whom would I like to start my business? | Elaborate on the business idea, building the entrepreneurial team; define consumer groups |
| (3) Pre Start-up | What do I require to start a viable business? | Determine the legal form, present the business plan to business angels in order to find financial support |
| (4) Start-up | How do I successfully implement my business idea? | Business registration, office work |
| (5) Business growth | How do I establish my business? | Bind the customers, develop new products |

provided supplementary examples. After 60 to 90 minutes, the trainers emphasized that the participants of this workshop started with the first phase and now transferred to the second stage, where market and sector analysis as well as a specification of customer groups becomes necessary. Participants received some detailed tasks that were to be completed as homework for the following two days until the workshop met again. Participants were encouraged to allot the tasks between group members. Three broad tasks were allocated: (1) Market and sector analysis, (2) Competitor analysis, and (3) Specification of customer groups.

## (K) Sharpen the Business Ideas (Module 3)

After two days without meeting the participants, the third day started with a welcome back and the presentation of the agenda. Then an external referent was welcomed at Workshop 1. At Workshop 2, one of the trainers took over this part. This session comprised alternating information inputs and group inputs. Participants were targeted as experts, as they provided information inputs resulting from their homework. For instance, they collected data based on interview investigations and internet search. In collaboration with referents' input and participants' input as well as with the adaptive reapplication of some creativity techniques the ideas were questioned, specified, and refined.

In general, the process resulted in a decrease of ideas, but there was an increase for particular aspects (e.g., inclusion of unused customer groups as a result of market analysis). It was mentioned that various creativity techniques could facilitate the refinement of ideas. For instance, abstraction exercises may advance the specification of a core idea (see Scherer, 2007, p. 99-100). Furthermore, the Osborn checklist (see e.g., Higgins, 1996, p. 377; cf. also Amabile, 1983, p. 192), also referred to as the SCAMPER technique (Scherer, 2007, p. 73-88), may be useful. Experiences suggest that it is difficult for some participants to keep a self-critical distance from their self-developed ideas. Here, techniques that objectify idea evaluation may be adequate (e.g., cost-utility analysis; see e.g., Scherer, 2007). This session was limited to 120 minutes. For the purposes of this workshop, this was enough time.

## (L) Final Presentations (Module 3)

At this point of the workshop the participants had a comparatively specific conception about their business idea, and the trainers provided a final, large task. Participants were prompted to prepare a group presentation to introduce their business ideas in front of financiers. The trainers took over the roles of financiers. The presentation was to contain the following components.

- Core business idea, Unique Selling Proposition
- List of products/ services
- Customer groups and customer benefits, based on market information
- Allocation of competencies and responsibilities to group members
- Milestones for idea implementation
- Persuasive pleading of 30 to 60 seconds

While participants worked on their presentations, trainers requested aspects of the idea and provided support with the further concretization of aspects, including a reapplication of Brainstorming and Brainwriting 635. Working time was limited to two hours. The trainers experienced a comprehensive working session from all the participants, and the trainers' support was requested several times.

At the presentation session, 15 minutes were pure presentation time, and 15 minutes were discussion time for each group. Presentations revealed some open questions regarding the business idea, but also provided an overview of the core ideas. The persuasive pleadings were contextualized; participants who prepared for it had to imagine they would meet a promising financier spontaneously (e.g., at a red traffic light), and that they would have just a few seconds to catch the attention of this person. The aim of this task was to stress the idea of target-group specificity when communicating own ideas.

## (M) Coming to an End (Module 3)

The presentations were very convincing in the light of the rather short time participants spent on the development of a business idea. The trainers provided positive feedback and then crossed over to creativity and the creativity process. The trainers emphasized the contribution that each participant, as well as the group as a whole, made. The ability to generate ideas and refine them was apparent in all participants, who were reminded of the time requirements of real-world idea generation.

Then some words were mentioned regarding the handling of the ideas developed within this workshop. Only ideas we were allowed to communicate were presented, including the explanations in this chapter. Later, some generated ideas were prepared for further market analysis. At the end of this workshop, participants were asked to fulfill the evaluation questionnaire. An open feedback circle finished the workshops.

## Post-Evaluation of the Workshop

Some issues the participants mentioned as "particularly positive" were: "consent to individual needs", "exercising time", "to experience how fast ideas evolve into business concepts", "increased awareness regarding things to be considered for business creation", and "exchange among participants". Also, some suggestions were made for improvements. Examples are: "shorten market analysis and discussion of results", "include a short talk of a real business founder as an experience report", and "information before the workshop should clarify that it is more about general training than about finding own business ideas".

40% noted an increase in entrepreneurial intention and nobody stated a decrease. In the light of the rather high proportion of high-intentional participants in the beginning, this evaluation result indicates that the workshop may be useful to build intentions within groups of interdisciplinary people. In terms of the specification of awareness-increasing aspects, the highest percentages of mentioned awareness increasing are stated concerning "the process of entrepreneurship" (80%), followed by "the occupational option of entrepreneurship" (78%) and "the multiple ways of entrepreneurship" (75%). The lowest percentages of stated awareness increasing are found for "awareness raising regarding the consequences of starting an own business" (25%). Thus, the intended issues were higher than other issues that were not the main objective, suggesting that this workshop met the main objectives. Of course, the trainers are aware of the need for deeper evaluation that would contain directly related issues as the awareness of own creative abilities and the perception of the role that creativity plays for entrepreneurs. Those issues will be discussed in the next and final section of this chapter, considering the scope of enhancement and future applications of this workshop concept.

## SCOPE OF FUTURE APPLICATIONS

The workshop illustrated in the previous paragraphs falls into the category of entrepreneurship-related trainings, defined as focusing on particular themes and representing rather small group behavioral interventions with a short duration of some hours to some days (Katz, 2007). The workshop represented a rather cognitive intervention in terms of Amabile (1983) but incorporated some socially important issues too (e.g., choice, modeling, stimulation; see Amabile, 1983, p. 189 ff.). Referring to considerations about curriculum development, it seems necessary that each adaptation regarding the target group requires reconsiderations of objectives, contents and didactics (e.g., sequencing of the modules, provision of examples and exercising time) (e.g., Gagne & Dick, 1983; Reigeluth, 1996; Salas & Cannon-Bowers, 2001; Silber, 1998). Thus, it seems inadequate to simply transfer the workshop to other objectives and other target groups without a sound reflection of specific conditions. For instance, it can be assumed the prevalent creativity techniques and approaches to creativity may differ in other countries (see Kaufman & Sternberg, 2006).

For the context of universities, this workshop may be useful to enhance individual awareness regarding entrepreneurship related aspects and to realize entrepreneurship as a generally valued competence cluster that goes beyond starting own businesses. Some principles of the workshop concept seem to be generally important and thus, transferable to other applications too. Such principles include (a) the connection of the contents to the particular experiences and backgrounds of participants (including the regional context), (b) the reflection of expectations of participants and objectives of trainers, (c) application of practiced techniques in authentic contexts (as the creation of business ideas), (d) action-orientation rather than knowledge-orientation, and (e) the general contextualization of broader competencies in entrepreneurship. Trainings applying such principles are very demanding for the trainers, and the involvement of multi-person interdisciplinary trainer-teams seems to be necessary to allow the appropriate assistance.

One of the main opportunities for improvement is the evaluation procedure. Referring to the recommendations of Henry et al. (2005b), that "the best means by which to evaluate training courses is to directly relate programme outcomes to objectives" (p. 161), we improved our procedure for subsequent workshops. The new pre- and post-measures now include a block of eight questions related to beliefs about the role and appearance of creativity and idea generation, as well as global beliefs about entrepreneurship (e.g., "I believe that I would be able to generate a good, convincing, and innovative-creative business idea if I wanted to."). Actually, a sound evaluation of the effectiveness of the training would contain longitudinal follow-up measurements too (cf. Henry et al., 2005b). Furthermore, a pre-post comparison using creativity tests may provide some insights into the strengths and weaknesses of such training.

There are some variations we had already proved. For instance, we prepared a short-term variation for the purposes of practising particular techniques and arousing students' curiosity about creative techniques. This short-term variation was established for the use in classes of between 10 to 25 students for duration of 60 to 90 minutes. Experiences underscore the applicability of these techniques even for short-term exercises, but they are actually limited for illustrating single techniques instead of increasing awareness of them. Nevertheless, such short-term applications may represent a measure for university lecturers from various fields to integrate some basics on creativity and entrepreneurship.

There are other variations not proved yet. For instance, the starting point selected here (working experiences, domains) can be varied and focus more on individual skills, attributes and attitudes. The techniques and exercises may be easily varied and adopted. But trainers should be familiar with techniques, and not every technique may be adequate for short-term training, as some techniques are not easily learned (McFadzean, 1998).

## CONCLUSION

This chapter presented a workshop concept for Interdisciplinary Idea Generation Processes, its theoretical, practical and empirical foundations, as well as its application and results. Practical background was a public-funded project at an East German University of Technology. The theoretical foundation came from the literature about entrepreneurship education and creativity. The empirical foundation resulted mainly from our own research conducted previously. The workshop was adjusted for interdisciplinary academics, both staff and students, who were not necessarily intent on starting an own business. Built for small groups of up to 12 participants, the workshop was applied twice. Reflections about adult learning led to the conceptualization of a workshop, where a multidisciplinary trainer team referred to the participants' relevant previous knowledge and experiences. Much emphasis was placed on metacognitive reflection about the creative idea generation process and the entrepreneurial process, as well as to specialization in respect of the typing approach to entrepreneurship. Multiple instructional strategies were applied. The workshop showed that by means of guided instruction, participants were able to generate a broad range of ideas and then deepen particular business ideas. Although duration was limited to three days plus two days for homework, an illustration and practice of a full creative idea generation process was realized. Having more time would enable a broader range of techniques to be utilized, and it would likely result in even broader and deeper ideas. Generated ideas were creative in terms of Dasgupta's (1996) psychological creativity. Subsequent applications of this workshop concept should produce some more scientifically based evaluation results, which may be one of our future tasks.

## REFERENCES


Ajzen, I. (1991). The theory of planned behavior. *Organizational Behavior and Human Decision Processes, 50*(2), 179–211. doi:10.1016/0749-5978(91)90020-T

Anderson, J. V. (1993). Mind mapping: A tool for creative thinking. *Business Horizons, 36*(1), 41-46. Retrieved October 9, 2008, from http://www.sciencedirect.com/science/article/B6W45-4FN1XTN-7/2/a433a975929052093fe8aee-1bf11b2e8

Baer, J., & Kaufman, J. C. (2006). Creativity research in English-speaking countries. In Kaufman, J. C., & Sternberg, R. J. (Eds.), *The international handbook of creativity* (pp. 10–38). N.Y.: Cambridge University Press.

Bandura, A. (2001). Social cognitive theory: An agentic perspective. *Annual Review of Psychology, 52*(1), 1–26. doi:10.1146/annurev.psych.52.1.1

Barrett, F. D. (1978). Creativity techniques: Yesterday, today, and tomorrow. *Advanced Management Journal, 43*(1), 25-35. Retrieved October 9, 2008, from http://search.ebscohost.com/login.aspx?direct=true&db=buh&AN=4606291&site=ehost-live

Barron, F., & Harrington, D. M. (1981). Creativity, intelligence, and personality. *Annual Review of Psychology*, *32*, 439–476. doi:10.1146/annurev.ps.32.020181.002255

Bygrave, W. D., & Hofer, C. W. (1991). Theorizing about entrepreneurship. *Entrepreneurship: Theory & Practice*, *16*(2), 13–22.

Cooper, P. (2005). A study of innovators' experience of new product innovation in organisations. *R & D Management*, *35*(5), 525–533. doi:10.1111/j.1467-9310.2005.00409.x

Dasgupta, S. (1996). *Technology and creativity*. New York: Oxford University Press.

Edmonds, G., Branch, R., & Mukherjee, P. (1994). A conceptual framework for comparing instructional design models. *Educational Technology Research and Development*, *42*(4), 55–72. doi:10.1007/BF02298055

Ericsson, K. A. (1996). The acquisition of expert performance: An introduction to some of the issues. In Ericsson, K. A. (Ed.), *The road to excellence. The acquisition of expert performance in the arts and sciences, sports, and games* (pp. 1–50). Mahwah, NJ: Erlbaum.

European Commission. (2004). *Final report of the expert group, "Education for Entrepreneurship". Making progress in promoting entrepreneurial attitudes and skills through primary and secondary education*. Brussels: European Commission. Retrieved February 8, 2008, from http://ec.europa.eu/enterprise/policies/sme/files/support_measures/training_education/doc/entrepreneurship_education_final_en.pdf

Faltin, G. (2001). Creating a culture of innovative entrepreneurship. *Journal of International Business and Economy*, *2*(1), 123–140.

Fasko, D. (2000-2001). Education and creativity. *Creativity Research Journal*, *13*(3&4), 317–327.

Federal Ministry of Economics and Technology (2007, August). Gründungsideen entwickeln - Weniger Glück als Verstand. *GründerZeiten Nr. 39*.

Federal Ministry of Education and Science (Ed.). (2005). *Wissens- und technologieorientiertes Gründungsgeschehen. Kienbaum-Bestandsaufnahme für eine Weiterentwicklung von EXIST.* Bonn, Berlin: Federal Ministry of Education and Science. Retrieved July 21, 2006, from www.bmbf.de/pub/wissens_technologieorientiertes_gruendungsgeschehen.pdf

Gagne, R. M., & Dick, W. (1983). Instructional psychology. *Annual Review of Psychology*, *34*(1), 261–295. doi:10.1146/annurev.ps.34.020183.001401

Garfield, M. J., Taylor, N. J., Dennis, A. R., & Satzinger, J. W. (2001). Research report: Modifying paradigms-individual differences, creativity techniques, and exposure to ideas in group idea generation. *Information Systems Research*, *12*(3), 322–333. doi:10.1287/isre.12.3.322.9710

Geschka, H. (2007). Kreativitätstechniken. In Köhler, R., Küpper, H.-U., & Pfingsten, A. (Eds.), *Handwörterbuch der Betriebswirtschaft* (6th ed., pp. 992–1005). Stuttgart: Schäffer-Poeschel Verlag.

Goldberg, J. C. (1980). Counseling the adult learner: A selective review of the literature. *Adult Education Quarterly*, *30*(2), 67–81. doi:10.1177/074171368003000201

Greene, P. G., Katz, J. A., & Johannisson, B. (2004). Entrepreneurship education. *Academy of Management Learning & Education*, *3*(3), 238–241.

Griffiths-Hemans, J., & Grover, R. (2006). Setting the stage for creative new products: Investigating the idea fruition process. *Journal of the Academy of Marketing Science*, *34*(1), 27–39. doi:10.1177/0092070305281777

Guilford, J. P. (1988). Some changes in the structure-of-intellect model. *Educational and Psychological Measurement*, *48*(1), 1–4. doi:10.1177/001316448804800102

Häcker, H., & Stapf, K. H. (Eds.). (1998). *Dorsch. Psychologisches Wörterbuch* (13th ed.). Bern, Göttingen, Toronto, Seattle: Huber.

Heilman, K. M., Nadeau, S. E., & Beversdorf, D. O. (2003). Creative innovation: Possible brain mechanisms. *Neurocase*, *9*(5), 369–379. doi:10.1076/neur.9.5.369.16553

Heller, K. A. (2004). Identification of gifted and talented students. *Psychological Science*, *46*(3), 302–323.

Henry, C., Hill, F., & Leitch, C. (2005a). Entrepreneurship education and training: Can entrepreneurship be taught? Part 1. *Education+ Training, 47*(2), 98–111.

Henry, C., Hill, F. & Leitch, C. (2005b). Entrepreneurship education and training: Can entrepreneurship be taught? Part 2. *Education + Training, 47*(3), 158-169.

Hewstone, M. (1989). *Causal attribution: From cognitive processes to collective beliefs*. Oxford: Blackwell.

Higgins, J. M. (1996). Innovate or evaporate: Creative techniques for strategists. *Long Range Planning*, *29*(3), 370–380. doi:10.1016/0024-6301(96)00023-4

Hockey, J. (2004). Practical creativity. *Medical Device Technology*, *15*(1), 20–23.

Holzbaur, U. (2007). *Entwicklungsmanagement. Mit hervorragenden Produkten zum Markterfolg*. Berlin, Heidelberg, New York: Springer.

Howard-Jones, P. (2002). A dual-state model of creative cognition for supporting strategies that foster creativity in the classroom. *International Journal of Technology and Design Education*, *12*(3), 215–226. doi:10.1023/A:1020243429353

Isaksen, S. G., Puccio, G. J., & Treffinger, D. J. (1993). An ecological approach to creativity research: Profiling for creative problem solving. *The Journal of Creative Behavior*, *27*(3), 149–170.

Jones, C. & English, J. (2004). A contemporary approach to entrepreneurship education. *Education + Training, 46*(8/9), 416-423.

Katz, J. A. (2007). Education and training in entrepreneurship. In Baum, J. R., Frese, M., & Baron, R. (Eds.), *The psychology of entrepreneurship* (pp. 209–235). Mahwah, NJ/ London: Lawrence Erlbaum Associates.

Kaufman, J. C., & Sternberg, R. J. (Eds.). (2006). *The international handbook of creativity*. New York: Cambridge University Press.

Kirby, D.A. (2004). Entrepreneurship education: Can business schools meet the challenge? *Education + Training, 46*(8/9), 510-519.

Koper, R., & Bennett, S. (2008). Learning design: Concepts. In H.H. Adelsberger, Kinshuk, J. M. Pawlowski & D. Sampson (Eds.), *Handbook on information technologies for education and training* (2nd ed., pp. 135-154). Heidelberg: Springer.

Kulicke, M., Stahlecker, T., Lo, V., & Wolf, B. (2006). *EXIST - Existenzgründungen aus Hochschulen. Bericht der wissenschaftlichen Begleitung zum Förderzeitraum 1998 bis 2005 (Kurzfassung)*. Berlin: Bundesministerium für Wirtschaft und Technologie. Retrieved September 7, 2007, from http://www1.isi.fraunhofer.de/p/download/EXIST/exist_kurzfassung.pdf

Lange, A. (2009a). Entrepreneurs as exceptional: Indications from giftedness and expertise research. In M. Rebernik, B. Bradac & M. Rus (Eds.), *The winning products: Proceedings of the 29th Conference on Entrepreneurship and Innovation Maribor - PODIM, Maribor, 25th - 26th March 2009* (pp. 89-105). Maribor: IRP Institute for Entrepreneurship Research.

Lange, A. (2009b). Gender-specific beliefs about entrepreneurship: Specifications and implications. In J. Braet (Ed.), *The Proceedings of the 4th European Conference on Entrepreneurship and Innovation, The University of Antwerp, Belgium, 10-11 September 2009,* (pp. 273-280). Antwerp: University of Antwerp.

Lange, A. (2010). TPB-basierte Überzeugungen bei Studierenden mit unterschiedlichen Gründungsintentionen: Ausprägungen und Schlussfolgerungen. In A. Rese, D. Baier, M. Mißler-Behr & M. Kaiser (Eds.), *Entrepreneurship Education: Symposium des Brandenburgischen Instituts für Existenzgründung und Mittelstandsförderung (BIEM e.V.) "Gründung und Innovation" vom 11.-12. Juni 2009, Brandenburgische Technische Universität Cottbus* (pp. 83-109). Lohmar: Eul.

Lentzy, J., & Weineck, G. (2009). *Linking findings in innovation management to entrepreneurship: Clustering potential entrepreneurs by innovator types.* Paper presented at the 1st International Conference on Strategic Innovation and Future Creation, March 2009, Malta.

Luczkiw, E. (2008). Entrepreneurship education in an age of chaos, complexity and disruptive change. In Potter, J. (Ed.), *Entrepreneurship and higher education* (pp. 65–93). OECD Publishing. doi:10.1787/9789264044104-5-en

McFadzean, E. (1998). The creativity continuum: Towards a classification of creative problem solving techniques. *Creativity and Innovation Management, 7*(3), 131. doi:10.1111/1467-8691.00101

Merriam, S. B. (2001). Andragogy and self-directed learning: Pillars of adult learning theory. *New Directions for Adult and Continuing Education, 89*(3).

Mouchiroud, C., & Lubart, T. (2006). Past, present, and future perspectives on creativity in France and French-speaking Switzerland. In J.C. Kaufman & R.J. Sternberg (Eds.), *The international handbook of creativity,* (pp. 96-123). New York: Cambridge University Press.

Mumford, M. D. (2000). Managing creative people: Strategies and tactics for innovation. *Human Resource Management Review, 10*(3), 313–351. doi:10.1016/S1053-4822(99)00043-1

Preiser, S. (2006). Creativity research in German-speaking countries. In Kaufman, J. C., & Sternberg, R. J. (Eds.), *The international handbook of creativity* (pp. 167–201). New York: Cambridge University Press.

Reigeluth, C. M. (1996). A new paradigm of ISD? *Educational Technology, 36*(3), 13–20.

Runco, M., & Chand, I. (1995). Cognition and creativity. *Educational Psychology Review, 7*(3), 243–267. doi:10.1007/BF02213373

Runco, M. A. (2004). Creativity. *Annual Review of Psychology, 55*(1), 657–687. doi:10.1146/annurev.psych.55.090902.141502

Runco, M. A., & Okuda, S. M. (1988). Problem discovery, divergent thinking, and the creative process. *Journal of Youth and Adolescence, 17*(3), 211–220. doi:10.1007/BF01538162

Salas, E., & Cannon-Bowers, J. A. (2001). The science of training: A decade of progress. *Annual Review of Psychology, 52*(1), 471–499. doi:10.1146/annurev.psych.52.1.471

Scherer, J. (2007). *Kreativitätstechniken. In 10 Schritten Ideen finden, bewerten, umsetzen.* Offenbach, Germany: Gabal.

Shane, S., & Venkataraman, S. (2000). The promise of entrepreneurship as a field of research. *Academy of Management Review, 25*(1), 217–226. doi:10.2307/259271

Silber, K. (1998). The cognitive approach to training development: A practitioner's assessment. *Educational Technology Research and Development, 46*(4), 58–72. doi:10.1007/BF02299674

Sim, E. W., Griffin, A., Price, R. L., & Vojak, B. A. (2007). Exploring differences between inventors, champions, implementers and innovators in creating and developing new products in large, mature firms. *Creativity and Innovation Management, 16*(4), 422–436. doi:10.1111/j.1467-8691.2007.00457.x

Smith, G. J. W., & Carlsson, I. (2006). Creativity under the northern lights. Perspectives from Scandinavia. In Kaufman, J. C., & Sternberg, R. J. (Eds.), *The international handbook of creativity* (pp. 202–234). New York: Cambridge University Press.

Sternberg, R. J. (1998). Principles of teaching for successful intelligence. *Educational Psychologist, 33*(2/3), 64–72.

Sternberg, R. J. (2006a). Introduction. In Kaufman, J. C., & Sternberg, R. J. (Eds.), *The international handbook of creativity* (pp. 1–9). New York: Cambridge University Press.

Sternberg, R. J. (2006b). The nature of creativity. *Creativity Research Journal, 18*(1), 87–98. doi:10.1207/s15326934crj1801_10

Sternberg, R. J., & Kaufman, J. C. (1998). Human abilities. *Annual Review of Psychology, 49*(1), 479–502. doi:10.1146/annurev.psych.49.1.479

Summers, I., & White, D. E. (1976). Creativity techniques: Toward improvement of the decision process. *Academy of Management Review, 1*(2), 99–108. doi:10.2307/257490

Urban, K. K. (2002). From creativity to responsible Createlligence® as future competence. *Journal of Educational Sciences, 1*(2), 200-170. Retrieved December 27, 2008, from http://qspace.qu.edu.qa/handle/10576/8482

Weisberg, R. W. (2006). Modes of expertise in creative thinking: Evidence from case studies. In Ericsson, K. A., Charness, N., Feltovich, P. J., & Hoffman, R. R. (Eds.), *The Cambridge handbook of expertise and expert performance* (pp. 761–787). New York: Cambridge University Press.

Woodman, R. W., Sawyer, J. E., & Griffin, R. W. (1993). Toward a theory of organizational creativity. *Academy of Management Review, 18*(2), 293–321. doi:10.2307/258761

Zampetakis, L., & Moustakis, V. (2006). Linking creativity with entrepreneurial intentions: A structural approach. *The International Entrepreneurship and Management Journal, 2*(3), 413–428. doi:10.1007/s11365-006-0006-z

## ADDITIONAL READING

Acs, Z. J., & Audretsch, D. B. (Eds.). (2005). *Handbook of Entrepreneurship Research: An Interdisciplinary Survey and Introduction*. N.Y.: Springer US.

Baum, J. R., Frese, M., & Baron, R. (Eds.). (2007). *The Psychology of Entrepreneurship*. Mahwah; New Jersey; London: Lawrence Erlbaum Associates.

Couger, J. D. (1995). *Creative Problem Solving and Opportunity Finding. Danvers*. Massachusetts: Boyd & Fraser Publishing.

De Bono, E. (1992). *Serious Creativity: Using the Power of Lateral Thinking to Create New Ideas*. London: Harper Collins.

Ericsson, K. A., Charness, N., Feltovich, P. J., & Hoffman, R. R. (Eds.). (2006). *The Cambridge Handbook of Expertise and Expert Performance*. New York: Cambridge University Press.

Guilford, J. P. (1950). Creativity. *The American Psychologist*, *5*, 444–454. doi:10.1037/h0063487

Herstatt, C., & Verworn, B. (2004). The “fuzzy front end” of innovation. In: Durand, T. et al. (Eds.), Bringing technology and innovation to the boardroom: strategy, innovation, and competences for business value (pp. 347–372). Houndmills: Palgrave Macmillan.

Higgins, J. M. (1994). *101 Creative Problem Solving Techniques: the Handbook of New Ideas for Business*. Florida: The New Management Publishing Company.

Hipp, C. (Ed.). (2002). *Lectures on Innovation Practices – Overview and Summary of Innovation Methods. Wireless Future Studies (WFS), No. 1*. München: Vodafone Group R&D.

Kirton, M. (Ed.). (1994). *Adaptors and Innovators: Styles of Creativity and Problem-solving*. London: Routledge.

Sternberg, R. J. (Ed.). (1988). *The nature of creativity: Contemporary psychological perspectives*. N.Y.: Cambridge University Press.

## KEY TERMS AND DEFINITIONS

**Entrepreneurship:** The processes concerning the management of opportunities by individuals.

**Entrepreneurship Education:** Any educational context that aims toward an enhancement of personal competencies to manage opportunities at any stage of their development.

**Extra-Curricular Entrepreneurship Education:** A special kind of entrepreneurship education; educational contexts outside the regular curriculum.

**Creativity:** Intentional cognitive activities, leading to (individually) new solutions.

**Creativity Techniques:** Operation guidelines with the aim to increase the likelihood of new, creative ideas and solutions.

**Intuitive Creativity Techniques:** Operation guidelines, mainly applying associative thoughts, with the aim to increase the likelihood of new, creative ideas and solutions.

**Systematic Creativity Techniques:** Operation guidelines, mainly applying systematic and logical thoughts, with the aim to increase the likelihood of new, creative ideas and solutions.

## ENDNOTES

1. See http://www.exist.de/englische_version/index.php.
2. This project, also public funded by the Federal Ministry of Education and Science and by the European Social Fund, is called “Mobilization of innovative entrepreneurial potential of female students (MobI$^{UP}$)”. For more information see http://www.tu-cottbus.de/fakultaet3/de/personalmanagement/zusatzangebot/mobiup.html.
3. See http://zab-brandenburg.de/; http://www.mwe.brandenburg.de/
4. Lusatia is a region in East Germany (parts of Saxony and Brandenburg) and Poland.

# Chapter 9
# The Structure of Idea Generation Techniques: Three Rules for Generating Goal-Oriented Ideas

**Stefan Werner Knoll**
*University of Magdeburg, Germany*

**Graham Horton**
*University of Magdeburg, Germany*

## ABSTRACT

*Idea generation techniques play an important role in the innovation process. Until recently, the space of techniques has been unstructured, and no clear guidelines have been available for the selection of an appropriate technique for a given innovation goal. This chapter uses an engineering approach to study and develop idea generation techniques with the aim of obtaining more structured and rigorous guidelines for generating ideas. One element of this approach was to identify and understand the fundamental mental principles underlying an idea generation technique. In this chapter, three such principles suffice to cover a large range of published idea generation techniques and can be used to improve the utility of idea generation within the innovation process.*

## INTRODUCTION

Suppose you are the innovation manager of a company that produces and sells a high-quality product. Business is good, but changing circumstances are influencing your competitive position. For example, competitors are introducing new or improved products or your patent protection will soon expire. In order to maintain a competitive position in the market, your company introduces an innovation process. This multi-stage process

DOI: 10.4018/978-1-60960-519-3.ch009

combines a variety of techniques and methods to analyse the market situation, define strategic goals, and generate and implement ideas, yielding new products and market strategies.

As innovation manager, you know that the biggest weakness in the innovation process is the so-called *Fuzzy Front End* (Herstatt, Verworn, & Nagahira 2004), which ranges from the generation of an idea to either its approval for development or its termination. Furthermore, different strategic goals of the company, such as maintaining market share, entering new markets, or achieving a particular growth rate require different types of innovations. These may include incremental and platform product innovations, process innovations or business model innovations. These in turn require different types of ideas. Therefore, the Fuzzy Front End needs to generate a variety of ideas, in order to increase the probability of obtaining ones that are suitable for the given innovation goal. One approach to achieve such a pool of ideas is to conduct an idea generation workshop which uses a variety of common idea generation techniques such as *Brainstorming* (Osborn, 1963), *Analogies* (VanGundy, 1988, pp. 82-84) and *Bionic Ideas* (VanGundy, 2005 pp. 229-231). *But how do you know that the chosen combination of idea generation techniques will yield ideas that are appropriate for your innovation goal?* This question has remained largely unanswered.

Since Alex Osborn (1963) introduced Brainstorming as a method to improve the creativity of groups, more than one hundred idea generation techniques have been published (Higgins, 1994; VanGundy, 1988, 2005). Each of them provides a step-by-step sequence of actions or instructions to support a group or an individual in an ideation process. Several studies have characterized and classified these techniques along several dimensions (Herring, Jones, & Bailey, 2009; Smith, 1998; VanGundy, 1988). For example, VanGundy (1988) subdivided idea generation techniques into group and individual techniques and used the dimensions: whether idea generation is "verbal" or "silent"; whether ideas are produced by "forced relationships" or "free association"; and whether the technique employs stimuli that are "related" or "unrelated" to the problem. However, most idea generation techniques are generic, i.e. they are presented in a non-problem-specific form. From the literature, we found no clear guidelines for the selection of a given technique with regard to the innovation goal. As a result, the innovation manager has to rely on experience for the selection of an appropriate technique.

This need for experience is increased by technological enhancements like Group Support Systems (GSS), which have changed the innovation process by enabling group work with virtual teams across geographical distances. GSS represent an information technology-based group meeting environment, which offers a variety of tools to assist the group in the structuring of activities, generating ideas, and improving group communication (Nunamaker Jr., Dennis, Valacich, Vogel, & George, 1991; Vreede, Vogel, Kolfschoten, & Wien, 2003). Examples for GSS include: *ThinkTank*™ ("ThinkTank" 2010), *Google Wave*™ ("Google Wave" 2010) and *StreamWork*™ ("Stream Work" 2010).

Most GSS implement ideation as a divergent thinking process called electronic brainstorming. In this process, the participants use the GSS functionalities to contribute ideas individually while also reading and elaborating on the ideas of others to improve the overall result. Research has studied electronic brainstorming to determine factors such as the impact of stimuli type and GSS structure (Hender, Dean, Rodgers, & Nunamaker Jr, 2002; Hender, Rodgers, Dean, & Nunamaker Jr., 2001) or the optimum group size for ideation using GSS (Valacich, Dennis, & Connolly, 1994; Gallupe, Dennis, Cooper, Valacich, Bastianutti, & Nunamaker Jr., 1992).

From the literature review, our conclusion is that under certain circumstances, groups using GSS for ideation can produce more ideas, and more good ideas than groups using pen-and-paper methods. But experience shows that the value of GSS depends on how purposefully and skillfully it is used (Briggs et al., 1999). Idea generation techniques were designed to support ideation in a face-to-face workshop. Most of these techniques recommend the use of tools or materials (e.g. balloons, newspapers or whiteboards) to support the creative process but it remains unclear how these could be mapped to a technological implementation such as a GSS. As a result, the innovation manager who uses the GSS functionalities for ideation will execute idea generation techniques already provided by the GSS and does not benefit from other techniques, which might be better suited to his or her goal.

Our research is aimed at developing a scientific basis for idea generation, which provides a model for the mental processes that lead to new ideas. This model should both explain the processes involved and at the same time enable specific idea generation techniques to be created for a given innovation goal. This research aims to answer the following questions:

1. What are the fundamental mental principles of idea generation techniques?
2. How do these principles determine the effectiveness of an idea generation technique for a given innovation goal?
3. How can these principles be used to define methods that enhance the selection of an appropriate idea generation technique and its implementation with a GSS?

In this chapter, we will give a comprehensive answer to the first question and provide partial answers for questions two and three.

One result of our research has been to show that the very large number of individual idea generation techniques that are in use can be reduced to just three underlying mental principles. An innovation manager can apply one of these principles to create an idea generation technique that can be used either in a traditional face-to-face workshop or in a GSS tool. This is much more efficient than trying to master all the known techniques and implement a large subset of them in a GSS.

In the next section, we present our perspective on idea generation, which is to treat it as an Engineering discipline. We then outline a cognitive model that attempts to explain how idea generation techniques affect the mental processes of an individual. We will show how these mental processes can be mapped onto just three cognitive principles, which each yield a different class of idea generation techniques. We also include some hints for the application of these insights. Finally, we present our conclusions and suggest some practical implications and possible future directions for research.

## AN ENGINEERING PERSPECTIVE ON CREATIVITY

Over several decades, a variety of approaches to creativity have been developed. Creativity research discriminates these approaches by the perspective they take (Bostrom, & Nagasundaram, 1998):

- The creative *person* (creativity as a property of an individual),
- The creative *product* (creativity as a property of a process outcome),
- The creative *press* (creativity as a result of the environment and context), or
- The creative *process* (creativity as a result of a defined process or technique).

Here, we are interested in the creative process, in particular the mental activities that lead to ideas and the techniques that correspond to them.

Like most researchers, we see the creative process as the exploration and transformation of conceptual spaces of an individual to generate new ideas (Mednick, 1962). Different stepwise models of creativity have been developed (Shneiderman,

2000; Amabile, 1983; Osborn, 1963; Wallas, 1926), which assume that the creative process occurs in various phases. A generic creativity model is given by Warr and O'Neill (2005), who combine common creativity models and divide the creative process into the following three phases:

- *Problem preparation* (building up knowledge about the problem from information resources),
- *Idea generation* (producing ideas through combination of existing knowledge) and
- *Idea evaluation* (assessing the ideas for their appropriateness as a solution to the given problem).

In the phase of idea generation, an *idea generation technique* represents a formalized protocol (Santanen, Briggs, & Vreede, 2004) that provides step-by-step sequences of actions or instructions to guide the individual.

Our research looks at idea generation techniques from an *engineering perspective* (Horton, 2006). According to one definition, Engineering includes the creative application of scientific principles to design or develop manufacturing processes (Encyclopedia Britannica Online, 2010). We view idea generation as a manufacturing process, and our goal is to develop both a theoretical understanding of this process and practical rules for practitioners to improve it. The result will be an improvement in the utility of idea generation for the innovation process.

The methods result from this approach should fulfill the following conditions:

- *Reliable* (they always perform in the same way),
- *Predictable* (they generate a given result with predefined characteristics),
- *Transparent* (they are comprehensible and learnable for anyone),
- *Efficient* (they can generate the required quantity of output using few resources),
- *Well-founded* (they are based on scientific principles),
- *Measurable* (their performance is quantifiable) and
- *Adaptable* (they can be tailored to meet a given goal).

These properties are all characteristic of traditional engineering methods and therefore represent the standard which idea generation techniques should ideally also be able to achieve.

## THREE INGREDIENTS OF IDEA GENERATION TECHNIQUES

Research has shown that idea generation techniques can be seen as a combination of different ingredients, which define the mental activity of the individual; the material or tools, which support the mental process and a set of requirements for the environment of the creative process (Smith, 1998). We denote these three ingredients by the terms *Algorithm*, *Format* and *Modifier* (Figures 1 and 2).

The *Algorithm* of an idea generation technique is a sequence of formal steps which guides the

*Figure 1. Examples of the three ingredients of Idea Generation techniques*

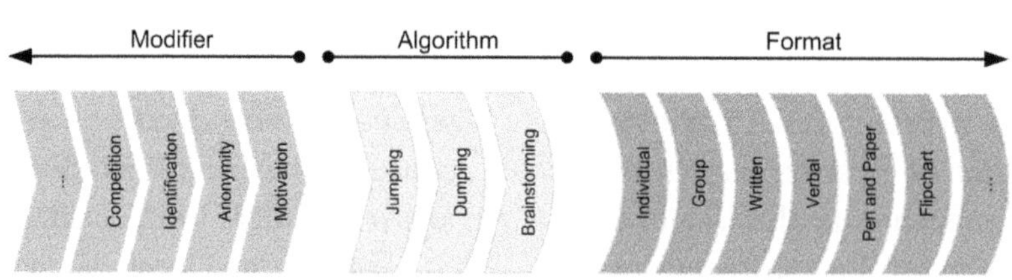

*Figure 2. Representation of different Idea Generation techniques using the three ingredients*

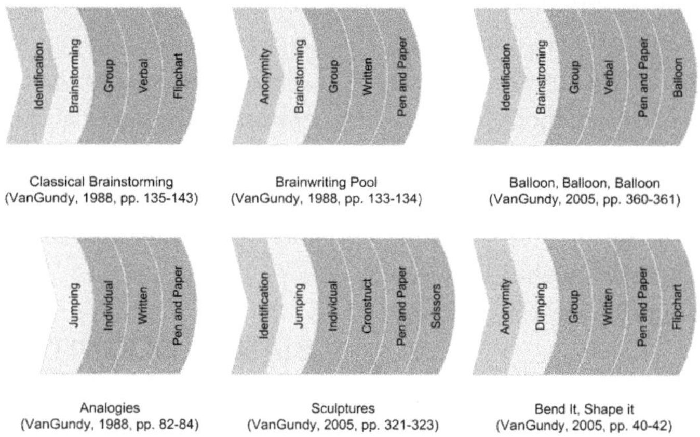


mental activities of the individual. In practice, the Algorithm is usually implemented as a sequence of instructions or questions given by the GSS or the facilitator to the participants. As will be shown in a later section, these algorithms can be categorised into just three principles which we call *Jumping, Dumping* and *Brainstorming*, which correspond to mental processes of the human cognitive network. Algorithms can use stimuli to lead the individuals to different areas of their knowledge. These stimuli are received through the five senses of the individual and can be categorised into goal- or task-related stimuli.

The *Format* of an idea generation technique refers to the organization of the Algorithm when carried out. It is primarily concerned with the organization of the participants, the role of the facilitator, and the material or tools used. Thus *Classical Brainstorming* (VanGundy, 1988, pp. 135-143) simply uses an Algorithm of the category Brainstorming and has the format of putting all participants in one group, who then call out their ideas to a facilitator, who in turn uses a flipchart to collect the ideas. Alternative formats include creating smaller groups which work in parallel, removing (or replicating) the facilitator and having the participants write down their ideas on sheets of paper or entering them into a GSS.

The *Modifier* of an idea generation technique designs a situation that supports the ideation process for the given group situation. In most cases, the ideation process is implemented as a group process, where individuals work with others as part of a formal or informal group to generate ideas. Under these conditions, the idea generation process of a group is both a *cognitive process* within individual group members and a *social process* as group members interact (Dennis, Aronson, Heninger, & Walker II, 1999). The resulting *social phenomena* influence the performance of an idea generation technique: e.g. *social loafing* (Karau & Williams, 1993): the tendency of participants to expend less effort when they believe their ideas to be dispensable and not needed for group success; or *evaluation apprehension* (Diehl & Stroebe, 1987): the fear of negative evaluation may cause participants to withhold their ideas. The Modifier of an idea generation technique changes the Format to create an environment that supports the ideation process for the given group situation. For example, *evaluation apprehension* can be reduced by the Modifier *Anonymity* that allows the participants to generate ideas in an anonymous form. Thus the format of *Classical Brainstorming* (VanGundy, 1988, pp. 135-143) would be changed to a non-verbal process, where participants write down their ideas anonymously on a sheet of paper.

To use the ideas of others as stimuli, the generated ideas are collected in the middle of a table and will be exchanged among the participants (occasionally referred to as *Brainwriting Pool* (VanGundy, 1988, pp. 133-134)). Other Modifiers like *Competition* or *Identification* influence the motivation of the participant and can reduce the phenomenon of *social loafing*.

Separation of idea generation techniques into these three ingredients supports our Engineering approach, for example:

- **Predictability:** By understanding and controlling the Algorithm and stimulus used, we are better able to predict the output of the technique.
- **Transparency:** The ingredients of idea generation techniques provide a simple approach to explain and teach creativity and problem solving (Horton, 2006).
- **Efficiency:** The Format and Modifier ingredients provide a mechanism for implementing an Algorithm in a way that most efficiently utilizes the given resources and group situation.
- **Well-Foundedness:** Basing the Algorithms on an established common cognitive model providing them with a theoretical foundation. Theories on group work and social processes can be used to define new elements or rules for the ingredients Format and Modifier.
- **Adaptability:** An understanding of the different Algorithms can help to define appropriate idea generation techniques for a given goal.

**Advice to the Practitioner:** By separating idea generation techniques into the ingredients *Algorithm*, *Format* and *Modifier*, the innovation manager can better select and adapt idea generation techniques for a given task and group situation. The Algorithm can be chosen to achieve a certain type of idea, while the Format and Modifier enable him or her to adapt the implementation to a given situation. For example, the presence of the supervisor of a group increases the possibility of *evaluation apprehension.* Thus the idea generation technique can be modified by using the Modifier *Anonymity,* which can be implemented, for example, by choosing a written Format, rather than a discursive one.

## SIAM: A COGNITIVE MODEL OF IDEA GENERATION

Lubart (2001) argues that in terms of a comprehensive understanding of the creative process, researchers need to specify the fundamental subprocesses of a creative process. Different studies have explored the mental activities of the individual. For example, Mumford and his colleagues (Mumford, Mobley, Reiter-Palmon, Uhlman, & Doares, 1991) specified the idea generation process as the combination and reorganization of information that involved reasoning, analogy use, and divergent thinking processes. The *geneplore model* (Finke, Ward, & Smith, 1992) combines generative and explorative processes to describe creativity. In this model, the individual uses generative processes like knowledge retrieval, idea association, synthesis, transformation and analogical transfer to construct loosely formulated ideas. These ideas will be elaborated and examined by explorative processes like the interpretation of the idea, hypothesis testing and searching for limitations.

We believe that the mental activities represent an interesting approach for the development of guidelines for the selection of an appropriate technique for a given innovation goal. In order to gain a better understanding of the *Algorithm* of an idea generation technique and the underlying mental activities of the individual, we used a creative cognition approach (Ward, 2006) whereby we analysed the mental activity of an individual using different idea generation techniques with

*Figure 3. Network of knowledge*

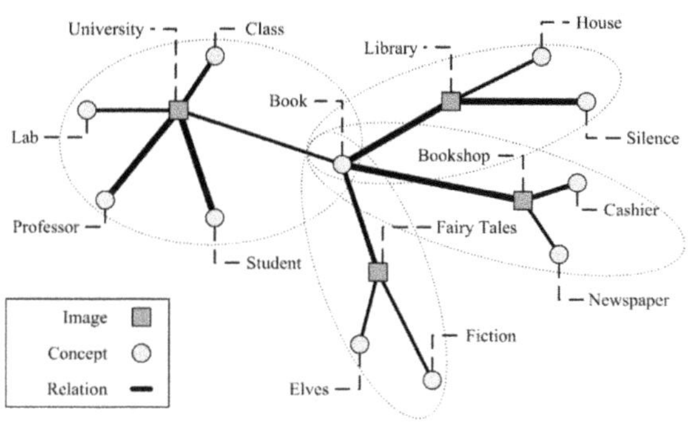


respect to a cognitive model of idea generation, called *Search for Ideas in Associative Memory* (SIAM).

Like most cognitive models, SIAM assumes that the individual has two memory systems, the *long-term memory* (LTM) and the *working memory* (WM). The WM is assumed to be a temporary storage system with limited capacity, which is used by the individual to execute conscious operations, such as rehearsal, recognition and decision making. The LTM is assumed to be an unlimited storage system, which is used to store previously acquired knowledge. This knowledge is stored as a richly interconnected network with numerous levels, categories and associations. This network is partitioned into *images*; knowledge structures that consist of a central concept and a number of features of that concept or associations with that concept. For example, the concepts *class*, *student*, *professor*, *lab* and *book* may be grouped together into the image called *university*. Images have fuzzy boundaries, may overlap to a considerable degree, and have mutual associations. For example the item *book* of the image *university* is strongly linked to the images for *library*, *bookshop* and *fairy tales* (see Figure 3). The strength of a link may be due to the frequency of its traversal, or to the relatedness among the images that it connects (Collins, & Loftus, 1975).

Based on SIAM, the idea generation process can be described as a controlled associative process (Nijstad, & Stroebe, 2006). In this process, the individual activates knowledge in the LTM depending on a search cue. This search cue is generated in the WM by external stimuli that are received through the five senses of the individual. Which images in the LTM will be activated depends on the strength of the association between the search cue and the images. The activated image will be temporarily stored in the WM, after which the concepts of the image will be accessible for the individual. The activated knowledge is used in the WM to generate ideas by forming new associations or by applying knowledge to a new domain (Mednick, 1962). Therefore, the individual combines the concepts of the image with one another or with elements of the search cue. A new image will be activated by adding previously generated ideas or external stimuli to the search cue. The process will be terminated, if the individual gets the impression that only few additional ideas can be generated. Nijstad and Stroebe (2006) assume that without any external stimuli, the individual will only modify the search cue by adding previously generated ideas. Thus, the generated ideas will be semantically related to each other. As a result, the individual will think primarily within limited areas of his or her knowledge network. The likelihood of forming new associations between previously unrelated images decreases and only a small area of the solution space will be considered (Gettys, Pliske, Manning, & Casey, 1987; Mednick, 1962).

According to Mednick (1962), we assume that unexpected associations between previously unrelated knowledge in the LTM lead to the formation of creative ideas. Idea generation techniques use external stimuli to lead the individual to different areas of their knowledge networks. Thus, several approaches exist to generate and use stimuli during the creative process. For example, *Combo Chatter* (VanGundy, 2005, pp. 127-129) combines different words related to the creative task and use the combination as semantic stimulation. In contrast, *Greeting Cards* (VanGundy, 2005, pp. 372-374) uses random pictures from magazines as stimuli to generate ideas. The resulting new perspective on a given creative task allows the individual to combine concepts of semantically unrelated images. Thus, the generated ideas will cover larger areas of the possible solution space. We see the use of external stimuli as a basic requirement for a creative process and call the resulting mental process of an individual a *change of perspective.* We define the change of perspective as a mental process which uses external stimuli to activate larger areas of the knowledge network of an individual, which would not be activated by an associative process. We assume that the change of perspective helps the individual to leave well-trodden thought paths and overcome occupational blindness. However, the influence of external stimuli on the mental process of the individual is not well understood. Therefore, we analysed well-known idea generation techniques to gain a theoretical understanding of the different ways how stimuli influence the mental processes (Knoll & Horton, 2010).

In a first research phase, the methodology focused on the relationship between the creative task, the instructions provided, the material used and the outcome of common idea generation technique like *Analogies* (VanGundy, 1988, pp. 82-84), *Assumption Reversal* (VanGundy, 1988, pp. 84-85), *Wishful Thinking* (VanGundy, 1988, pp. 127-128) and *Classical Brainstorming* (VanGundy, 1988, pp. 135-143). We analysed the intended activities for the given instructions and formalised abstract activities, which form a specific sequence of steps for a given technique. We will call this sequence of steps the algorithm of an idea generation technique. Each algorithm was subsequently normalised: Recurring steps were deleted, similar ones were consolidated and complex steps were divided into basic steps like *add*, *select* and *move*. The resulting algorithm represents the mental and physical activities of the individual during the execution of an idea generation technique.

We used the SIAM model to further analyse the mental activities. We discovered that the mental activities involved can be mapped onto just three mental principles. Two of these principles create a change of perspective by modifying the given associations in the knowledge network or activating unrelated knowledge areas. We call these two principles *Dumping* and *Jumping*. In contrast, the third mental principle *Brainstorming* represents a sequence of mental activities similar to the idea generation technique Brainstorming (Osborn, 1963) and uses only knowledge strongly related to the given creative task as a stimulus. The three mental principles will be explained in more detail with the help of SIAM in the next section.

In the second research phase, 101 idea generation techniques (VanGundy, 2005) were again reviewed to determine the completeness of the mental principles defined. We analysed and clustered the mental procedure of each of the idea generation techniques with respect to the defined mental principles. Furthermore, we focused on the relationship between the stimuli used and the tasks situation. We distinguished the techniques into those using stimuli related to the tasks situation, those using stimuli related

*Figure 4. Underlying questions of Idea Generation techniques with respect to their mental principle and the type of stimulus used*

| Stimulus / Mental Principle | **None:** Using the Creative Task as a Stimulus | **Task Situation:** Stimulus related to the Task Situation | **Goal:** Stimulus related to the Task Goal |
|---|---|---|---|
| **Brainstorming:** Activate Neighbouring Knowledge Areas | *How can we solve the task?* | *If we focus on the task situation, how can we solve the task?* | *If we focus on the task goal, how can we solve the task?* |
| **Jumping:** Activate Distant Knowledge Areas | *Be inspired by a random concept?* | *How would others in the same situation solve the task?* | *How would others with the same goal solve the task?* |
| **Dumping:** Challenge Assumptions of the Knowledge Areas | *If we challenge something, how can we solve the task?* | *If we challenge the task situation, how can we solve the task?* | *If we challenge the task goal, how can we solve the task?* |

to the tasks goals and those using no stimuli at all. The result of this analysis was that we could map all analysed idea generation technique to a combination of underlying stimulus and mental principle. We designed a matrix that categorises idea generation techniques with respect to their mental principle and the type of stimulus used (see Figure 4). Each cell of the matrix is described by an abstract question of the underlying mental process of an idea generation technique that will be categorised in this cell.

In the next chapter, we will explain the mental principles *Jumping*, *Dumping* and *Brainstorming* in more detail. Based on our experience with different innovation workshops of a consulting company ("Zephram" 2010), we will give a comprehensive answer how the provided matrix can be used to support the selection of an appropriate technique for a given innovation goal.

## THE THREE MENTAL PRINCIPLES

A change of perspective is defined as a mental process that allows individuals to cover large areas of their knowledge networks which they usually would not cover in an associative thought process. According to this definition, only the principles *Jumping* and *Dumping* are changes of perspective. We found that 36 idea generation techniques use the principle Jumping and 18 of the 101 idea generation techniques (VanGundy, 2005) use the principle Dumping to enhance the creative process (see Figure 5). In contrast, the mental principle *Brainstorming* represents an associative process that activates areas of the knowledge network that have a strong association to the image of the creative task. The analysis has shown that 47 of the 101 idea generation techniques use the mental principle Brainstorming. Each of the given mental principles will be further described by a formal sequence of steps for applying the mental principle using different type of stimuli.

For each principle/stimulus combination in the following sections, we give an example based the innovation task "Invent an innovative wristwatch."

## THE MENTAL PRINCIPLE JUMPING

The mental principle Jumping uses external stimuli to activate distant knowledge areas that have no or only a weak association to the image of the creative task (we jump to a distant location in the associative network). Idea generation techniques which use this mental principle request the individual to think of a random situation and use the knowledge about this situation to generate ideas for the creative task. Most idea generation techniques use random elements as stimuli, which have no relation to the given creative task. These

*Figure 5. Summary of the analysis of 101 Idea Generation techniques (VanGundy, 2005) with respect to their mental principle*

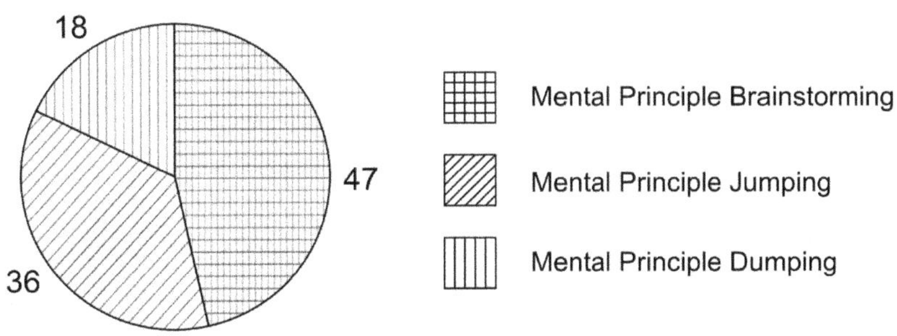

random elements are mostly provided as words (e.g. *Picled Brains*, *Say What* or *Fairy Tale Time*) or physical elements like sculptures or inkblots (e.g. *Sculptures* or *Rorschach Revisionist*) (VanGundy, 2005).

Our experience with innovation workshops has shown that the mental principle Jumping supports the generation of innovative ideas by prompting the individual to form new associations between distant knowledge areas. However, we see a weakness in the use of random elements that could generate a high cognitive load for the individual (Sweller, van Merrienboer, & Paas, 1998). With increasing distance between the image invoked by the stimulus and the image of the creative task, the individual will need more cognitive resources to form new associations between the images (Santanen, Briggs, & Vreede, 2002). Furthermore, the individual can generate different associative chains to connect the concepts of the both images. By using random elements, the idea generation technique provides no starting point for the individual. It remains unclear which concepts of the activated images should be used to generate new associations. As a result, the individual will jump between different concepts as well as different association chains. During this process, the activated knowledge will be stored into the limited storage system of the working memory. We assume that individuals will terminate the association process after a short time, if they do not find any ideas or reach the capacity of their working memory.

The following list presents the mental principle Jumping as a formal sequence of steps using different kinds of stimuli.

## Mental Principle: Jumping/ Stimulus: None

*Be inspired by a random concept*! A random stimulus is provided. Solutions or associations are collected and applied to the original task.

1. Choose a random stimulus.
2. What does this stimulus bring to mind? How the task might be solved in this situation?
3. Apply this solution to the original task.

Example:

1. Traffic light.
2. A sequence of changing colours.
3. Display the time using coloured LEDs.

## Mental Principle: Jumping / Stimulus: Task Situation

*How would others in the same situation solve the task?* The individual uses characteristic attributes of the task to find analogies. Existing solutions for the analogous problem are collected and applied to the original creative task.

1. Choose a characteristic attribute of the task situation.
2. Choose an analogous situation with the same attribute.
3. Imagine how the task might be solved in this situation.
4. Apply this solution to the original task.

Example:

1. A wristwatch is wrapped around my arm.
2. A blood pressure monitor is also wrapped round my arm.
3. The blood pressure monitor tells me my blood pressure.
4. Create a wristwatch that can measure blood pressure.

## Mental Principle: Jumping/ Stimulus: Goal

*How would others with the same goal solve the task?* The individual uses characteristic attributes of the goal to find an analogous situation. Existing solutions for the analogous problem will be collected and applied to the original creative task.

1. Choose a characteristic attribute of the task solution.
2. Choose an analogous situation with the same attribute.
3. Imagine how this solution might be realised in this situation.
4. Apply this solution to the original task.

Example:

1. A wristwatch is used to express individuality.
2. A car can be used to express individuality.
3. Cars can be configured to my personal taste and requirements.
4. Create a watch that can be configured (perhaps online) by the customer.

**Advice to the Practitioner:** A large proportion of idea generation techniques use the random Jump algorithm, varying only in the Format and Modifier used. Many commonly quoted methods differ only in the source of the random input (for example magazines, dictionaries or catalogues.) The algorithm is appropriate for very open problems, such a finding a theme for an article. By contrast, it does not work well when the ideation task has many boundary conditions or must meet specific criteria.

The Jump algorithm with focus on the given task corresponds to the classical analogy technique. It is very versatile, since by choosing the attribute to focus on, the style of analogy can be controlled. In a conservative form, it amounts to little more than observing or benchmarking the competition, while more exotic attributes yield more exotic analogies.

The jump algorithm with stimulus goal is seldom found in the literature. However, it can be very useful, since it makes use the goal of the ideation task, while at the same time providing a change of perspective.

## THE MENTAL PRINCIPLE DUMPING

The mental principle *Dumping* challenges the assumptions of the creative task to generate a new perspective on the creative task (we discard or dump an assumption). Most techniques request the individual to think of a random situation which describes the creative task and to then challenge this situation. The individual analyses the modified situation for consequences or processes and uses them to generate ideas for the creative task.

Our experience has shown that the mental principle Dumping supports the generation of innovative ideas by challenging the given associations in the knowledge network. Like *Jumping*, the mental principle *Dumping* might generate a high cognitive load for the individual. A given situation can be modified in different ways to generate new images. Most idea generation techniques provide no starting point for the individual, describing which part of the situation should be challenged as well as how the situation should be changed. As a result, the individual will jump between different concepts as well as different association chains.

The following list presents the mental principle Dumping as a formal sequence of steps using different kind of stimuli.

### Mental Principle: Dumping/ Stimulus: None

*If we challenge something, how can we solve the task?* A random belief or observation is challenged. Consequences of this challenge are collected and used as inspiration for solving the original task.

1. Challenge a random assumption.
2. Find the consequences that result from this challenge.

3. Apply these consequences to the original task.

Example:

1. I can change the colour of the sky.
2. Perhaps I would change the colour of the sky according to my mood.
3. Create a "mood watch" that allows colour changes (for example by using LEDs).

## Mental Principle: Dumping/ Stimulus: Task Situation

*If we challenge the given task situation, how can we solve the task?* A characteristic attribute of the task situation is challenged. Consequences of this challenge are collected and used as inspiration for solving the task.

1. Choose a characteristic attribute of the task situation.
2. Challenge this attribute.
3. Find consequences that result from the challenged attribute.
4. Apply these consequences to the original task.

Example:

1. The wristwatch alarm goes off at a predetermined time.
2. What if the alarm went off at a random time?
3. The wristwatch would interrupt me unpredictably.
4. A "self-discipline" wristwatch, whose random alarm causes me to reflect on how I am currently using my time.

## Mental Principle: Dumping/ Stimulus: Goal

*If we challenge the task goal, how can we solve the task?* A characteristic attribute of the task goal is challenged. Consequences of this challenge are collected and used as inspiration for solving the task.

1. Choose a characteristic attribute of the task goal.
2. Challenge this attribute.
3. Find consequences that result from this challenge.
4. Apply these consequences to the original task.

Example:

1. A wristwatch should work for everybody.
2. What if the wristwatch only worked for certain wearers?
3. For example a voice-activated watch.
4. Create a watch that requires voice recognition to display the time (or perhaps perform other functions.)

**Advice to the Practitioner:** The first algorithm corresponds to the common admonition to "just think out of the box" when looking for ideas. We do not believe it is of much use as an algorithm for generating ideas *per se*, since it is both extremely arbitrary and also difficult to execute.

The Dump algorithm based on the given situation corresponds to VanGundy's methods *Law Breaker* and *Turn Around* (VanGundy, 2005) and De Bono's method of *Provocation/Escape* (De Bono, 1993). Since it challenges assumptions about the world, it is useful for overcoming occupational blindness. It is therefore appropriate when familiarity with a problem is a hindrance or when situations can benefit from a radical reinterpretation.

The third algorithm is used, for example, by the *Problem Reversal* method of VanGundy (VanGundy, 2005). As in the first case, this algorithm seems both difficult to execute and arbitrary in its results.

## THE MENTAL PRINCIPLE BRAINSTORMING

The mental principle *Brainstorming* produces no change of perspective. Brainstorming represents a sequence of mental activities which activates areas of the knowledge that have a strong association to the image of the creative task. Idea generation techniques (i.e. Brainstorming) which use this mental principle instruct the individual to think about the creative task and use their knowledge about this task to generate ideas. Furthermore, most techniques suggest a group of individuals to share their resulting ideas as stimuli to inspire one another and activate knowledge unrelated to the creative task.

Our experience has shown that ideas generated by a homogenous group (individuals with similar knowledge about the creative task) cannot inspire a change of perspective in this group. Therefore, sharing ideas will activate unrelated knowledge only by accident. In this case, the innovation manager has no significant influence on the cognitive processes of the participants. As a result, each individual considers only a small area of the solution space and cannot overcome occupational blindness. The analysis has shown that 47 of the 101 idea generation techniques use the mental principle Brainstorming. An explanation for the large number of techniques can be given by Smith (1998) who regards an idea generation technique as a combination of different ingredients Strategy, Tactics and Enabler. With regard to this framework, most of the analysed techniques use similar mental Strategies (mental activities of the individual) and only differ in their Tactic (tools that support mental activities) or Enabler (conditions that support the creative process). For example, the technique *Idea Pool* (VanGundy, 2005, pp. 340-341) instructs the group members to write down ideas on a sheet of paper, to collect them in the middle of a table and to use the idea of others to generate new ideas. In contrast, the technique *Museum Madness* (VanGundy, 2005, pp. 342-343) defines the procedure to write ideas individually on a sheet of flip-chart paper; walk around and read each other's ideas and to use these ideas as stimuli for new ideas. Therefore, several new idea generation techniques can be developed simply by changing the Tactic or the Enabler of existing techniques. However, all of these apparently new techniques used the same mental process.

The following list presents the mental principle Brainstorming as a formal sequence of steps using different kind of stimulus.

### Mental Principle: Brainstorming/ Stimulus: None

*How can we solve the task?* The individual generates ideas by thinking about the creative task.

1. Imagine how the task can be solved.

### Mental Principle: Brainstorming/ Stimulus: Task Situation

*If we focus on the task situation, how can we solve the task?* The individual focuses on characteristic attributes of the task situation. Solutions will be generated by using these attributes to solve the task.

1. Choose a characteristic attribute of the task situation.
2. Imagine how the task might be solved by using or referencing the characteristic attribute.

### Mental Principle: Brainstorming/ Stimulus: Goal

*If we focus on the task goal, how can we solve the task?* The individual focuses on characteristic attributes of the goal. Solutions will be generated using these characteristic attributes of the goal.

1. Choose a characteristic attribute of the task solution.

2. Imagine a solution for the task that attains this characteristic attribute.

**Advice to the Practitioner:** The innovation manager can improve the association process of an individual by requesting or providing stimuli having a relation to the given creative task (Santanen et al., 2004). Thereby, the idea generation technique uses attributes or goals as the creative task. An example is the idea generation technique Direct-Brainstorming (Santanen et al., 2004), which decomposes the solution space of a given problem and presents a series of stimuli that are derived from the criteria for an effective solution.

When the stimulus is the task itself, attributes can be generated systematically by using attribute categories. One such set of categories is formed by the "eight P's" (Parts, Properties, Problems, People, Processes, Places, Parameters, and Purposes.) By asking questions such as "What parts does a wristwatch have?", "What properties does a wristwatch have?" or "What problems are associated with wristwatches?", a large set of task-related attributes can be quickly generated. The association process of an individual can be improved further by a checklist for the modification of the creative task. A generalized checklist is *Osborn's Checklist* (Osborn, 1963) that includes the following verbs: *put to other uses*, *adapt*, *modify*, *magnify*, *minify*, *substitute*, *rearrange*, *reverse* and *combine.*

## PRACTICAL APPLICATION IN THE INNOVATION PROCESS

The practical use of the structure of idea generation techniques for the innovation process is given in different ways. With regard to the Fuzzy Front End, the innovation manager can use the principles to select or create an idea generation technique that fits his strategic goal. The ingredient Algorithms is the skeleton of an idea generation technique that can be adapted to the given resources and group situations by the ingredients Format and Modifier. An algorithm of an idea generation technique could be adapted to the *technology push* and *market pull* approaches to innovation. The technology push approach generates ideas based on existing resources in the organisation. Current sources of ideas like marketing tools focus on the collection of customer data. The resulting customer data represents task situation stimuli, which can be used in an idea generation workshop. Thus the innovation manager can use the provided ingredients to select idea generation techniques, which use task situation stimuli. The generation of multifaceted ideas will be supported by the combination of idea generation techniques with different mental principles. A market pull-oriented innovation approach on the other hand generates ideas which fulfil customers' demands. Corresponding idea generation techniques use goal stimuli representing these customer demands.

The structure is also helpful to implement idea generation techniques using information technology like Group Support Systems (GSS). Current idea generation techniques were designed for a face-to-face workshop and their technological implementation often remains unclear. Our research suggests that the mental principles of an idea generation technique can be described by a formal sequence of steps (Knoll & Horton, 2010). Each of the mental principles defined can be implemented as a question-and-answer application with given GSS technology. With regard to the provided technology, the innovation manager may have a choice of media (video, image or audio file) for presenting the stimuli, and how the participants will contribute their ideas (mind map, wiki or electronic brainstorming). As a result, an idea generation workshop could benefit from techniques which were designed for a face-to-face workshop and GSS technology.

## CONCLUSION

Our goal is to develop both a theoretical understanding and practical rules for innovation managers to improve the utility of idea generation for the innovation process. Therefore, we look at idea generation techniques from an engineering perspective. Like most researchers, we believe that idea generation techniques improve the creative process by using external stimuli. We define a change of perspective as a mental principle that stimulates the ideation process. This is done by supporting individuals in visiting locations in their knowledge network which they would not otherwise become aware of. One-hundred-and-one idea generation techniques were analysed to gain a theoretical understanding of the different ways how ideation can be stimulated. The analysis showed that there are three mental principles which can be extended by different types of stimuli. Creativity research can use these mental principles to categorize idea generation techniques.

A new approach to describing idea generation techniques is introduced by the three ingredients:

- **Algorithm:** is a sequence of formal mental activities of the individual
- **Format:** defines the organization of the participants, the role of the facilitator, and the use of material or tools
- **Modifier:** defines the adaption of the Format with regard to the social process of the group.

The formalisation can be used to define guidelines and methods for the selection of an appropriate idea generation technique for a given creative task or a specific group setting. We believe that the resulting methods for the selection will fulfil the requirements for an engineering discipline. Further research is needed to analyse given idea generation techniques with regard to their Modifiers and Format. We will need further guidelines and methods that support the innovation manager in designing, configuring and selecting an appropriate idea generation technique for an innovation goal. In this manner, we hope to make the ideation processes for the Fuzzy Front End of innovation more efficient.

## ACKNOWLEDGMENT

This chapter is based on the results of cooperative research between the University of Magdeburg and the consulting company Zephram. We would like to thank Jana Görs, René Chelvier, Wolf Brüning and Falko Werner for their contributions to our discussions. Thanks are also due to Claudia Krull of the University of Magdeburg for supporting us during the preparation of this chapter and to the reviewers for their valuable suggestions.

## REFERENCES

Amabile, T. M. (1983). *The social psychology of creativity*. New York: Springer-Verlag.

Bostrom, R. P., & Nagasundaram, M. (1998). Research in creativity and GSS. *Proceedings of the 31st Hawaii International Conference on System Sciences*. Los Alamitos: IEEE Computer Society Press.

Briggs, R. O., Adkins, M., Mittleman, D., Kruse, J., Miller, S., & Nunamaker, J. F. Jr. (1999). A technology transition model derived from field investigation of GSS use aboard the USS Coronado. *Journal of Management Information Systems, 15*(3), 151–195.

Collins, A. M., & Loftus, E. F. (1975). A spreading-activation theory of semantic processing. *Psychological Review, 82*(6), 407–428. doi:10.1037/0033-295X.82.6.407

De Bono, E. (1993). *Serious creativity: Using the power of lateral thinking to create new ideas*. HarperBusiness.

de Vreede, G. J., Vogel, D. R., Kolfschoten, G. L., & Wien, J. S. (2003). Fifteen years of GSS in the field: A comparison across time and national boundaries. *Proceedings of the 36th Hawaii International Conference on System Sciences*. IEEE Computer Society Press.

Dennis, A. R., Aronson, J. E., Heninger, W. G., & Walker, E. D. II. (1999). Structuring time and task in electronic brainstorming. *Management Information Systems Quarterly*, *23*(1), 95–108. doi:10.2307/249411

Diehl, M., & Stroebe, W. (1987). Productivity loss in brainstorming groups: Toward the solution of a riddle. *Journal of Personality and Social Psychology*, *53*, 497–509. doi:10.1037/0022-3514.53.3.497

Diehl, M., & Stroebe, W. (1991). Productivity loss in idea-generating groups: Tracking down the blocking effects. *Journal of Personality and Social Psychology*, *61*(3), 31–63. doi:10.1037/0022-3514.61.3.392

Encyclopædia Britannica Online. (2010). *Engineering*. Retrieved April 01, 2010, from www.britannica.com/EBchecked/topic/187549/engineering

Finke, R. A., Ward, T. B., & Smith, S. M. (1992). *Creative cognition: Theory, research, and applications*. Cambridge, MA: MIT Press.

Gallupe, R. B., Dennis, A. R., Cooper, W. H., Valacich, J. S., Bastianutti, L. M., & Nunamaker, J. F. Jr. (1992). Electronic brainstorming and group size. *Academy of Management Journal*, *35*(2), 350–369. doi:10.2307/256377

Gettys, C. F., Pliske, R. M., Manning, C., & Casey, J. T. (1987). An evaluation of human act generation performance. *Organizational Behavior and Human Decision Processes*, *39*(1), 23–51. doi:10.1016/0749-5978(87)90044-6

Google Wave. (2010). *Homepage information*. Retrieved May 9, 2010, from http://wave.google.com

Hender, J. M., Dean, D. L., Rodgers, T. L., & Nunamaker, J. F. Jr. (2002). An examination of the impact of stimuli type and GSS structure on creativity: Brainstorming versus non-brainstorming techniques in a GSS environment. *Journal of Management Information Systems*, *18*(4), 59–85.

Hender, J. M., Rodgers, T. L., Dean, D. L., & Nunamaker, J. F., Jr. (2001). Improving group creativity: Brainstorming versus non-brainstorming techniques in a GSS environment. *Proceedings of the 36th Hawaii International Conference on System Sciences*. IEEE Computer Society Press.

Herring, S. R., Jones, B. R., & Bailey, B. P. (2009). Idea generation techniques among creative professionals. *Proceedings of the 42nd Hawaii International Conference on System Sciences*. Los Alamitos, CA: IEEE Computer Society Press.

Herstatt, C., Verworn, B., & Nagahira, A. (2004). Reducing project related uncertainty in the "fuzzy front end" of innovation — a comparison of German and Japanese product innovation projects. *International Journal of Product Development*, *1*(1), 43–65. doi:10.1504/IJPD.2004.004890

Higgins, J. M. (1994). *101 creative problem solving techniques: The handbook of new ideas for business*. Winter Park, FL: New Management Publishing Company.

Horton, G. (2006, July). *Idea engineering: Teaching students how to generate ideas*. Paper presented at the 9th International Conference on Engineering Education, San Juan, Puerto Rico.

Karau, S., & Williams, K. (1993). Social loafing: A meta- analytic review and theoretical integration. *Journal of Personality and Social Psychology*, *65*, 681–706. doi:10.1037/0022-3514.65.4.681

Knoll, S. W., & Horton, G. (2010) Changing the perspective: Improving generate thinkLets for Ideation. *Proceedings of the 43rd Hawaii International Conference on System Sciences*. Los Alamitos, CA: IEEE Computer Society Press.

Lubart, T. I. (2001). Models of the creative process: Past, present and future. *Creativity Research Journal*, *13*(3/4), 295–308. doi:10.1207/S15326934CRJ1334_07

Mednick, S. A. (1962). The associative basis of the creative process. *Psychological Review*, *69*(3), 220–232. doi:10.1037/h0048850

Mumford, M. D., Mobley, M. I., Reiter-Palmon, R., Uhlman, C. E., & Doares, L. M. (1991). Process analytic models of creative capacities. *Creativity Research Journal*, *4*(2), 91–122. doi:10.1080/10400419109534380

Nijstad, B. A., & Stroebe, W. (2006). How the group affects the mind: A cognitive model of ideation in groups. *Personality and Social Psychology Review*, *10*(3), 186–213. doi:10.1207/s15327957pspr1003_1

Nunamaker, J. F. Jr, Dennis, A., Valacich, J., Vogel, D., & George, J. F. (1991). Electronic meeting systems to support group work. *Communications of the ACM*, *34*(7), 40–61. doi:10.1145/105783.105793

Osborn, A. F. (1963). *Applied imagination: Principles and procedures of creative thinking* (2nd ed.). New York: Scribner's & Sons, Inc.

Santanen, E. L., Briggs, R. O., & de Vreede, G. J. (2002). Toward an understanding of creative solution generation. *Proceedings of the 35th Hawaii International Conference on System Sciences*. IEEE Computer Society Press.

Santanen, E. L., Briggs, R. O., & de Vreede, G. J. (2004). Causal relationships in creative problem solving: Comparing facilitation interventions for ideation. *Journal of Management Information Systems*, *20*(4), 167–197.

Shneiderman, B. (2000). Creating creativity: User interfaces for supporting innovations. *ACM Trans. On Computer-Human Interaction*, *7*(1), 114-138.

Smith, G. F. (1998). Idea-generation techniques: A formulary of active ingredients. *The Journal of Creative Behavior*, *32*(2), 107–133.

Stream Work. (2010). *Features*. Retrieved May 9, 2010, from http://www.sapstreamwork.com/features/

Sweller, J., van Merrienboer, J. J. G., & Paas, F. (1998). Cognitive architecture and instructional design. *Educational Psychology Review*, *10*(3), 251–296. doi:10.1023/A:1022193728205

Think Tank. (2010). *Group systems solutions*. Retrieved May 9, 2010, from http://www.groupsystems.com/solutions/

Valacich, J. S., Dennis, A. R., & Connolly, T. (1994). Idea generation in computer based groups: A new ending to an old story. *Organizational Behavior and Human Decision Processes*, *57*(3), 448–467. doi:10.1006/obhd.1994.1024

VanGundy, A. B. (1988). *Techniques of structured problem solving*. New York: Van Nostrand Reinhold Company Incorporated.

VanGundy, A. B. (2005). *101 activities for teaching creativity and problem solving*. San Francisco: Wiley & Sons, Inc.

Wallas, G. (1926). *The art of thought*. New York: Harcourt, Brace & World.

Ward, T. B. (2006). Creative cognition as a window on creativity. *Methods (San Diego, Calif.)*, *42*(1), 28–37. doi:10.1016/j.ymeth.2006.12.002

Warr, A., & O'Neill, E. (2005). Understanding design as a social creative process. *Proceedings of the 5th conference on Creativity & Cognition*, ACM.

Zephram. (2010). *Home*. Retrieved May 9, 2010, from http://www.zephram.de/?lang=en

## ADDITIONAL READING

Cooper, R. B., & Jayatilaka, B. (2006). Group creativity: The effects of extrinsic, intrinsic, and obligation motivations. *Creativity Research Journal*, *18*(2), 153–172. doi:10.1207/s15326934crj1802_3

Gabora, L. (2002). Cognitive mechanisms underlying the creative process. *Proceedings of the 4th conference on Creativity & Cognition*, ACM.

Garfield, M. J., Taylor, N. J., Dennis, A. R., & Satzinger, J. W. (2001). Research report: Modifying paradigms—individual differences, creativity techniques, and exposure to Ideas in group idea generation. *Information Systems Research*, *12*(3), 322–333. doi:10.1287/isre.12.3.322.9710

Jung, J. J. H., Schneider, C., & Valacich, J. S. (2005) The influence of real-time identifiability and evaluability performance feedback on group electronic brainstorming performance. *Proceedings of the 38th Hawaii International Conference on System Sciences*, IEEE Computer Society Press.

Knoll, S. W., Chelvier, R., & Horton, G. (2007). Formalised online creativity using thinXels. *Proceedings of the 10th European Conference on Creativity and Innovation*.

Mueller, S. Ch., Diehl, M., & Ziegler, R. (2006) Cross-cuing versus self-cuing: What enhances performance in a brainstorming task? *Proceeding of the 28th Annual Conference of the Cognitive Science Society*.

Nagasundaram, M., & Dennis, A. R. (1993). When a group is not a group: The cognitive foundation of group idea generation. *Small Group Research*, *24*(4), 463–489. doi:10.1177/1046496493244003

Nijstad, B. A., Stroebe, W., & Lodewijkx, H. F. M. (2002). Cognitive stimulation and interference in groups: Exposure effects in an idea generation task. *Journal of Experimental Social Psychology*, *38*(6), 535–544. doi:10.1016/S0022-1031(02)00500-0

Reinig, B. A., & Briggs, R. O. (2006). Measuring the quality of ideation technology and techniques. *Proceedings of the 39th Hawaii International Conference on System Sciences*, IEEE Computer Society Press.

Santanen, E. L. Briggs, R.O. & Vreede, G.J. de (2000). The cognitive network model of creativity: A new causal model of creativity and a new brainstorming technique. *Proceedings of the 33rd Hawaii International Conference on System Sciences*, IEEE Computer Society Press.

Santanen, E. L. (2005). Resolving ideation paradoxes: Seeing apples as oranges through the clarity of thinkLets. *Proceedings of the 38th Hawaii International Conference on System Sciences*, IEEE Computer Society Press.

Santanen, E. L., Briggs, R. O., & de Vreede, G. J. (2003). The impact of stimulus diversity on creative solution generation: An evaluation of the cognitive network model of creativity. *Proceedings of the 36th Hawaii International Conference on System Sciences*, IEEE Computer Society Press.

Santanen, E. L., & de Vreede, G. J. (2004) Creative approaches to measuring creativity: comparing the effectiveness. *Proceedings of the 37th Hawaii International Conference on System Sciences*, IEEE Computer Society Press.

Shepherd, M. M., Briggs, R. O., Reinig, B. A., & Yen, J. (1995). Social loafing in electronic brainstorming: invoking social comparison through technology and facilitation techniques to improve group productivity. *Proceedings of the 28th Annual Hawaii International Conference on System Sciences*, IEEE Computer Society Press.

Welling, H. (2007). Four mental operations in creative cognition: The importance of abstraction. *Creativity Research Journal*, *19*(2/3), 163–177. doi:10.1080/10400410701397214

## KEY TERMS AND DEFINITIONS

**Algorithm:** A sequence of formal mental activities of the individual which defines an idea generation technique.

**Change of Perspective:** A mental process which uses external stimuli to activate larger areas of the knowledge network of an individual, which would not be activated by an associative process.

**Format:** Organization of the participants, the role of the facilitator, and the use of material or tools used by an idea generation technique.

**Idea Generation Technique:** A method for systematically producing ideas, usually in a form of a sequence of simple questions or instructions.

**Mental Principle Brainstorming:** A mental process which uses external stimuli to activate areas of knowledge that have a strong association to the creative task.

**Mental Principle Dumping:** A Change of Perspective which uses external stimuli to challenge the assumptions of the creative task.

**Mental Principle Jumping:** A Change of Perspective which uses external stimuli that have no or only a weak association to the image of the creative task.

**Modifier:** Adaption of the Format with regard to the social process of the group to support the idea generation technique.

Section 3

# Tools for Creativity

# Chapter 10
# Experience with Self-Guiding Group Support Systems for Creative Problem Solving Tasks

**Gwendolyn L. Kolfschoten**
*Delft University of Technology, The Netherlands*

**Calvin Lee**
*TeamSupport, The Netherlands*

## ABSTRACT

*Many teams and groups use brainstorming to improve their creativity. Brainstorming can be supported with Group Support Systems (GSS). However, GSS are most successful when offered in combination with facilitation or at least training. Unfortunately, facilitation or training will impose a barrier to use such systems. In this chapter the use of a GSS for a multi-step creative problem solving task was evaluated. The groups using this GSS got no training, had no GSS experience and got no support, other than a 1 page log-in instruction. With this limited instruction and no training all participating groups handed in a report with the results of their brainstorm, using the tool. This chapter will report the process, the way it is embedded in the tool, and the results of our exploratory questionnaire among the participants.*

## INTRODUCTION

Creativity is a critical competence in organizations. Organizations need to improve their services and products continuously in order to remain competitive. To foster creativity, it is important that people in organizations collaborate, as creative solutions often are the result of multiple perspectives and interdisciplinary problem solving. Frost and Sullivan surveyed 946 decisions makers globally, using a collaboration index, and found that collabora-

DOI: 10.4018/978-1-60960-519-3.ch010

tion is a key driver of performance in organizations, its impact is twice the impact of strategic orientation, and five times the impact of market and technological turbulence (Frost & Sullivan, 2007). Given the importance of collaboration and creativity it is important to develop and support these competences in organizations.

A well known technique for creativity is brainstorming. Brainstorming is a method in which a group collectively shares ideas to resolve a problem. Originally brainstorming was developed as a face to face group process, where participants share ideas and write them on a flipchart. Osborn (1953) set four key rules to further stimulate creativity: (1) don't criticize, (2) freewheel, (3) combine and improve, and 4) the wilder the better. These rules are intended to prevent the individual participant from withholding specific ideas for fear of being chastised by other group members. To further support creativity, electronic brainstorming with GSS has been introduced. GSS enable parallel input which increases the efficiency of a collaborative creativity or brainstorming. Furthermore, they offer tools to reduce information overload and, through anonymity of participants, dominance and fear of contributing is diminished (Nunamaker, Briggs, Mittleman, Vogel, & Balthazard, 1997). Santanen et al (Santanen & Vreede, 2004) found that using GSS, the need for some of Osborn's rules is reduced, for instance, the rule 'don't criticize.' Since GSS are anonymous, the negative (blocking) effects of criticizing are reduced, and critique can even motivate participants to sharpen their ideas in this context (Santanen & Vreede, 2004). Further, creativity can be stimulated giving the group directions and triggering different perspectives (Knoll & Horton, 2010).

GSS therefore could potentially help organizations to increase their creative capacities. However, collaboration is also challenging, and the use of GSS requires additional procedural support from experts such a facilitators, trainers or at least technical assistants (Dennis & Wixom, 2001; Kolfschoten, Niederman, Vreede, & Briggs, 2008; Nunamaker, et al., 1997). This creates a significant barrier to sustainably implement collaboration support in organizations (Briggs, Vreede, & Nunamaker, 2003). We therefore looked for a way to guide groups through a brainstorming process without the need for procedural support. This resulted in the development of a GSS that does not require any additional support. In this chapter we will discuss the role and purpose of collaboration support, and its challenges. Next we will present the tool developed, called TeamSupport. Finally we will present an experiment with the tool to evaluate its role in creativity, and the extent to which the tool is self-guiding, enabling its use without additional support.

## BACKGROUND

Collaboration support can in some circumstances enable groups to accomplish their goals more efficient and effective (Fjermestad & Hiltz, 2001; Vreede, Vogel, Kolfschoten, & Wien, 2003b). Collaboration support technology offers mostly tools to collect and combine input from participants in activities such as brainstorming and voting (Nunamaker, et al., 1997). However, collaboration support is often used in combination with training or facilitation, which poses an additional barrier to its use and implementation.

While collaboration support such as GSS has proven to increase efficiency and effectiveness of groups, it is challenging to implement such collaboration support in organizations (Vreede & Briggs, 2005; Vreede & Bruijn, 1999; Vreede, Davison, & Briggs, 2003a; Vreede, et al., 2003b). Lab and field studies in collaboration support show conflicting results (Fjermestad & Hiltz, 1999, 2001; Santanen, 2005) with respect to the effectiveness and efficiency of GSS. Research has indicated that collaboration support often depends on a single champion, and when this person leaves the facilities are abandoned (Munkvold & Anson, 2001). Further the training of a facilitator can take a significant amount of time and effort, and is often offered in a master-apprentice style (Ackermann, 1996). Finally, it can be challenging to create a business case for collaboration support (Agres, Vreede, & Briggs, 2005; Post, 1993), especially because of the costs of hardware and human resources.

Fjermestad and Hiltz (Fjermestad & Hiltz, 1999, 2001) found (see Table 1) that in 200 experimental studies with GSS, 63% of the studies reported their subjects received training. In the field, only 37% got training however, 63% was supported by a facilitator, and another 17% got support from a "chauffeur" (a technical facilitator). In the experimental studies, this was very different. Only 30% of the studies in experimental setting reported offering facilitation support. While the difference between sessions in the context of research experiments and field studies are fairly large, both seem to generally offer their groups support or guidance in using the technology. Tasks in experimental setting are often more simple and take less time, and thus do not need facilitation support. However, apparently they do need training to operate the technology for the specific task, as 63% reported this. A significant amount of studies did not report at all on the use of training, and therefore we cannot infer the percentage that got neither training nor facilitation support, but we expect that this is small, as the use of these tools is not very intuitive.

To overcome the need for training, researchers have collected scripts for the use of GSS called thinkLets (Briggs, et al., 2003). Originally, thinkLets described a script for a specific tool, and its precise configuration. In this way a clear instruction on how to use GSS was created. Nowadays we conceptualized thinkLets more tool-independent. For this purpose, we describe for each thinkLet the rules and capabilities that need to be afforded to the group using technology (Kolfschoten, Briggs, Vreede, Jacobs, & Appelman, 2006). This increases the applicability of thinkLets, as they can be instantiated with different tools. However, it also makes the translation from thinkLets to tool instruction more challenging, as thinkLets now needed to be instantiated for the specific tool with which they are used. This again requires a group in need of GSS support to receive training or process support.

*Table 1. Training and facilitation support in GSS sessions*

| | Experiment | Field |
|---|---|---|
| training reported | 63,50% | 37,04% |
| no training | 3,00% | 5,56% |
| not reported | 33,50% | 57,41% |
| | | |
| facilitator | 30,00% | 62,96% |
| no facilitator | 70,00% | 20,37% |
| chauffeur | not mentioned | 16,67% |

Researchers have been exploring possibilities to support facilitation and the appropriation of GSS technology (Antunes, Ho, & Carriço, 1999; Kolfschoten, Briggs, & Vreede, 2009; Kolfschoten & Veen, 2005). Further, it is reported that restriction of functionality to only the tools that are used for the specific task might be a way to offer guidance in the use of GSS (Dennis, Wixom, & Vandenberg, 2001). Also, a large project by Briggs et al. is aiming to create a suite in which thinkLet based custom made collaboration processes can be developed to support specific tasks, which can be guided by non-experts (Briggs, Kolfschoten, Lukosch, Vreede, & Dean, 2010). In this chapter we report on an experiment with a GSS to support a 4 step creativity task, which offers only functionality for the task, and offers build-in guidance for both the group and the person initiating the session.

## THE TEAM SUPPORT GSS

To support our brainstorming task a GSS named TeamSupport (www.teamsupport.net) was used. The tool offers an online anonymous GSS environment for creativity tasks. Anonymity helps to reduce barriers for participation and removes the need for Osborns rule on judgment (Santanen & Vreede, 2004). When people are anonymous, ideas are judged on content, not on author, and people are less reluctant to share wild ideas. To support the group in sharpening their ideas, the GSS offers not only a brainstorming step, but also several steps to converge the set of ideas in a more concise set. For this purpose the tool has a build in process of four steps; brainstorming, clustering, grouping (within clusters) and discussion of the resulting ideas. The process is build with four thinkLets. First an OnePage brainstorm is performed. Participants can add ideas to a shared page. Next, a ChauffeurSort is done, where the appointed group leader has to cluster the ideas, based on a discussion with the group. Next, the Concentration thinkLet is performed to merge double ideas in the clusters. Finally a LeafHopper thinkLet is used to add comments to the final set of ideas, and these are discussed in the group. The Facilitation process model of this collaboration process can be found in Figure 1.

This process thus helps the group not only to share creative ideas, but also to converge these ideas to a small set for further consideration. It is therefore a very effective problem solving process. The tool requires the group to appoint a "leader" who is afforded more capabilities than the other group members by the tool, and has a coordinating role. This leader has however, no experience, no facilitation skills, and no support from professional facilitators, and thus in no way needs the training or skills of a professional facilitator as described in the literature (Clawson, Bostrom, & Anson, 1993; Kolfschoten, et al., 2008). The person fulfilling the leader role is thus a complete novice. The leader sets up the brainstorm session, invites the other participants (verbal, through chat or e-mail, by giving them a URL and code) and enters the brainstorm topic or question. Participants can add ideas. The leader can move the group to the next step in which they discuss the

*Figure 1. Creative collaboration process*

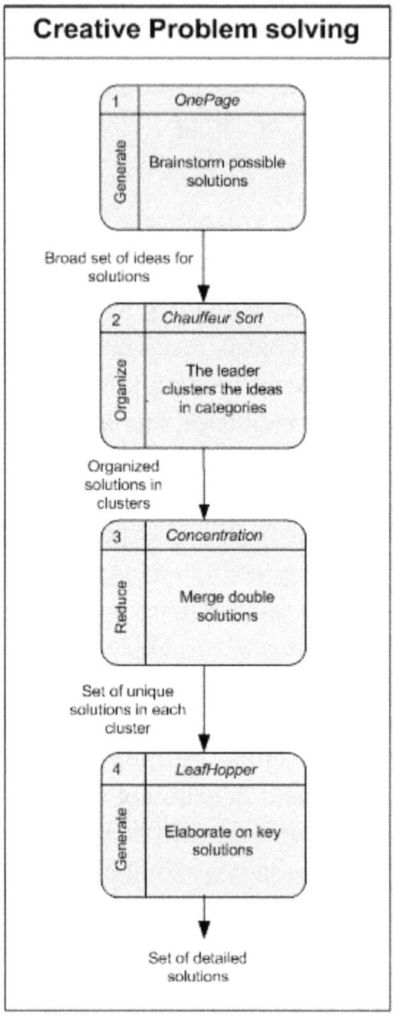


ideas and cluster them. There is a next step button, and the leader can close the previous activity entirely when finished. The leader has the ability to clusters ideas while the participants can follow along and give suggestions. In the next phase the group combines ideas in each cluster that are the same or similar into so called groups. These ideas are grouped and re-labeled to rephrase them in a way that is more precise in capturing the key idea of the group. In the last step, these groups of ideas can be discussed and comments can be added, to capture the discussion. In Figure 2a-d, several screenshots of the tool are visible. As shown, the

*Figure 2. a) Brainstorm; b) Clustering in buckets; c) Grouping within a cluster; d) Adding remarks*

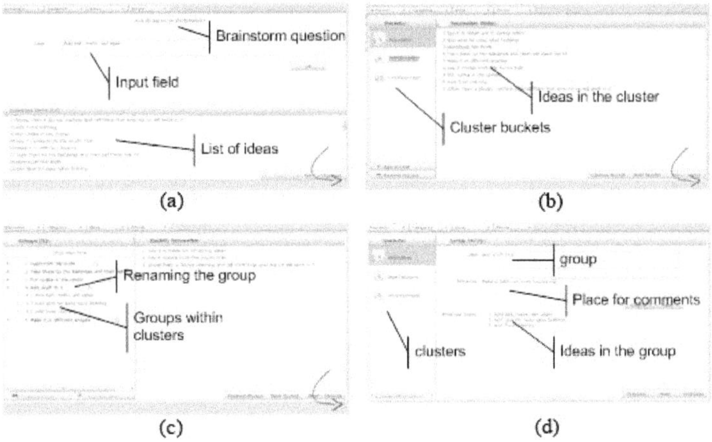


participants see only the brainstorm question and a field where they can enter their ideas. This makes the step very intuitive, and restricts functionalities to only the functions required for the task. The leader also has a button "next" which will move the group to the next step.

## Experiment

To see if this way of restriction enables groups to use the GSS without training or the support of a professional or experienced facilitator, we did an experiment in educational setting.

## Method

We asked student groups to use the GSS tool to see if they could use it without facilitation or training. In 2008 and 2009 first year students of a bachelor program in 'policy making in engineering' participated in a project course. The students had to analyze a problem case presented by a problem owner from business. In 2008 this was a hospital department manager who presented a problem in effectively using the Deming cycle for business process improvements, 98 students participated. In 2009 this was a consultant/accountant presenting a problem of improving the financial administration of a ministry in a developing country, 119 students participated. Students worked for 8 weeks on the project going through a process of problem analysis, modeling, solution finding, evaluation and reflection. Half way in the project they are instructed to brainstorm with their group of 4-6 students to identify solutions for the problem case.

Their assignment was to use the Team Support tool and to brainstorm at least 30 ideas. They got a 1 page instruction on how to acquire an account for the tool, how to log in and how to start the session. No instructions were provided on how to go through the creative problem solving process. In the second year the vendor offered a video on their website with instruction on how to use the tool. One student indicated that one of them watched the video. We do not expect that many students used this video, and it was not referred to in the instruction.

To explore the feasibility of GSS without facilitation or training, we asked all students to fill out an exploratory evaluation questionnaire. A limitation is that some students filled this out alone, while others filled it out with their group. The questionnaire contained questions about the process, the ideas generated, and the way in which they used the tool. The questionnaire also enabled students to give feedback on the tool. The questionnaire was in Dutch.

In the questionnaire we asked the students if the tool was easy or difficult to use and how this was for the leader. We used a five point scale: very easy, a little easy, easy, a little difficult, very difficult. Next, we asked if the tool supported the task, and whether it was useful (yes/no and open answer). We also asked if the students considered their ideas as creative, we used a four point scale: very creative, creative, a little creative, not creative. Further, we asked the students how they used the tool, the time they spent, whether it helped them, and if they had suggestions for improvement. These questions were open questions.

The questionnaire is exploratory and was not validated. The objective was to see if students could use the tool without support. Therefore we wanted to see how they would appropriate the tool. For this reason, we did not design the study as an experiment, and thus did not set strict guidelines on how the students should use the tool. While this introduced some limitations as indicated above, it also revealed some interesting patterns in how students used the tool, which might not have emerged in a controlled experiment. The results of the evaluation are discussed in the next section.

## Results

We received 53 questionnaires in total. All student groups used the tool successfully and handed in the brainstorming report, 42 groups, and 217 students used the tool in total.

The results for ease of use are listed in table 2. The score for the leader was not always filled out as some participants did not fulfill this role. The scores for the leader are somewhat lower than for the participants. This is not surprising, as the leader had more tasks, and needed to learn to understand more functionalities of the system. However, most leaders (22) scored the tool as a little easy, and some even scored it very easy, which seems to indicate that the difficulty was acceptable.

Next, we asked if the tool supported the task. 73% reported that the tool was supporting their task. We consider this high, given that students are generally critical about the tools they are offered. We also asked if the students considered their ideas as creative. 4% evaluated their ideas very creative, 52% evaluated their ideas as creative, 40% considered them a little creative and 4% considered them not creative. This indicates again that the tool has supported the students in identifying creative ideas. On average the groups brainstormed 38 ideas (30 was asked in the assignment), from those they created on average 5 clusters, and 11 groups representing converged ideas. Note that we removed 4 outliers who brainstormed less than 30 ideas and probably interpreted the question as the number of ideas they eventually had after grouping the ideas, as an end result.

*Table 2. Ease of use for participants and leaders*

| | **For participants n=52** | | **For leader n=41** | |
|---|---|---|---|---|
| Very difficult | 1 | 1,92% | 1 | 2,38% |
| A little difficult | 8 | 15,38% | 6 | 14,29% |
| A little easy | 19 | 36,54% | 22 | 52,38% |
| Easy | 1 | 1,92% | 0 | 0% |
| Very easy | 23 | 44,23% | 12 | 30,95% |

We did not instruct the groups on how to use the tool. 55% of the respondents decided to use it in a face to face setting, usually sitting co-located in the computer room at the university. 28% reported that they used msn or another chat tool and worked distributed. 7% reported a combination of these and 6 percent used no additional communication tools. One group used Skype. We found it very remarkable that groups used the tool distributed, and combined it with a chat tool. Research indicates various additional challenges of online facilitation (Macaulay, Alabdulkarim, & Kolfschoten, 2006; Romano, Nunamaker, Briggs, & Mittleman, 1999). The number of students using it in a distributed setting indicates that the tool has overcome several of these challenges. The groups spend on average 2 hours on the task and some reported the time for the leader which was usually the same, but sometimes slightly longer, on average 30 min longer.

We finally compared the groups working face to face with those working through chat. We found that the average time spend when working through chat was longer, (chat 2 hours and 36 min, face to face 2 hours), the groups working with chat however ended up with on average 13 grouped ideas, while the face to face group ended up with only 10. Face to face 17% scored the use of the tool as participant to be difficult, for leaders this was 8%. Others scored a little easy to very easy. For the chat group, 14% of the participants scored difficult or very difficult and 30% of the leaders indicated this score. Leading the group process in a distributed way, thus was more difficult. However, these groups also handed in their reports, and managed to go through the entire process. Last, we asked about the anonymity of the activity. Some reported that they discussed ideas and therefore anonymity was reduced or removed. This occurred more in the face to face setting than in the chat setting. For small groups, a distributed session thus improves anonymity, which as discussed earlier, can improve creativity.

## Tool Suggestions and Improvements

Besides the quantitative evaluation we also asked the respondents to reflect on the tool and to offer suggestions for improvement.

The most prominent feedback was that the tool was very easy to use. One student mentioned "the tool forced us to really think our solutions through." Some students indicated that they struggled to understand the tool in the beginning and that the tool did not offer enough 'overview.' However, most groups indicated that they understood it after a while. Some indicated that it would be much easier to use a second time. Also the students mentioned the characteristics of GSS (anonymity, parallel communication, automatic minutes) to be supportive. Several requested to add a manual or tutorial with more instruction on how to use the tool. Some students indicated that they did not see the added value of the exercise. In some occasions this was because they had already identified solutions with their group. A final difficulty was in the discussion phase. Some groups did not succeed in this phase. The reason for this is not entirely clear.

Several improvement suggestions were made. A suggestion has been made about the fact that currently only the leader can categorize while group members cannot move ideas into the categories, and instead have to instruct the leader to do this. It would make the process even more efficient and faster if the participants could also do this task. It is also suggested that drag & dropping multiple items into a bucket would be appreciated. Other suggestions for improvement include the ability to comment on the ideas from others in the brainstorm phase, a more intuitive submission interface, and to incorporate chat functionality in the TeamSupport tool to eliminate the need to work in two applications. Further the tool needs a help button and an overview of the rights of the participants and the leader. One person suggested building in a voting functionality. An overall suggestion on the tool is to make it look more attractive and to provide more of a structure and overview. The developer implemented some of these features, and accommodated the suggestions of the students in a new release of TeamSupport, presented in the next section.

## Revised Self-Guiding GSS

Based on the suggestions and the outcomes of the evaluation the following modifications were made to the tool:

1. A voting functionality has been added to rate ideas.
2. A new layout with a better overview and structure has been realized by implementing a navigation panel at the top, removing the old ability to navigate to the next step in three different ways.
3. To improve the intuitiveness of the interface, the ability to select multiple ideas for categorizing and grouping, by holding the CTR-button has been implemented, and the submission of ideas by pressing ENTER has been implemented.
4. The session leader can send messages/instructions to all participants, to partly replace separate chat tools
5. Template management functionality has been added that enables the session leader to (re-)order the process steps to his liking, for more advanced users.

The new process also slightly changed the sequence in the collaboration process and the thinkLets used. In this process, again the group starts with an OnePage to brainstorm ideas, next a ChauffeurSort to sort ideas in categories. After the ChauffeurSort, again Concentration is used and next instead of LeafHopper, the group performs a StrawPoll. (The LeafHopper can be inserted as an extra step in the process via the Template Management functionality.) In this way the group has a clearer basis to select a final set of ideas to consider further in their problem solving process. The facilitation process model of the revised process is shown in Figure 3.

*Figure 3. Creative problem solving process*

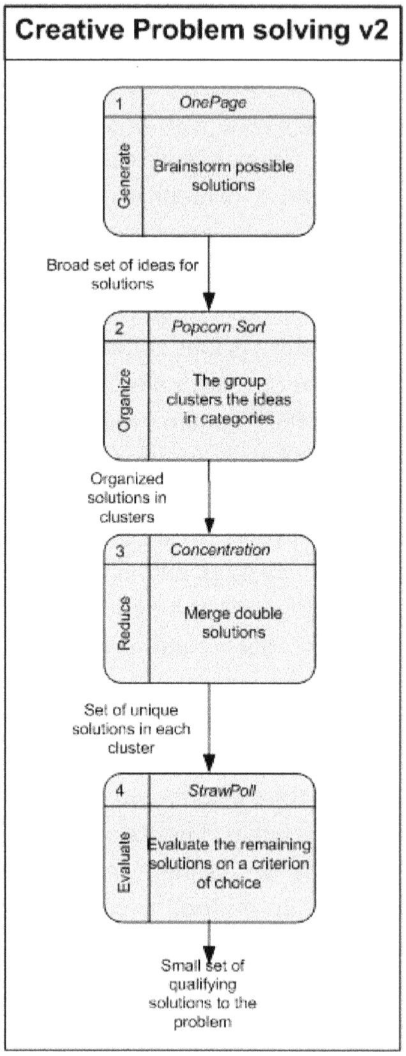


The session flow navigation pane (1) in Figure 4a shown below is the main mechanic the group leader has in order to guide the participants through the process. It is located prominently at the top of every window. When you start the session, the text field for the brainstorm topic will be highlighted (2) so the group leader knows to enter the brainstorm topic or question first.

*Figure 4. a) Brainstorm; b) Clustering in buckets; c) Merging similar ideas together; d) Rate the ideas in the Voting step*

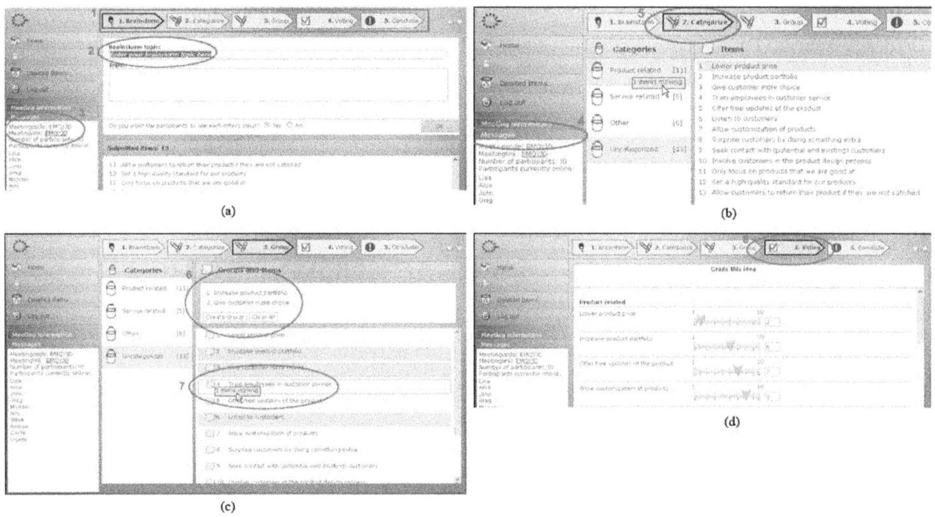

The group leader invites the participants to the session by providing them with the meeting link (3) through for example e-mail or another digital medium. Participants get into the session through the meeting link. When entering the meeting, participants will see the brainstorming topic, and will start generating ideas.

In the messages tab (Figure 4b), the group leader can send instructions to the participants whenever necessary to guide them through the process. (4) Together with the participants the group leader can categorize the generated ideas by navigating to the "Categorize" (5) step and drag and drop the items into buckets.

The next step is to group the items together that are similar or related to each other. (Figure 4c) This can be done in several ways, by dragging & dropping the items together in the drop area (6) and pressing "Create Group", or by dragging an item onto another item to drop it in order to group them together similar to managing folders and files in Windows systems (7). Both ways are very intuitive. Next (Figure 4d), the group leader can move the group to the "Voting" (8) step and press "Start voting" to let the participants start grading their ideas. This will help the group to set priorities among the ideas they identified. The voting results are presented neatly in both graph and table form. The last step is to conclude the meeting and to generate the report. This digital report can be send to the participants through regular e-mail.

## CONCLUSION AND FUTURE RESEARCH DIRECTIONS

We consider the results from the first evaluation highly encouraging. All groups managed to use the tool as intended, and found out how it worked by themselves. The tool offered all benefits of GSS and the students reported that these benefits helped them in their task. Unlike other GSS, this tool does not offer any configurable functionality. This restriction of course limits the applicability of the tool in some ways, but it ensures that the participants follow the intended process, without the need for training and facilitation. This enables groups to use a GSS for smaller and less critical tasks, as the organization of a brainstorm requires less effort, and the costs are significantly lower. The use of the step buttons at the top of the screen, and the labels of the activities (brainstorm, cluster, group, discuss) offered most groups sufficient instruction to perform the task.

Further, we expect the tool can be configured to offer even more guidance especially for the leader role. This could include clear and more specific build-in instructions that can be customized like the brainstorm question. For instance, an editable text box could be added to each functionality, to say e.g. "Please cluster the ideas based on our core competencies," instead of the current generic clustering instruction. This would help the leader to customize the tool to refer to specific organizational processes and templates for instance. Further, the creativity of the solutions might be improved if the tool would motivate participants with a target (minimum number of ideas) or with comments and feedback. We would in addition like to experiment with a timed instruction to the leader to motivate the group to come up with better ideas (e.g. after 10 minutes). The new version of TeamSupport shows that improvements in the user interface contribute highly into an even more self-guiding GSS. With the new trends in Web 2.0 techniques and usability focus, much can still be improved to accommodate the users of such tools.

Limitations of the study are numerous, as it was an explorative survey. We tried to compensate for the fact that some groups handed in multiple questionnaires, while other groups handed in one, but this was not always clear. Further we used a categorical scale for feedback. Also we did not get much input on the time difference between leaders and participants and what leaders did in this time difference.

Research indicated that a flexible/adaptive facilitation style is beneficial (Dickson, Limayem, Lee Partridge, & DeSanctis, 1996; Nunamaker, et al., 1997). However, we found that when groups want to use GSS without support, restriction increases the usability and enables groups to follow the process. We will further explore how we can support the appointed novice 'facilitator' in supporting the group and increasing the quality of their results. One of the directions for this research is to create intelligent collaboration support, to offer guidance to the leader, based on the computer's understanding of progress and activities. Also, we are interested in studying different types of users such as elderly and children, to see if they also can use these types of processes to collaborate and learn from each other.

## ACKNOWLEDGMENT

We thank TeamSupport for providing the software for this research.

## REFERENCES

Ackermann, F. (1996). Participants perceptions on the role of facilitators using group decision support systems. *Group Decision and Negotiation*, *5*(1), 93–519.

Agres, A., de Vreede, G. J., & Briggs, R. O. (2005). A tale of two cities: Case studies of GSS transition in two organizations. *Group Decision and Negotiation*, *14*(4), 256–266. doi:10.1007/s10726-005-0315-6

Antunes, P., Ho, T., & Carriço, L. (1999). *A GDSS agenda builder for inexperienced facilitators.* (pp. 1-15). Paper presented at the 10th Euro GDSS Workshop, Copenhagen, Denmark.

Briggs, R. O., de Vreede, G. J., & Nunamaker, J. F. Jr. (2003). Collaboration engineering with thinkLets to pursue sustained success with group support systems. *Journal of Management Information Systems*, *19*(4), 31–63.

Briggs, R. O., Kolfschoten, G. L., Lukosch, S. G., de Vreede, G. J., & Dean, D. (2010). *Facilitator in a box: CACE and PSS for rapid development of collaborative applications for high-value tasks.* (pp. 1-10). Paper presented at the Hawaii International Conference on System Science. Waikoloa, HI.

Clawson, V. K., Bostrom, R., & Anson, R. (1993). The role of the facilitator in computer-supported meetings. *Small Group Research, 24*(4), 547–565. doi:10.1177/1046496493244007

de Vreede, G. J., & Briggs, R. O. (2005). *Collaboration engineering: Designing repeatable processes for high-value collaborative tasks.* Paper presented at the Hawaii International Conference on System Science, Kauai HI.

de Vreede, G. J., Davison, R., & Briggs, R. O. (2003a). How a silver bullet may lose its shine-learning from failures with group support systems. *Communications of the ACM, 46*(8), 96–101. doi:10.1145/859670.859676

de Vreede, G. J., & de Bruijn, J. A. (1999). Exploring the boundaries of successful GSS application: Supporting inter-organizational policy networks. *Database, 30*(3-4), 111–131.

de Vreede, G. J., Vogel, D. R., Kolfschoten, G. L., & Wien, J. S. (2003b). *Fifteen years of in-situ GSS use: A comparison across time and national boundaries.* Paper presented at the Hawaii International Conference on System Science, Waikoloa HI.

Dennis, A. R., & Wixom, B. H. (2001). Investigating the moderators of the group support systems use with meta-analysis. *Journal of Management Information Systems, 18*(3), 235–257.

Dennis, A. R., Wixom, B. H., & Vandenberg, R. J. (2001). Understanding fit and appropriation effects in group support systems via meta-analysis. *Management Information Systems Quarterly, 25*(2), 167–183. doi:10.2307/3250928

Dickson, G. W., Limayem, M., Lee Partridge, J., & DeSanctis, G. (1996). Facilitating computer supported meetings: A cumulative analysis in a multiple criteria task environment. *Group Decision and Negotiation, 5*(1), 51–72.

Fjermestad, J., & Hiltz, S. R. (1999). An assessment of group support systems experimental research: Methodology and results. *Journal of Management Information Systems, 15*(3), 7–149.

Fjermestad, J., & Hiltz, S. R. (2001). A descriptive evaluation of group support systems case and field studies. *Journal of Management Information Systems, 17*(3), 115–159.

Frost & Sullivan. (2007). *Meetings around the world: The impact of collaboration on business Performance.* Retrieved on March 30, 2010, from http://www.verizonbusiness.com/us/resources/conferencing/impactcollab.pdf

Knoll, S. W., & Horton, G. (2010). *Changing the perspective: Improving generate thinkLets for ideation.* (pp. 1-10). Paper presented at the Hawaii International Conference on System Science, Kauai HI.

Kolfschoten, G. L., Briggs, R. O., & de Vreede, G. J. (2009). A technology for pattern-based process design and its application to collaboration engineering. In Rummler, S., & Ng, K. B. (Eds.), *Collaborative technologies and applications for interactive information design: Emerging trends in user experiences* (pp. 1–18). Information Science Reference.

Kolfschoten, G.L., Briggs, R.O., & Vreede, G.J., de, Jacobs, P.H.M. & Appelman, J.H. (2006). Conceptual foundation of the thinkLet concept for collaboration engineering. *International Journal of Human Computer Science, 64*(7), 611–621. doi:10.1016/j.ijhcs.2006.02.002

Kolfschoten, G. L., Niederman, F., de Vreede, G. J., & Briggs, R. O. (2008). *Roles in collaboration Support and the effect on sustained collaboration support.* (pp. 1-10). Paper presented at the Hawaii International Conference on System Science, Waikoloa HI.

Kolfschoten, G. L., & Veen, W. (2005). *Tool support for GSS session design.* Paper presented at the Hawaii International Conference on System Sciences, Kauai HI.

Macaulay, L. A., Alabdulkarim, A., & Kolfschoten, G. L. (2006). *An analysis of the role of the facilitator and alternative scenarios for collaboration support.* Paper presented at the First HICSS Symposium on Case and Field Studies of Collaboration, Waikoloa HI.

Munkvold, B. E., & Anson, R. (2001). *Organizational adoption and diffusion of electronic meeting systems: A case study.* (pp. 279–287). Paper presented at the Proceedings of ACM GROUP01, New York.

Nunamaker, J. F. Jr, Briggs, R. O., Mittleman, D. D., Vogel, D., & Balthazard, P. A. (1997). Lessons from a dozen years of group support systems research: A discussion of lab and field findings. *Journal of Management Information Systems, 13*(3), 163–207.

Osborn, A. F. (1953). *Applied imagination.* New York: Scribners.

Post, B. Q. (1993). A business case framework for group support technology. *Journal of Management Information Systems, 9*(3), 7–26.

Romano, N. C., Jr., Nunamaker, J. F., Jr., Briggs, R. O., & Mittleman, D. D. (1999). *Distributed GSS facilitation and participation: Field action research.* Paper presented at the Hawaii International Conference on System Sciences, Waikoloa. HI.

Santanen, E. L. (2005). *Resolving ideation paradoxes: Seeing apples as oranges through the clarity of ThinkLets.* Paper presented at the Hawaii International Conference on System Sciences, Kauai HI.

Santanen, E. L., & de Vreede, G. J. (2004). *Creative approaches to measuring creativity: Comparing the effectiveness of four divergence Thinklets.* Paper presented at the Hawaiian International Conference on System Sciences, Waikoloa HI.

## ADDITIONAL READING

Briggs, R. O., de Vreede, G. J., & Nunamaker, J. F. Jr. (2003). Collaboration Engineering with ThinkLets to Pursue Sustained Success with Group Support Systems. *Journal of Management Information Systems, 19*(4), 31–63.

de Vreede, G. J., Briggs, R. O., & Kolfschoten, G. L. (2006). ThinkLets: A Pattern Language for Facilitated and Practitioner-Guided Collaboration Processes. *International Journal of Computer Applications in Technology, 25*(2/3), 140–154.

Knoll, S. W., & Horton, G. (2010). *Changing the Perspective: Improving Generate ThinkLets for Ideation.* Paper presented at the Hawaii International Conference on System Science, Kauai HI. 1-10.

Kolfschoten, G. L., Briggs, R. O., & de Vreede, G. J. (2009). A Technology for Pattern-Based Process Design and its Application to Collaboration Engineering. In Rummler, S., & Ng, K. B. (Eds.), *Collaborative Technologies and Applications for Interactive Information Design: Emerging Trends in User Experiences* (pp. 1–18). Information Science Reference.

Kolfschoten, G.L., Briggs, R.O., & Vreede, G.J., de, Jacobs, P.H.M., & Appelman, J.H. (2006). Conceptual Foundation of the ThinkLet Concept for Collaboration Engineering. International. *Journal of Human Computer Science, 64*(7), 611–621. doi:10.1016/j.ijhcs.2006.02.002

Kolfschoten, G. L., & de Vreede, G. J. (2009). A Design Approach for Collaboration Processes: A Multi-Method Design Science Study in Collaboration Engineering. *Journal of Management Information Systems*, *26*(1), 225–256. doi:10.2753/MIS0742-1222260109

Niederman, F., Beise, C. M., & Beranek, P. M. (1996). Issues and Concerns about Computer-Supported Meetings: The Facilitator's Perspective. *Management Information Systems Quarterly*, *20*(1), 1–22. doi:10.2307/249540

Nunamaker, J. F. Jr, Briggs, R. O., Mittleman, D. D., Vogel, D., & Balthazard, P. A. (1997). Lessons from a Dozen Years of Group Support Systems Research: A Discussion of Lab and Field Findings. *Journal of Management Information Systems*, *13*(3), 163–207.

Pinsonneault, A., Barki, H., Gallupe, R. B., & Hoppen, N. (1999). Electronic brainstorming: The illusion of productivity. *Information Systems Research*, *10*(2), 110–133. doi:10.1287/isre.10.2.110

Reinig, B. A., & Briggs, R. O. (2008). On the Relationship Between Idea-Quantity and Idea-Quality During Ideation. *Group Decision and Negotiation*, *17*(5), 403–420. doi:10.1007/s10726-008-9105-2

Romano, N. C., Jr., Nunamaker, J. F., Jr., Briggs, R. O., & Mittleman, D. D. (1999). *Distributed GSS Facilitation and Participation: Field Action Research.* Paper presented at the Hawaii International Conference on System Sciences, Waikoloa HI. 1-10.

Santanen, E. L. (2005). *Resolving Ideation Paradoxes: Seeing Apples as Oranges through the Clarity of ThinkLets.* Paper presented at the Hawaii International Conference on System Sciences, Kauai, Hi. 1-10.

Santanen, E. L., de Vreede, G. J., & Briggs, R. O. (2004). Causal Relationships in Creative Problem Solving: Comparing Facilitation Interventions for Ideation. *Journal of Management Information Systems*, *20*(4), 167–197.

## KEY TERMS AND DEFINITIONS

**Brainstorming:** Generating new ideas or solutions in a participative group process.

**Collaboration Engineering:** "An approach to designing collaborative work practices for high-value recurring tasks, and deploying those designs for practitioners to execute for themselves without ongoing support from professional facilitators (Briggs et al. 2003)."

**Divergence:** Having a group generating a shared set of contributions, such as ideas, issues, problems, risks, solutions, etc.

**Facilitation:** Offering groups process and/or technology guidance to help them in achieving their collaborative goals.

**Group Support Systems (GSS):** A class of collaboration software used to move groups through the steps of a process toward their goals.

**Restriction:** Ensuring that users can only do those activities that are intended according to the work process designed.

**ThinkLet:** "A named, scripted collaborative activity that gives rise to a known pattern of collaboration among people working together toward a goal (Briggs et al. 2003)."

# Chapter 11
# The OLC Questionnaire:
## A Measure to Assess an Organization's Cultural Orientation towards Learning

**Teresa Rebelo**
*University of Coimbra, Portugal*

**A. Duarte Gomes**
*University of Coimbra, Portugal*

## ABSTRACT

*This chapter is centered on the psychometric qualities of the OLC questionnaire, which has the objective of measuring the orientation of organizational culture towards learning – a kind of culture that promotes creativity and innovation in organizations. Hence, it includes description and discussion of its conception, assessment of content validity and the main construct validity studies already carried out. Its bi-dimensionality in terms of internal integration and external adaptation processes and its potentialities for research and intervention are also discussed, as well as future research directions to continue its journey of validation.*

## INTRODUCTION

In organizational learning and learning organization literature (especially in the latest), organizational culture is mostly seen as a facilitating factor for learning in (and by) organizations. This orientation of culture towards learning is called, in the literature, oriented learning culture or simply learning culture and, in short, it is the type of culture that a learning organization should have. It can be described as an organizational culture that is oriented towards the promotion and facilitation of

DOI: 10.4018/978-1-60960-519-3.ch011

workers' learning, and its share and dissemination, in order to contribute to organizational development and performance (Rebelo & Gomes, 2009).

Concerning the characteristics that distinguish this kind of culture from other cultures, the convergence points between authors are easily seen. Among them, we can highlight learning as one of the organization's core values, focus on people, concern about all stakeholders, stimulation of experimentation, encouraging an attitude of responsible risk, readiness to recognise errors and learn from them, the promotion of open and intense communication, as well as promotion of cooperation, interdependence and knowledge share (e.g., Ahmed Loh & Zairi, 1999; Baetz, 2003; Hill, 1996; Kandemir & Hult, 2005; López, Péon & Órdas, 2004; Marquardt, 1996; Marsick & Watkins, 2003; McGill & Slocum, 1993; Reeves, 1996; Schein, 1992, 1994; Yeung, Ulrich, Nason & Glinow, 1999).

From its definition emerges the central idea underlying this kind of culture, that is to say, the organization, through culture, promotes and values individual learning with the objective that good, or productive, individual learning, through sharing processes, is turned into group learning or organizational learning and, in this way, contributes to organizational performance.

This central idea is responsible for the importance and relevance of learning culture from the nineties to today. In fact, facing more and more global, dynamic and uncertain environments, an organizational culture oriented to productive learning that leads to new and useful knowledge which, in turn, leads to innovative ways to solve problems or optimize processes, increases the probability of an organization being successful.

Therefore, conceived as a relevant facilitator of learning in, and by, organizations, the consequences of this kind of cultural orientation have been studied in the literature, namely its effects on innovation (e.g., Baetz, 2003; Bates & Khasawneh, 2005; Kontoghiorghes, Awbrey, & Feurig, 2005; Sta. Maria, 2003). And, in fact, the majority of studies concerning this issue point out a positive relationship between the existence of a learning culture and innovation in organizations.

Based on the aforesaid, it becomes important to assess an organization's cultural orientation towards learning. Besides, a valid and reliable measure of learning culture endows managers with a way to assess it in their organizations, enabling them to see and use it as a management tool.

In this respect, and in the area of instrumental research, centred on the development and validation of data collection instruments (Drenth, 1998), this chapter aims to present the OLC (Organizational Learning Culture) questionnaire, describing its development process and giving an overview of the validity studies which have supported satisfactorily its psychometric qualities.

In the scope of this book, the main goal of the present chapter is to present a valid measure of a contextual factor that positively impacts on creativity and innovation in organizations, providing, in this way, a tool for research and organizational assessment.

According to its objective, the chapter has two main sections: the first aims to introduce the central theme – the learning culture -, presenting its definition, some proposals about its critical characteristics and also some assumptions around the concept present in the literature. This section finishes with a discussion, based on some previous research, of the role of this kind of culture in relation to creativity and innovation in organizations. The second main section presents the OLC questionnaire and also its conception, development and validity studies carried out. A discussion about some strengths, shortcomings and application procedures of OLC ends this section. The chapter concludes with some indications for future research and reinforcement of the importance of instrumental research, namely around OLC.

## ORGANIZATIONAL LEARNING CULTURE: DEFINITION, CHARACTERISTICS AND ASSUMPTIONS

A better understanding of the concept of learning culture is facilitated with a short overview of the concept of organizational culture itself and the link between organizational culture and learning in organizations.

Hence, we can briefly define organizational culture using the well-known statement by Deal & Kennedy (1982): it is the way people do things in their organization. For a deeper understanding of what organizational culture is, we can resort to the definition by Schein (1992), as well-known as the former. For this author, organizational culture can be defined as:

*A pattern of shared basic assumptions that the group learned as it solved its problems of external adaptation and internal integration, that has worked well enough to be considered valid and, therefore, to be taught to new members as the correct way to perceive, think, and feel in relation to those problems. (p. 12)*

In addition, knowing Schein's proposal, concerning the levels of cultures which encompass the various cultural elements, contributes to better understanding of this concept. The Schein model comprises three distinct levels of elements, depending on the observer's visibility. However, they are interrelated. The most visible level is the level of artifacts, which comprises all the phenomena that an observer sees, hears and feels when in contact with a group (Schein, 1992). It includes cultural elements, such as architecture, decoration, language in use and jargon, myths and stories, observed rituals and ceremonies. Next, we encounter a second level, considered by Schein as espoused values, which comprises espoused strategies, goals values, rules and norms of behaviour "that members of the culture use as a way of depicting the culture to themselves and others" (op. cit, p. 16). If the values at this conscious level (what people say) do not match what people effectively do and how they behave, we witness incongruence between the espoused theories and the theories-in-use, that is to say, the implicit assumptions that really guide behaviour (Argyris & Schön, 1978). Thus, some of these espoused justifications could be simply rationalizations or future aspirations. However, if the elements of this espoused level are in fit with the next level of culture, the least visible – the basic underlying assumptions -, then the "articulation of those values into a philosophy of operating can be helpful in bringing the group together, serving as a source of identity and core mission" (Schein, 1992, p. 21). Due to the frequently registered incongruence between espoused values and basic assumptions, a part of observed behaviours and routines are sometimes hard to interpret and explain. So, for a thorough understanding of a group's culture, we need to look deeper, we need to look to the basic assumptions, which comprise shared taken-for-granted beliefs, perceptions, thoughts and feelings. A group creates these "taken-for-granted" beliefs when a hint, a value, a rule, that was a mere hypothesis for solving something, when implemented, worked repeatedly and successfully. When this happens, it comes gradually to be treated by the group as a guide for action and, in time, becomes the way of doing things in the organization.

So, basic assumptions bound what is "true", what is "right" to do, think and feel in a group of people. This is why organizational culture is a construction of people who work together, and functions as an organizational differentiating factor. Indeed, an organization could be conceived as a mini-society, endowed with symbols and rituals, its own language, a framework of shared understanding of reality, and a journey that makes itself unique and distinct from other organizations (Gomes, 2000).

Therefore, different organizational cultures have different focuses and orientations. Depending on shared assumptions, beliefs, values, routines, norms and patterns of behaviours that exist in the organization, culture can stimulate or, on the contrary, create barriers for other organizational processes. And organizational learning is one of these processes (Brown, 1998). As we have already mentioned, organizational culture is mostly seen as a potential facilitator for learning in (and by) organizations (e.g., Ahmed et al., 1999; Baetz, 2003; Campbell & Cairns, 1994; Conner & Clawson, 2004; Hill, 1996; Jones, 1996; Maccoby, 2003; Marquardt, 1996; Marsick & Watkins, 2003; Pedler, Burgoyne, & Boydell, 1997). In fact, when talking about learning culture we are speaking about the positive relationship between an organizational culture (with some specific characteristics) and learning.

However, other relationships between culture and learning appear in the literature, despite having a weaker presence. Actually, culture also assumes the role of an organizational metaphor for better understanding of the organizational learning process (e.g., Cook & Yanow, 1996; Weick & Westley, 1996). Other authors, such as Lundberg (1985) or even Schein (1992), emphasize the fact that culture itself is a learning process, and, in this context, Fiol and Lyles (1985), in their seminal article, state that these concepts have a circular relationship, because culture is a factor that influences learning and also something that people learn. Hedberg (1981), Mahler (1997), Levitt and March (1988) and Walsh and Ungson (1991), among others, highlight that organizational culture also constitutes a learning repository, as a storehouse of lessons learned or acquired knowledge.

Therefore, we could describe the relationship between culture and learning as a multiform one. Among these multiple shapes of association, this chapter is centred on the largely cited relationship that sees organizational culture as a factor that impacts on learning in, and by, organizations. Specifically, our focus is on a specific kind of culture called, in the literature, learning culture, which by definition, is a type of culture that facilitates learning and that an organization should develop if it wants to be a learning one. As Marquardt (1996, p. 70) stated "a successful corporate learning culture has a system of values that is supportive of learning", and Brown and Gray (2004) state that the transformation of an organization into a learning one necessarily implies a cultural change.

## Characteristics of a Learning Culture

A concrete way to look closely at the learning culture concept is to focus on its distinctive characteristics.

Ahmed et al. (1999), Hill (1996), Marquardt (1996), Marsick and Watkins (2003), and Schein (1992, 1994) have focused on this topic and systematised its characteristics in a way that serves as a helpful framework for understanding this type of culture in depth.

For Ahmed et al. (1999), learning and continuous improvement are intrinsically linked. These authors argued that organizational learning and continuous improvement required more than speeches or allocated financial resources, it requires an organizational culture that guides and commits workers to learning. Hence, they proposed a learning culture framework based on the literature, namely on O'Reilly (1989), Pinchot and Pinchot (1996), Schneider, Brief, and Guzzo (1996), Judge, Fryxell, and Dooley (1997), and Picken and Dess (1997). Inspired by Schein's definition of organizational culture, this framework split characteristics in two dimensions: external adaptability and internal consistency. External adaptability comprises: a safe environment to try out, make mistakes and fail; risk-taking and the expectation that innovation is part of the job; future orientation; openness, mainly in terms of communication and information sharing; autonomy; and external orientation and customer perspective. Internal consistency includes: time to experiment, opportunities and time for learning; reward learning; investment on training; leadership committed to learning; shared common goals; orientation to action and empowerment.

Hill (1996) proposes an intervention methodology for managing an organization's cultural change, aiming for transformation to a learning one. Based on Pedler et al. (1991), and also on action-research studies she carried out, the author presents a set of nuclear attributes of a learning organization, which must be reflected in the organizational culture. Thus, these attributes take the place of the characteristics of a learning culture and according to the author, they are: focus on the customer; valuing inquiry and challenging the norm; valuing creativity, experimentation and action orientation; tolerance of mistakes; valuing detection and correction activities as a learning experience; a shared vision; integration of learning opportunities for all in organizational strategy; encouraging, valuing and recognizing learning and self-development; and open and free communication with stakeholders.

Marquardt (1996) based his model of the learning organization on acquired experience in more than 50 recognised learning organizations, as well as on the literature. He presents nine characteristics of a learning culture, corresponding to key-points of learning organization management. The first is, obviously, to value learning (*learners are the heroes*), achieved through a climate that stimulates learning and an efficient compensation system, where learning is highly rewarded, for instance through award ceremonies or in the pay-check. The second is related to the fact that responsibility for learning should be shared by all, and according to the organization's objectives. The third is about mutual trust and autonomy, that is to say, a learning culture must allow and encourage experimentation and workers' autonomous decision-making in situations that require autonomous action. Consequently, giving feedback to staff is one of the main roles of leaders in this type of culture. As the fourth characteristic, the author emphasises the incentive for innovation, experimentation and responsible risk-taking and, in this context, highlights the value of mistakes as an opportunity to improve and innovate. An organization that wishes to become a learning one must be financially committed to staff training and development, to improve the quality of learning. This financial commitment is the fifth element. As the sixth element, Marquardt, just as Schein (1992, 1994), defends the need for variety and diversity in organizations, as a way to promote creativity and innovation. To value and be committed to continuous improvement of products and services is the seventh attribute of a learning culture presented by the author. The eighth refers to facing changes as a challenge, instead of being afraid of them. Responsiveness to change leads to attitudes of determination and strength, in order to meet new challenges with creativity. So, change and chaos should be conceived as opportunities to unlearn, learn, and innovate. Finally, the ninth characteristic concerns quality of working life and focuses on the commitment that learning organizations should have regarding the well-being of their workers, in physical and psychological terms, and regarding respect for the whole person.

Marsick and Watkins (2003) claimed that much of the positive learning produced in organizations happens in an informal and non-structured way, i.e., in the workplace or in group conversations. Thus, organizations need to develop a learning culture to promote and support this learning. Just as Hill (1996), these authors placed organization culture as central to a learning organization. Hence, the contents proposed for a learning culture correspond to the dimensions they propose to define a learning organization. This correspondence justifies the use of the DLOQ (Dimensions of the Learning Organization Questionnaire) to audit an organization's cultural orientation towards learning. The model comprises three levels: the individual, group and organizational. These levels are distributed in two main components: people and structure. The component related to people is essential since a learning culture must be focused on people (the primary learning agents), and comprises the individual and the group levels. However, this is not enough to promote organizational learning. So, the other component – structure – related to the organizational level, is also important and necessary.

Concerning the content of a learning culture, the authors argue that, at an individual level, the creation of continuous learning opportunities and promotion of inquiry and dialogue are elements that this type of culture should deal with. At the group level, encouragement of collaboration and team learning must also be present in this type of culture. At the organizational level, aspects like the creation of systems to capture and share learning, empowerment of people towards a collective vision, connection between the organization and its environment, and leadership oriented to learning and promoting staff learning, are all elements that a learning culture must support.

Schein (1994) departs from the assumption that organizational culture is central in the learning processes and it could act as either a facilitator or as an inhibitor. To act as a facilitating factor, the author proposes seven main elements of characterization which must be part of an organizational culture oriented towards learning. The origin of these seven characteristics is based on the cumulative knowledge provided by various organizations studied by Schein.

The first element is related to orientation towards people and a similar concern for all stakeholders. The second is a shared belief in the organization that people can, and want to learn, and that they also value learning and change. The third is about a shared belief that the external environment is malleable, in the sense that organizations have the capacity to produce changes in the environment, to control their own destiny. The key-point is that if people believe the contrary, that the environment cannot be changed, learning to improve will be worthless. As the fourth element, Schein (1994) points out the need for some time to be available in organizations to think and experiment, as well as the existence and promotion of sub-cultural diversity to allow generation of creative solutions. The fifth characteristic is centred on communication, and, to promote learning, open and extensive communication in the whole organization is required. The focus should be on development of a shared vocabulary and on promotion of open and intense communication about tasks and problems to be solved. The sixth refers to a systemic approach to situations, that is to say, it refers to people's involvement in thinking and learning, considering the multiple forces that affect an event, as well as in terms of the consequences a decision could have. Finally, the seventh element emphasises the need for teamwork and interdependence. A learning culture has to contain the shared belief that work teams really work and that individual competition should co-exist with interdependence to face an increasingly complex and specialised world.

This model of learning culture has some differences in relation to the model presented by Schein in 1992. Here, the author presents 10 elements, or dimensions, of a learning culture. A comparison of the two proposals shows that Schein adds, withdraws and fuses dimensions to build the 1994 proposal. Summarising the main differences, in the 1994 model, Schein adds orientation towards people. However, he does not mention the elements concerning the conception or nature of time, in terms of planning and orientation for the near future, task vs. relationship orientation, and the dimension relating to the nature of reality and truth. Finally, he fuses the dimensions regarding the conception of human activity and the conception of human nature in the dimension of the 1994 model related to the shared belief about people's willingness to learn.

Analysis of the frameworks proposed by these authors highlights great convergence on the characteristics a learning culture should have, as well as the idea that this kind of cultural orientation aims to promote and support individual and group learning and also allow organizational learning, through sharing new knowledge and new ways of doing things and solving problems through staff.

## Assumptions about Learning Culture

The analysis of these frameworks, as well as the literature, leads to some assumptions about learning culture. The first, and the most evident, is that learning culture is nuclear to a learning organization, because learning has a strong social component. In fact, "(...) successful learning happens with and through other people and what we choose to learn depends on who we are, who we want to become, what we care about, and which communities we wish to join" (Brown & Gray, 2004, p. 3). Also because a large amount of successful solutions of day to day working situations are based on informal and improvised ways of solving problems and dealing with circumstances (Marsick & Watkins, 2003). Consequently, it must be guaranteed by an appropriate organizational culture, which shapes the way people think, feel and act regarding learning, its use and share (Popper & Lipshitz, 2000).

Apart from the assumption that learning culture is central to a learning organization, another assumption prevailing in the literature is that learning culture should be a strong corporate culture, that is to say, a culture developed by top management, strategically disseminated through all the organization, in order to become uniformly shared. However, authors like Cook and Yanow (1996), DiBella, Nevis, and Gould (1996), Nevis, DiBella, & Gould (1995), and Schein (1993, 1996, 1997) point out the existence of different learning styles, different learning orientations in the same organization, which could originate the coexistence of different subcultures with different involvements and visions about learning. Regarding this issue, Schein (1997) highlights that different subcultures are usually misaligned and this contributes to the failure of organizational learning. Schein (1996) also states that many learning and organizational development programmes have difficulties in succeeding because they do not pass through the organization's hierarchical and functional boundaries.

The close link between learning culture and positive organizational outcomes, leading to organizations' greater capacity to face the challenges of a global and dynamic environment successfully, is another assumption underlying this type of culture. As already mentioned, this has even been responsible for its popularity, in conjunction with the notion of a learning organization, since the nineties. In this context, research about the consequences of a learning culture has been carried out, aiming to understand its effect on some outcomes, such as organizational performance (López et al., 2004; Yang, 2003; Škerlavaj, Štemberger, Škrinjar, & Dimovski, 2007), satisfaction (Egan, Yang, & Bartlett, 2004), or participation in decision making (Thompson and Kahnweiler, 2002). Among them, the relationship between learning culture and innovation has warranted attention by some researchers (e.g., Bates & Khasawneh, 2005; Kontoghiorghes, et al., 2005; Sta. Maria, 2003). The following section is dedicated to exploring the role this kind of culture has in the promotion of creativity and innovation in organizations.

## Learning Culture as a Contextual Factor for Creativity and Innovation in Organizations

The recognized centrality of organizational culture in organizational life and the maturity reached by the concept in the literature led to moving research from the concept itself towards the study of the role it plays in other organizational issues (Pettigrew, 2000).

In relation to the role organizational culture plays in creativity and innovation, it influences these processes through socialisation processes in organizations, when individuals learn what behaviours are acceptable. According to shared norms and values, individuals make assumptions about whether creative and innovative forms of behaviour are accepted and welcome in the organization. Since organizational culture is reflected in structure, daily practices, management practices, leadership and procedures, individuals come to perceive what is considered not only acceptable, but valuable, in organizations, and how they should act in the workplace (Martins & Terblanche, 2003).

Therefore, by definition, organizational culture is a contextual factor that could enable or block other organizational factors, such as creativity or innovation, and its impact depends on its characteristics. According to Angle (1989), innovation (and also creativity) occurs in organizations that have a context containing enabling and motivating conditions for it, the reverse being also true. Hence, the central question about organizational culture and creativity and innovation is which characteristics an organizational culture should have in order to promote these two processes in organizations?

Martins and Terblanche (2003) carried out a literature review aiming to develop a model that comprises the specific determinants of an organizational culture that support creativity and innovation. The elements highlighted by these authors as critical for enabling these processes in organizations (such as autonomy, empowerment, mistake handling, risk taking) match the characteristics of a learning culture. Based on several authors, Martins and Terblanche go even further, indicating that a culture that supports a continuous learning orientation encourages creativity and innovation: "By focusing on being inquisitive, encouraging personnel to talk to another (…), keeping knowledge and skills up to date and learning creative thinking skills, a learning culture can be created and maintained." (2003, p. 72).

The literature review by McLean (2005), based on the work of authors like Theresa M. Amabile, Rosabeth M. Kanter and the Minnesota Innovation Research Program of Van Ven, Angle and Poole, revealed that some characteristics of organizational culture and dimensions of organizational climate, such as organizational encouragement, supervisory encouragement, work group encouragement, freedom/autonomy, and resources, appear consistently in the literature as supports of creativity and innovation. As impediments, control is the characteristic that has been identified as decreasing organizational creativity and innovation. As can be seen from the characteristics of the learning culture mentioned above, this type of culture encompasses the characteristics that have emerged from McLean's literature review, reinforcing the promoting role of a learning culture for creativity and innovation in organizations.

This author also emphasizes that "it appears that organizational culture and climate characteristics that support creativity are similar, if not the same, as those that support innovation" (p. 241). This result is not unexpected, in so far as creativity and innovation in organizations are interrelated processes, although distinct ones. Indeed, based on Amabile, Conti, Coon, Lazenby, and Herron (1996, p. 1155), "we define creativity as the production of novel and useful ideas in any domain. We define innovation as the successful implementation of creative ideas within an organization". By these definitions, its relationship emerges naturally, that is to say, as Amabile et al. claim, creativity is the starting point for innovation, being a necessary but not sufficient condition for innovation to occur. In fact, without the generation of new and useful ideas, innovation in organizations will not occur. However, as McLean (2005, p. 226) stated "in the life of an organization, many brilliant ideas never see the light of the day". So, the process of recognizing, developing and successfully implementing a new idea - the innovation process -, while depending on the generation of new and useful ideas, also depends on other factors, such as availability of different types of resources (time, funds), technology and environmental demands. As McLean (2004, p. 226) states "creativity without innovation is of significantly diminished value".

Due to the bond between creativity and innovation, and being aware that if creativity is the seed of innovation, the psychological perceptions of innovation within an organization, the implementation of workers' ideas, motivate the generation of new ideas (Amabile et al., 1996), in other words, feed creativity. Then, organizational culture, as a contextual factor, must support both processes. However, since innovation is a more related outcome of organizational performance facing turbulent environments, many researchers have centred those studies on the impact culture has on innovation, rather than on creativity.

Among them, Hurley and Hunt (1998) carried out a study based on a sample of 9648 employees from 56 organisations of a large agency of the US federal government, centred on the relationships between cultural aspects that support a cultural orientation to innovation, that is to say, cultural initiation and receptivity to innovation (which the authors named innovativeness), and the relationship between innovativeness and capacity to innovate (operationalized as the number of new ideas adopted and successfully implemented). The four cultural characteristics considered as aspects impacting on innovativeness were power sharing, participative decision making, learning and development, and support and collaboration). Results indicated that innovativeness is associated with a greater capacity to innovate, but in addition, innovativeness is associated with cultures that emphasize learning and development (defined as the degree to which learning and development are encouraged in the organization), and also participative decision making.

Bates and Khasawneh (2005), in an exploratory examination of the relationship between organizational learning culture, learning transfer climate and organizational innovation, viewing learning culture as an antecedent that influences learning transfer which, in turn, affects organizational innovation, have also reinforced the importance of a learning culture for innovation. With a sample of 450 employees in 28 different Jordanian organizations, the findings supported a partially mediated model in which learning culture had a direct effect on organizational innovation and learning transfer climate, and that learning transfer climate, in turn, affects organizational innovation (measured by a five-item scale to assess the perceived ability of an organization to adopt or create new ideas and successfully implement them in the development of products, services, processes and procedures).

Recently, Škerlavaj, Song, and Lee (2010), using data from 201 South Korean companies, tested a model of innovation improvement based on the impact of organizational learning culture. The results show that learning culture has a strong positive direct effect (0.60, $t = 5.53$) on innovations in organizations (measured by a questionnaire regarding product, services and process innovations), as well as a moderate positive indirect impact via cultural elements supportive of innovation.

Other authors have found significant and positive relationships between learning culture and innovation, such as Kontoghiorghes et al. (2005). With a sample made up of 579 employees from four different organizations in service and manufacturing companies, they found that the most important learning organization dimensions for change adaptation, innovation (defined as the extent to which the organization can introduce new products or services quickly and easily), and organizational performance are those pertaining to organizations' structural, cultural and communication systems. Specifically, the statistical analysis identified resource availability, open communications and information sharing, as well as risk taking and new idea promotion, as the strongest predictors of innovation in organizations. Faced with these results, the authors concluded "that organizational interventions that focus on the structural, cultural, and communication system characteristics of the organization will be more likely to produce higher levels of performance, change adaptation, and innovation than those that strictly focus on learning and its application." (p. 205).

In turn, Sta. Maria (2003), based on data collected from 628 individuals in 11 Malaysian public-sector organizations, found that the perception of a learning culture has a significant and positive impact on the use of innovation, and that this effect is higher than the effect produced by the employees' concern about innovation. According to the author, "this may suggest that the overall learning culture of the organization has more to do with the use of innovation than what the individuals feel about the particular innovation" (p. 209). However, as Sta. Maria points out, the analysis across the 11 organizations shows some variability in the relationship between the variables under study. It suggests that other organizational variables also impact on innovation issues and therefore should be taken into account.

Despite the awareness of the impact of various organizational variables on the creativity and innovation processes, literature reviews and the research that we have briefly presented offer evidence of the positive effect that an organizational culture oriented to learning has on these processes. Moreover, as Bates and Khasawneh (2005) state, organizational learning and innovation appear to be based on the same related processes and both influenced by several contextual variables, such as culture, climate, structure, environment, or management practices. As a learning culture is, by definition, the culture a learning organization possesses, and this kind of organization aims to adapt successfully to the dynamics of a turbulent external environment, learning culture should be strongly linked to the promotion of creativity and innovation. In fact, we agree with Bates and Khasawneh (2005, p. 98) when they state that learning culture "emphasizes the open exchange of information and ideas in ways that facilitate learning and its creative application. In effect, learning organization culture can be seen as a critical facilitator of creativity and innovation because it supports inquiry, risk-taking, and experimentation", among other aspects. Learning culture has elements that together make this type of organizational culture a contextual factor supportive of creativity and innovation, enabling them to occur.

## THE OLC QUESTIONNAIRE

As already said, due to the recognised importance of a cultural orientation towards learning in organizations, namely as a contextual factor for promoting creativity and innovation in organizations, valid and reliable measures are needed to assess it in order to contribute to the internal validity of research, and to help managers to have a better grasp of their organizations' culture. In this context, our work with the development of OLC scale aims precisely to contribute to providing researchers and practitioners with a valid measure, that is to say, with an instrument that "does what it is intended to do" (Nunnally, 1978). This section is devoted to OLC presentation in terms of construction, development and its validity.

### Construction of the OLC and Assessment of its Content Validity

The OLC scale was created in 2000. We based it on data collected from six semi-structured interviews and learning culture frameworks proposed by Schein (1992, 1994), Hill (1996), Marquardt (1996) and Ahmed et al. (1999), presented above, namely in the characteristics where they converge. Managers were the target of the semi-structured interviews and the objective was to comprehend how they defined and characterised a cultural orientation towards learning, in order to make a comparison with the literature and, using both sources, create the scale items with more assurance. In this way, the items would be in tune with the literature and with practitioners' understanding.

Regarding its construction and the assessment of content validity, the phases suggested by Hill and Hill (2000) and Benson and Clarke (1982) were followed.

After the interview and the literature review phase, we developed a previous version of the questionnaire. The items and answer scale were elaborated based on the recommendations of Czaja and Blair (1996), Fink and Kosecoff (1985), Ghiglione and Matalon, (1977/1997), and Moreira (1997). In order to diversify item content, the following topics guided construction: customer orientation, orientation to the environment, learning value, training and human resources development, learning from mistakes and promotion of experimentation, leadership support, communication, autonomy, trust and empowerment, information and knowledge share, and interdependence. The first version of OLC comprised 40 items (9 reversed), with a five-point answer scale (In this organization… 1- *doesn't apply*, 2 – *applies little*, 3 – *applies moderately*, 4 – *applies very much*, and 5 – *totally applies*). The items were ordered by random assignment.

This previous version was refereed by two expert researchers on the topic, to evaluate the degree of appropriateness and representativeness of the items to the construct they intended to measure.

After refereeing, the questionnaire was administrated to a sample of 60 employees in a Portuguese manufacturing firm to accomplish the pre-test or pilot study phase. According to Ghiglione and Matalon (1977/1997), the pilot study was carried out in two steps: first, the OLC was administrated to 10 subjects, with different jobs and belonging to different departments, to assess the degree of suitability and comprehension of each item and the answer scale; secondly, 50 subjects (in small groups of six) filled in the questionnaire, to test its suitability with a large set of individuals and also to allow preliminary descriptive data analysis.

The results of the first step led to changing the extremes of the answer scale, as respondents said it was very difficult for them to choose these options because they were formulated in an absolute manner (1- *doesn't apply* and 5 – *totally applies* to my organization), leading to avoiding this option. Therefore, the extreme options of the scale were modified to 1- *almost doesn't apply* and for 5- *almost totally applies*.

The updated questionnaire was well received by the other 50 respondents, who considered the items easy to understand and appropriate to their organizational situation. The descriptive analysis carried out revealed a satisfactory distribution of responses through all the five options, no items had significant missing values and each subject showed variability of responses over the questionnaire. However, this preliminary analysis showed some misunderstanding in answers to reverse items. Some employees expressed difficulty in understanding the reverse items during the administration sessions, not due to the contents, but due to the type of reasoning these items require to mark the desired option correctly. Despite this information, as we were at the beginning of the OLC development process, we decided to keep these reverse items in the version that would be used in the first assessment of construct validity.

## A Synthesis of OLC Construct Validity Studies

Construct validity is related with the assessment of the degree to which the instrument reflects the construct that intends to measure. Construct validity of OLC was centred on analysis of its dimensionality. Reliability was also estimated, using the Cronbach *alpha*. As can be seen in Table 1, various studies concerning OLC construct validity have been carried out up to now.

The first study was carried out in 2001, with 224 workers from seven Portuguese manufacturing firms. Due to the fact that it was the first, an exploratory technique was chosen for data analysis. Therefore, principal component analysis, in conjunction with theoretical reasons, led to a structure, with *varimax* rotation, composed of four dimensions. This structure explained 44%

*Table 1. OLC main construct validity studies*

| | Main studies |
|---|---|
| Rebelo (2001) | Preliminary study of a sample of 224 workers from seven Portuguese manufacturing firms (use of exploratory factorial analysis) – Reduction from 40 to 30 items. |
| Rebelo, Gomes & Cardoso (2003) | Study of a sample of 627 workers from 15 Portuguese manufacturing firms (use of exploratory factorial analysis). A structure composed of 23 items divided in two factors (Internal Integration and External Adaptation) emerged. |
| Rebelo, Gomes & Cardoso (2005) | Study using confirmatory factor analysis (CFA) with a sample of 619 individuals from 35 Portuguese manufacturing firms. This supported the bi-dimensional model of OLC (23 items) suggested by previous studies. |
| Rebelo (2006) | Study using CFA with a sample of 1122 workers from 107 Portuguese companies (manufacturing and consultancy industries). The bi-dimensional "Internal Integration/ External Adaptation" is supported. Reduction from 23 to 20 items. |

of total variance and all the items loaded the pertaining factor above .30, with communalities, in general, also above .30. Based on the contents of the items of each dimension, we named them: (1) External orientation (9 items; $\alpha$ =.82); (2) Autonomy and empowerment (8 items; $\alpha$ =.78); (3) Leadership support (7 items; $\alpha$ =.78); and (4) Learning incentive (6 items; $\alpha$ =.67). This preliminary study allowed a reduction from 40 to 30 items. Nevertheless, as the previous steps concerning content analysis had warned, all the reverse items were eliminated, due to their behaviour in data analysis.

The sample of the second study (Rebelo, Gomes, & Cardoso, 2003) was made up of 627 employees in 15 Portuguese manufacturing firms. In this study, we decided to continue using principal component analysis, since we still did not have a solid enough background to move up to a confirmatory technique. This time, a structure composed of 23 items, divided in two factors, emerged as the best solution. All the statistic indicators were satisfactory (48.3% of total explained variance; the large majority of communalities above .40; all loadings above .50). Here, we used an *oblimin* rotation, because the dimensions were inter-correlated ($r$ = .53).

Comparing the former structure to this one, analysis of the item distribution in the two dimensions showed that the previous dimension of "external orientation" was replicated in one of the new dimensions. The other three dimensions ("autonomy and empowerment", "leadership support", and "learning incentive") were combined in the other new dimension. So, what seemed so different at first glance was not so different at all. In fact, this new structure only polarised the two central aspects of organizational life present in Schein's definition of culture: external adaptation and internal integration. In consequence, we decided to use these terms for the two dimensions of the OLC questionnaire. The Internal Integration dimension comprised 14 items, with a Cronbach *alpha* of .92, whereas the External Adaptation dimension comprised 9 items, with an *alpha* of .86.

Due to the fact that this bi-dimensional structure has a high interpretability in the literature (e.g., Schein, 1992), even being used in one of the learning culture frameworks where OLC is based (Ahmed et al., 1999), the statistical indicators were better than the former structure, and the size of the sample gave some confidence regarding the stability of the structure, we decided to test it in the next study.

Hence, the third study, with a sample of 619 individuals from 35 Portuguese manufacturing firms was carried out (Rebelo, Gomes, & Cardoso, 2005), using confirmatory factorial analysis (CFA) as statistical technique. This technique, as opposed to exploratory factor analysis (appropriate to situations where relationships between latent and observed variables are uncertain or even unknown), is suitable for use in situations where the objective is to test particular relationships between latent and observed variables, towards the assessment of the fit between data and the structure that constitutes the hypothesis (Byrne, 1994).

In this context, we tested the bi-dimensional structure of the previous study using EQS/Windows software package. Maximum Likelihood (ML) method was used for parameter estimation. Since sample statistics showed infringement of the multivariate normality assumption, the recommendation of Byrne (1994) for using Santorra-Bentler Scaled Statistic (S-B $\chi^2$) as fit index was followed.

The results indicated a satisfactory fit of the model to the data, S-B $\chi^2$ (228, $N$=619) = 504.11, $p$ <.001, CFI=.95, RMSEA =. 053, 90% confidence interval ranging from .048 to .058. All the parameters estimated were statistically significant and all the items' respective loading factor above .50. The two factors remained inter-correlated ($r$ = .69), as did the good reliability of both ($\alpha$ = .91 for Internal Integration factor and $\alpha$ = .84 for External Adaptation factor).

As we have already mentioned, from a conceptual point of view, this bi-dimensional structure has a high level of interpretation and, what is more, the results of this third study suggest its stability across manufacturing firm samples, inasmuch as this sample is made up of workers who differ from the workers in earlier studies in some characteristics, such as level of education, the dimension of the firms they belong to, and their job.

A second test of this bi-dimensional model, using CFA technique, was carried out by Rebelo (2006). This study used a sample of 1122 individuals from 107 companies (70% manufacturing and 30% consultancy firms). The test of the model was also with ML estimation method (Kline, 2005), but this time AMOS software was used. The results showed, once again, a good adjustment between the model and the data, $\chi^2$ (169, N = 1120) = 644.72, p =.000, $\chi^2$/df = 3.82, CFI =.95, RMSEA = .05, 90% confidence interval of .046-.054, p > .05. All parameters estimated were statistically significant and items loaded in respective factors above .50. The reliability estimated for both factors remained good ($\alpha$ = .91 for Internal Integration with 12 items, and $\alpha$ = .83 for External Adaptation with 8 items), and the factors were inter-correlated at $r$ =.73. In the process of data analysis, we made three re-specifications to the initial model of 23 items, and one item was removed in each re-specification, leading to a light version of the instrument, now with 20 items.

These results reinforce the bi-dimensionality "internal integration / external adaptation" of this learning culture measure, already present in previous studies, indicating that we are faced with a factorial matrix of considerable stability, as it remained in this sample which includes business consulting services, a context where this scale had never been administered.

Table 2 presents the 20 items, divided in Internal Integration/External Adaptation dimensions, of this last version of OLC questionnaire [1].

Until now, this 20 item version of OLC was used in the Rebelo and Almeida (2008) research, on a sample of 164 employees at four plants of a Portuguese ceramic group. Based on previous validation studies and because they were collecting data also from the manufacturing industry, they decided to forgo a re-examination of dimensionality. However, the reliability of two dimensions of the scale was estimated. The values obtained are indicative of good internal consistency ($\alpha$ = 0.92 for internal integration and $\alpha$ = 0.90 for external adaptation), similar (and even a little higher) than the values found in previous studies.

Carvalho, Lourenço, and Dimas (2010) have adapted the OLC scale to the group level. For this adaptation, the main changes the authors have introduced in the scale were in the vocabulary, changing organization to group and workers to members in some items. With a sample of 403 respondents, belonging to 73 work teams in 24 manufacturing and service firms, the results supported the bi-dimensionality of OLC at group level. In fact, the estimated reliability for the factors remained good ($\alpha$ = 0.90 for internal integration and $\alpha$ = 0.89 for external adaptation).

*Table 2. OLC dimensions and items (20 items version)*

| Dimensions | Items |
|---|---|
| Internal Integration | People are also paid for thinking.<br>Leaders are available and interested in listening to workers' suggestions for improvement.<br>Failures are seen as an opportunity to experiment new ways of doing things.<br>Contact between top management and any collaborator is easy.<br>We have the habit of sharing information and knowledge.<br>Those who contribute to ideas and solutions towards the improvement of work processes are considered the best employees.<br>Leaders encourage the search for solutions by their subordinates.<br>People are informed about organizational objectives.<br>This organization stimulates the professional development of its workers.<br>We have the habit of discussing in groups ways of solving problems when they appear.<br>Leaders agree to and support implementation of some suggestions by their subordinates.<br>There is an environment of trust and respect, where people listen to what the others say, even if they are criticisms. |
| External Adaptation | We know that if we work to quality standards, we will assure organizational success.<br>We know that a good relationship with suppliers is important.<br>We are aware that without clients there is no salary or stability.<br>We are aware that the work of one department depends on the work of other departments, and vice-versa.<br>There is a belief that people can, and want, to learn to improve.<br>Customer complaints are carefully analysed in order to improve.<br>We know it is important to contribute with innovative ideas for the improvement of work processes.<br>We recognize it is important to know our competitors so as to do better than them. |

## Discussion

All the main construct validation studies described above point to a bi-dimensional structure that reflects two kinds of problems that organizations and groups are faced with, and which they have to deal with to ensure viability: adaptation to the external environment and integration of internal processes (Schein, 1992). Internal integration consists of the structuring and coordination of internal processes in the organization (e.g., leadership style, the way work is arranged – with or without flexibility, communication structure) whereas external adaptation is related to the organization's orientation to the exterior (e.g., clients, competitors and other stakeholders), in order to correspond successfully to the demands of the environment.

Therefore, the role of culture is stressed in managing a paradox inherent to organizational life: the contrast between the processes of change (necessary for adapting organizations to the external environment), and the need to preserve internal consistency (which requires mechanisms of integration). Also the correlation between the two factors makes sense, because the processes of external adaptation and internal integration must be understood as interdependent, their mutual influence being necessary to assure survival and even organizational success.

So as already mentioned, and from the conceptual point of view, the structure has a high interpretability, since thinking about learning culture in terms of these two processes has implications for the way we consider how organizations deal with learning issues. In fact, the identified structure emphasizes the importance of a close relationship between both dimensions, leading to the idea that the organization and its employees, internally, should be actively structured, committed and learning oriented to respond effectively to the external environment, customers, and other stakeholders, in order to ensure organizational viability, as suggested by the literature on learning organizations (e.g., Salaman, 2001).

These two interrelated dimensions of learning culture suggest it is not enough to have a culture of adaptability to the external environment, which values the customer and is aware of competitors, trying to correspond to demands and expectations. If an organization wants to ensure high performance, a learning approach must be reflected within the organization, through promotion and facilitation of learning. Learning must be perceived as an integral part of any job in the organization and, therefore, should be supported, valued and rewarded, in order to establish itself as a value (a cultural element or ingredient) internalized and shared by employees. That is to say, as something acquired or established by experience and updated by organizational actors, as something similar to the process Weick (1995) called enactment.

In short, the bi-dimensionality "Internal Integration / External Adaptation", as well as the interdependence of the two factors, reinforce that the organization must learn to be flexible internally so that, based on constant surveillance of environmental conditions (customers, competition, etc.) it is able to respond effectively to their requests (for example, through innovation or better quality of products and services).

Thus, it is important to emphasize that adaptation to the environment, according to the results obtained with the OLC scale, is only one dimension to consider. Internal integration should be given equal attention, as well as their relationship (Rebelo et al., 2005).

From the point of view of intervention, this two-factor structure can also be useful, inasmuch as it can provide distinct assessment of these two main organizational processes. Based on differentiated scores for internal integration and external adaptation, managers could observe which area needs more attention regarding the facilitation of learning, and strategically decide what measures should be taken.

Apart from theoretical support, we have good statistical indicators for this factorial matrix, which seems to be stable across different samples and even at different levels of analysis. In addition, some studies already carried out using OLC questionnaire suggest it has predictive validity, since it was found that learning culture (measured by OLC) impacts on organizational profitability and customer satisfaction (Rebelo, 2006; Rebelo, Gomes & Cardoso, 2007) and also that it is able to discriminate between different learning subcultures within organizations (Rebelo, Gomes & Cardoso, 2002). Therefore, these results, to some extent, reassure us concerning the non-existence of reverse items, a shortcoming of this scale. However, in spite of being a recommendation in the literature, the reverse items of this scale did not work sufficiently well to be kept in its latest versions.

Finally, we want to make some recommendations about the use of OLC questionnaire. This measure is a quantitative technique of data collection, and being aware of the multiple levels of organizational culture, the use of questionnaires to assess organizational culture hardly captures its essence, that is to say, the basic underlying assumptions of Schein (1992). Thus, if we are aiming for a deep picture of any organizational learning culture, we must triangulate results obtained with a questionnaire (OLC or another) with qualitative methods of collecting data.

Nevertheless, a satisfactory way of increasing the confidence in results obtained with a culture questionnaire is its administration to a representative sample of employees within an organization. As Wilderom, Glunk, and Maslowsky (2000) noticed, in a review of studies concerning the culture-performance relationship, one of the shortcomings detected is the deficit of workers' representativeness in culture assessment, concluding that some of the studies reviewed only questioned top management. If organizational culture is a construction of organizational actors, it should be assessed based on investigating them, to obtain a more accurate picture of the situation. So, as OLC is a cultural measure, it should be administrated following these rules, in order to increase the confidence of researchers and practitioners about the results produced.

What is more, the choice of samples within organizations depends on the objectives of the research or the organizational diagnosis. If, for instance, the aim is to assess the strength of cultural orientation towards learning, the sample must comprise representative subsamples of departments and jobs, because they are potential *loci* of subcultures (Louis, 1985).

## FUTURE RESEARCH DIRECTIONS

We can state that the validation process is like "a never-ending story". Hence, many paths are open to carry on with OLC validation.

Although previous studies concerning construct validity present interesting and promising results, a common shortcoming of these studies is that they were performed in the same national context – that of Portugal. One research-in-progress concerning OLC is being developed in Brazil, but we know these two countries have some cultural similarities. Therefore, we think it would be important to spread the validation of this measure to other national contexts, with different national cultures from the Portuguese and Brazilian ones.

Until now, the samples used have only included organizations for profit. This limitation of OLC validity is another indication for future research, i.e., to extend its administration to non-profit organizations, to compare results and assess the stability of the bi-factorial matrix on a new kind of organizations. In this context, another research-in-progress being carried out by our research team is the adaptation of OLC to organizations in the Portuguese public sector.

Specifically regarding construct validity, other studies apart from analysis of its dimensionality should be developed, such as analysis of convergent-discriminant validity.

Besides content and construct validity studies, it is also relevant to move towards deeper assessment of OLC predictive validity, namely with some outcomes where the literature claims learning culture has an effect, such as organizational innovation or quality. Other interesting areas to include in future research to assess the predictive power of learning culture, using OLC as a measure, would be the use of information technology in organizations and also employees' well-being and satisfaction.

## CONCLUSION

We agree with Williams, Ford, and Nguyen (2002) when they state that organizational researchers "typically use less-than-perfect measures of variables that represent the substantive constructs of interest, given a theory and/or set of hypotheses being tested" (p. 367). Indeed, we are sometimes confronted with studies presenting complex and sophisticated data analysis, but with poor information about the validity of the data collection measures. And if the research design is poor and/or the measures are insufficiently valid, no data analysis, however it is elaborated, can guarantee reliable results. The same rationale could be applied to the field of intervention. In fact, a rigorous organizational diagnosis is only possible if the measures for collection of information are valid.

This is why instrumental research, aiming to develop and validate instruments for data collection is so important and useful, as much for researchers as for practitioners. In this context, the objective of our work on OLC is effectively to provide those who are interested in organizational learning and learning organizations with a consistent questionnaire for assessing cultural orientation towards learning in organizations. Our main objective, in making a good learning culture measure available, is to contribute to the quality of future research on this topic and provide an assessment tool for those managers who are concerned about learning, continuous improvement, creativity and innovation in their organizations.

The OLC has been proven to have good psychometric qualities, and its bi-dimensionality highlights the dynamics between internal integration and external adaptation processes, where learning plays a central role. In fact, nowadays, a cultural orientation towards learning should be present on a daily basis in the organization, in order to allow assimilation of new information by organizational members, support creativity and innovation, and appropriate internal reorganization. This is what organizations ultimately search for to gain competitiveness (Easterby-Smith, Lyles & Tsang, 2008). These are more than sufficient arguments for pursuing research around the improvement and validity of the Organizational Learning Culture questionnaire.

## REFERENCES

Ahmed, P., Loh, A., & Zairi, M. (1999). Cultures for continuous improvement and learning. *Total Quality Management*, *10*(4/5), 426–434.

Amabile, T. M., Conti, R., Coon, H., Lazenby, J., & Herron, M. (1996). Assessing the work environment for creativity. *Academy of Management Journal*, *39*(5), 1154–1184. doi:10.2307/256995

Angle, H. L. (1989). Psychology and organizational innovation . In Van deVen, A. H., Angle, H. L., & Poole, M. S. (Eds.), *Research on the management of innovation: The Minnesota studies* (pp. 135–170). New York: Harper & Row.

Argyris, C., & Schön, D. (1978). *Organizational learning: A theory of action perspective*. Reading, MA: Addison-Wesley.

Baetz, W. G. (2003). *Organizational learning practices.* Doctoral dissertation, University of Albany, State University of New York. ProQuest Information and Learning Company.

Bates, R., & Khasawneh, S. (2005). Organizational learning culture, learning transfer climate and perceived innovation in Jordanian organizations. *International Journal of Training and Development*, *9*(2), 96–109. doi:10.1111/j.1468-2419.2005.00224.x

Benson, J., & Clark, F. (1982). A guide for instrument development and validation. *The American Journal of Occupational Therapy.*, *36*(12), 789–800.

Brown, A. (1998). *Organisational culture* (2nd ed.). Harlow, UK: Prentice Hall.

Brown, J. S., & Gray, E. S. (2004). Introduction . In Conner, M. L., & Clawson, J. G. (Eds.), *Creating a learning culture: Strategy, technology, and practice* (pp. 1–16). Cambridge, U.K.: Cambridge University Press.

Byrne, B. M. (1994). *Structural equation modeling with EQS and EQS/Windows: Basic concepts, applications and programming*. Thousand Oaks, CA: Sage.

Campbell, T., & Cairns, H. (1994). Developing and measuring the learning organization from buzz words to behaviours. *Industrial & Commercial Training*, *26*(7), 10–15. doi:10.1108/00197859410064583

Carvalho, Lourenço, and Dimas. (2010). Cultura y eficacia: El papel de la orientación cultural para el aprendizaje en la determinación de la eficacia del grupo/equipo. *Ciencia Y Trabajo, 12*(36), 306-311.

Conner, M. L., & Clawson, J. G. (2004). Afterword . In Conner, M. L., & Clawson, J. G. (Eds.), *Creating a learning culture: Strategy, technology, and practice* (pp. 327–338). Cambridge, U.K.: Cambridge University Press.

Cook, S., & Yanow, D. (1996). Culture and organizational learning . In Cohen, M. D., & Sproull, L. E. (Eds.), *Organizational learning* (pp. 430–459). Thousand Oaks: Sage.

Czaja, R., & Blair, J. (1996). *Designing surveys: A guide to decisions and procedures*. Thousand Oaks, CA: Pine Forge Press.

Deal, T., & Kennedy, A. (1982). *Corporate cultures*. Reading, MA: Addison-Wesley.

DiBella, A., Nevis, E., & Gould, J. (1996). Organizational learning style as a core capability . In Moingeon, B., & Edmondson, A. (Eds.), *Organizational learning and competitive advantage* (pp. 38–55). London: Sage.

Drenth, P. J. (1998). Research in work and organizational psychology: Principles and methods. In P.J. Drenth, H. Thierry & C. J. de Wolff (Eds.), *Handbook of work and organizational psychology, (vol. 1), Introduction to work and organizational psychology (2nd ed.),* (pp. 11-46). Hove, East Sussex, UK: Psychology Press.

Easterby-Smith, M., Lyles, M., & Tsang, E. (2008). Inter-organizational knowledge transfer: Current themes and future prospects. *Journal of Management Studies, 45*(4), 678–690. doi:10.1111/j.1467-6486.2008.00773.x

Egan, T. M., Yang, B., & Bartlett, K. R. (2004). The effects of organizational learning culture and job motivation to transfer learning and turnover intention. *Human Resource Development Quarterly, 15*(3), 279–301. doi:10.1002/hrdq.1104

Fink, A., & Kosecoff, J. (1985). *How to conduct surveys: A step-by-step guide*. Newbury Park: Sage.

Fiol, C., & Lyles, M. (1985). Organizational learning. *Academy of Management Review, 10*(4), 803–813. doi:10.2307/258048

Ghiglione, R., & Matalon, B. (1997). *O inquérito, teoria e prática* (3ª ed.). Oeiras: Celta Editora.

Gomes, A. D. (2000). *Cultura Organizacional: Comunicação e Identidade*. Coimbra: Quarteto Editora.

Hedberg, B. (1981). How organizations learn and unlearn . In Nystrom, P. C., & Starbuck, W. H. (Eds.), *Handbook of organizational design* (*Vol. 1*, pp. 3–27). Oxford: Oxford University Press.

Hill, M., & Hill, A. (2000). *Investigação por questionário*. Lisboa: Sílabo.

Hill, R. (1996). A measure of the learning organization. *Industrial & Commercial Training, 28*(1), 19–25. doi:10.1108/00197859610105440

Hurley, R. F., & Hult, G. T. (1998). Innovation, market orientation, and organizational learning: An integration and empirical examination. *Journal of Marketing, 62*, 42–54. doi:10.2307/1251742

Jones, S. (1996). *Developing a learning culture: Empowering people to deliver quality, innovation and long-term success*. London: McGraw-Hill.

Judge, W., Fryxell, G., & Dooley, R. (1997). The new task of R & D management: Creating goal directed communities for innovation. *California Management Review, 39*, 72–84.

Kandemir, D., & Hult, G. T. (2005). A conceptualization of an organizational learning culture in international join ventures. *Industrial Marketing Management, 34*, 430–439. doi:10.1016/j.indmarman.2004.10.002

Kline, R. B. (2005). *Principles and practice of structural equation modelling* (2nd ed.). New York: The Guilford Press.

Kontoghiorghes, C., Awbrey, S. M., & Feurig, P. L. (2005). Examining the relationship between learning organization characteristics and change adaptation, innovation, and organizational performance. *Human Resource Development Quarterly, 16*(2), 185–211. doi:10.1002/hrdq.1133

Levitt, B., & March, J. (1988). Organizational learning. *Annual Review of Sociology, 14*, 319–340. doi:10.1146/annurev.so.14.080188.001535

López, S. P., Péon, J. M., & Órdas, C. J. (2004). Managing knowledge: The link between culture and organizational learning. *Journal of Knowledge Management, 8*(6), 93–104. doi:10.1108/13673270410567657

Louis, M. R. (1985). An investigator's guide to workplace culture . In Frost, P. J., Moore, F. L., Louis, M. R., Lundberg, C. C., & Martin, J. (Eds.), *Organizational culture* (pp. 73–93). Beverly Hills, CA: Sage.

Lundberg, C. C. (1985). On the feasibility of cultural intervention in organizations . In Frost, P. J., Moore, F. L., Louis, M. R., Lundberg, C. C., & Martin, J. (Eds.), *Organizational culture* (pp. 169–185). Beverly Hills, CA: Sage.

Maccoby, M. (2003). The seventh rule: Create a learning culture. *Research Technology Management, 46*(3), 59–60.

Mahler, J. (1997). Influences of organizational culture on learning in public agencies. *Journal of Public Administration: Research and Theory, 7*(4), 519–540.

Marquardt, M. (1996). *Building the learning organization*. New York: McGraw-Hill.

Marsick, V. J., & Watkins, K. E. (2003). Demonstrating the value of an organization's learning culture: The dimensions of learning organization questionnaire. *Advances in Developing Human Resources, 5*(2), 132–151. doi:10.1177/1523422303005002002

Martins, E. C., & Terblanche, F. (2003). Building organisational culture that stimulates creativity and innovation. *European Journal of Innovation Management, 6*(1), 64–74. doi:10.1108/14601060310456337

McGill, M., & Slocum, J. Jr. (1993). Unlearning the organization. *Organizational Dynamics, 22*(2), 67–79. doi:10.1016/0090-2616(93)90054-5

McLean, L. (2005). Organizational culture's influence on creativity and innovation: A review of the literature and implications for human resource development. *Advances in Developing Human Resources, 7*(2), 226–246. doi:10.1177/1523422305274528

Moreira, P. (1997). *A Aprendizagem Organizacional, a Performance Organizacional e a Vantagem Competitiva Sustentável. Dissertação de Mestrado não publicada.* ISCTE - Instituto Superior das Ciências do Trabalho e da Empresa.

Nevis, E., DiBella, A., & Gould, J. (1995). Understanding organizations as learning systems. *Sloan Management Review, 36*(2), 73–85.

Nunnally, J. (1978). *Psychometric theory*. New York: Mc Graw-Hill.

O'Reilly, C. (1989). Corporations, culture and commitment: Motivation and social control in large organizations. *California Management Review, 31*(4), 9–25.

Pedler, M., Burgoyne, J., & Boydell, T. (1991). *The learning company*. Maidenhead: McGraw-Hill.

Pettigrew, A. M. (2000). Foreword . In Ashakanasy, N. M., Wilderom, C. P., & Peterson, M. F. (Eds.), *Handbook of organizational culture and climate* (pp. xiii–xv). Thousand Oaks: Sage.

Picken, J., & Dess, G. (1997). Out of (strategic) control. *Organizational Dynamics*, (Summer): 35–48. doi:10.1016/S0090-2616(97)90026-7

Pinchot, E. & Pinchot, G. (1996). Seeding a climate for innovation. *Executive Excellence, June,* 17-18.

Popper, M., & Lipshitz, R. (2000). Organizational learning: Mechanisms, culture, and feasibility. *Management Learning, 31*(2), 181–196. doi:10.1177/1350507600312003

Rebelo, T. (2001). *Organização, aprendizagem e cultura: Estudo sobre a homogeneidade/heterogeneidade da orientação cultural para a aprendizagem*. Unpublished master dissertation, University of Coimbra, Portugal.

Rebelo, T. (2006). Orientação cultural para a aprendizagem nas organizações: Condicionantes e consequentes. Unpublished doctoral dissertation, University of Coimbra, Portugal.

Rebelo, T., & Almeida, N. (2008). Cultura de aprendizagem: O papel das lideranças intermédias. *Psychologica, 47*, 125–143.

Rebelo, T., & Gomes, A. (2009). Different types of organization, different cultural orientations towards learning: What factors explain this? In Fanti, K. A. (Ed.), *Applying psychological research to understand and promote the well-being of clinical and non-clinical populations* (pp. 175–186). Athens: ATINER.

Rebelo, T., Gomes, A., & Cardoso, L. (2007). *Is learning culture a good bet? A cultural orientation and its impact on firm performance.* Paper presented at the 13th European Congress of Work and Organizational Psychology, Stockholm, Sweden.

Rebelo, T., Gomes, A. D., & Cardoso, L. (2002). Orientações culturais para a aprendizagem nas organizações: homogeneidade e/ou heterogeneidade. *Psychologica, 30*, 345–363.

Rebelo, T., Gomes, A. D., & Cardoso, L. (2003). Cultures d'apprentissage: L'échelle OCA. In C. Vandenberghe, N. Delobbe & G. Karnas (Eds). *Développement des compétences, investissement professionnel et bien-être des personnes – Actes du 12e Congrès de Psychologie du Travail et des Organisations: Vol 2. Dimensions individuelles et sociales de l'investissement professionnel,* (pp. 497-507). Louvain–la-Neuve: Presses Universitaires de Louvain.

Rebelo, T., Gomes, A. D., & Cardoso, L. (2005). Cultura de aprendizagem: A (bi)dimensionalidade da escala OCA. *Psychologica, 38*, 191–208.

Reeves, T. (1996). Rogue learning on the company reservation. *The Learning Organization, 3*(2), 20–29. doi:10.1108/09696479610113774

Salaman, G. (2001). A response to Snell: The learning organization: Fact or fiction? *Human Relations, 54*(3), 343–359. doi:10.1177/0018726701543004

Schein, E. H. (1992). *Organizational culture and leadership* (2nd ed.). San Francisco: Jossey-Bass.

Schein, E. H. (1993). On dialogue, culture and organizational learning. *Organizational Dynamics*, (Winter): 40–51. doi:10.1016/0090-2616(93)90052-3

Schein, E. H. (1994). *Organizational and managerial culture as a facilitator or inhibitor of organizational learning.* M.I.T. - Sloan School of Management.

Schein, E. H. (1996). Culture: The missing concept in organization studies. *Administrative Science Quarterly, 41*(2), 229–241. doi:10.2307/2393715

Schein, E. H. (1997). *Three cultures of management: The key to organizational learning in the 21st century*. M.I.T. - Sloan School of Management.

Schneider, B., Brief, A., & Guzzo, R. (1996). Creating a climate and culture for sustainable change. *Organizational Dynamics*, (Summer): 7–20. doi:10.1016/S0090-2616(96)90010-8

Škerlavaj, M., Song, J.H. & Lee, Y. (2010). Organizational learning culture, innovative culture and innovations in South Korean firms. *Expert Systems and Applications*.

Škerlavaj, M., Štemberger, M., Škrinjar, R., & Dimovski, V. (2007). Organizational learning culture–the missing link between business process change and organizational performance. *International Journal of Production Economics, 106*, 346–367. doi:10.1016/j.ijpe.2006.07.009

Sta. Maria, R. F. (2003). Innovation and organizational learning culture in the Malaysian public sector. *Advances in Developing Human Resources*, *5*(2), 205–214. doi:10.1177/1523422303005002008

Thompson, M. A., & Kahnweiler, W. M. (2002). An exploratory investigation of learning culture theory and employee participation in decision making. *Human Resource Development Quarterly*, *13*(3), 271–288. doi:10.1002/hrdq.1031

Walsh, J. P., & Ungson, G. R. (1991). Organizational memory. *Academy of Management Review*, *16*(1), 57–91. doi:10.2307/258607

Weick, K. (1995). *Sensemaking in organizations*. Thousand Oaks, CA: Sage.

Weick, K., & Westley, F. (1996). Organizational learning: Affirming an oxymoron . In Clegg, S., Hardy, C., & Nord, W. (Eds.), *Handbook of Organization Studies* (pp. 440–458). London: Sage.

Wilderom, C. P. M., Glunk, U., & Maslowsky, R. (2000). Organizational culture as a predictor of organizational performance . In Ashakanasy, N. M., Wilderom, C. P., & Peterson, M. F. (Eds.), *Handbook of organizational culture and climate* (pp. 193–209). Thousand Oaks: Sage.

Williams, L. J., Ford, L. R., & Nguyen, N. (2002). Basic and advanced measurement models for confirmatory factor analysis . In Rogelberg, S. G. (Ed.), *Handbook of research methods in industrial and organizational psychology* (pp. 366–389). Oxford: Blackwell.

Yang, B. (2003). Identifying valid and reliable measures for dimensions of a learning culture. *Advances in Developing Human Resources*, *5*(2), 152–162. doi:10.1177/1523422303005002003

Yeung, A. K., Ulrich, D. O., Nason, S. W., & Glinow, M. A. (1999). *Organizational learning capability: Generating and generalizing ideas with impact*. Oxford: Oxford University Press.

## KEY TERMS AND DEFINITIONS

**Creativity in Organizations:** The generation of new and useful ideas.

**Innovation in Organizations:** The successful implementation of creative ideas related to products, services, processes or procedures.

**Instrumental Research:** Research centered on the development and validation of data collection instruments.

**Learning Culture:** An organizational culture that is oriented towards the promotion and facilitation of workers' learning, and towards its share and dissemination.

**Organizational Culture:** A shared system of assumptions, beliefs, values and artifacts that shapes the behavior of individuals in an organization, and gives organization a particular identity.

**Validity:** The study process of a measure or a data collection instrument, in order to guarantee that it really does what it is intended to do.

## ENDNOTE

1 For further information regarding OLC questionnaire please contact Teresa Rebelo (terebelo@fpce.uc.pt).

# Chapter 12
# Knowledge Management and Innovation:
## Mapping the Use of Technology in Organizations

**Leonor Cardoso**
*University of Coimbra, Portugal*

**A. Duarte Gomes**
*University of Coimbra, Portugal*

## ABSTRACT

*Nowadays it is generally recognized that technology has a not inconsiderable role in organizational knowledge management, inasmuch as it provides new forms for holding and exchanging information and knowledge in intra and inter-organizational contexts. The potential of technological means is therefore emphasized as tools supporting the various organizational processes related to knowledge, and questions arise concerning their comparative or relative relevance for increased innovation and creativity.*

*In a sample of 1275 individuals belonging to 50 Portuguese organizations, the use of technology plays an important part in terms of the organizational processes related to knowledge management, but this is limited above all to those which are formally instituted and based on knowledge of a mainly explicit nature. In addition, this chapter highlights the importance of management of organizational processes related to tacit knowledge, which emerges essentially from processes of social and discursive interaction involving organizational actors.*

DOI: 10.4018/978-1-60960-519-3.ch012

## INTRODUCTION

Organizations considered as advanced and competitive base their activities, the communication between people working in them and the relationship they form with other organizations, on information generated and managed by technology. Considering how organizational knowledge in this context is seen and managed, it is above all American organizations that tend to be perceived and presented as pioneers or leaders in the way they manage it through recourse to new information and communication technology (Takeuchi, 2001). In other words, there is selective attention to the way organizations have valued their knowledge resources and the diverse technological means they have available, from a perspective of joint, interactive and optimized management. Those that have been considered as adopting the best practices are found, in most cases, in the area of consultancy organizations – where knowledge corresponds to the product. In these organizations, there is indeed great attention given to "integration of organizational knowledge", which means the development of activities directly related to the elaboration, structuring and ordering of databases, with categorization and formatting of documents and with the destruction of material considered obsolete. In this way an effort has been made to use information and communication technology (ICT) effectively and efficiently to acquire, document, make available, share and use knowledge in organizations (Sarmento et al., 2000). According to the organizational systemic perspective (O'Brien, 1993; Stoner & Freeman, 1992), importance is given to the availability and use of communication channels linking the diverse participants and through which processed information circulates. This will flow through the whole organization, forming a complex network, frequently referred to as the organization's information system (Zorrinho, 1995). Therefore, as long as the appropriate infrastructure is established, information and knowledge can be easily distributed or transferred, both intra and inter-organizationally. As Davenport and Prusak (1998) highlight, as organizations interact with their environment, "they absorb information, turn it into knowledge, and take action based on it in combination with their experiences, values, and internal rules" (p. 52). This means that all organizations generate and manage their knowledge.

However, with knowledge being of vital importance for organizational functioning, it is not enough for its creation to occur spontaneously or for its management to be occasional, haphazard and non-deliberate. On the contrary, this implies conscious, systematic and intentional behaviour which must be operationalized within a set of formally instituted activities and initiatives. Therefore, according to Sousa (1999), knowledge management, as a management attitude, forms a process that integrates in organizational strategy the management of people and information and communication technology, with a view to promoting integrated organizational learning, using information gathered from colleagues, clients, suppliers, competitors, etc., so as to use the results of its treatment and synthesis at the right time and more quickly than the competition. In this sense, it is up to top management to make the first commitment regarding knowledge, and this should be reflected in the development of a set of processes aiming for and stimulating the acquisition (internal and external), systemization, retention and share of knowledge within its structure, in order to accelerate and improve problem-solving and decision-making. These processes, which are more cultural than technological, should stimulate a working environment that emphasizes and rewards the total commitment of all organizational actors to knowledge and the essential need to share it.

In this chapter, we seek to describe the contexts of the emergence of technology (and its use) in organizations, with a view to promoting innovation and organizational performance in the area of which knowledge management is particularly relevant. We stress the role played by human resources as the main catalyzing elements in the process of joint, interactive and optimized management of knowledge and the technological means available. Finally, and turning to an empirical study, we try to understand the role played by technology in the processes supporting organizational knowledge management. To do so, we investigate the existence of different perceptions of the various organizational actors concerning the degree of applicability of knowledge management processes in the organizations we studied (assessed through the Knowledge Management Questionnaire (Cardoso, 2007), both globally and in each of the four factors considered therein) according to the use (or non-use) of information and communication technology. A diversified set of variables is also considered relating to the use of this type of means, to be able to investigate if there are differences in the perception of the organizational actors who use them, concerning the knowledge management processes in question.

## BACKGROUND

### Organizations and Information and Communication Technology

Quite frequently, when the expression "information society" is used to designate today's social context, it conveys more or less explicitly the idea of the emergence of a new stage of historical development of those societies considered most developed or advanced, inasmuch as they experienced throughout the twentieth century a set of transformations directly related to information and technology. Indeed, it is commonly accepted that the combination of computer science with telecommunications sparked off the emergence of a new era, greatly marked by the appearance and massive use of information and communication technology. As the maximum exponent of the so-called "information society", this acquired such a status that it is difficult to imagine the time when it was something limited to the field of computer science and its specialists, and nowadays it is extremely difficult to imagine the functioning of organizations and society in general without it (Neto & Figueiredo, 2001; Vandijck, 1998), as it forms a basic requisite for the prosperity of organizations, when inserted in a context where competitiveness predominates.

However, despite the fact that technology greatly mediates the actions we perform in daily life, the weight and significance of the expression "information society" tends to diminish. Faced with the permanent social, cultural, political and economic changes that can form a threat to organizations' survival, perception of the context in which they operate, as a social context based on knowledge – commonly designated as the knowledge society (Toffler, 1970) –, despite having less clear or evident aspects, seems to have greater potential than that which was focused exclusively on information. Gradually, the idea came to the fore that possession of knowledge functions as a critical element of organizational success. In the words of Prusak (1996), "the only thing that gives an organization a competitive edge – the only thing that is sustainable – is what it knows, how it uses what it knows and how fast it can know something new" (p. 6).

We have witnessed the exponential valuing of knowledge, reflected in greater concern about its use, storage, perfection, share and use of everything an organization and its collaborators know. In this context, organizations come to be perceived as sets of people who act in a disciplined way to produce something or provide any service, objectives that are only fulfilled in circumstances where they are equipped with consolidated knowledge. In other words, the competitive advantages of a given organization depend to a great extent on the way it creates, acquires, makes available, stores, uses and applies new knowledge; they depend on the degree to which it knows where the knowledge is, as well as its capacity to identify those who possess it (Murray, 2000). The organizational process of decision-making itself, which is essential and inherent to any management activity, is highly dependent on it, as besides routine decisions (which consequently are not subject to great reflection), it involves others of a strategic nature which inevitably imply the existence of a set of knowledge that guarantees decision-making in accordance with organizational progress.

Despite the high position technology has occupied in society in general, the truth is that organizations have been the context *par excellence* where its impact has been felt most, given the acquired awareness of its high instrumentality. Therefore, Sá-Soares (1998) argues that a given organization's information system represents a fundamental component for its success, forming an area of essential intervention in terms of management, comparable to that of any other organizational area. Amaral (1994), in turn, states that the said information system's mission should be the improvement of individuals' performance in organizational processes, through effective use of the technological means available and, as in any other organizational system it should contribute to achieving the organization's mission. By its innovative nature, its practical usefulness and by the added value it provides, information and communication technology became an operational necessity in any organization oriented towards change (e.g., Kiesler & Sproull, 1982; Knights, Murray & Hugh, 1993; Perry, 1993; Sá-Soares, 1998; Tushman & Moore, 1988).

Besides contributing to increased productivity, to improved quality of products and services, to promotion of the organization's image, to reducing human involvement in repetitive tasks, to greater autonomy and collaborator satisfaction, technology also has a crucial role in organizational activities related to reducing the cost of action, coordination and production processes, allowing in addition greater agility and speed of response (Hall, 1986). It also serves as a gelling and catalyzing element of organizational communication, since it is the only means of communication able to support and digitalize all types of information, namely text documents, mathematical and financial analyses, pictures, audio and video supports, etc. (Sousa, 1997).

New information and communication technology is also responsible for decentralization and relocation of some organizations (both services and production) to regions which, without it, would not be advisable in terms of investment. However, that relocation is limited to production means, since in most cases the technology remains at the headquarters in the home country and does not contribute therefore to reducing inequality in possession of knowledge, as in spite of a certain degree of transfer to the host country, the fact is that the core of the knowledge remains in the country of origin.

## Human Resources and Technology: Implications for Tacit and Explicit Knowledge

Knowledge can be characterized according to a vast variety of dimensions. Polanyi (1966) classified knowledge as explicit and tacit. McAdam, Mason and McCrory (2007) analyzed the relative interpretations of tacit knowledge, from which emerged a recurring set of associated topics. Setting out from these authors' analysis, tacit knowledge can be conceptualized as typically individualistic and practical, associated with skills and experience, centred on the organizational domain and referring to specific contexts. In addition, it emerges as non-codifiable, intuitive, associated with personal competences commonly referred to as know-how and acquired with little or no environmental support. The terminological evolution associated with the definition of tacit knowledge can be synthesized in the classic bi-dimensionality proposed by Polanyi (1966): technical, related to a type of knowledge rooted in a subject's action and performance for a specific context; and cognitive, which includes elements such as intuition, emotions, schemes, values, beliefs, attitudes and competences. In turn, explicit knowledge can be codified, forming a repository of formal organizational knowledge, and is easily transferred from one subject to another, since it is expressed articulately and acquired through education. Although these descriptions are centred on the individual, both tacit and explicit knowledge can be observed at the collective/organizational level (Cook & Brown, 1999; Leiponen, 2006; McAdam et al., 2007; Nelson & Winter, 1982; Spender, 1996). For example, collective knowledge materialized in group and organizational routines and processes is tacit knowledge, whereas objectivized knowledge (e.g., intellectual properties, products, services) is explicit knowledge. The level of ownership of knowledge determines who accesses and uses it in an organization: cooperation is a necessary condition for organizational use of knowledge held individually, whereas knowledge held collectively makes organizations less dependent on key people and more able to maintain their structure (Leiponen, 2006; Nelson & Winter, 1982).

McEvily and Chakravarthy (2002) argue that performance is affected by the complexity (taken to be the difficulty in understanding an organization's functioning or the way to produce results) and by the tacitness and explicitness (seen as the difficulty or ease of articulating the knowledge that influences performance) of knowledge. The proposal of Garyd and Kumaraswamy (1995) is centred on the modularity of knowledge which, through specification of interface patterns, allows separation of production processes. Introduction of the concepts of complexity and modularity contributes to greater intelligibility of organizational knowledge since together with tacitness they form the main dimensions of knowledge (Ju, Li & Lee, 2006). Therefore, for example, high complexity associated with high levels of modularity and predominantly tacit knowledge hinders processes of integrating knowledge in organizations.

Extension of the separability of tacit and explicit knowledge is currently being debated (Leiponen, 2006; McAdam et al., 2007). Some authors (e.g., Cook & Brown, 1999) claim the association of functionalities distinct to each form of knowledge, arguing for example that practices based on the combination of tacit and explicit knowledge are distinct from those based only on tacit knowledge (Leiponen, 2006), and that sharing explicit knowledge promotes the emergence of tacit knowledge through internalization, in the same way that tacit knowledge is shared through processes of socialization, becoming explicit through externalization (Nonaka & Takeuchi, 1995). On the other hand, Orlikowski (2002) and also Tsoukas (1996) claim the impracticality of separating tacit and explicit knowledge, and focus studies on practice, taking both types of knowledge together. In this study, we consider the existence of two types of knowledge which, although interdependent, have specific characteristics and functionalities which allow us to distinguish formally instituted knowledge (i.e., explicit) and knowledge of a fundamentally informal nature, associated with processes of social and discursive management (i.e., tacit).

Indeed, according to Malhotra (2001), knowledge management concerns critical aspects related to the capacity for adaptation, survival and competence in organizations in the face of irregular transformations in their environment. It incorporates essentially organizational processes that seek to combine, synergetically, the capacity for data and information processing supported by technology, and individuals' ability to create and innovate. However, mere access to technological means does not guarantee good organizational performance. Success – and not just access – concerns questions to do with its use and appropriation (Davenport & Prusak, 1998; Santana, 1999). Indeed, we can see a considerable division between possession of new technology and the ability to use and apply it, and so it becomes crucial to carry out appropriate and integrated human resource management, giving value to questions related to training and qualifications.

As Soliman and Spooner (2000) argue, the process of human resource management translates as follows: "the means by which value is added to raw knowledge (inputs) in order to create processed knowledge (outputs), i.e., adding value for clients" (p. 338). In the same context, Kovács (1993) argues that the functionality of information and communication technology depends to a great extent on the quality of the human factor in the organization, as well as on effective information circulation and correct organization and coordination of its work teams. This requires a fluid, participative, cooperative organizational structure adaptable to external demands and changes, meaning a anthropocentric perspective reflected in the organization and operationalization of the production systems themselves (from its planning to placing products or services on the market), stimulating, using and valuing the capacities of the diverse organizational actors (in terms of creativity, initiative, responsibility and innovation), with a view to drawing profit from the technological means available. As this author mentions, the truth is that many organizations, after introducing new technology, adopt a technocentric perspective from which they perceive it as a likely means to substitute all individual potential, which results in a less flexible, more centralized and deterministic organization as regards the tasks that are normally performed mechanically. In the same connection, development of an identity and culture forming the organizational social structure facilitates and promotes communication of tacit knowledge (Zander & Kogut, 1995), making share and transfer of knowledge between individuals and groups better intra-organizationally than inter-organizationally (Kogut & Zander, 1996). On the other hand, explicit knowledge, which is easily codified and taught, can be transmitted as much intra-organizationally as inter-organizationally (Zander & Kogut, 1995; Kogut & Zander, 1993).

Following on from what we have stated, information and communication technology cannot be conceived only as technological systems, as it has implications for the social system, or in other words, it is a matter of optimized management of technical and social components (Gomes, 1992). Even though an organization that intends to be innovative and competitive cannot ignore the progressive technological developments characterizing current society (Hall, 1986), the fact is that it is important to assess its indispensability with a view to competitive advantage, and also important, from the point of view of the knowledge Organizational Psychology has accumulated concerning the introduction of any change, not to ignore other factors of innovation which can be equally or more important, namely the human factor.

## Knowledge Management, Technology and Innovation: Limited Potential

Knowledge management programmes intend basically to give organizations a set of means able to catalyze fulfilment of their objectives, as well as the effective creation and use of knowledge, through structuring its contents, using available technological means for the purpose (e.g., Blackler, Reed & Whitaker, 1993; Davenport, De Long & Beers, 1999; Davenport, Eccles & Prusak, 1992; Davenport, Hammer & Metsisto, 1989; Davenport & Prusak 1998; Scarborough, 1993; Winograd & Flores, 1986; Zuboff, 1988). According to De Long (1999), any knowledge management strategy conceived to improve organizational performance should be directed to three components: people or work activities that create or increase organizational knowledge; technological infra-structure to support the retention, transfer and use of knowledge; behavioural norms and practices – set in an "organizational culture" – which are essential for effective use of knowledge.

Setting out from the analysis of thirty one programmes implemented in twenty four organizations, Davenport and collaborators (1999) propose a categorization of the objectives underlying them, considering that to some extent, they form the guiding elements of the great majority of knowledge management systems: (a) creation of repositories to store it; (b) perfecting methods giving access to it; (c) promotion of an atmosphere conducive to its development; and (d) its management as an organizational resource.

While the objective related to creation of repositories of knowledge has to do with the need for its capture, other systems emphasize making it more accessible and interpersonally transferable, despite recognizing the inherent difficulties. Promotion of an environment conducive to knowledge development is related to creating a suitable atmosphere for its effective creation, transfer and use. Various systems aim to stimulate awareness and cultural receptiveness to knowledge, others promote initiatives to transform behaviours related to it and others again aim to perfect management processes themselves. In the end, knowledge management as an organizational resource aims above all for its inclusion in the organization's accounting balance sheets, which is common practice in European organizations focused on the possibility of measuring their resources of intellectual capital.

Davenport and Prusak (1998) point out three essential sub-processes in a global process of knowledge management, which they consider to be at present the major driving force of organizational change: knowledge generation, knowledge codification and coordination, and knowledge transfer, highlighting the important role played by new information and communication technology in developing this entire process. Reinforcing this perspective, Ruggles (1999) concludes that knowledge management consists of three major processes: creation (how new knowledge is discovered), codification (capture and representation of knowledge) and transfer (movement of knowledge between people, functions and places).

According to Davenport and Prusak (1998), the main objective guiding the use of technology in the field of knowledge management concerns precisely the possibility of transferring individual, group and organizational knowledge. Accordingly, in the opinion of Hibbard (1997), knowledge management consists of the technological combination, indexing and search to facilitate organization of data stored in multiple sources and distribute relevant information to its users. In this context, Domínguez (2001) warns of the fact that in the knowledge management process based on technological tools, what is at stake is not knowledge management itself, but rather the use of tools able to capture, organize, store and transmit it. This alone is justified by the fact that, even in circumstances where the sources are the same, individuals acquire different knowledge since the knowledge absorbed will depend on their training, the context and the way they access these sources.

Davenport and Prusak (1998) even argue that the concept of knowledge management would be much less appealing in the absence of the possibility of using the technology directed towards it, its main instrumentality arising from its power to increase the reach and speed of the transfer process, enabling extraction and structuring of the knowledge of an individual or work team and its later use by other members of the organization or its external stakeholders. In addition, these authors maintain that even though organizations implant technological systems in relatively restricted areas of knowledge and considerable development is expected in the near future, individuals will continue to play a major role, rather than being relegated to mere passive users of technology oriented towards knowledge.

The majority of authors concentrating on the subject of knowledge management consider the storage process to be of prime importance (e.g., Ash, 1998; Bassi, 1997; Blake, 1998; Cole-Gomolski, 1997; 1997a, 1998; DiMattia & Oder, 1997; Finerty, 1997; Hibbard, 1997; Laberis, 1998; LaPlante, 1997; Mårtensson, 2000; Mayo, 1998; Nerney, 1997; Ostro, 1997; Symoens, 1998; Watson, 1998; Yeh, Chang & Qyang, 2000). Setting out from analysis of various knowledge management programmes, Davenport and collaborators (1999), just as Sveiby (1997) identify three basic types of repositories: (a) of external knowledge, for example of competitive intelligence; (b) of structured internal knowledge, namely investigation reports, market studies, techniques and methodologies; and (c) of informal internal knowledge, such as debates and discussions rich in know-how. The dominant view in the literature considers that share of tacit knowledge mediated by information and communication technology is very limited (e.g., Lundvall, 2001; Salter & Gann, 2003). In this context, authors such as Wenger and Snyder (2000) defend communities based on practice as central vectors in the development and share of tacit knowledge. There are two types of communities (Brown & Duguid, 1999). In fact, "communities of practice" are centred on face-to-face collaborative participation of groups of individuals in developing new knowledge. These practices have a double function: training newcomers and creating new knowledge. On the other hand, "networks of practice", which take place fundamentally via the internet, allow interaction between different "communities of practice", facilitate refinements of existing knowledge and promote the share of knowledge between people in a given specialization and greater motivation to acquire new knowledge in an area where an individual is already a specialist. In this connection, there are various overlaps and complementarities between the two communities (Coenen, Moodyson & Asheim, 2004; Gibson & Gibbs, 2006; Hildrum, 2009). Each community has specific potentialities which are limited by the other. As an example, we highlight that although "communities of practice" depend mainly on face-to-face interaction, coordination and support of that interaction depends heavily on information and communication technology. Besides, share of tacit knowledge is correlated with the creation of social environments that can take on a local or electronic character (Hildrum, 2009). Indeed, in the circumstances where in an organization there are interests, aptitudes and cultural orientation to matters of knowledge, technology allows wider access, making it available to the person who needs it, at the right time. Davenport and Prusak (1998) warn of valuing technology in implementing knowledge management systems limitations, emphasizing that technical systems do not substitute the role performed by people in the sphere of this process. The same idea is reinforced by Freire (2001), alleging that for the present at least, no technology replaces human thought, inasmuch as knowledge management goes far beyond the acquisition of advanced hardware and software equipment. Illustrating these arguments, Malhotra (1998) states that knowledge management joins diverse processes that ideally should aim for synergy between technology's ability to process data and information and the creative and innovative ability of the various organizational actors. That is to say, technology alone is incapable of making someone share knowledge, or encouraging him to look for it. Therefore, the mere presence of technology does not create an organization which learns or which generates and manages knowledge.

To sum up, and turning to the words of Davenport and Prusak (1998), we will say there can be no confusion between "information – or knowledge – with the technology that delivers it" (p. 4). Or again, in the appropriate expression by Martiny (1998), "technology is an enabler, not a driver of knowledge management" (p. 76). Therefore, it is important to enable its use in an interactive way and recognize the essential role of the various organizational actors at all times of the knowledge management process. They are irreplaceable, above all when it comes to creation, this being an essential process for organizational innovation and development, as well as for reaching and holding on to sustainable competitive advantages.

The countless definitions of innovation that appear in the literature share one basic idea: tacit knowledge can be transformed into new products, processes and services and explicit knowledge promotes improvement of that which already exists, increasing organizational competitive advantage and satisfying clients' constantly changing needs (Carnegie & Butlin, 1993; Gloet & Terziovski, 2004; Nystrom, 1990). Gieski (1999) goes as far as to point out four mechanisms that contribute to continuous innovation: capacities, behaviours, contingencies and levers, knowledge management being transversal to the four mechanisms.

Studies on knowledge management and innovation are based on four major topics (Van de Ven & Engleman, 2004). The first topic relates to the human component, focused on innovation through the valuing and exploitation of new knowledge rather than recourse to knowledge that already exists (e.g., Taminiau, Smit & Lange, 2009). The second concerns processes and centres on developing processes that allow management and implementation of new ideas (e.g., Darroch, 2005; Lundvall & Nielsen, 2007; Huang & Li, 2009). The third topic refers to construction of infra-structure that allows firstly development of knowledge and secondly facilitation, support and promotion of innovation (e.g., Scarbrough, 2003). The present study, although not exclusively, is set in this structural topic. Finally, study of leadership seeks to develop an appropriate context for innovation (e.g., Jong & Hartog, 2007). The four topics have been studied taking into account their influence on knowledge management and on innovation of internal factors (e.g., organizational structures, communication channels, organizational cultures) and external ones (e.g., the role of governments, use of technology for innovation based on inter-organizational collaboration) (Lu, Tsang & Peng, 2008). Generically, studies consider that the processes and types of knowledge management explored above function as coordinating and previous mechanisms for greater effectiveness and innovation.

## EMPIRICAL STUDY

The main goal of this chapter is to determine the role of technology in the organizational processes related to knowledge management, specifically in those related to more formal processes (supported by mainly explicit knowledge) or more informal processes (set in essentially tacit knowledge and potentially more relevant in processes of creation and innovation).

### Hypotheses

In the light of previous considerations we formulated the following two general hypotheses:

*Hypothesis 1: Differences in users and non-users of technology are centered on knowledge management of a mainly explicit nature.*

*Hypothesis 2: The use of technology plays an important role in knowledge management of a mainly explicit nature.*

## METHOD

### Participants and Procedure

To carry out the empirical study, multivariate analyses of the variance (MANOVA) were used. Data were gathered from 50 industrial organizations in the district of Viseu, Portugal, after distribution of 1824 questionnaires of which 1547 were taken in and the answers of 1275 participants analyzed (response rate = 69.9%). This is above the average response rate of published research in the managerial and behavioral sciences, so the comparison is favorable (55.6% overall, see Baruch, 1999, for extensive review).

The sample was predominately male (55.5%). Considering the sample as a whole, most of the subjects were aged between 20 and 30 (41.9%), followed by subjects aged between 30 and 35

(18%), 40 and 50 (16.2%) and between 35 and 40 (13.5%); 7.9% of participants were over 50 and the remaining 2.5% were under 20. The majority of respondents (62.7%) had prior experience of work in other companies. The vast majority (60.2%) worked in the "production department" of the companies they belong to. 12.8% of respondents worked in administrative departments, 8.8% in trade, 4.1% in quality, 2.4% in human resources, 1.6% in financial departments and 1.0% in research and development. It should be noted that 9.0% belong to "other functional areas".

## Measures

Knowledge management was measured through Knowledge Management Questionnaire (Cardoso, 2007), a questionnaire made up of 32 items ($\alpha$=.93) and which shows a tetra-dimensional structure, with: Factor 1 – Knowledge management practices (groups organizational actions developed around formally instituted processes, centered on knowledge of a mainly explicit nature – 10 items; $\alpha$=.88); Factor 2 – Cultural orientation towards knowledge (reflects a framework serving as a guide for instituted practices, rules, norms and procedures – 11 items; $\alpha$=.86); Factor 3 – Social and discursive knowledge management (translates the informal interaction occurring in the organization which facilitates social construction of knowledge – 6 items; $\alpha$=.79); and Factor 4 – Strategic knowledge management (reflects the organization's orientation towards the exterior – 5 items; $\alpha$=.76). Answers were given on a Likert-type scale of 5 points, ranging from 1 (*Almost never applies*) to 5 (*Applies almost completely*).

Use was also made of the UTI, a questionnaire directed to employees who use technology to carry out their work, and therefore its first question separates the sample subjects concerning this aspect, collaboration in completing it fully being only asked of those who used this type of resource in their functions. It is in two parts: the first contains nine questions aiming to determine if the work carried out by respondents is or is not dependent on using technological means, if they dispose of the means they need, if they share them with other colleagues or not, what is the degree of importance they give these means, how often they use them and to what extent, what are the underlying reasons for using less than their full potential, as well as the purpose they are used for. The second part is made up of eight statements aiming to assess the role played by technology in management of data, information and knowledge - technological orientation to the management of explicit knowledge. Always bearing in mind their own organizational context, respondents were asked to choose one of five optional answers (1 – *Almost never applies*, 2 – *Applies little*, 3 – *Applies moderately*, 4 – *Applies a lot* and 5 – *Almost completely applies*). The reliability of this scale was high ($\alpha$=.91).

# RESULTS

## Knowledge Management and Use of Information and Communication Technology

The purpose of this section was to investigate the existence of different perceptions of respondents about the degree of applicability of knowledge management processes in the organizations studied, according to use or non-use of information and communication technology, as well as subsequent assessment of the existence of perceptive differences between organizational actors who use it in carrying out their daily tasks, according to a set of variables related to use of this type of means. This being so, we will assess the influence of a set of variables integrating the Questionnaire on use of information technology in knowledge management (assessed through Knowledge Management Questionnaire, both globally and in each of the four factors considered).

## Use of Computers

The first question of the UTI (UTI01) let us discriminate which respondents used (or did not use) computers in carrying out their work, allowing us now to check for differences of perception between each group concerning the degree of applicability of knowledge management processes. To this end, we resorted to a MANOVA, taking as independent variable use of computers and as dependent variables the four factors of Knowledge Management Questionnaire (KMQ). The result of the Wilks Lambda is 0.872 [$F(4, 1210) = 44.47$, $p < .001$] and the eta ($\eta$) partial squared is 0.128, indicating that 12.8% of the variability in knowledge management is found to be associated with use of computers. The subsequent univariate tests (cf. Table 1) point to differences in Factors 1 and 4 of KMQ. In these, the average scores indicate that organizational actors who use computers to carry out their work perceive organizational processes related to practices of knowledge management and

*Table 1. Average scores and standard deviations of KMQ according to use of computers, dependence on use of computers to carry out work and the availability of the necessary computing means to carry out work: univariate tests*

| **Do you use computers to do your work?** | | | | | | | |
|---|---|---|---|---|---|---|---|
| | **No (*n* = 568)** | | **Yes (*n* = 647)** | | **Total (*N* = 1215)** | | **F (1, 1213)** |
| | **M** | **DP** | **M** | **DP** | **M** | **DP** | |
| **KMQ (TA)** | *3.22* | *0.58* | *3.35* | *0.55* | *3.29* | *0.56* | |
| **Factor 1** | *2.60* | *0.83* | *3.03* | *0.75* | *2.83* | *0.82* | *89.77*** |
| **Factor 2** | *3.67* | *0.61* | *3.60* | *0.57* | *3.63* | *0.59* | *3.39* |
| **Factor 3** | *3.30* | *0.70* | *3.34* | *0.64* | *3.32* | *0.67* | *1.63* |
| **Factor 4** | *3.36* | *0.75* | *3.45* | *0.66* | *3.41* | *0.70* | *4.63** |
| **Can you do your work without using a computer?** | | | | | | | |
| | **No (*n* = 508)** | | **Yes (*n* = 130)** | | **Total (*N* = 638)** | | **F (1, 636)** |
| | **M** | **DP** | **M** | **DP** | **M** | **DP** | |
| **KMQ (TA)** | *3.33* | *0.53* | *3.42* | *0.61* | *3.35* | *0.55* | |
| **Factor 1** | *3.02* | *0.74* | *3.08* | *0.80* | *3.03* | *0.76* | *0.53* |
| **Factor 2** | *3.57* | *0.56* | *3.72* | *0.62* | *3.60* | *0.57* | *6.70** |
| **Factor 3** | *3.34* | *0.62* | *3.36* | *0.69* | *3.34* | *0.64* | *0.12* |
| **Factor 4** | *3.42* | *0.65* | *3.55* | *0.69* | *3.45* | *0.66* | *3.66* |
| **Do you have available the computers you need to do your work?** | | | | | | | |
| | **No (*n* = 66)** | | **Yes (*n* = 566)** | | **Total (*N* = 632)** | | **F (1, 630)** |
| | **M** | **DP** | **M** | **DP** | **M** | **DP** | |
| **KMQ (TA)** | *3.16* | *0.53* | *3.37* | *0.55* | *3.35* | *0.55* | |
| **Factor 1** | *2.71* | *0.73* | *3.07* | *0.75* | *3.03* | *0.76* | *13.65*** |
| **Factor 2** | *3.45* | *0.58* | *3.63* | *0.57* | *3.61* | *0.57* | *5.56** |
| **Factor 3** | *3.24* | *0.64* | *3.35* | *0.64* | *3.34* | *0.64* | *1.69* |
| **Factor 4** | *3.30* | *0.69* | *3.47* | *0.66* | *3.45* | *0.66* | *3.55* |

* $p < .05$ ** $p < .001$

**Legend:** KMQ (TA) – Total average of KMQ; Factor 1 – Knowledge management practices; Factor 2 – Cultural orientation towards knowledge; Factor 3 – Social and discursive knowledge management; Factor 4 – Strategic knowledge management.

its strategic management as applying to a greater extent to the organizations studied.

We conclude that as regards the other factors of KMQ, namely cultural orientation towards knowledge (Factor 2) and its social and discursive management (Factor 3), there are no differences of perception found as to their applicability to these organizations between organizational actors who use or do not use computers to fulfil the tasks inherent in their function.

After investigating the existence of different perceptions by respondents regarding the degree of applicability of knowledge management processes to the organizations studied according to use (or non-use) of information and communication technology, we will now consider exclusively organizational actors who use it in carrying out their work, seeking to assess if among them there are differences of perception of the degree of applicability of knowledge management processes according to a diversified set of variables related to use of these means.

## Dependence on Use of Computers for Work

We took now as VI the variable related to dependence on computers by organizational actors who use them in carrying out their work (UTI02), applying it to a MANOVA. The multivariate test obtained reveals that in the set of the four factors of KMQ, there are significant differences [Wilks Lambda = 0.985; F (4, 633) = 2.42 $p<.05$], which the univariate tests indicate as being due exclusively to the second, cultural orientation towards knowledge (cf. Table 1). Although Table 1 shows average scores that are higher in all the factors of KMQ in respondents who depend on computers for their work, only in Factor 2 does that superiority emerge as significant.

## Availability of the Necessary Computing Means for Work

It is fundamental to make available to the diverse organizational actors the computers they need to carry out efficiently the tasks inherent in their duties. Therefore, our objective was to check for differences of perception of the degree of applicability of knowledge management processes as a function of that availability (UTI04). We carried out a MANOVA and found the existence of a statistically significant multivariae effect [Wilks Lambda = 0.977, F (4, 627) = 3.63, $p<.01$], which the univariate tests identified as being due to differences in Factors 1 and 2 of KMQ. Table 1 shows the average scores obtained in these factors, which allow us to conclude it is participants who have available the computers they need to fulfil their duties who perceive practices of knowledge management and cultural orientation to knowledge as applying to a greater extent to the organizations studied.

## Importance Attributed to the Computer to Carry Out Work

Only in Factor 1 do we find differences of perception regarding the degree of applicability of organizational processes related to knowledge management according to the importance attributed to the computer in performing tasks (UTI03). In fact, the multivariate test from the MANOVA indicates a Wilks Lambda of 0.960 [F (16, 1940.60) = 1.65, $p=.05$], which the univariate tests point out as being due to knowledge management practices. Carrying out Tukey multiple comparison tests, for this factor, let us see that these practices are perceived as applying to a greater extent by participants who consider it important to use a computer to perform their tasks than by those who consider it as merely useful (cf. Table 2).

*Table 2. Average scores and standard deviations of KMQ according to the importance attributed to the computer to carry out tasks: univariate and multiple comparison tests*

| What importance do you give to use of a computer to carry out your tasks? | | | | | | | |
|---|---|---|---|---|---|---|---|
| **Measures** | **Harmful ($n$ = 2)** | **Irrelevant ($n$ = 4)** | **Useful ($n$ = 59)** | **Important ($n$ = 175)** | **Indispensable ($n$ = 403)** | **Total ($N$ = 643)** | **F (4, 638)** |
| | **Average scores** | | | | | | |
| **KMQ (TA)** | *3.69* | *3.36* | *3.24* | *3.40* | *3.35* | *3.35* | |
| **Factor 1** | *2.90* | *2.90* | *2.76* | *3.11* | *3.04* | *3.04* | *2.47** |
| **Factor 2** | *4.45* | *3.59* | *3.58* | *3.64* | *3.59* | *3.60* | *1.43* |
| **Factor 3** | *3.50* | *3.50* | *3.27* | *3.34* | *3.34* | *3.34* | *0.30* |
| **Factor 4** | *3.80* | *3.60* | *3.38* | *3.50* | *3.44* | *3.45* | *0.67* |
| **Standard deviations** | | | | | | | |
| **KMQ (TA)** | *0.35* | *0.76* | *0.59* | *0.53* | *0.55* | *0.55* | |
| **Factor 1** | *0.28* | *1.04* | *0.74* | *0.71* | *0.77* | *0.75* | |
| **Factor 2** | *0.00* | *0.71* | *0.66* | *0.53* | *0.58* | *0.57* | |
| **Factor 3** | *0.94* | *0.89* | *0.77* | *0.60* | *0.63* | *0.64* | |
| **Factor 4** | *0.57* | *1.02* | *0.73* | *0.64* | *0.66* | *0.66* | |
| | **Differences between averages (Factor 1)** | | | | | | |
| **Irrelevant** | *0.00* | - | | | | | |
| **Useful** | *-0.14* | *-0.14* | - | | | | |
| **Important** | *0.21* | *.021* | *0.35** | - | | | |
| **Indispensable** | *0.14* | *0.14* | *0.28* | *-0.07* | - | | |

$p < .05$

**Legend:** KMQ (TA) – Total average of KMQ; Factor 1 – Knowledge management practices; Factor 2 – Cultural orientation towards knowledge; Factor 3 – Social and discursive knowledge management; Factor 4 – Strategic knowledge management.

## Frequency of Use of Available Computer Resources

We carried out a MANOVA using as dependent variables the four factors of KMQ and as independent variable the frequency of use of available computer resources (UTI05). Although the multivariate test result does not reach the threshold of statistical significance [Wilks $\Lambda$ = 0.974, F (12, 1685.64) = 1.41, $p$ >.10], univariate tests indicated a significant effect in terms of Factor 4 (cf. Table 3). Indeed, Tukey multiple comparison tests indicated that participants using the available IT tools a few times and often, compared to those who seldom use them, perceive strategic knowledge management processes as applying to a greater degree.

## Exclusive Use of Computers

We aim to find out if exclusive use of computers as opposed to sharing (UTI06) reveals differences in perception of the degree of applicability of knowledge management processes. The MANOVA carried out, taking now as VI the dichotomic information made available by this variable, does not present any significant effect, whether from the multivariate test [Wilks $\Lambda$ = 0.993, F (4, 624) = 1.13, $p$ >.30], or from the univariate tests considered individually (cf. Table 4). Unlike the availability of the necessary computers to carry out work, we found that sharing does not mean differences in respondents' perception of knowledge management.

*Table 3. Average scores and standard deviations of KMQ according to the frequency of use of available computer facilities: Univariate and multiple comparison tests*

| Measures | Rarely $n = 6$ | Few Times $n = 15$ | Sometimes $N = 55$ | Frequently $n = 568$ | Total $N = 644$ | F (3, 640) |
|---|---|---|---|---|---|---|
| | **Average scores** | | | | | |
| **KMQ (TA)** | *3.20* | *3.05* | *3.34* | *3.36* | *3.35* | |
| **Factor 1** | *2.63* | *2.81* | *3.03* | *3.04* | *3.03* | *1.05* |
| **Factor 2** | *3.67* | *3.35* | *3.59* | *3.61* | *3.60* | *1.09* |
| **Factor 3** | *3.06* | *3.08* | *3.34* | *3.35* | *3.34* | *1.27* |
| **Factor 4** | *3.47* | *2.88* | *3.43* | *3.46* | *3.45* | *3.87*** |
| | **Standard deviations** | | | | | |
| **KMQ (TA)** | *0.55* | *0.50* | *0.53* | *0.55* | *0.55* | |
| **Factor 1** | *0.83* | *0.67* | *0.64* | *0.76* | *0.75* | |
| **Factor 2** | *0.66* | *0.61* | *0.52* | *0.58* | *0.57* | |
| **Factor 3** | *0.60* | *0.52* | *0.77* | *0.63* | *0.64* | |
| **Factor 4** | *0.69* | *0.63* | *0.60* | *0.66* | *0.66* | |
| | **Differences between averages (Factor 4)** | | | | | |
| **Few times** | *-0.59* | - | | | | |
| **Sometimes** | *-0.03* | *0.55** | - | | | |
| **Frequently** | *-0.00* | *0.58*** | *0.03* | - | | |

* $p < .05$ ** $p < .01$

**Legend:** KMQ (TA) – Total average of KMQ; Factor 1 – Knowledge management practices; Factor 2 – Cultural orientation towards knowledge; Factor 3 – Social and discursive knowledge management; Factor 4 – Strategic knowledge management.

*Table 4. Average scores and standard deviations of KMQ according to exclusive use of computers: univariate tests*

| Exclusive use of computers | | | | | | | |
|---|---|---|---|---|---|---|---|
| | Shared equipment ($n = 356$) | | Exclusive use of equipment ($n = 273$) | | Total ($N = 629$) | | F (1, 627) |
| | M | DP | M | DP | M | DP | |
| **KMQ (TA)** | *3.35* | *0.54* | *3.35* | *0.53* | *3.35* | *0.54* | |
| **Factor 1** | *3.03* | *0.74* | *3.05* | *0.76* | *3.04* | *0.75* | *0.07* |
| **Factor 2** | *3.61* | *0.56* | *3.60* | *0.57* | *3.60* | *0.57* | *0.07* |
| **Factor 3** | *3.36* | *0.62* | *3.31* | *0.64* | *3.34* | *0.63* | *0.98* |
| **Factor 4** | *3.42* | *0.69* | *3.47* | *0.62* | *3.44* | *0.66* | *1.06* |

**Legend:** KMQ (TA) – Total average of KMQ; Factor 1 – Knowledge management practices; Factor 2 – Cultural orientation to knowledge; Factor 3 – Social and discursive knowledge management; Factor 4 – Strategic knowledge management.

## Perception of the Degree of Harm from Sharing Computers

After ascertaining the existence of different perceptions of knowledge management according to exclusive use or share of computers, we decided to investigate, in the latter circumstances, if the perception of harm (or not) arising from this (UTI07) has implications for participants' perception of the degree of applicability of knowledge management processes. In this context, turning once more to a MANOVA, we found a significant multivariate effect [Wilks $\Lambda = 0.936$, $F(12, 913.08) = 1.93$, $p < .03$], which the univariate tests identify as being attributed to Factors 1 and 2 of KMQ, knowledge management practices and cultural orientation to knowledge (cf. Table 5). Observation of the Tukey multiple comparison tests shows the existence of significant linear contrasts for these factors, as well as for Factor 3, social and discursive knowledge management. Concerning the first two factors, those collaborators who state that share of computers has little or no negative effect on the fulfilment of their duties perceive knowledge management practices and cultural orientation to knowledge as applying to a greater extent in their organizations, compared to those stating that sharing is very harmful to their work. As for the third factor, differences of perception are only found between those for whom sharing is perceived to be very harmful to their work and those for whom sharing has no negative effect.

## Percentage Use of Computers

From the point of reference that 100% corresponds to making full use of computers, we asked participants about the degree of use they make of them (UTI08), determining afterwards to what extent there are differences of perception among participants as to the degree of application of organizational processes related to knowledge as a function of that degree of use. We calculate the Pearson correlation coefficients, taking as variables in the analysis the percentage use in question and each of the four factors of the Knowledge Management Questionnaire. The coefficients found, despite not reaching the threshold of statistical significance for the KMQ as a whole ($r = .06$, $p > .10$) and for Factors 2, 3 and 4 (Pearson correlation coefficients of -.01,.05 e.03, respectively, $p > .20$), are seen to be significant for Factor 1 ($r = .11$, $p < .01$), although the proportion of shared variability is reduced. We conclude that the higher the percentage of computer use by participants, the higher the perception of the degree of applicability of knowledge management practices occurring in the sample organizations.

## Determinant Factors of Less Use of Computers

From participants who indicated a percentage of computer use under 100%, we gathered information concerning the reasons for this lower use (UTI09), each of them as follows being considered as a variable influencing perception of knowledge management. Carrying out MANOVAs, the results of which appear in Table 6, let us identify those causing differences in perception of the degree of application of knowledge management processes in the sample organizations. Therefore, participants who did not tick out-of-date hardware and software as reasons for less use (UTI09_1 E UTI09_2) have a perception corresponding to higher applicability of KMQ in all its component factors, than those who ticked those reasons. We also found that for Factor 1, knowledge management practices, perception is of less applicability by collaborators indicating reasons for less use of computer potential as being lack of training and sharing equipment with colleagues (UTI09_3 E UTI09_5).

## Purposes of Using Computers

The UTI contemplates question 10 (UTI10), which groups items relating to the diverse purposes

*Table 5. Average scores and standard deviations of KMQ according to perception of the degree of harm caused by sharing computers: univariate and multiple comparison tests*

| Degree to which sharing computers is harmful to the fulfilment of work | | | | | | |
|---|---|---|---|---|---|---|
| | Not at all ($n$ =127) | Little ($n$ = 114) | Moderate ($n$ = 72) | Very much ($n$ = 39) | Total ($N$=352) | F (3, 348) |
| | **Average Scores** | | | | | |
| KMQ (TA) | 3.39 | 3.40 | 3.34 | 3.07 | 3.35 | |
| Factor 1 | 3.04 | 3.12 | 3.03 | 2.68 | 3.03 | 3.55* |
| Factor 2 | 3.69 | 3.64 | 3.56 | 3.33 | 3.61 | 4.47** |
| Factor 3 | 3.40 | 3.39 | 3.35 | 3.10 | 3.35 | 2.52 |
| Factor 4 | 3.44 | 3.42 | 3.47 | 3.27 | 3.42 | 0.80 |
| | **Standard deviations** | | | | | |
| KMQ (TA) | 0.54 | 0.51 | 0.56 | 0.46 | 0.55 | |
| Factor 1 | 0.78 | 0.73 | 0.75 | 0.64 | 0.75 | |
| Factor 2 | 0.60 | 0.51 | 0.59 | 0.48 | 0.56 | |
| Factor 3 | 0.67 | 0.53 | 0.58 | 0.74 | 0.62 | |
| Factor 4 | 0.75 | 0.69 | 0.63 | 0.61 | 0.69 | |
| | **Differences between averages (Factor 1: Knowledge management practices)** | | | | | |
| Not at all | - | -0.08 | 0.01 | 0.36* | | |
| Little | | - | 0.09 | 0.45** | | |
| Moderate | | | - | 0.36 | | |
| | **Differences between averages (Factor 2: Technological orientation to knowledge)** | | | | | |
| Not at all | - | 0.05 | 0.14 | 0.36** | | |
| Little | | - | 0.08 | 0.31* | | |
| Moderate | | | - | 0.22 | | |
| | **Differences between averages (Factor 3: Social and discursive knowledge management)** | | | | | |
| Not at all | - | 0.00 | 0.05 | 0.29* | | |
| Little | | - | 0.05 | 0.29 | | |
| Moderate | | | - | 0.25 | | |

* $p < .05$ ** $p < .01$

**Legend:** KMQ (TA) – Total average of KMQ; Factor 1 – Knowledge management practices; Factor 2 – Cultural orientation towards knowledge; Factor 3 – Social and discursive knowledge management; Factor 4 – Strategic knowledge management.

pursued by organizational actors when using computers in carrying out their duties. We tried to determine to what extent there is an association between the purpose focused in each item and the perception held about the degree of applicability of knowledge management processes. The Pearson correlation analyses made, the coefficients of which appear in Table 7, allow identification of the non-existence of relationships with the items related to extranet and programming languages. In the other items analyzed, we find positive relationships between the items referring to:

*Table 6. Average scores of KMQ according to reasons determining less use of computers: multivariate and univariate tests*

| Items | Average scores | | | | | | | | Wilks Λ | F (4, 472) | Univariate scores F (1, 475) | | | |
|---|---|---|---|---|---|---|---|---|---|---|---|---|---|---|
| | Factor 1 | | Factor 2 | | Factor 3 | | Factor 4 | | | | Factor 1 | Factor 2 | Fac-tor 3 | Factor 4 |
| | NT | T | NT | T | NT | T | NT | T | | | | | | |
| **UTI09_1** | *3.07* | *2.51* | *3.63* | *3.28* | *3.35* | *3.15* | *3.49* | *3.08* | *.940* | *7.59*** | *23.29*** | *16.16*** | *4.21** | *17.25*** |
| **UTI09_2** | *3.06* | *2.56* | *3.63* | *3.26* | *3.36* | *3.12* | *3.48* | *3.14* | *.952* | *5.93*** | *18.43*** | *17.54*** | *5.43** | *11.40*** |
| **UTI09_3** | *3.08* | *2.92* | *3.60* | *3.59* | *3.36* | *3.29* | *3.47* | *3.41* | *.983* | *2.02** | *5.12** | *0.03* | *1.12* | *1.13* |
| **UTI09_4** | *3.02* | *3.03* | *3.58* | *3.70* | *3.34* | *3.33* | *3.45* | *3.44* | *.985* | *1.83* | *0.03* | *3.18* | *0.01* | *0.02* |
| **UTI09_5** | *3.06* | *2.88* | *3.63* | *3.51* | *3.36* | *3.27* | *3.48* | *3.35* | *.986* | *1.66* | *5.64** | *3.76* | *1.52* | *3.60* |
| **UTI09_6** | *3.02* | *3.01* | *3.60* | *3.62* | *3.34* | *3.30* | *3.45* | *3.40* | *.999* | *0.15* | *0.00* | *0.03* | *0.08* | *0.18* |
| **UTI09_7** | *2.97* | *3.07* | *3.61* | *3.58* | *3.31* | *33.6* | *3.44* | *3.46* | *.986* | *1.69* | *2.38* | *0.35* | *0.58* | *0.19* |
| **UTI09_8** | *3.03* | *2.98* | *3.59* | *3.62* | *3.34* | *3.31* | *3.46* | *3.40* | *.994* | *0.73* | *0.39* | *0.15* | *0.19* | *0.65* |
| **UTI09_9** | *3.02* | *2.98* | *3.60* | *3.59* | *3.34* | *3.32* | *3.45* | *3.40* | *1.00* | *0.03* | *0.03* | *0.01* | *0.01* | *0.09* |
| **UTI09_10** | *3.02* | *3.01* | *3.59* | *3.63* | *3.33* | *3.34* | *3.47* | *3.40* | *.990* | *1.13* | *0.03* | *0.44* | *0.00* | *1.27* |

* $p < .05$ ** $p < .001$

**Legend:** *NT* – not ticked; *T* – ticked; UTI09_1 – Out-of-date equipment; UTI09_2 – Out-of-date software; UTI09_3 – Lack of training to make good use of it; UTI09_4 – Lack of practice; UTI09_5 – Share of equipment with colleagues; UTI09_6 – Preference for manual work; UTI09_7 – Lack of time to explore programs; UTI09_8 – More time is wasted in carrying out some tasks; UTI09_9 – Lack of motivation and interest in the technology; UTI09_10 – Not enough knowledge to make efficient use of it.

*Table 7. Correlation coefficients between purposes of computer use and knowledge management*

| Purposes of computer use | Knowledge management | | | | | |
|---|---|---|---|---|---|---|
| | *N* | KMQ (TA) | Factor 1 | Factor 2 | Factor 3 | Factor 4 |
| UTI10_1 | 581 | 0.12* | 0.08 | 0.11* | 0.07 | 0.09* |
| UTI10_2 | 558 | 0.19** | 0.30** | 0.03 | 0.17** | 0.07 |
| UTI10_3 | 549 | 0.10* | 0.19** | 0.00 | 0.02 | 0.07 |
| UTI10_4 | 549 | 0.08 | 0.17** | - 0.02 | 0.06 | - 0.01 |
| UTI10_5 | 555 | 0.05 | 0.04 | 0.05 | 0.04 | 0.06 |
| UTI10_6 | 563 | 0.09* | 0.17** | - 0.01 | 0.05 | 0.08 |
| UTI10_7 | 570 | 0.07 | 0.20** | - 0.03 | - 0.04 | 0.03 |
| UTI10_8 | 557 | 0.22** | 0.29** | 0.12* | 0.15** | 0.10* |
| UTI10_9 | 539 | 0.20** | 0.27** | 0.09* | 0.12* | 0.13* |
| UTI10_10 | 547 | 0.12* | 0.15** | 0.06 | 0.09* | 0.06 |
| UTI10_11 | 536 | 0.05 | 0.08 | 0.03 | 0.02 | 0.00 |
| UTI10_12 | 579 | 0.10* | 0.12* | 0.06 | 0.12* | 0.03 |
| UTI10_13 | 579 | 0.19** | 0.14** | 0.17** | 0.18** | 0.18** |
| UTI10_14 | 574 | 0.13* | 0.13* | 0.07 | 0.12* | 0.12* |
| UTI10_15 | 572 | 0.10* | 0.08* | 0.05 | 0.14** | 0.07 |

* $p < .05$ ** $p < .001$

**Legend:** KMQ (TA) – Total average of KMQ; Factor 1 – Knowledge management practices; Factor 2 – Cultural orientation towards knowledge; Factor 3 – Social and discursive knowledge management; Factor 4 – Strategic knowledge management; UTI10_1 – Entering data; UTI10_2 – Use of e-mail; UTI10_3 – Use of the Internet; UTI10_4 – Use of intranet; UTI10_5 – Use of extranet, UTI10_6 – Word processing; UTI10_7 – Spreadsheet; UTI10_8 – Graphic presentations; UTI10_9 – Project management; UTI10_10 – Database management system; UTI10_11 – Programming languages; UTI10_12 – To obtain essential data for own work and that of colleagues; UTI10_13 – To keep/supply essential data for own work and that of colleagues; UTI10_14 – To distribute essential data for own work and that of colleagues; UTI10_15 – To retrieve essential data for own work and that of colleagues.

- Graphic presentations, project management and register/supply of essential data for work (items 10_8, 10_9 and 10_13), with all the factors of KMQ, the magnitude of the relationship being greater with the first, knowledge management practices;
- Distribution of essential data for work (UTI10_14), with Factors 1, 3 and 4 of KMQ, the magnitude of the relationship also being greater with the first factor;
- Use of e-mail, database management system, obtaining and retrieval of essential data for work (UTI10_2, UTI10_10, UTI10_12 and UTI10_15), with knowledge management practices and its social and discursive management (Factors 1 and 3 of the Knowledge Management Questionnaire);
- Entering data (UTI10_1), with Factors 2 and 4, cultural orientation to knowledge and strategic knowledge management;
- Use of the Internet and intranet, word processing and spreadsheet (UTI10_3, UTI10_4, UTI10_6 and UTI10_7), with Factor 1, knowledge management practices.

In general terms, we can conclude that increased frequency of technology use with a determined purpose by participants corresponds to a perception of greater applicability of the various processes related to knowledge management. It is worth mentioning that in the set of analyses

*Table 8. Correlation coefficients between technological orientation to explicit knowledge management and KMQ: test of the difference between two independent correlation coefficients*

| | Technological orientation to explicit knowledge management | | | | |
|---|---|---|---|---|---|
| | r (N =646) | Z scores | | | |
| | | *Factor 1* | *Factor 2* | *Factor 3* | *Factor 4* |
| KMQ (TA) | *.54*** | - | - | - | - |
| ***Factor 1***: Knowledge management practices | *.52*** | - | *2.20** | *3.04** | *3.50*** |
| ***Factor 2***: Cultural orientation to knowledge | *.43*** | - | - | *0.84* | *1.30* |
| ***Factor 3***: Social and discursive knowledge management | *.39*** | - | - | - | *0.46* |
| ***Factor 4***: Strategic knowledge management | *.37*** | - | - | - | - |

* $p < .05$ ** $p < .001$
**Legend:** KMQ (TA) – Total average of KMQ.

made, associations stand out between the items assessing the purposes of computer use and Factor 1 of KMQ (knowledge management practices).

## Technological Orientation towards Explicit Knowledge Management and Knowledge Management

Finally, we decided to evaluate associations between the results obtained on the scale of "technological orientation to explicit knowledge management" (UTI11) and with KMQ. We intended to find out specifically if given the existence of significant relationships between this scale and each of the factors of KMQ, in which of them, compared to the others, the proportion of shared variability is significantly greater. First, we calculated the Pearson correlation coefficients presented in Table 8, which indicate that the average proportion of shared variability between the variables analyzed is of the order of 18.28%. In this way, we conclude that the greater the perception of the degree of application of technological orientation to explicit knowledge management in these organizations, the greater the perception of the degree of applicability of the four factors of KMQ.

Secondly, we carried out the test of the difference between two independent correlation coefficients (Alferes, 2002), in order to assess the existence of associations of different magnitudes between technological orientation to explicit knowledge management and each factor of KMQ. A reading of the $Z$ values, presented in Table 8, indicates that the association between that technological orientation and Factor 1 is stronger than the relationship between that orientation and Factors 2, 3 and 4. In turn, we do not find differences between the magnitudes of the associations of technological orientation to explicit knowledge management and any of the other factors. Therefore, our conclusion is the existence of significantly stronger associations between the technological orientation in question and knowledge management practices.

# DISCUSSION AND CONCLUDING REMARKS

## Summary and Discussion of Results

Throughout this chapter we presented the results obtained from a set of analyses aiming to assess if there were differences in respondents' perceptions, concerning the degree of applicability of organizational processes related to knowledge, according to use of information and communication technology, also seeking to ascertain if there were differences between those who use it as a function of a set of variables regarding use. Next, we summarize the main results reached, attempting to draw some conclusions.

As for the results obtained in the other analyses made, we point out firstly those which allowed us to ascertain the existence of perceptive differences between organizational actors who used technological means to carry out their work and those who did not use them. These differences indicate that the former perceive greater applicability in the organizations studied of organizational processes related to knowledge management practices (Factor 1) and to strategic management (Factor 4). We also conclude that perceptions of the applicability of processes related to the other factors of KMQ (namely cultural orientation to knowledge and its social and discursive management) are not seen to be influenced by the independent variable in question. These results seem relevant since, by restricting the influence of use of technological means to perceptions of greater applicability of factors 1 and 4, they support the theoretical principle that availability and use of this type of means does not satisfy organizations' needs concerning the creation and maintenance of an organizational culture oriented to learning and knowledge, or of management of organizational processes related to knowledge of a mainly tacit nature, which emerge essentially from processes of social and discursive interaction by the various organizational actors.

Focusing now exclusively on organizational actors who use information and communication technology to carry out their work, the results obtained from the analyses made suggest on the influence of the diverse variables referring to use of technological means on perception of the degree of applicability of the different knowledge management processes in the organizations studied. It stands out, however, that this influence is felt particular strongly in terms of Factor 1 (knowledge management practices). Indeed, we find the existence of differences in these participants' perception of the applicability of organizational processes related to practices of management of knowledge that is mainly explicit (Factor 1) as a function of almost all the variables considered (except for variables UTI02, UTI 05 and UTI 06). As for the remaining factors, we find that perception regarding the degree of applicability to these organizations (a) of cultural orientation towards knowledge (Factor 2) varied as a function of the perception of dependence on and possession of these means to carry out work, the perception of harm arising from sharing, as well as a restricted number of purposes pursued with their use; (b) processes related to social and discursive knowledge management (Factor 3) varied as a function of the perception of harm arising from share of technological means and some of the purposes originating their use, while (c) strategic knowledge management (Factor 4) was influenced by the frequency with which technological resources were used and some of the purposes they were used for.

It seems therefore we can state that the perception of participants who use technological means in their daily tasks regarding the degree of applicability of knowledge management practices in these organizations is likely to vary according to a wide range of variables, finding a perception of greater applicability of practices related to this factor when the diverse organizational actors consider that: technological means are important to carry out their duties, the means they need are available to them (a variable which revealed itself to be among the most influential), sharing them has little or no harmful effect on carrying out functions, they make appropriate use of computers' potential, also considering that they do so because they possess the necessary training for optimized use and because of the fact that sharing the available means does not mean taking less advantage of their potential. We also find the influence of almost all the purposes listed in question 10 of UTI on perception of the degree of application of knowledge management practices. Indeed, from the set of fifteen mentioned purposes, twelve were seen as likely to influence participants' perception regarding this factor, and nine of them particularly so. Finally, it is important to mention that despite finding that the greater the perception of respondents about the degree of applicability of "technological orientation to explicit knowledge management" in these organizations, the greater the perception of the degree of applicability of the four factors of KMQ, we also found more strongly significant associations between that technological orientation and knowledge management practices. These results as a whole suggest that, besides technology use revealing itself to be influential in organizational actors' perception of the applicability of knowledge management practices in these organizations, when we take exclusively the perceptions of subjects who use it to carry out their work and find differences as a function of a wide range of variables related to its use, these are found to be mainly limited to this same factor. It seems a conclusion can be made about the strong relationship existing between organizational processes related to knowledge management practices (Factor 1) and use of technology (assessed through the diverse variables considered).

Compared to what happens in terms of Factor 1, we found that perception of the degree of applicability to these organizations of cultural orientation towards knowledge (Factor 2) by respondents who use technological means in fulfilling their duties was shown as likely to vary according to a more reduced range of variables. In this context, we identified a perception of greater applicability of organizational processes related to this factor when participants consider that: they depend on computers to do their work, they dispose of the means they need, sharing has little or no harmful effect on carrying out the work, concluding finally on the influence of a reduced number of purposes listed in question 10 of UTI on perception of the degree of application of this factor to the organizations studied. Indeed, only four of these fifteen purposes were revealed to be influential and only one (UTI_13) revealed an influence we can consider high.

Regarding the processes related to social and discursive knowledge management (Factor 3), we confirmed a dominant perception corresponds to its greater applicability when respondents consider that sharing technological means harms their work greatly (compared to those for whom sharing is harmless), also finding the influence of eight of the fifteen purposes listed in question 10 of UTI concerning perception of the degree of this factor's application to the organizations studied, the influence exercised by four of these variables being quite high.

Finally, considering the results obtained for strategic knowledge management (Factor 4), these indicate a perception that reflects a greater degree of this factor's application to the organizations studied when participants who use technology in performing their duties consider they use it sometimes and frequently (as opposed to those who rarely use it). Five of the fifteen purposes listed in question 10 of UTI were seen to influence perception of the degree of this factor's application to the sample organizations, only one showing an influence we can consider as high.

It is also important to mention that participants who did not tick the optional answers about the existence of out-of-date software and hardware as reasons for using less than the full potential of the technological means available perceived any of the four factors of KMQ as applying to a greater degree to these organizations. However, the values obtained for Factor 3 (social and discursive knowledge management), although significant, were not as high as those found in the other factors.

Distributed over eleven questions, we considered a total of thirty six variables related to use of information and communication technology in the analyses made. The first (UTI01) allowed discrimination of respondents who used technology and those who did not, finding that the former perceived the organizational processes related to Factors 1 and 4 as applying to a greater degree in these organizations. As for the other variables (where data refers only to respondents who use technological means in their work), we found that exclusive use or share of equipment (UTI06) was not seen to influence the perceptions of organizational actors in any of the four factors of KMQ, the same occurring in relation to six (UTI09_4, UTI09_6, UTI09_7, UTI09_8, UTI09_9, UTI09_10) of the ten variables included in the question referring to the reasons for less use of the computer's potential, as well as two (UTI10_5, UTI10_11) of the fifteen concerning the purposes that determine use of these means. Some of the variables considered were seen to influence a single factor of knowledge management. The influence of the perception of dependence on technological means to carry out work (UTI02) was only felt in Factor 2, the importance given to use of equipment to carry out tasks (UTI03), the extent to which it is used (UTI08) in Factor 1 and the frequency with which it is used (UTI05) in Factor 4. The influence of having (or not having) the necessary means to carry out work (UTI04) was felt in two factors, namely 1 and 2, while perception of the harm (or not) arising from share of these means (UTI07) influenced perception of the degree of application of organizational processes related to factors 1, 2 and 3. Some variables were shown to influence the perception of participating organizational actors who used technological means in all factors of KMQ. We refer specifically to the scale of technological orientation to explicit knowledge management, to two of the variables related to the reasons for less use of the means available [out-of-date hardware (UTI09_1) and software (UTI09_2)] and to three of the purposes determining use of these means [graphic presentations (UTI10_8), project management (UTI10_9) and save/supply essential data for own work and that of colleagues (UTI10_9)]. It should also be mentioned that in the sphere of reasons for less use of the means available, having (or not having) training to use them well (UTI09_3) and sharing (or not) equipment with colleagues (UTI09_5) only influenced respondents' perception of the applicability of organizational processes related to Factor 1. Focusing now on the purposes for using technological means, we found that four of these purposes only influenced the perception held regarding Factor 1 [use of the Internet (UTI10_3), use of the intranet (UTI10_4), word processing (UTI10_6) and spreadsheet (UTI10_7)], while another four were shown to influence perception related to two factors. We refer specifically to entering data (UTI10_1) which was seen to influence Factors 2 and 4, as well as use of e-mail (UTI10_2), database management system (UTI10_10), obtaining (UTI10_12) and retrieving (UTI10_15) essential data for own work and that of colleagues, the influence of which was felt in perception related to Factors 1 and 3. Finally, we found the existence of different perceptions relating to the degree of applicability of Factors 1, 3 and 4 as a function of the purpose related to distribution of essential data for work (UTI10_14).

## Discussion of Results and Managerial Implications

We conclude that in these organizations, the technological means available are effectively used by the organizational actors who need them to carry out their work. We also conclude on the existence of global differences in participants' perception about the degree of application to the organizations studied of knowledge management processes as a function of use of information and communication technology, as well as on the existence of different perceptions between organizational actors who use it to carry out their work as a function of a diversified set of variables related to its use. In particular, we emphasize the fact that when we evaluated the differences between users and non-users, the latter were centred on Factors 1 and 4 of KMQ, whereas exclusive evaluation of the former originated differences in the four factors of the Knowledge Management Questionnaire. This being so, respondents who use technological means to carry out their work are distinguished from those who do not, in the way they perceive knowledge management practices and strategic knowledge management, but are not distinguished in the perception they hold of these organizations' cultural orientation knowledge and their social and discursive knowledge management. The former (those who use computers daily) presents perceptive differences in any of the four factors of KMQ.

However, in each case, it was in knowledge management practices that those differences were found as a function of a greater number of variables and more significantly. Therefore, in our opinion, the main conclusion to be drawn from the investigation carried out is that use of information and communication technology plays an important part in organizational processes related to knowledge management, but this is restricted above all to those that are formally instituted and focused on knowledge of a mainly explicit nature.

In the same connection, these results point clearly to a conceptual distinction between explicit knowledge and tacit knowledge. Indeed, the emergence of the dimensions of knowledge management practices and social and discursive knowledge management points towards specific processes associated with the two types of knowledge. In addition, for communities of practice to reach their potential, it is necessary to take into consideration not only the questions related to tacit and explicit knowledge but also their integration in the strategy of organizational knowledge management supported by use of information and communication technology and by cultural orientation to knowledge. The data point towards complex organizational dynamics, and therefore we consider that valid interventions in terms of knowledge management, innovation and use of information and communication technology should focus simultaneously on multiple organizational aspects: on the one hand, an intervention can, for example, solve specific aspects, but acting simultaneously in other fields helps the organization to create solid bases to remain innovative. The mapping of technology use in organizations related to knowledge management provides managers with a useful tool, in that it facilitates focused interventions bearing in mind the organizational systemic complexity. When the combination of knowledge management, information and communication technology and innovation are aligned and managed strategically, the organization increases its competitiveness in a sustained way integrating different combinations of information with the skills of human resources.

## REFERENCES


Alferes, V. R. (2002). *Programas e rotinas complementares do SPSS.* Retrieved January 20, 2010, from http://www.fpce.uc.pt/nucleos/niips/

Amaral, L. (1994). *Um referencial para o planeamento de sistemas de informação.* Unpublished doctoral dissertation, University of Minho, Portugal.

Ash, J. (1998). Managing knowledge gives power. *Communication World, 3*(15), 23–26.

Baruch, Y. (1999). Response rate in academic studies: A comparative analysis. *Human Relations, 52*, 421–438. doi:10.1177/001872679905200401

Bassie, L. J. (1997). Harnessing the power of intellectual capital. *Training & Development, 12*(51), 25–30.

Blackler, F., Reed, M., & Whitaker, A. (1993). Editorial introduction: Knowledge workers and contemporary organizations. *Journal of Management Studies*, (November): 851–862. doi:10.1111/j.1467-6486.1993.tb00469.x

Blake, P. (1998). The knowledge management expansion. *Information Today, 1*(15), 12–13.

Brown, J. S., & Duguid, P. (1999). *The social life of information.* Boston: Harvard Business School Press.

Cardoso, L.M. (2007). *Gerir conhecimento e gerar competitividade. Estudo empírico sobre a gestão do conhecimento e seu impacto no desempenho organizacional.* Penafiel: Editorial Novembro.

Carnegie, R., & Butlin, M. (1993). *Managing the innovative enterprise: Australian companies competing against the world's best.* Melbourne: Business Council of Australia.

Coenen, L., Moodyson, J., & Asheim, B. T. (2004). Nodes, networks and proximities: On the knowledge dynamics of the Medicon Valley biotech cluster. *European Planning Studies, 12*(7), 1003–1018.

Cole-Gomolski, B. (1997). Chase uses new apps to ID best customers. *Computerworld, 35*(31), 49–50.

Cole-Gomolski, B. (1997a). Users loath to share their know-how. *Computerworld, 46*(31), 6.

Cole-Gomolski, B. (1998). Vendors cram knowledge-ware market. *Computerworld, 5*(31), 55–56.

Cook, S. D., & Brown, J. S. (1999). Bridging epistemologies: The generative dance organizational knowledge and organizational knowing. *Organization Science, 10*(4), 381–400. doi:10.1287/orsc.10.4.381

Darroch, J. (2005). Knowledge management, innovation and firm performance. *Journal of Knowledge Management, 9*(3), 101–115. doi:10.1108/13673270510602809

Davenport, T., Eccles, R. G., & Prusak, L. (1992). Information politics. *Sloan Management Review*, (Fall): 53–65.

Davenport, T., Hammer, M., & Metsisto, T. (1989). How executives can shape their company's information systems. *Harvard Business Review*, (March-April): 130–134.

Davenport, T. H., De Long, D. W., & Beers, M. C. (1999). Successful knowledge management projects. In Cortada, J. W., & Woods, J. A. (Eds.), *The knowledge management yearbook 1999-2000* (pp. 89–107). Boston: Butterworth-Heinemann.

Davenport, T. H., & Prusak, L. (1998). *Working knowledge: How organizations manage what they know.* Boston: Harvard Business School Press.

De Long, D. (1999). *Building the knowledge-based organization: How culture drives knowledge behaviors.* Center for Business Innovation: Ernst & Young LLP.

DiMattia, S., & Oder, N. (1997). Knowledge management: Hope, hype, or harbinger? *Library Journal, 15*(122), 33–35.

Domínguez, M. S. (2001). Importancia de la gestión del conocimiento: Homo sapiens versus maquina sapiens. *Proceedings of the 2nd Conference of the Portuguese Association of Information Systems.* Évora: Portuguese Association of Information Systems.

Finerty, P. (1997). Improving customer care through knowledge management. *Coastal Management, 9*(71), 33.

Freire, P. (2001). *Gestão do conhecimento e conhecimento da gestão.* Retrieved January 20, 2010, from http//:www.kmol.online.pt

Garyd, R., & Kumaraswamy, A. (1995). Technological and organizational designs for realizing economies of substitution. *Strategic Management Journal, 16*, 93–109. doi:10.1002/smj.4250160919

Gibson, C. B., & Gibbs, J. L. (2006). Unpacking the concept of virtuality: The effects of geographic dispersion, electronic dependence, dynamic structure, and national diversity on team innovation. *Administrative Science Quarterly, 51*(3), 451–495.

Gieski, J.F. (1999). *Continuous improvement and global innovation management trial project*. Euro-Australian Cooperation Centre Newsletter, No. 2.

Gloet, M., & Terziovski, M. (2004). Exploring the relationship between knowledge management practices and innovation performance. *Journal of Manufacturing Technology Management, 15*(5), 402–409. doi:10.1108/17410380410540390

Gomes, A. D. (1992). Formas de organização e técnicas de gestão: Análise comparativa e níveis de análise. *Revista Portuguesa de Gestão, 2*, 29–47.

Hall, S. C. (1986). Thoughts on emerging technology. *Information Technology Quarterly, 5*(1), 13–20.

Hibbard, J. (1997). Knowing what we know. *Information Week, 653*, 49–54.

Hildrum, J. M. (2009). Sharing tacit knowledge online: A case study of e-learning in Cisco's network of systems integrator partner firms. *Industry and Innovation, 16*(2), 197–218. doi:10.1080/13662710902764360

Huang, J., & Li, Y. (2009). The mediating effect of knowledge management on social interaction and innovation performance. *International Journal of Manpower, 30*(3), 285–301. doi:10.1108/01437720910956772

Jong, J. P., & Hartog, D. N. (2007). How leaders influence employees' innovative behaviour. *European Journal of Innovation Management, 10*(1), 41–64. doi:10.1108/14601060710720546

Ju, T. L., Li, C. Y., & Lee, T. S. (2006). A contingency model for knowledge management capability and innovation. *Industrial Management & Data Systems, 106*(6), 855–877. doi:10.1108/02635570610671524

Kiesler, S., & Sproull, L. (1982). Managerial response to changing environments: Perspectives on problem sensing from social cognition. *Administrative Science Quarterly, 27*, 548–570. doi:10.2307/2392530

Knights, D., Murray, F., & Hugh, W. (1993). Networking as knowledge work: A study of strategy development in the financial services industry. *Journal of Management Studies, 30*(6), 975–995. doi:10.1111/j.1467-6486.1993.tb00475.x

Kogut, B., & Zander, U. (1993). Knowledge of the firm and the evolutionary theory of the multinational corporation. *Journal of International Business Studies, 24*(4), 625–646. doi:10.1057/palgrave.jibs.8490248

Kogut, B., & Zander, U. (1996). What firms do? Coordination, identity, and learning. *Organization Science, 7*(5), 502–518. doi:10.1287/orsc.7.5.502

Kovács, I. (1993). Mudança tecnológica e organizacional: Análise de tendências na Indústria. *Socius Working Paper, 2*(94). 1-54.

Laberis, B. (1998). One big pile of knowledge. *Computerworld, 5*(32), 97.

LaPlante, A. (1997). Sharing the wisdom. *Computerworld, 22*(41), 73.

Leiponen, A. (2006). Managing knowledge for innovation: The case of business-to-business services. *Journal of Product Innovation Management, 23*, 238–258. doi:10.1111/j.1540-5885.2006.00196.x

Lu, Y., Tsang, E. W., & Peng, M. W. (2008). Knowledge management and innovation strategy in the Asia Pacific: Toward an institution-based view. *Asia Pacific Journal of Management, 25*, 361–374. doi:10.1007/s10490-008-9100-9

Lundvall, B., & Nielsen, P. (2007). Knowledge management and innovation performance. *International Journal of Manpower, 39*(3/4), 207–223. doi:10.1108/01437720710755218

Lundvall, B. A. (2001). Innovation policy in the globalising learning economy. In Archibuigi, D., & Lundvall, B. A. (Eds.), *The globalising learning economy*. Oxford: Oxford University Press.

Malhotra, Y. (1998). Tools @ work: Deciphering the knowledge management hype. *Journal for Quality and Participation*, *4*(21), 58–60.

Mårtensson, M. (2000). A critical review of knowledge management as a management tool. *Journal of Knowledge Management*, *4*(3), 204–216. doi:10.1108/13673270010350002

Martiny, M. (1998). Knowledge management at HP Consulting. *Organizational Dynamics*, *2*(47), 71–77. doi:10.1016/S0090-2616(98)90025-0

Mayo, A. (1998). Memory bankers. *People Management*, *2*(4), 34–38.

McAdam, R., Mason, B., & McCrory, J. (2007). Exploring the dichotomies within the tacit knowledge literature: Towards a process of tacit knowing in organizations. *Journal of Knowledge Management*, *11*(2), 43–59. doi:10.1108/13673270710738906

McEvily, S., & Chakravarthy, B. (2002). The persistence of knowledge-based advantage: An empirical test for product performance and technological knowledge. *Strategic Management Journal*, *23*(4), 285–305. doi:10.1002/smj.223

Murray, P. (2000). Designing for business benefits from knowledge management. In Despres, C., & Chauvel, D. (Eds.), *Horizons knowledge: The present and the promise of knowledge management*. Boston: Butterworth Heinemann.

Nelson, R. R., & Winter, S. G. (1982). *An evolutionary theory of economic change*. Cambridge, MA: Harvard University Press.

Nerney, C. (1997). Getting to know knowledge management. *New World (New Orleans, La.)*, *39*(14), 101.

Neto, E. D., & Figueiredo, A. D. (2001). The risks of systems and information technology: A multidimensional approach. *Proceedings of the 2nd Conference of the Portuguese Association of Information Systems*. Évora: Portuguese Association of Information Systems.

Nonaka, I., & Takeuchi, H. (1995). *The knowledge-crating company: How Japanese companies create the dynamics of innovation*. New York: Oxford University Press.

Nystrom, H. (1990). *Technological and market innovation: Strategies for product and company development*. London: John Wiley & Sons.

O'Brien, J. A. (1993). *Management information systems: A managerial end user perspective*. Homewood, IL: Dow Jones Irwin.

Orlikowski, W. J. (2002). Knowing in practice: Enacting a collective capability in distributed organizing. *Organization Science*, *13*(3), 249–273. doi:10.1287/orsc.13.3.249.2776

Ostro, N. (1997). The corporate brain. *Chief Executive*, *123*, 58–62.

Perry, N. (1993). Scientific communication, innovation networks and organization structures. *Journal of Management Studies*, *30*(6), 957–973. doi:10.1111/j.1467-6486.1993.tb00474.x

Polanyi, M. (1966). *The tacit dimension*. London: Routdedge & Kegan Paul.

Prusak, L. (1996). The knowledge advantage. *Planning Review*, *2*(24), 6–8.

Ruggles, R. L. (1999). The state of the notion: Knowledge management in practice. In Rutkowski, M. L., & Stasko, S. R. (Eds.), *Knowledge management: People make the difference*. The Knowledge Chronicle, Walsh College.

Sá-Soares, D. (1998). *Planeamento de sistemas de informação: Estudo das variáveis que condicionam a sua estratégia de execução*. Unpublished master thesis, University of Minho, Portugal.

Salter, A., & Gann, D. (2003). Sources of ideas for engineering design. *Research Policy*, *32*, 1309–1324. doi:10.1016/S0048-7333(02)00119-1

Santana, S. (1999). *Tecnologias de informação e da comunicação e pequenas e médias empresas: Uma abordagem centrada na aprendizagem organizacional*. Unpublished doctoral dissertation, University of Aveiro, Portugal.

Sarmento, A., Batista, J., Cardoso, L., Lousã, M., Rebelo, T., Babo, R., et al. (2000). Gestão do Conhecimento e Sistemas de Informação: Que relação? *Proceedings of the VIII Accounting and Auditing Congress*, Aveiro: ISCAA. CD-ROM Edition.

Scarborough, H. (1993). Problem-solutions in the management of information systems expertise. *Journal of Management Studies*, *30*(6), 939–955. doi:10.1111/j.1467-6486.1993.tb00473.x

Scarbrough, H. (2003). Knowledge management, HRM and the innovation process. *International Journal of Manpower*, *24*(5), 501–516. doi:10.1108/01437720310491053

Soliman, F., & Spooner, K. (2000). Strategies for implementing knowledge management: Role of human resources management. *Journal of Knowledge Management*, *4*(4), 337–345. doi:10.1108/13673270010379894

Sousa, C. (1999). A gestão do conhecimento ou as virtudes da partilha. *Recursos Humanos Magazine*, *1*(4), 20–24.

Sousa, S. (1997). *Tecnologias de informação: O que são? Para que servem?*Lisboa: Editora de Informática.

Spender, J. C. (1996). Making knowledge the basis of a dynamic theory of the firm. *Strategic Management Journal*, *17*, 45–62.

Stoner, J., & Freeman, R. (1992). *Management*. Englewood Cliffs, NJ: Prentice Hall.

Sveiby, K. E. (1997). *The new organizational wealth: Managing and measuring knowledge-based assets*. San Francisco: Berret-Koehler Publishers, Inc.

Symoens, J. (1998). Site server is a fine set of tools for web site building. *InfoWorld*, *4*(20), 128.

Takeuchi, H. (2001). Towards a universal management of the concept of knowledge. In Nonaka, I., & Teece, D. (Eds.), *Managing industrial knowledge: Creation, transfer and utilization* (pp. 315–329). London: Sage Publications.

Taminiau, Y., Smit, W., & Lange, A. (2009). Innovation in management consulting firms through informal knowledge sharing. *Journal of Knowledge Management*, *13*(1), 42–55. doi:10.1108/13673270910931152

Toffler, A. (1970). *Future shock*. London: Bodley Head.

Tsoukas, H. (1996). The firm as a distributed knowledge system: A constructionist approach. *Strategic Management Journal*, *17*, 11–25.

Tushman, M., & Moore, W. (1988). *Readings in management of innovation*. New York: Harper Collins Publishers.

Van de Ven, A., & Engleman, R. (2004). Central problems in managing corporate innovation and entrepreneurship. *Advances in Entrepreneurship. Firm Emergence and Growth*, *7*, 47–72. doi:10.1016/S1074-7540(04)07003-5

Vandijck, E. (1998). Information and communication technology in the 21st century. *Revista Portuguesa de Gestão*, *2*, 15–22.

Watson, S. (1998). Getting the 'aha'! *Computerworld*, *4*(32), 1–2.

Wenger, E. C., & Snyder, W. M. (2000). Communities of practice: The organisational frontier. *Harvard Business Review*, (January–February): 139–145.

Winograd, T., & Flores, F. (1986). *Understanding computers and cognition: A new foundation of design*. Norwood, NJ: Ablex.

Yeh, J.-C., Chang, J.-Y., & Qyang, Y.-J. (2000). Content and knowledge management in digital library and museum. *Journal of the American Society for Information Science American Society for Information Science*, *4*(51), 371–379. doi:10.1002/(SICI)1097-4571(2000)51:4<371::AID-ASI6>3.0.CO;2-6

Zander, U., & Kogut, B. (1995). Knowledge and the speed of the transfer and imitation of organizational capabilities: An empirical test. *Organization Science*, *6*(1), 76–92. doi:10.1287/orsc.6.1.76

Zorrinho, J. (1995). *Gestão da informação: Condição para vencer*. IAPMEI, Ministry of Industry and Energy.

Zuboff, S. (1988). *In the age of the smart machine: The future of work and power*. New York: Basic Books.

## ADDITIONAL READING

Adam, E. R., & Freeman, C. (2000). Communities of practice: Bridging technology and knowledge assessment. *Journal of Knowledge Management*, *1*(4).

Amidon, D. M. (1997). *Innovation strategy for the knowledge economy*. Boston, M.A.: Butterworth-Heinemann.

Anderson, P., & Tushman, M. (1991). Managing through cycles of technological change. *Research Technology Management*, May-June, 26-31.

Aramburu, N., Sáenz, J., & Rivera, O. (2006). Fostering innovation and knowledge creation: the role of management context. *Journal of Knowledge Management*, *10*(3), 157–166. doi:10.1108/13673270610670920

Bohn, R. (1994). Measuring and managing technological knowledge. *Sloan Management Review*, (Fall): 61–73.

Boisot, M. H. (1998). *Knowledge assets: Securing competitive advantage in the information economy*. Oxford: Oxford University Press.

Boisot, M. H. (1999). Is your firm a creative destroyer? Competitive learning and knowledge flows in the technological strategies of firms. In Zack, M. H. (Ed.), *Knowledge and strategy* (pp. 251–274). Boston: Butterworth-Heinemann. doi:10.1016/B978-0-7506-7088-3.50017-6

Brown, J. S., & Duguid, P. (1991). Organizational learning and communities of practice: Towards a unified view of working, learning and innovation. *Organization Science*, *1*(2), 40–57. doi:10.1287/orsc.2.1.40

Carneiro, A. (2000). How does knowledge management influence innovation and competitiveness? *Journal of Knowledge Management*, *2*(4), 87–98. doi:10.1108/13673270010372242

Cohen, W. M., & Levinthal, D. A. (1990). Absorptive capacity: A new perspective on learning and innovation. *Administrative Science Quarterly*, *35*(1), 128–152. doi:10.2307/2393553

Darroch, J., & McNaughton, R. (2002). Examining the link between knowledge management practices and types of innovation. *Journal of Intellectual Capital*, *3*(3), 210–222. doi:10.1108/14691930210435570

Earl, M. J. (1989). *Management strategies for information technology*. New Jersey: Prentice Hall.

Edvinsson, L., Dvir, R., Roth, N., & Pasher, E. (2004). Innovations: The new unit of analysis in the knowledge era. The quest and context for innovation efficiency and management of IC. *Journal of Intellectual Capital*, *5*(1), 40–58. doi:10.1108/14691930410512914

Handfield, R., & Pannesi, R. (1994). Managing component life cycles in dynamic technological environments. *International Journal of Purchasing and Materials Management*, *2*(30), 20–27.

Jang, S., Hong, K., Bock, G. W., & Kim, I. (2002). Knowledge management and process innovation: the knowledge management transformation path in Samsung SDI. *Journal of Knowledge Management*, *6*(5), 479–485. doi:10.1108/13673270210450582

Jonscher, C. (1994). An economic study of the information technology revolution. In Allen, T. J., & Morton, M. S. (Eds.), *Information Technology and Corporation of the 1990s*. Oxford: Oxford University Press.

Kogut, B., & Zander, U. (1997). Knowledge of the firm: Combinative capabilities and the replication of technology. In Prusak, L. (Ed.), *Knowledge in organizations* (pp. 17–36). Boston: Butterworth-Heinemann. doi:10.1016/B978-0-7506-9718-7.50005-6

Leonard-Barton, D. (1995). *Wellsprings of knowledge: Building and sustaining the sources of innovation*. Boston, MA: Harvard Business School Press.

Maqsood, T., Walker, D. H., & Finegan, A. D. (2007). Facilitating knowledge pull to deliver innovation through knowledge management. A case study. *Engineering, Construction, and Architectural Management*, *14*(1), 94–109. doi:10.1108/09699980710717007

McDermott, R. (1999). Why information technology inspired but cannot deliver knowledge management. *California Management Review*, 103–117.

Nonaka, I., Reinmöller, P., & Toyama, R. (2001). Integrated information technology systems for knowledge creation. In Dierkes, M., Berthoin Antal, A., Child, J., & Nonaka, I. (Eds.), *Handbook of organizational learning and knowledge* (pp. 827–848). Oxford: Oxford University Press.

Plessis, M. (2007). The role of knowledge management in innovation. *Journal of Knowledge Management*, *11*(4), 20–29. doi:10.1108/13673270710762684

Stata, R. (1989). Organizational learning: The key to management innovation. *Sloan Management Review*, *12*(1), 63–74.

Teece, D. (1981). The market for know-how and the efficient international transfer of technology. *The Annals of the Academy of Political and Social Science*, November, 81-96.

von Krogh, G., Ichijo, K., & Nonaka, I. (2000). *Enabling knowledge creation: How to unlock the mystery of tacit knowledge and release the power of innovation*. Oxford: Oxford University Press.

Winter, S. G. (1987). Knowledge and competence as strategic assets. In Teece, D. (Ed.), *The competitive challenge: Strategic for industrial innovation and renewal* (pp. 159–184). New York: Ballinger.

## KEY TERMS AND DEFINITIONS

**Cultural Orientation Towards Knowledge:** Points out aspects framed in a cultural perspective, because it depicts common, known, assumed and shared values that will guide co-workers in terms of the way they are, behave and work.

**Explicit Knowledge:** Is purely formal, systematic, detailed and based on rules. This type of knowledge is easily quantifiable and tasks associated with its encoding, decoding, transfer, communication and sharing are fluid in various ways.

**Innovation:** Development of new products, processes and services through the transformation of tacit knowledge and improvement of what already exists through explicit knowledge, increasing organizational competitive advantage and satisfying clients' constantly changing needs.

**Knowledge Management Practices:** Reflects concern for the creation and acquisition of knowledge through a set of knowledge sharing practices provided by training, attending conferences, or solving problems in groups, as well as use of knowledge transfer circles.

**Knowledge Management:** Knowledge management can be defined as the creation and development of internal organizational conditions that catalyze all processes related to knowledge. Through the scope of its intervention, KM highlights the active and constructive role performed by the human element in the context of organizational knowledge management.

**Social and Discursive Knowledge Management:** Translates the importance that a discursive practice assumes in knowledge creation and sharing. It also represents the informal organization and the network of interactions that characterizes and facilitates social construction of knowledge and creation of a common language and meaning.

**Strategic Knowledge Management:** Is related to the organizational environment and to oriented behavior focused on organizations' external context, highlighting the importance of clients and competitors for organizations directed towards competitiveness.

**Tacit Knowledge:** Knowledge of a tacit nature, which in many cases is not easily interpretable or shared, and in order to be understandable, it requires a collective and shared situation that distinguishes what is relevant within the organization.

Section 4

# Best Practices

# Chapter 13
# Exploring Alternative Assessments to Support Digital Storytelling for Creative Thinking in Primary School Classrooms

**Lee Yong Tay**
*Beacon Primary School, Singapore*

**Siew Khiaw Lim**
*Beacon Primary School, Singapore*

**Cher Ping Lim**
*Hong Kong Institute of Education, China*

## ABSTRACT

*This chapter documents the use of digital storytelling as a teaching approach to facilitate the learning of creative thinking among students (aged 7 and 8) in a primary school setting. A constructive teaching approach is adopted to allow students to create their own digital stories based on an authentic experience and express their thoughts. The focus of this chapter is to show how a shift from traditional classroom assessment to more flexible, alternative assessment format facilitates higher level thought processes (e.g., creative thinking) and range of skills. Several issues and challenges of using alternative assessment in digital storytelling are explored and discussed. Findings suggest that digital storytelling may be effectively used as an approach to foster creative thinking. They also suggest that refinements to the assessment process are needed to make it more formative in nature.*

DOI: 10.4018/978-1-60960-519-3.ch013

## INTRODUCTION

Research studies in education have shown that information and communication technologies (ICT) coupled with the necessary pedagogical strategies could engage students in critical, creative, and higher-order type of thinking (Hollingworth & McLoughlin, 2001; Jonassen & Carr, 2000; Jonassen, Peck, & Wilson, 1999; Kearney & Treagust, 2001). However, whether these ICT tools could be used to engage students in such higher-order type of thinking depends on how they were being used in the classrooms (Lim & Tay, 2003). Like any tool in the learning environment, ICT may be used well or poorly, and care and experience are needed when using it (Healy, 2001).

In the recent years, there has been a shift from traditional methods of classroom assessment to more flexible, alternative assessment formats that capture higher level thought processes (e.g., creative thinking), range of skills, and capabilities. This chapter is our preliminary exploration (from 2008 to 2009) on the use of digital storytelling (Lambert, 2006) as a constructive tool to engage and facilitate the learning of language skills and the acquisition of creative thinking skills for the lower primary students. It focuses on how alternative assessments have been employed to support its use and discusses the challenges faced in the implementation of such assessments in English Language primary classrooms.

## LITERATURE REVIEW

### Digital Storytelling

*"Clearly we are increasingly visual in our orientation towards communications. Reading and writing literacies are adversely affected, as each generation sees the screen as a preferable expressive medium. However, the inherent values of textual communication begin to become stronger by contrast, just as the strength of recitation and musical orality became stronger with the dissemination of literacy" (Lambert, 2007, p. 27).*

Bull and Kajder (2005) and Lambert (2006, 2007) propose the use of digital storytelling as a strategy to engage learners in writing. According to Bull and Kajder (2005), a digital story consists of a series of still images combined with a narrated soundtrack to tell a story. They reiterate that the focus in the language arts classroom should be on the writing and communication process rather than technical effects. Digital storytelling could be used as a constructive tool to engage students in higher order type of thinking skills as it could help students to 'visualise' and 'verbalise' their writing, develop creativity, and facilitate the learning of technological literacy, including the embedding of the voices of the students. In this creation process, students need to synthesise their writing and personal thoughts with clarity and organisation. However, it would also be necessary to design an appropriate approach to assess students' learning and acquisition of higher order thinking skills through the use of digital storytelling.

### Digital Storytelling for Creative Thinking

The major components of creative thinking are synthesising, imagining, and elaborating (Jonassen, 2000).

Synthesising involves skills such as analogical thinking, summarising, hypothesising, and planning. Imagining, as the name implies, is something that is less concrete and not as easily taught as other skills but it is important for the generation of new ideas. It also involves intuition and fluency of thinking, and requires students to visualize some form of objects or actions. Imagining includes skills such as fluency, predicting, speculating, visualising, and intuition. Elaborating on information involves adding personal meaning and information by relating it to personal experiences or building on an idea. Elaborating includes skills such as expanding, modifying, extending, shifting categories, and concretising.

Jonassen (2000) proposes the use of computers as mind-tools for teaching and learning. He puts forward the idea of using of computers as tools for students to construct and create their own learning and knowledge. Computers as mind-tools could be used to facilitate the acquiring of higher-order type of thinking skills (e.g., creative thinking).

## Alternative Assessment to Support Digital Storytelling

According to Abrami and Barrett (2005), Kubiszyn & Borich (2007), and Gipps (1994), there has been a shift from traditional methods of classroom assessment (e.g., Multiple Choice Questions, True and False) to more flexible assessment formats that measure higher level of thought processes (e.g., Problem Solving). Zacharis (2010) calls for innovative assessment for learning to enhance the learning processes. Digital storytelling used for assessment purpose parallels the shift from a quantitative tradition of assessment to a more qualitative approach. The digital storytelling approach may be supported by an alternative mode of assessment to evaluate language acquisition and also study students' higher-order thinking skills and various other skills. For example, in creating the digital story, students were required to plan and write creatively, select/create images and record their reading. These assessable artefacts provide teachers with an insight into students' higher order thinking skills and other skills.

In this chapter, the use of alternative assessment in digital storytelling for both formative and summative purposes will be examined for its fitness of purpose, validity, reliability, and equity/fairness in the assessment of higher-order type of thinking and creative skills. Fitness of purpose, validity, reliability, and equity/fairness are common themes that are being used to evaluate the effectiveness of assessment approaches (Black, 1998; Crooks, Kane, & Cohen, 1996; Gipps, 1994; Kubiszyn & Borich, 2007; Messick, 1989; & Stiggins, 2002)

### Fitness for Purpose

The fundamental questions in assessment design are the purpose of assessment and whether the assessment methods fit the purpose of assessment. As mentioned by Black (1998, p.24), "it is important to match the selection and use of assessment methods to a particular purpose which the assessment is meant to serve".

The majority of assessment methods fulfil two assessment purposes: assessment of learning (summative assessment) or assessment for learning (formative assessment). According to Stiggins (2002), "the crucial distinction is between assessment to determine the status of learning and assessment to promote greater learning" (p. 761). In the more recent time, there is another approach, which is the *assessment as learning*. For this approach, learners and instructors set learning goals, share learning intentions and success criteria, and evaluate their learning through dialogue and self and peer assessment (Henderson 2003). However, for this chapter, the focus is on the assessment for learning (i.e., formative assessment) to facilitate the learning of creative thinking.

Digital storytelling could be designed as assessment for both formative and summative evaluation of learning. There is a need to make a distinction between digital storytelling to support assessment of learning and/or for learning. This will determine the content, the creation process, and the design of the evaluation rubrics so it is important to have a clear sense of intended purpose from the start.

Once the purpose is identified, it is important to have the assessment to fit its purpose because of this direct impact on teaching and learning. According to Stiggins (2002), if the purpose of assessment is of learning to inform decisions or driven by the needs of policy makers, one consequence might be that schools challenge students to work harder to rise test scores. This may cause teachers to teach to achieve good grades rather for understanding and enhance learning for some students while discouraging others and causing them to give up. On the other hand, Stiggins (2002) mentioned that if the purpose of assessment is for learning, teachers use assessment information about student achievement as a means to advance their teaching and students' learning and not merely checking on student learning.

## Validity

Validity is the most important consideration in the use of assessment procedure as affirmed by most literature on assessment (Crooks, Kane, & Cohen, 1996; Kubiszyn & Borich, 2007; Gipps, 1994; Messick, 1989). Valid assessment can minimise unintended negative consequences such as reduced motivation and exclusion from further learning opportunities as mentioned by Crooks et al. (1996).

The evidence of validity in assessment can vary depending on its purpose. It was also surfaced in the literature that consequential validity argument does differ for formative and summative assessment. According to Stobart (2006), in summative assessment it is the trustworthiness of the inferences that are drawn from the results, and in formative assessment, it is about consequences on whether further learning has taken place as a result of the assessment.

In formative assessment, feedback is an important concept and the consequential validity issue here is whether feedback leads to further learning (Stobart, 2006). Black (1998) believes that there is action based on feedback in formative assessment towards integration of learning and development in understanding. However, Stobart (2006) pointed out that feedback does not necessary help learners to close gap in learning as many current feedback practices may not lead to further learning. As such, these may not be valid formative assessment.

## Reliability

Reliability is a major consideration in evaluating an assessment procedure as it can affect validity especially on the inferences that can be drawn based on the literature reviewed. Gipps (1994, p.171) summarizes the definition of reliability by indicating that it relates to consistency of assessment tasks being presented in the same way to all students assessed; assessment criteria being interpreted in the same way by all teachers; and students performance being evaluated according to the same rubric and standards by all markers. In the assessment of digital stories created, the reliability question of whether there is consistency of approach to the assessment tasks, consistency in interpretation of rubrics as well as consistency of standards in assessment would be examined. This is especially important for the summative mode of assessment.

Threats to reliability identified in the literature reviewed are inter-rater consistency and consistency of approach to assessment tasks. One way to achieve consistency in assessment in order to enhance quality is for group moderation to occur as recommended by Gipps (1994).

Reliability is also interpreted differently in formative assessment. According to Stobart (2006), as formative purposes judgements are essentially student-referenced and the focus is on providing feedback to "close the learning gap" so there is no need for consistency in formative assessment.

## Equity/Fairness

All assessments should allow for students of all backgrounds to do equally well and provide equal opportunity for all students to demonstrate the skills and knowledge being assessed. This is the equity or fairness aspect of assessment. According to Gipps (1994), equity is a complex area in assessment, and it is not just to do with the design of tests and assessment material. Equity issues include equality of access to learning, 'biased' curriculum, and inhibiting classroom practices. Gipps (1994) views bias in assessment to mean that the assessment is unfair to one particular group or another and thus will jeopardise the fairness of the assessment.

In this context, fairness could be viewed as no bias of using digital storytelling as a form of assessment towards any group of students, allowing for equitable treatment in the assessment process, and providing opportunity for all to learn the necessary skills and knowledge required for a satisfactory creation.

The issues related to fairness would need to be examined. These are: Will students with good keyboard or writing skills have an advantage in digital storytelling? Will ready access to computers outside the school favour particular social groups over others?

## RESEARCH DESIGN

A multiple case studies approach (Stake, 2006) is used to report the implementation of the digital storytelling and how it has been used as an alternative mode of assessment to promote higher-order type of learning (i.e., creative thinking) among the students. In this multiple case study, each of the class constitutes a case of how digital storytelling is implemented and used as assessment tool to promote creative thinking. A total of 8 classes of primary school students were involved in this study. These students were in Primary 1 in 2008 (aged 7) and Primary 2 in 2009 (aged 8). The issues and challenges in using digital storytelling as an assessment tool are highlighted. The cases are triangulated within each case and between the cases to enhance the validity of the study.

The main purpose of this paper is to explore how alternative assessments were used to support digital storytelling for creative thinking in the primary school classrooms. The tracing of the challenges and issues faced could provide insights for future implementation of such an innovation in an educational setting so that further refinements or improvements could be made.

### Data Collection Methods

Data were collected from multiple sources to provide a more complete picture on the issues explored and increase the validity of the study. In this case study, data were gather from the students, the teachers involved and the authors. Bush (2002) termed this way of comparing many sources of evidence to determine the accuracy of information or phenomena as triangulation and he viewed such cross-checking data as a means of establishing validity.

The data collection procedures of this study included the followings:

1. **Observation notes by the authors:** The observation notes focused on the school developments and programs. Observation notes were audio-taped using a digital audio recorder for further analysis. These notes were kept to have systematic records of significant developments and events during the period of implementation.
2. **Lesson observations:** Formal lesson observations were conducted by the authors. Lesson observation notes on the activities carried out in the classrooms were recorded for further analysis. The main purpose was to look into how classroom lessons were conducted on how the digital storytelling process was carried out. The observations were done in a non-intrusive manner with the researcher sitting behind the classroom recording the activities that were being carried out.
3. **Meetings with teachers and the authors:** Meetings were conducted to monitor the progress of the school programs and also to further understand teachers' intent and assumptions when they were conducting

their lessons. Meetings were more official channel where teachers and the researchers met up to review the program.

4. **Student produced artefacts, reflections, and feedback:** The digital stories (work assigned to them) were collected for evaluation and analysis. These artefacts were collected by all teachers and submitted to the authors to assess and examine the learning of the students. The students' works were assessed according to an in-house rubrics developed by the teachers. In addition, students' written reflections and verbal feedback were also taken into account.

## FINDINGS & DISCUSSIONS

### Implementation of Digital Storytelling

This section discusses how digital storytelling was used to facilitate the acquiring of creative thinking among the students. This is then followed by a discussion on how the digital storytelling approach has been used to meet the summative and formative purpose and whether it has fulfilled the objective of using it as an assessment tool, in particular, the measurement of higher-order and creative thinking.

The main objective of the digital storytelling was to allow students to apply their higher order type of thinking skills (e.g., creative thinking skills) through the creation of the digital stories through a constructivistic approach. The digital stories created by the students would be assessed both formatively and summatively, that is, to facilitate further learning and also assess the outcomes of their learning, respectively (see Appendix: Digital storytelling rubrics).

The lesson plan/idea was a very simple one, that is, to use a presentation software to create a digital story by the lower primary students (Primary 1 and 2, aged 7 and 8) with text, digital images, and sound recordings. Students were given a series of tasks to guide them to the completion of their digital storytelling assignment. The tasks included brainstorming for ideas in groups or pairs for profiles of characters, drafting of story outlines and finally recording their narration of the stories. The completed digital stories were then "published', as in shared on the school network. This mode of sharing completed work facilitates ease of access among students, and a plaftorm to appreciate and evaluate one another's work.

With clear objectives stipulated in the series of lessons, these students created their own digital stories with not only texts but also digital images and sound recordings, it took only about 15 to 18 one-hour lessons for the lower primary students to learn how to create their first very own digital stories. More important, students might be more motivated to learn the English language.

This process of organising and translating their ideas into text and colourful visuals engaged them. Students applied creative thinking skills such as synthesising and imagine planing their stories. In addition, it spurred them on to further elaborate and develop a story about their created characters. With the use of ICT tools to present their stories creatively, students could learn from each others' work while work was still in progress. During those lessons where their works were shared in class through the school network, students were clearly motivated to improve upon their pieces of work. Students were constantly encouraged to improve on their draft stories. In the fine tuning process of their stories, their writing skills were sharpened. In the process of recording their narration of their stories, students did multiple readings and recordings until they were satisfied with their work. This process encouraged students to critically think about their reading, self evaluate and thus improve upon their reading skills. The completed digital stories gave students a sense of satisfaction and accomplishment as they readily shared their stories with peers, teachers and even visitors to the school.

On the whole, students were observed to be very engaged during the lessons with the use of the Tablet PCs and the presentation software. It provided students the platform to showcase their creativity as no two end products are the same.

## Using Digital Storytelling to Facilitate the Learning and Acquiring of Creative Thinking

The use of digital storytelling as an approach to engage students in the learning of language and to facilitate the development of higher-order and creative thinking skills were very well received by teachers and administrators of the school. This initiative started off with just one class piloting its implementation in early 2008. The success and impact of this approach on students was very quickly cascaded to the other classes, with a total of 16 classes (all of Primary 1 and 2) adopting it in their teaching of the various languages (i.e., English, Chinese, Malay, and Tamil) by the end of year 2009.

It was undoubted that students were given the opportunity to creatively construct and create their own digital stories and hence, this process facilitated their learning and development of their creative thinking.

In the following year of implementation (i.e., 2009), this approach was further refined by some teachers to further use ICT tools to better engage the students and nurturing their creative thinking skill. For instance, in one of the cases, one of the English Language teachers explored the teaching of simple animation using commonly available presentation software in her English lessons. In addition, she also tried to incorporate art and music into the creation of the animated digital stories by the students. The teacher recognised the benefits of using the concept of animation to engage young learners in the learning of the English language. She work with the Art and Music teacher to further incorporated elements of art, music, and technology into the students animated stories.

The teacher designed a series of lessons based on fables books the students have read. She started by providing her students with background ideas to develop their own animated, multimedia stories. The students were required to plan and write a script based on the selected fable. The script involved speeches of the narrator and two more characters in the story. Students used the Sound Recorder software that is available on their computer desktop to record their voices to be embedded into the animated, multimedia stories. Students also drew their own characters for their stories using simple paint software. In addition, students also used available music software to mix and match musical loops to create the concept music that could bring out the moods of various scenes of their stories. To complete their animated digital stories, students took up the role of the characters and recorded their voice to be incorporated into the animated digital story. Students peer evaluated their voice recordings and further improved on them before embedding them into their animated digital story productions. The creation of the animated stories captivated the attention of the students, engaged them in the learning of the English language, and stretched their imagination and creativity. It was basically a series of English writing lessons that were further facilitated with the use of technologies, animation, art, and music. The key elements in the assessment were usage of language, the content of the story, higher order thinking skills such as imagining, elaborating and synthesising.

In another case of a Chinese language class, the teacher recognised the benefits of collaborative learning and group work in producing a news report and also the application of Chinese Language in real life situation. She further incorporates this collaborative element with authentic learning environment, leverages on technologies. She designed a year-long project work called the "Young Journalists" for all her Primary 2 students to engage them in collaborative type of learning. First, she taught her students basic reporting and interviewing skills. Her students used the sound recorder software to record the interviews. Students paired up to do their peer interviews with the sound recorder software and these audio files were subsequently uploaded to the online platform for sharing. Following this initial exposure to the basic interviewing skills and the handling of the technologies, students were formed into groups of journalists ready to cover a sports event in the school which is organised in line with the Asian Youth Games in 2009 where students also learned more about sportsmanship. Roles were assigned to the students who formed into groups of four (i.e., one editor, one photographer, one reporter, and one interviewee). Basically the groups went around interviewing their schoolmates who were having their co-curricular activities in the afternoons. The groups then created their sports news using online free presentation and video software to create their multimedia stories. To further interest her students, she enlisted help from fellow teachers to rewrite the Asian Youth Games related sports news (so to make the news more easily understandable by lower primary students) and uploaded them to online free educational portal (i.e., the cloud computing concept). Students will access this online portal to retrieve the sports news to get ideas and information so that the students could create their individual multimedia news report using online free presentation software.

## Alternative Assessment in Digital Storytelling

In this section, the issues and challenges faced during the implementation of alternative assessment (for both formative and summative aspects) in digital storytelling would be discussed in relation to the purpose of fit, validity, reliability, and fairness/equity as outlined in the literature review section. In addition, assessment issues pertaining to the facilitation of creative thinking among students would also be discussed.

### Purpose of Fit

The intended purpose was to use both formative and summative approaches towards the assessment of the digital stories created by the students, that is, both the assessment for learning and the assessment of their learning. The focus was to use the assessment to further facilitate the learning of the students and also the evaluation of what they have learned. Although the use of the digital story telling approach managed to engage students and allowed them to continuously improve their works and facilitated their learning of creative thinking, the assessment approach used by the teachers was generally more summative than formative in nature.

From our observations, the interviews with the teachers involved, and the triangulation among the authors, the findings seem to suggest that the current assessment process and rubrics support more summative purpose than formative purposes. Nichols, Meyers, and Burling (2009) also caution the importance of the use of formative assessment. The above authors feel that "for an assessment use to be appropriately labelled "formative," both empirical evidence and reasoned arguments must be offered to support the claim that improvements in student achievement can be linked to the use of assessment information (Nichols, Meyers, & Burling, 2009, p. 14). The teachers involved also reflected that they were not technically very proficient to carry out the formative assessment approach. Through the interviews, teachers generally believe the benefits and potentials of taking a more formative approach to assess the learning of their students but in terms of practice they were observed to be otherwise. One main objective of using DST for assessment was to be able to measure HOT and range of skills. However, through the interviews, teachers also felt that there was a diverse spectrum of skills to be assessed and the assessment of creative thinking would inevitably be watered down vis-à-vis the other aspects to be assessed at the same time (see Appendix).

## Validity

As mentioned in the literature review section, consequential validity is an important consideration in formative assessment. That is, whether further learning has taken place as a result of the formative assessment. Feedback is an important step in formative type of assessment and when used the right way, it could lead to further learning (Stobart, 2006). The teachers involved provided feedback to the students but they felt that it was too time consuming and that they would require more training in how to provide useful feedback that would lead to the students' further learning. Some found it cumbersome as this was not within their repertoire of teaching practices.

The rubrics developed by the teachers seemed to be a good instrument in bringing out the validity for the more summative assessment of the digital stories. However, as the assessment process usually ended around end of the term, the teachers did not made use of the results to draw inferences on students' progress in learning.

## Reliability

Reliability was another major issue during the summative assessment of the digital stories created by the students. Through the examination of the artefacts (i.e., the digital stories), lesson observations, and the interview with teachers, there were discrepancies on how the assessment tasks were being presented to the students and ultimately affected the consistency of the final product (i.e., the digital stories). There were vast differences in the digital stories presented by the students at the end of the project among the eight classes in the first year of implementation. The teachers had different ideas, concept, and approach towards the implementation of the digital storytelling process despite meetings throughout the period of the project. In addition, the assessment criteria (the rubric) were also interpreted differently among the eight teachers. Although the same rubric was used, standards of evaluation between the teachers varied. Hence, inter-rater consistency was being compromised. The assessment process was refined subsequently to address these issues of inconsistencies.

Although Stobart (2006) argued that formative assessment is student-referenced and there is no need for consistency, the teachers had different interpretation of the assignment task and the rubrics as their mindset was mainly summative assessment in nature and consistency was one of the important criteria to be considered. As the teachers have perceived assessment was more summative in nature, the formative aspect of nurturing creative thinking and skills might have been compromised.

## Equity/Fairness

In order to allow for students of all background (e.g., social economic) to do equally well and provide equal opportunity for all students to demonstrate the skills and knowledge being assessed, all information communicative technological tools were provided by the school during the curriculum hours. Students who needed more time to complete their work could always stay back to use the computers in school. Hence, the issue of a lack of ICT tools did not pose as an issue.

Students without prior computer skills and knowledge were not disadvantaged as the software application was easy to use and they were given ample time to practise their computer skills. The rubrics also ensured the overall grade will not be affected too much by students' competency in their technical skills.

As story creation hinges on students' language abilities, students with better writing skills would be able to perform better in the language component such as in the content, grammar, and flow of the stories they created. However, students who are not strong in language will not be penalised if they can perform well in the other components measured by the rubrics such as reading and team work.

## IMPLICATIONS AND RECOMMENDATIONS

The above findings from the various cases observed, it seems to suggest that the digital storytelling pedagogical approach could be used as an approach to engage students in the acquisition of creative and higher-order type of thinking. However, there remain issues and challenges with regard to the use of digital storytelling as an assessment tool to facilitate the learning of creative thinking.

From the findings and discussions, there are five interrelated implications that need further considerations:

1. Iterative approach towards the implementation of alternative assessment in digital storytelling for the acquisition of creative thinking;
2. Understanding of the essence of formative assessment approach and creative thinking;
3. Teachers' professional development / building teachers' capacity;
4. Standardisation and moderation for inter-rater consistency; and
5. Balanced approach: formative and summative assessment.

The implications were derived from the following data collection methods as shown in Table 1. The data collected were triangulated within each method and between methods for a valid and balanced presentation of the findings.

### Iterative Approach towards the Implementation of Alternative Assessment in Digital Storytelling for the Acquisition of Creative Thinking

As indicated in the findings from the observation notes by the authors and the meeting sessions between the teachers and authors, the socio-cultural context where the digital storytelling approach was used as an assessment tool for the learning and acquisition of creative thinking might not be epistemologically and systemically ready, especially in its initial months of implementation.

General teaching practices in schools, the focus of the society, and parents' expectations have been leaned more towards summative grades and

*Table 1. Implications derived from the various data collection methods*

| No. | Implications | Data Collection Methods |
|---|---|---|
| 1. | Iterative approach towards the implementation of alternative assessment in digital storytelling for the acquisition of creative thinking | a) Observation notes by authors &<br>b) Meeting sessions with teachers and authors. |
| 2. | Understanding of the essence of formative assessment approach and creative thinking | a) Observation notes by authors &<br>b) Lesson Observations. |
| 3. | Teachers' professional development / building teachers' capacity | a) Meeting sessions with teachers and authors. |
| 4. | Standardisation and moderation for inter-rater consistency | a) Observation notes by authors,<br>b) Lesson observations,<br>c) Meeting sessions with teachers and authors, &<br>d) Student produced artefacts, reflections, and feedback. |
| 5. | Balanced approach – formative and summative assessment | a) Observation notes by authors. |

test scores. Stobart (2008), also echoes similar sentiments, "one of the most persistent threats to formative assessment is that it is conscripted to support the demands of summative assessment or neglected once 'real' assessment is imminent" (p. 5). Time and continuous effort would be necessary for any substantive and significant change to materialise. Sustaining such initiative would also require substantial effort. An iterative approach might be necessary for the continuous fine-tuning of the assessment process for implementation of using digital storytelling as an approach to facilitate the learning of creative thinking and also for it to gain its significance in assessment approaches in schools. Hence, for any innovative approaches to be sustained and have an impact on the system (i.e., teaching and learning practices in schools), time for continuous improvements would have to be factored in for any significant change to take place.

### Understanding of the Essence of Formative Assessment Approach and Creative Thinking

From the observation notes by the authors and the lesson observations conducted, another important aspect is for teachers and school administrators to understand the essence of formative assessment for further learning rather than just for assessment of learning. Although the concept on assessment for learning is not new, this concept needs to be constantly communicated and reminded to those who are implementing such educational initiative. This essence of formative assessment could easily be diluted by the everyday busyness of teaching and administrative duties as it requires more time for teachers to customise the feedback for each student. Hence, constant discussions and dialogues would be necessary to revisit the fundamental nature of formative assessment.

In addition, the concept of creative thinking would still need to be further explored and discussed among the teachers and also the school administrators. The definition of creative thinking and creativity still varies among the teachers. Some felt that creativity could be explicitly taught whereas some teachers thought otherwise. There is a need for a whole school approach towards implementing programmes to explicitly promote creativity in students and develop their creative thinking.

From the current socio-cultural context of assessment and the expectations by the society, reliability in terms of uniformity in the assignment tasks, assessment criteria, and inter-rater consistency are still very necessary to maintain fairness. This socio-cultural context has affected how schools and teachers approach the use of formative type of assessment. The approach was still very summative biased.

## TEACHERS' PROFESSIONAL DEVELOPMENT/BUILDING TEACHERS' CAPACITY

Feedback from teachers through the meeting sessions with them reflected that teachers who were involved in the project reflected that they were not well prepared to carry out more formative type of assessment and also the assessment of creative thinking as they lacked the necessary skills and knowledge. One of the learning focuses for professional development of teachers would be to build their assessment literacy through attending courses, sharing sessions, and seminars/ conferences.

### Standardisation and Moderation for Inter-Rater Consistency

From the observation notes by the authors, lesson observations, meetings with teachers, and student produced artefacts, reflections, and feedback, it was reflected that there was a need for consistency in interpreting and communicating the assignment tasks and assessment rubrics, standardisation through consensus grading of sample digital stories, and moderation after evaluation to enhance inter-rater consistency. The fundamentals to better standardisation and moderation lie with good planning and constant dialogue among teachers who are involved in the implementation. Enhancing the reliability aspect of the assessment would inevitability increase the fairness of assessment and confidence levels of the students, parents, and other stakeholders.

### Balanced Approach: Using Digital Storytelling as both Forms of Formative and Summative Assessment

From the observation notes by the authors, this education innovation is situated in a highly structured and highly efficient educational system with internationally renowned summative high-stake assessment system, the implementation of a more formative mode of assessment would be like a vessel moving against the currents. A balanced approach of blending both formative and summative purposes may be a more workable and practical move towards the implementation of a more holistic/comprehensive assessment in schools. In short, a more balanced assessment approach of and for learning especially in the development of creative thinking, is called for.

## CONCLUSION

This preliminary multiple case study of the use of alternative assessment in digital storytelling to foster creativity highlighted several interesting and critical issues for further considerations for research and curriculum innovations in the area of assessment and the use of information communication technology in schools. First, this multiple case study highlighted the importance and influence of the socio-cultural context. The decades of highly structured and efficient system of mainly summative approach of assessment has a considerable impact on how formative type of assessments could be carried out in the schools. Second, a longer timeline would be necessary for the schools, teachers, and students to learn and get used to the more formative forms of assessment through an iterative implementation approach and a whole school approach is necessary for the shift. In addition, teachers would not only need to have a change in their epistemological beliefs towards a more formative nature of assessment, they also require the necessary knowledge and skills so that their practices could be transformed to bring out the essence of formative assessment, that is, to promote further learning among our students, especially in the development of the students' creative thinking and creativity.

With the introduction of any new practices, be it in educational settings or otherwise, it is essential to consider the socio-cultural factors that could possibly facilitate or impede such initiatives. Another important consideration is that a longer timeline should be factored for more iterations for the resolution of the challenges surfaced for the new innovation to be assimilated into existing practices.

## REFERENCES


Abrami, P. C., & Barrett, H. (2005). Directions for research and development on electronic portfolios. *Canadian Journal of Learning and Technology, 31*(3). Retrieved on February 5, 2006, from http://www.cjlt.ca/content/vol31.3/abrami.html

Black, P. (1998). *Testing: Friend or foe? Theory and practice of assessment and testing*. London: RoutledgeFalmer.

Bull, G., & Kajder, S. (2005). Digital storytelling in the language arts classroom. *Learning and Leading with Technology, 32*(4), 46–49.

Bush, T. (2002). Authenticity–reliability, validity and triangulation. In Coleman, M., & Briggs, A. (Eds.), *Research methods in educational leadership and management*. London: Paul Chapman.

Crooks, T. J., Kane, M. T., & Cohen, A. S. (1996). Threats to the valid use of assessment. *Assessment in Education*, *3*(3), 265–285. doi:10.1080/0969594960030302

Fontana, L. A., Dede, C., White, C. S., & Cates, W. M. (1993). *Multimedia: A gateway to higher-order thinking skills*. Center for Interactive Educational Technology: George Mason University, VA.

Gipps, C. V. (1994). *Beyond testing: Towards a theory of educational assessment*. London: The Falmer Press.

Healy, J. M. (1998). *Failure to connect: How computers affect our children's minds–for better and worse*. New York: Simon & Schuster.

Henderson, A. J. (2003). *The e-learning question and answer book: A survival guide for trainers and business managers*. Broadway, NY: American Management Association.

Herrington, J., & Oliver, R. (1998). *Using situated learning and multimedia to promote higher-order thinking*. Western Australia: Edith Cowan University.

Hollingworth, R. W., & McLoughlin, C. (2001). Developing science students' metacognitive problem solving skills online. *Australian Journal of Educational Technology*, *17*(1), 50–63.

Jonassen, D. H. (2000). *Computers as mindtools for schools: Engaging critical thinking* (2nd ed.). Upper Saddle River, NJ: Merrill/Prentice Hall.

Jonassen, D. H., & Carr, C. S. (2000). Mindtools: Affording multiple knowledge representations for learning. In Lajoie, S. P. (Ed.), *Computers as cognitive tools, vol. two: No more walls* (pp. 165–196). Mahwah, NJ: Lawrence Erlbaum.

Jonassen, D. H., Peck, K. L., & Wilson, B. G. (1999). *Learning with technology–a constructivist perspective*. Upper Saddle River, NJ: Merrill/Prentice Hall.

Kearney, M., & Treagust, D. F. (2001). Constructivism as a referent in the design and development of a computer program using interactive digital video to enhance learning in physics. *Australian Journal of Educational Technology*, *17*(1), 64–79.

Kubiszyn, T., & Borich, G. (2007). *Educational testing and measurement–classroom application and practice* (8th ed.). USA: John Wiley & Sons, Inc.

Lambert, J. (2006). *Digital storytelling-capturing lives, creating community*. Berkeley, CA: Digital Diner Press.

Lambert, J. (2007). Digital storytelling–how digital media help preserve cultures. *The Futurist*, *41*(2), 25.

Lim, C. P., & Tay, L. Y. (2003). Information and communication technologies (ICT) in an elementary school: Students' engagement in higher order thinking. *Journal of Educational Multimedia and Hypermedia*, *12*(4), 425–451.

Messick, S. (1989). Validity. In Linn, R. L. (Ed.), *Educational management* (3rd ed., pp. 13–103). London: Collier Macmillan.

Neo, K. T. K., & Neo, M. (2001). A constructivist learning experience: Reconstructing a web site using web based multimedia authoring tools. *Australian Journal of Educational Technology*, *17*(3), 330–350.

Nichols, P. D., Meyers, J. L., & Burling, K. S. (2009). A framework for evaluating and planning assessments intended to improve student achievement. *Educational Measurement: Issues and Practice*, *28*(3), 14–23. doi:10.1111/j.1745-3992.2009.00150.x

Oliver, K., & Hannafin, M. (2000). Student management of web-based hypermedia resources during open-ended problem solving. *The Journal of Educational Research*, *94*(2), 75–92. doi:10.1080/00220670009598746

Sarapuu, T., & Adojann, K. (1999). *Usage of educational web pages to develop students' higher-order thinking skills*. Science Didactics Department, University of Tartu.

Stake, R. E. (1995). *The art of case study research*. Thousand Oaks, CA: SAGE.

Stake, R. E. (2006). *Multiple case study analysis*. New York: The Guilford Press.

Stiggins, R. J. (2002). Assessment crisis: The absence of assessment for learning. *Kappan Professional Journal, 83*(10), 758-765. Retrieved on March 16, 2006, from http://electronicportfolios.org/afl/Stiggins-AssessmentCrisis.pdf

Stobart, G. (2005). Fairness in multicultural assessment systems. *Assessment in Education, 12*(3), 275–287. doi:10.1080/09695940500337249

Stobart, G. (2006). The validity of formative assessment. In Gardner, J. (Ed.), *Assessment and learning* (pp. 133–146). London: Sage Publications.

Stobart, G. (2008). *Validity in formative assessment*. Paper presented at AEA-Europe ninth annual conference Hisar, Bulgaria, 6-8 November 2008.

Zacharis, N. Z. (2010). Innovative assessment for learning enhancement: Issues and practices. *Contemporary Issues in Education Research*, *3*(1), 61–69.

## ADDITIONAL READING

Abrami, P. C. (2001). Understanding and promoting complex learning using technology. *Educational Research and Evaluation*, *7*(2-3), 113–136. doi:10.1076/edre.7.2.113.3864

Barab, S. A., Hay, K., & Duffy, T. (1998). Grounded constructions and how technology can help. *TechTrends*, *43*(2), 15–23. doi:10.1007/BF02818171

Black, P. (1998). *Testing: Friend or Foe? Theory and Practice of Assessment and Testing*. London: RoutledgeFalmer.

Black, P., Harrison, C., Lee, C., Marshall, B., & William, D. (2002). *Working inside the black box. Assessment for learning in the classroom*. London: King's College.

Black, P., & William, D. (1998). Inside the Black Box: Raising Standards through Classroom Assessment. *Phi Delta Kappan*, Oct 1998. Retrieved 12 Dec 2006 from http://www.pdkintl.org/kappan/kbla9810.htm

Bloom, B. S., Engelhart, M. D., & Furst, E. J., Hill, W. H., & Krathwohl, D. R. (1956). *Taxonomy of educational objectives, Handbook I: Cognitive domain*. New York: McKay.

Bush, T. (2002). Authenticity – reliability, validity and triangulation. In Coleman, M., & Briggs, A. (Eds.), *Research Methods in Educational Leadership and Management*. London: Paul Chapman.

De Bono, E. (1993). The direct teaching of thinking as a skill. *Phi Delta Kappan*, *64*(10), 703–708.

Driscoll, M. P. (2000). *Psychology of learning for instruction* (2nd ed.). Needham Heights, Massachusetts: Allyn & Bacon.

Driscoll, M. P. (2001). Computers for what? Examining the roles of technology in teaching and learning. *Educational Research and Evaluation*, 7(2-3), 335–349. doi:10.1076/edre.7.2.335.3861

Gagne, R., Briggs, L., & Wager, W. (1992). *Principals of instructional design* (4th ed.). Fort Worth, TX: Harcourt Brace Jovanovich College Publishers.

Herrington, J., & Oliver, R. (1998). *Using situated learning and multimedia to promote higher-order thinking*. Edith Cowan University: Western Australia. (ERIC Document Reproduction Service No. ED 428 672)

Iowa Department of Education. (1989). *A guide to developing higher-order thinking across the curriculum*. Des Moines, IA (ERIC Document Reproduction Service No. ED306 550)

Klenowski, V. (1996). *Connecting Assessment and Learning*. Paper presented at the British Educational Research Association Annual Conference. Retrieved on 14 Jan 2007 from http://www.leeds.ac.uk/educol/documents/000000190.htm.

Noonan, B., & Duncan, C. R. (2005). 'Peer and Self-Assessment in High Schools'. *Practical Assessment. Research Evaluation, 10*(17).

Paris, S. G., & Ayres, L. R. (1994). *Becoming Reflective Students and Teachers*. Washington, DC: American Psychological Association. doi:10.1037/10158-000

Rolheiser, C., & Ross, J. A. (2000). *Student Self-Evaluation: What Research Says and What Practice Shows*. Article published on Centre for Developing and Learning Website. Retrieved on 2 Jan 2007 from http://www.cdl.org/resource-library/articles/self_eval.php?type=subject&id=4

Stiggins, R. J. (2002), 'Assessment Crisis: The Absence of Assessment for Learning'. *Kappan Professional Journal*, V83/10, 758-765. Retrieved on 16 Nov 2006 from http://electronicportfolios.org/afl/Stiggins-AssessmentCrisis.pdf

Wells, G. (1990). Talk about text: Where literacy is learned and taught. *Curriculum Inquiry, 20*, 369–405. doi:10.2307/1179876

Young, A. C. (1997). Higher-order learning and thinking: what is it and how is it taught? *Educational Technology, 37*(4), 38–41.

## KEY TERMS AND DEFINITIONS

**Digital Storytelling:** A process where students or individuals use digital media to produce and relate a story of their personal experience with the use of their voice, text, and digital images.

**Formative Assessment:** This form of assessment aims to facilitate further learning, commonly known as assessment ***for*** learning.

**Higher-Order Thinking:** In simple terms, the forms of thinking that require more than simple recall of content.

**Reliability:** Consistency in measurement; achieving the same results or outcomes after repeated trials or measurements.

**Standardisation:** An important process in assessment where teachers or assessors meet to discuss how to grade or assess so to achieve inter-rater consistency.

**Summative Assessment:** This form of assessment aims to evaluate the effectiveness ***of*** learning.

**Validity:** The measuring of what that is supposed or intended to be measured.

## APPENDIX 1.

*The digital storytelling rubrics*

| DST Rubrics | | | | | | Grades | | |
|---|---|---|---|---|---|---|---|---|
| Marks | Area of Assessment | | Assessment Items | | | 0–1 | 2–3 | 4–5 |
| | | | | | | Needs Improvement | Good | Excellent |
| 20 | GROUP WORK / WRIT-ING | TEAM-WORK & GROUP PROD-UCT (20) | Responsibility | 5 | Team-work | Task assigned requires close supervision to complete | Task assigned requires some supervision to complete | Task assigned requires mini-mal supervision to complete |
| | | | Teamwork | 5 | | Makes little or no effort to work cooperatively | Makes some effort to work cooperatively | Makes considerable effort to work cooperatively |
| | | | Contribution | 5 | | Seldom contributes or no con-tribution to project work | Some contributions to project work | Contributes actively to project work |
| | | | Group Writing Product | 5 | Product | Main ideas are not very clear | Main ideas are generally clear | Main ideas are consistently clear |
| | | | | | | | | |
| 30 | INDI-VDUAL WRIT-ING | LAN-GUAGE (15) | Grammar & Punctua-tion | 5 | | Many mistakes in grammar and punctuation | 4 to 6 mistakes in grammar and punctuation | 3 or less mistakes in grammar and punctuation |
| | | | Spelling | 5 | | Many mistakes in spelling | 4 to 6 mistakes in spelling | 3 or less mistakes in spelling |
| | | | Vocabulary | 5 | | Limited use of vocabulary | Some or good use of appropri-ate vocabulary | Very good or excellent use of vocabulary |
| | | CON-TENT & OR-GANI-SATION (15) | Relevancy of Ideas | 5 | | Ideas have little relevance to story | Ideas are somewhat relevant to story | Ideas are largely relevant to story |
| | | | Adequacy of Ideas | 5 | | Little expression of ideas in story | Adequate expression of ideas in story | Creative expression of ideas in story |
| | | | Organisation & Devel-opment of Ideas | 5 | Organi-sation | Little organisation of ideas | Some or good organisation of ideas | Effective organisation of ideas |
| | | | | | Devel-opment | Little development of ideas | Some development of ideas | Good development of ideas |

*continues on following page*

*The digital storytelling rubrics continued*

| | | | | | | | | |
|---|---|---|---|---|---|---|---|---|
| 30 | READING | ORAL PRE-SENTA-TION (30) | Clarity | 5 | | Little clarity in reading due to mumbling most of the time | Reads with a little mumbling some of the time | Reads clearly most of the time |
| | | | Fluency | 5 | | Jerky delivery | Generally smooth delivery | Smooth delivery |
| | | | | | | Reads with hesitation most of the time | Reads with hesitation some of the time | Reads every word without hesitation |
| | | | Expression | 5 | | Reads with little expression, like a robot, or does not change the sound of voice | Changes voice some or most of the time | Changes voice to show feel-ing or to sound like different characters |
| | | | Prominence | 5 | | Little attempt to place promi-nence | Prominence placed appropri-ately some of the time | Prominence placed appropri-ately most of the time |
| | | | Pronunciation | 5 | | Mispronounces most of the words | Mispronounces some of the words | Clear pronunciation most of the time |
| | | | Volume | 5 | | Voice is very soft | Voice is quite soft | Volume is adequate |
| | | | | | | | | |
| 20 | FINAL PROD-UCT | ICT skills (5) | Independent Creation of DST | 5 | | Needs teacher assistance all the time | Adequate use of functions in software | Proficient use of functions in software |
| | | | | | | Limited use of functions in software | Good use of functions in software | Creative use of functions in software |
| | | Creativ-ity (5) | Originality of Ideas | 5 | | Able to produce products that are mundane, predictable or with great support | Able to produce/adapt products that are unique with support | Able to produce product that are original, unique indepen-dently |
| | | Aes-thetics Presen-tation - Music (5) | Selection of Sound/ music | 5 | Quantity | Few sound effects/composi-tions are used to illustrate the story | Adequate sound effects/com-positions are used to illustrate the story | Sound effects/compositions are used to enhance the story |
| | | | | | Appro-priate-ness | Chosen sound effects/composi-tions match a small part the story | Chosen sound effects/composi-tions match some parts of the story | Chosen sound effects/composi-tions match most parts of the story |
| | | Aes-thetics Presen-tation - Art (5) | Selection of Images | 5 | Quantity | Few images are used to illus-trate the story | Adequate images are used to illustrate the story | Images are used to enhance the story |
| | | | | | Appro-priate-ness | 1–2 chosen images are relevant to the story/theme | 3–4 chosen images are relevant to the story/theme | All chosen images are relevant to the story/theme |
| | | | | | | | | |
| **100** | TOTAL | | | | | | | |

# Chapter 14
# Creative Management, Technology and the BBC

**Nicholas Nicoli**
*University of Nicosia, Cyprus*

## ABSTRACT

*The past decade has witnessed increased attempts by managers, scholars and policy-makers to stimulate the creativity of organisations. The practice of stimulating organisational creativity has led to a paradigm shift known as creative management, the focus of which is to use these practices to achieve competitive advantages. Such creative stimulation can come in a variety of forms. These include identifying and influencing environmental conditions that can increase the chances for creating new and significant products or services. In order to stimulate creativity, current creative management literature proposes the use of technology as a disseminator of knowledge and ideas. This chapter offers a literature review of creative management and technology use for creativity. It next introduces a case study of how technology is used as a creative management tool at the BBC. The findings of the study indicate that although the BBC's yearly revenues are under sustained pressure, the organisation has invested heavily in technology in order to maintain its high creative standing. In conclusion, supported by the findings of the case study, this chapter corroborates and further advocates the use of technology as a significant component of creative management practices.*

## INTRODUCTION

There is currently a massive proliferation in how organisations in developed societies deploy and manage creativity in order to attain a competitive advantage. Yet due to varying perspectives on how creativity can best be fostered and put into use, creative management discourses could be mistaken for an amalgam of all that is confusing in the study of creativity. Nevertheless, what seems to be endorsed by scholars, regardless of their discipline (e.g. Nonaka & Takeuchi, 1995 for knowledge management studies; Florida, 2002 for Urban Studies; Bilton, 2007 for communica-

DOI: 10.4018/978-1-60960-519-3.ch014

tions studies; Andriopoulos & Dawson, 2009 for organisational studies), is the use of technology and how it can best stimulate creativity.

This chapter highlights the growing significance of technology to how the British Broadcasting Corporation (BBC), the UK's main public service broadcaster, stimulates and manages creativity in television production. The significance of choice in the BBC is twofold. Firstly, due to the BBC's role as a bastion of public service organisations across the western world (Born, 2004), many of its practices are often duplicated, studied and implemented from across many academic disciplines and within a wide range of diverse organisations. Secondly, the BBC is a media organisation, and while many sectors are only now beginning to grasp the importance of managing creativity, media organisations have always regarded creativity as an essential element in their survival and everyday function (Davis & Scase, 2000). The making of a television or radio programme– the main functions of the BBC-implicitly entails a creative process that is paramount for the programme to be considered successful.

The chapter is divided in three sections. In the first, it addresses the major theoretical frameworks of organisational creativity management and the role technology plays in the creative management paradigm. This section begins with a brief introduction of the role creativity currently plays in organisations. It follows a series of four previous paradigms in organisational studies that have ultimately led to creative management. It next addresses the social context of creativity and the theoretical underpinnings that support creativity as a phenomenon that can be fostered through proper environmental measures. The second part of this section reviews some of the environmental changes that can be made in organisations in order to foster creativity. The final part of this section examines the technological tools and methods that support the creative management paradigm. It is noted here that there are no limits to how information technology and communicative technological tools can be used to allow creativity to flow, not only from within the organisation but across a broader social context beyond the organisation.

The second section focuses its attention on how the BBC has attempted to use technology as part of its creative management strategy. This section discusses the various changes implemented at the BBC since Mark Thompson was appointed Director General in 2004. It describes the BBC's *Creative Future* strategy and the significance given to technology by creating a larger, more inclusive technology department that impacts the whole organisation. The section looks in detail at two aspects of the BBC and the role technology plays in both. These are, the changing nature of how audiovisual audiences interact with media through different platforms, and the commissioning process of audiovisual content.

Finally, before moving on to the conclusion, the chapter offers a smaller third section on future research directions of the creative management paradigm.

## THEORETICAL OVERVIEW OF CREATIVE MANAGEMENT

It is important to acknowledge that interest in creativity derives from numerous academic disciplines such as communications, organisational studies, psychology, sociology and urban studies. There are two main reasons for the present attention afforded to creativity. The first is that a growing number of knowledge-driven societies are determined on becoming *'creative societies'*. A 'creative societies' theory, to a large extent, stems from the broader theoretical strand regarding information societies (Webster, 2006). Researchers from this approach prefer to concentrate on the study of how contemporary societies as a whole aim to shift creatively (Bilton, 2007), and how creativity has become an integral part of their lives. In Richard Florida's well-known work on the rise of the creative class (2004) he notes,

*Society is changing in large measure because we want it to... lately a number of diverse and seemingly unconnected threads are starting to come together... that driving force is the rise of human creativity as the key factor in our economy and society. Both at work and in other spheres of our lives, we value creativity more highly than ever and cultivate it more intensely. (p. 4)*

The second reason why interest in creativity has grown – as implied in the above citation - is the transformation of many societies into '*creative economies'*. Researchers here choose to concentrate solely on macro and micro economic policy, rather than study societies more generally (Asimakou, 2009). Cohesively, 'creative economies' thrive on new ideas through intellectual property (such as those using patents and copyrights) and research and development. For this reason Bilton (2007) argues that creativity 'should not be the privilege of a few officially designated businesses and missing from everywhere else' (p.3). Within this context managing creativity is just as vital for an automotive engineering firm or a pharmaceutical firm as it is for an advertising agency or an independent television production firm (Howkins, 2001). Undoubtedly, the literature on creative economies is of particular importance for this chapter, since organisations practicing 'creative management' approaches, together make up the sum of a creative economy.

In spite of the current embryonic state of creative management, Xu and Rickards (2007) note that it represents the most recent in a series of five paradigms that have been incorporated across organisations during the past century or so. The previous four are as follows:

1. **Rational Management:** E.g. Theories of Bureaucracy, Fordism, Taylorism
2. **Total Quality Management:** E.g. the scholarly works of W. Edwards Deming and Kaoru Ishikawa
3. **Humanistic Management:** E.g. Kurt Lewin's work on human relations within the organisation
4. **Organisational Creativity:** E.g. Lean production, knowledge management

Henry (2001) similarly states, 'there has been a paradigm shift in the nature of management and organisation, in which creativity, innovation and knowledge now have a central role' (p.22), while Bilton and Leary (2002) note that creativity management techniques, 'are seen to have replaced operational efficiency and strategic planning as the primary source of competitive advantage' (p.49). 'Creative management' as an organisational studies paradigm therefore involves the contextualised view regarding how creativity can be managed and used to increase levels of productivity and give organisations a competitive advantage (Henry, 2006; Xu and Rickards, 2007). Moultrie and Young (2009) note that for today's businesses, creativity is used 'to describe processes and outputs rather than inherent traits of individuals' (p. 300).

As the 'creative management' paradigm gathers momentum across the economy as a whole, an understanding of how the 'creative industries' manage creativity can significantly contribute towards the creative management paradigm (Bilton and Leary, 2002; Bilton, 2007; Andriopoulos & Dawson, 2009). Citing Jeffcutt and Pratt (2002), Moultrie and Young (2009) aver that, 'there is still much to be learned from this sector on encouraging and managing creativity (p. 299). The importance of creativity within 'creative industries', as the phrase itself implies, has *always* existed, since creativity is imperative for producing successful symbolic products of expressive value, such as television programmes, building designs, software design or advertising campaigns (Davis and Scase, 2000; Jeffcutt and Pratt 2002; Hesmondhalgh, 2007). This is mainly because creativity in such industries has an explicit function that results in the 'generation and exploitation of intellectual property' (DCMS, 2007; Bilton, 1999; Bilton 2007). 'Routine is the exception rather than the norm' for creative organisations, according to Moultrie and Young (2009, p. 299). The BBC represents a significant constituent of the creative industries, since the organisation produces and broadcasts television and radio programmes and web content across a wide range of genres and formats on a daily basis, impacting the lives of millions of people.

## The Social Context of Managing Creativity

Although creative management practices continue to spread, identifying "particular social and environmental conditions that can positively or negatively influence the creativity of most individuals" (Amabile, 1996, p. 5), has been studied for many years within the field of psychology. This is because psychologists understood early on that individual creativity should be placed within a socio-cultural context; this in turn has helped demystify many of the myths surrounding creativity.

The reason why the socio-cultural perspective offers a more viable way of understanding creativity is because it better explains the difficulties endemic to analyzing originality and value - these two concepts are considered fundamental attributes of an idea if it is considered to be creative. Both originality and value must 'pass social evaluation' (Csikszentmihalyi 1997, p. 23). Eysenck writes (1995, p. 37),

*When Shakespeare broke the rule of the consistencies (of time and place) in his plays he was certainly original, but was his originality relevant? What would relevance mean in this context? Who would decide? He was certainly condemned for his breach of convention by continental critics, until Lessing argued in his favour.*

In order to conceptualise creative achievement, Eysenck further argues that a combination of *cognitive*, *environmental* and *personality* variables as shown in Figure 1 be used. The *environmental* variable illustrates the significance of factors outside the realm of individual achievement to succeed in a creative act.

Csikszentmihalyi (1997) argues that individual creativity can flow stronger if the surroundings are managed appropriately; as he notes (p.1),

*An idea or product that deserves the label 'creative' arises from the synergy of many sources and not only from the mind of a single person. It is easier to enhance creativity by changing conditions in the environment than by trying to make people think more creatively.*

The social-system model shaped by Csikszentmihalyi is used increasingly by scholars as a theoretical framework in order to study different aspects of creativity (see for example, Gardner, 1993 for the creative lives of 'geniuses'; Bergh and Stuhlfaut, 2006 for advertising,). The theory is based on three mechanisms: the *domain*, the *field* and the *person* (Csikszentmihalyi, 1988, 1997). Each of the three mechanisms plays a pivotal role in the creative process. The *domain*, according to Csikszentmihalyi, is a set of rules, such as mathematics, music, television production. The *field* consists of all the gatekeepers in a certain domain. These gatekeepers decide whether the idea is creative enough, and therefore if it should be included in the domain – the members of a field might include collectors, commissioning editors, critics, teachers, television producers, et cetera. Finally, the *person* or creator uses the symbols of the domain to create a novel idea that is consequently analysed by the field before its members decide whether or not to include it as part of that same domain. Using the above elements, the definition of creativity, according to Csikszentmihalyi *(p. 28)* is:

*Figure 1. Eysenck's creative achievement model (adapted from Eysenck, 1995, p. 39)*

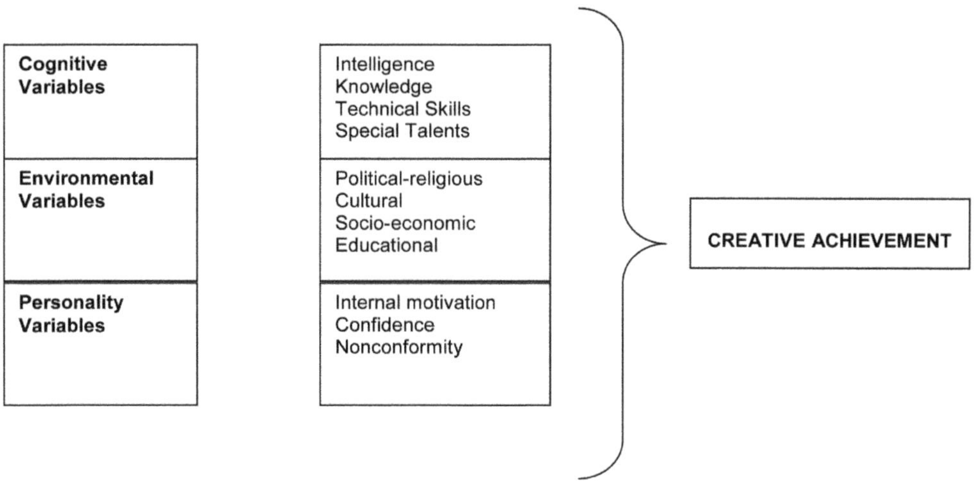


Any act, idea, or product that changes an existing domain or that transforms an existing domain into a new one…

By and large, organisations practicing creative management approaches must verbally and formally communicate to employees the importance of idea generation. The concept of creativity within organisations after all, is based on the generation of ideas that may ultimately give the organisation a competitive advantage (Bilton 2007).

Organisational creativity can be stimulated if the climate is managed in such a way that employees are encouraged to generate and implement ideas for the overall good of the organisation. With this in mind, Bilton has put forward seven hypotheses of how organisations should start managing creativity (p. 6):

1. Organisations should tolerate diversity, complexity and contradiction
2. Creativity derives from a combination of different types of thinking
3. Communicating the best creative ideas is as important as the creative idea itself
4. Creative thinking takes place neither inside the box nor outside the box, but at the edges of the box
5. Boundaries and constraints are a necessary part of the creative process
6. Diversity, compromise and collaboration are no less important to creative work than singularity of vision and purpose
7. Creativity is embedded in a cultural context

Psychologist Mark Runco (2007) also highlights the significance of organisational climate[1] in stimulating creativity. Organisations need to completely reform themselves toward a philosophical belief system based solely on creativity (pp. 168-171). Attempting to offer a more scientific approach, Ekvall's model of the creative climate of organisations (1996) acknowledges ten factors for establishing a creative climate:

1. Idea time
2. Risk taking
3. Challenge
4. Freedom
5. Idea support
6. Conflict
7. Debates
8. Playfulness / humour
9. Trust / openness
10. Dynamism / liveliness

Another often used scientific approach to stimulate organisational creativity draws from a large-scale study of 129 research and development scientists from different sectors (Amabile, 1996). The study identifies nine stimulants and nine obstacles of creativity (pp. 231 – 232):

### Stimulants to Creativity

1. Freedom
2. Good project management
3. Sufficient resources
4. Encouragement
5. Organisational characteristics promoting cooperation and collaboration
6. Recognition
7. Sufficient time
8. Challenge
9. Pressure (a general desire to accomplish something important)

### Obstacles to Creativity

1. Organisational characteristics with excessive red tape, lack of cooperation, inappropriate reward system
2. Lack of freedom and constraints
3. Insufficient resources
4. Poor project management
5. Inappropriate evaluation
6. Time pressure
7. Overemphasis on the status quo
8. Competition (interpersonal or intergroup competition within the organisation fostering a self-defensive attitude)
9. Organisational disinterest

Based on these points, Amabile (1997) brought forward what is called the 'componential theory of organisational creativity'. Amabile's theory seeks to comprehend, assess (through an organisational framework called KEYS) and ultimately stimulate creativity.

Moultrie and Young (2009) have themed the work on organisational creativity by Amabile and Ekvall into the following categories (p. 301):

1. Time
2. Rewards
3. Risks
4. Conflicts
5. Challenge
6. Debate
7. Freedom

With the exception of time and rewards, which are based solely on management decision-making, the remaining five points can all be particularly enhanced by the implementation of information systems and networked technologies.

## Creative Management and Technology

Creative organisations that wish to make the most of their creative stakeholders (whether internal or external) need to use the techniques discussed above by fostering networks and clusters of creative collaboration (see Sonnenburg, 2004), predominately through the use of online technologies. There are a number of information technology tools that can be used to make organisations more creative. Butler and Murphy (2007, p. 143) identify the following: data mining and learning tools, knowledge repositories, databases, electronic bulletin boards, discussion forums, intranets, extranets, emailing, calendaring tools, collaboration tools, knowledge directories, decisions support tools, expert systems, workflow systems, social network analysis tools and knowledge codification tools. While these tools inspire creativity to flow within an organisation by stimulating employees to take part in the creative process, one should not dismiss the contribution that can be made by the public at large with these technologies. It is here where a creative *economy* and a creative *society* collide. Public service organisations like the BBC can use the knowledge pool of literally millions of citizens and not only from its own human resources. Mass online innovation and creativity is burgeoning. Both users and creators act in order to achieve recognition and support, and are hence more intrinsically motivated to take part in the creative process of what it is they are interested in creating, or using. This leads to the indistinct lines between those that produce and those that consumer, and is a significant characteristic of what is generally known as the Web 2.0, social media landscape. The blurriness between producers and consumers is particularly apparent in media creative processes (Leadbeater, 2009). Web-users exchange ideas, share information, become journalists, and innovate without having to physically meet each other, while laypeople using Web 2.0, increasingly generate ideas for television and radio programmes and find opportunities to broadcast them. Accordingly, Leadbeater (2009) makes an appealing comparison of the industrial age with the information age,

*The factory encouraged us to see everything through the prism of the orderly production line delivering products to waiting consumers. The web will encourage us to see everyone as a potential participant in the creation of collaborative solutions through largely self-organizing networks (p. 8).*

Risk, conflict, challenge, debate and freedom, the factors in which technology can play a significant role, when combined effectively, can stimulate organisational creativity, particularly in small team meetings (see West & Sacramento, 2008; Runco, 2007).

## Team Meetings

Team meetings that are used to stimulate creativity can be divided into either *optimal team meetings* or *brainstorming sessions* (Runco, 2007). Optimal teams require heterogeneous and fairly small groups of people convening to find the best creative solutions, while brainstorming is a specific method of team-meeting for the generation of a multiplicity of ideas. The technique was developed in 1938 by Alex Osborn of the advertising agency Batten, Barton, Durstine and Osborn (known today as BBDO) (Stein, 1975; Andriopoulos & Dawson 2009). Stein (p.25), citing Osborn, notes that the sessions were called *brainstorming* because 'brainstorming means using the brain to storm a problem' (Osborn, 1963, p. 151). Osborn was apparently dissatisfied with the way meetings were held at his advertising agency, particularly how junior executives were not given enough opportunity to express their ideas. According to Andriopoulos and Dawson, 'this triggered him to come up with rules designed to improve team problem-solving, by giving people the freedom to speak their minds (p. 151). Osborn created four rules involved in the brainstorming session: (1) Avoid judgment (criticism is ruled out); (2) focus on the quantity of ideas and not their quality, producing the most amount of ideas possible (quantity is wanted because through it quality is achieved); (3) try finding ideas via 'piggybacking' or 'hitchhiking' as Runco notes (2007) (combination and improvement are sought) and finally, (4) participants should welcome 'free-wheeling' in which the more out of the ordinary the idea, the better.

Although teams can contribute towards creative solutions there are at least three disadvantages of team sessions of which managers need to be aware of. These are *social loafing, groupthink* and *blind conformity* (Andriopoulos & Dawson, 2009). *Social loafing,* whereby individuals within the team are inhibited from expressing ideas because they lack motivation or because they do not feel appreciated (Jackson and Williams, 1985) is cause for serious attention when brainstorming or when given group tasks. *Groupthink* is when groups become 'victimized' by knowing each other too well and are overly friendly, which results in 'group cohesion' and the homogeneity in the ideas put forth in the meeting. Similarly, *blind conformity* is when members in a group agree with other members in order to feel accepted by the rest of the group. *Blind conformity* differs from *groupthink* in that the group members might not be well-acquainted with one another (Andriopoulos & Dawson 2009, pp. 143-144). Some ways of avoiding these drawbacks are to use techniques such as *virtual teams* (Runco, 2007) or what Andriopoulos and Dawson call *electronic brainstorming* (p. 159). Electronic brainstorming is the implementation of online meetings in which the participants can all take part in a non-hierarchical manner. The rationale behind virtual teams / electronic brainstorming is that all the team members remain anonymous, 'thereby lowering the potential costs to original ideas' (Runco 2007, p. 169).

## THE BBC, CREATIVE MANAGEMENT AND TECHNOLOGY

Throughout its history the BBC has consistently attempted to apply the most current and fashionable managerial techniques. Born (2004, p.69), points out that the BBC is 'increasingly influenced by management theories'. The 'creative management' paradigm is no exception. Not only does the BBC wish to implement creativity management strategies at the level of its main function (that of creating programmes), but it is also aspiring to drive creativity throughout all areas of the BBC. Greg Dyke, former Director General of the BBC, seemed to have recognised the importance of managing creativity from his time at the helm of the BBC, from 2000 to 2004. In an interview he gave Georgina Born in 2002, after the organisation had gone through a decade of intense bureaucracy, self-regulation, high amounts of auditing and accountability, he stated:

*In the nineties, one of the stated aims of the BBC was "to be the best managed organisation in the public sector". I have to admit that wouldn't have got me out of bed in the morning. So let me offer you a new vision. We want the BBC to become the most creative organisation in the world, and I don't just mean in the production and programme areas, I mean right across the BBC, everywhere – including the commercial parts, finance, strategy, public policy and HR. Creativity is as important in those areas as in the obvious areas like production. (p.467)*

### Creative Future

In 2004 Mark Thompson took over from Greg Dyke as Director General of the BBC, while Michael Grade was appointed Chairman[2]. The two men took the BBC through a plethora of organisational changes, with creativity management a focal point of their strategy. In fact, the BBC's 2005/06 annual report states that its vision is to be 'the most creative organisation in the world' and to make creativity 'the lifeblood of the organisation' (BBC, 2006, p.1). Shortly after the two men were appointed, they published their manifesto for change, *Building Public Value*. The manifesto highlighted the need to pursue techniques that stimulate creativity in programme-making, from idea implementation to the finished programme. In tandem with the manifesto, the BBC also established a detailed blueprint, appropriately called *Creative Future*. Through *Creative Future* the BBC would pursue creativity across all parts of the organisation. In a seminal speech Thompson gave to all BBC staff regarding *Creative Future*, he noted, 'everywhere in the BBC creativity has got to become a priority for us – it's vital if we are serious about *Creative Future*' (Thompson, 2006). That speech kicked off a wave of 'creative' changes at the BBC. Following the speech, management effected significant organisational changes (see below), placed more emphasis on technology and audience involvement to enhance the creative process, and has created two divisions to promote creativity, *Creative Network* and *The*

*Creativity And Audiences Training Board.* Other *Creative Future* initiatives included opening the lines of communications amongst BBC staff with regular Q&A's, audience festivals, storytelling day and other creative initiatives.

The most important outcome of *Creative Future* is the restructuring of the whole BBC. As the organisational chart shows (Figure 2), the BBC is now structured in a straightforward, circular and decentralised manner.

The BBC's revised structure reflects how contemporary 21st Century organisations are transforming. According to Leadbeater (2009), in 21st Century organisations, "hierarchies are now flatter, job descriptions vaguer, the working day more flexible, the working week more extendable, careers more unpredictable…getting more productivity, innovation and quality seems to require ever greater pain to make organisations leaner and meaner" (pp. 88-89). The BBC's new structure places technology at a premium, and the way technology now affects creativity at the BBC is complex and nonlinear. Due to the new structure, technology plays an influential role in all other departments of the organisation - the BBC's technology division, named *Future[3] Media & Technology* now dominates all of the BBC's creative divisions (see below). The BBC's technological investments have come at a cost, however. Although the BBC has invested much of its

*Figure 2. BBC organisational chart (source: bbc.com: Used with Permission)*

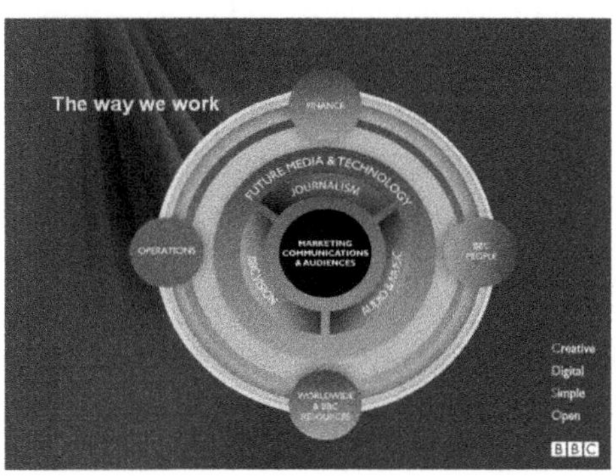


resources in ambitious technological initiatives such as *BBC iplayer, BBCi, BBC HD, the BBC website* and more recently *Project Canvas,* the organisation has also been forced to downsize (since 2004 the BBC has cut its annual budget by £350m and reduced its headcount by 7000, Rushton, 2008).

*BBC iplayer* (the organisation's on-demand service via the internet), and the BBC's website, bbc.co.uk, have enjoyed huge success and continue to draw an increasing number of users. The BBC is now trying to broaden the scope of *BBC iplayer* by allowing audiences to have access to it through their television sets, so that viewers who do not have internet access will be able to use the service via an 'online' set-top-box - the initiative is known as *project canvas* (Sweney, 2009). In order to follow through on such ambitious projects, *Future Media & Technology* has been assigned an undeniably high budget (approximately quarter of a million pounds), and all BBC technologists have been relocated to the division.

Since *Future Media & Technology* overarches all creative divisions of the BBC, the borders between technologists and other departments are becoming increasingly blurry. As illustrated in the new structure and as exemplified in various interviews with key BBC personnel, *Journalism, BBC Vision, Audio & Music*, the other divisions of the BBC, are all required to work closely with BBC technologists. Consequently, *Future Media & Technology* has three controllers that are assigned to each of the three cross-media content groups (for example, the Technology Controller of *Journalism*). Employees with significantly different backgrounds are now required to work together for the first time in 'creative' meetings on how content is produced. The previous head of *Future Media & Technology,* Ashley Highfield, noted "Creativity, ingenuity and innovation don't just start and end with the content" (2008, p. 29). Bilton (2007) emphasises the significance of such collaboration by pointing to the fact that creativity is 'brokered' when:

*connecting disparate talents not only with each other but also with co-creators, recourse providers, critics and audiences ... the brokering function extends beyond the initial phase of assembling the team or group, into an ongoing readjustment and interaction, reinjecting diversity and manipulating creative tensions in the group (p.36).*

Additionally, diverse meetings between professionals with media backgrounds and technologists are increasing, as dictated by the changing nature of how media are consumed. Commissioning editors and genre and channel controllers are the two occupational groups that deal directly with what audiovisual content gets produced. The primary data gathered for this case study clearly shows the added value of technologists during the creative process. Commissioning editors and controllers indicate that while technologists are not in all meetings regarding possible programmes (these meetings are usually made up of writers, independent producers and other forms of programme-makers), others require their presence. The increased participation of technologists in such meetings is justified due to the more active role of audiences in what is fast becoming an extremely advanced technological environment. Programmes are shown on multiple platforms and not just television and technologists are better placed at understanding what kind of audiovisual content can work on different platforms. The following section indicates in more detail the added value of technology during the creative process.

## Technology, the Creative Process and Different Media Platforms

Since audience consumption habits continuously shift and increasingly depend on new technologies (for example, see AppleTV by *Apple Inc.*), programmes are commissioned with cross-platform in mind (360-degree as it is known inside the BBC). What makes a programme a creative one is changing and developing creative content is becoming more challenging, on account of technological shifts in audiovisual consumption and revised audience expectations. Cross-platform commissioning is intended for producers wanting to pitch ideas for programmes that will be watched not only on television but also online and mobile devices. Television in other words, has become but one of several audiovisual platforms. BBC television commissioners, whose jobs are to commission specific ideas of television programmes from BBC producers or independent producers, need to bear in mind other platforms in which audiences view audiovisual content. These include 'mobisodes' (on mobile telephones), podcasting, live streaming and other such technological innovations regarding media consumption. Because of these changes, commissioners, controllers *and* technologists are all involved in the pursuit of both new formats and new programmes. For this reason, these divisions increasingly rely on each other to achieve their goals. According to a technologist at the BBC, commissioners (responsible for BBC content) are now more mindful of both 'time-shifting and place-shifting users'.

It seems technology plays a major role in the creation of original audiovisual production at the BBC. Before audiovisual content was distributed digitally on many platforms, programmes were made for large, passive audiences that 'helplessly' waited for their programmes to be broadcast when the television scheduler found a place for them – this is no longer the case. All BBC television programmes now have websites of their own and online technologies are frequently used as complementary devices to audiovisual versions of programmes (e.g., the documentary *Planet Earth* has a vast array of clips and information online which complement the programme). Furthermore, online communities can now actively become involved with the development of BBC content. For example, Simon Nelson (2007), Controller of Portfolio & Multimedia at *BBC Vision,* discussing the cross-platform project *Signs of Life*, pointed out that audiences are progressively interacting with original content and pushing cross-platform higher up on *BBC Vision's* agenda:

*It's a ground-breaking online drama where players move between watching the drama unfold and interacting with the plot. Where the emotional themes of each episode can be explored via interactive personality tests and the results saved by players and shared via widgets on their own social networking pages.*

The *Signs of Life* example demonstrates how important sharing and collaboration has become to a newer generation of online users, and why cross-platform is of such significance to the BBC as it pertains to the production of original content. Audiences, due to the availability of networked communications technologies are becoming a part of the creation process itself.

## E-Commissioning

The BBC receives hundreds of ideas for potential television programmes in every genre. These ideas arrive at the BBC from different organisations and individuals, and not just from BBC studios. The main reason why ideas arrive at the BBC from elsewhere, particularly from independent television producers, is that since 2003, the BBC is obliged by law to commission 25 percent of its total programmes from external suppliers and not just from its own studios and production units. In 2007 this figure changed again when the BBC began producing more commissions from independent, external suppliers through a system called the Window of Creative Competition (WoCC). The WoCC means the BBC is now further obliged to commission up to an additional 25 percent of its total programming from independent producers. This might mean that up to 50 percent of all BBC productions could be commissioned from independent producers. With more than 400 active independent production companies in the UK, a system needs to be in place that makes it possible for the BBC to take advantage of new ideas when they are presented to the organisation. The BBC achieves this through an extranet called E-commissioning. Proposals must now be submitted electronically on the BBC's E-commissioning website by both in-house and independent producers; this, according to the BBC, will enable *BBC Vision* – the department responsible for commissioning programmes – to filter the thousands of ideas it receives and will give the department members more time to discuss promising proposals (see below). E-commissioning is an example of how technology is playing a more significant role in the conception of original programmes. New ideas can be filtered, processed and tracked faster, and there is now more transparency in how ideas are managed from one genre to another. The E-commissioning website has gradually lead the BBC to move away from rounds of commissions, in which various slots are made public for producers to match with specific ideas, to a rolling commissioning policy, whereby producers can send in their proposals and ideas any time throughout the year.

If ideas are filtered well enough, E-commissioning is particularly useful in levelling the playing field for the evaluation of proposals and ideas for programmes. Good ideas are in short supply and since ideas can be generated by anyone, the E-commissioning website simplifies the submissions process for all suppliers, regardless of who they are. By filtering ideas this way, BBC commissioners and controllers can spend more time in the meetings that matter, rather than having to hunt for the best proposals, since it is in these meetings that ideas are harnessed and moulded into a form ready for development. The E-commissioning website states (BBC, 2009):

*Rather than replacing any of the creative conversations BBC Commissioning teams have with producers, the system is designed simply to make the process of filtering and comparing ideas much easier. As a result, creating more time for creative discussion around the best of those ideas.*

In addition to the independent production companies and internal BBC studios that submit ideas via the E-commissioning website, the BBC also helps normal citizens with little or no television background who present good ideas to the organisation, sometimes matching them with in-house or independent producers. Finally, E-commissioning can increase the organisational memory of the BBC since nothing can be lost this way.

## FUTURE RESEARCH DIRECTIONS

Long-term organisational success is based on the ability of an organisation to create environmental mechanisms that stimulate creativity in order to generate goods, services, strategies, processes and other organisational factors that contribute towards the organisations' competitive advantages. These abilities are translated into what is known as creative management. Creative management is the most recent in a series of organisational paradigms that have taken hold in many developed societies. While creative management is undoubtedly growing as an organisational paradigm, it is still in its infancy. As it burgeons, it continues to draw from a wide variety of disciplines and sectors, notably from psychology, creative industries strategies (such as those applied by media and advertising organisations), and the information technology (IT) sector. IT tools used to stimulate organisational creativity are embedded around communicative online networks. This is because research has shown that creativity happens in socio-cultural surroundings in which the internet in particular has a major influence in allowing people to collaborate.

In order for creative management to flourish into an established paradigm, more empirical research needs to be conducted in organisations that have applied creative management techniques and have truly invested in prioritising creativity. Case studies such as this one offer a more pragmatic interpretative stance since the BBC – as shown above - is so conspicuously 'pro-creativity'. Future studies therefore need to draw on more similar institutions. As different models for measuring organisational creativity grow (e.g. Amabile's KEYS and Ekvall's model of the creative climate), it will become easier to apply these models when researching organisations. It will consequently become more straightforward to conduct research on how organisations which apply creative management techniques compare against organisations which do no.

A cautionary note however needs to be made in that creative industries – the sector possibly used more than any other when studying organisational creativity - have very detailed sector-specific attributes. In media organisations for example, many creators are motivated intrinsically and are often trained outside the realm of commerce, and therefore require specific methods of management to stimulate their creativity (e.g. an environment allowing a more 'nonconformist' approach). Furthermore, this sector is laden with ideological biases that need to be considered, regardless of the subject matter.

## CONCLUSION

From a media management perspective, this chapter posits that the line between the production of media content – those within the walls of traditional media institutions – and online users, is blurring. The creative process has therefore changed to include input from a truly socio-cultural collaborative perspective (see Figure 3).

The BBC has embraced these changes and has strengthened its technology department to prepare for a future that is increasingly becoming unstable and ominous for media organisations. The indivisible relationship between creative management and audiovisual production methods practiced at

*Figure 3. The use of technology in the creative process at the BBC*

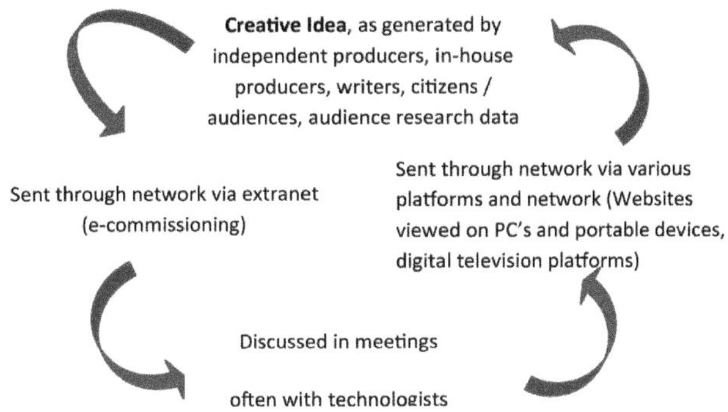


the BBC can be seen as a catalyst for change at other organisations in knowledge-driven societies. This is because while the BBC is under extreme pressure to downsize and cut its budgets, it has made significant investments in technological tools for creative management. The restructuring of the BBC gives technologists an active role in the creative process of the BBC, which means more of the organisation's funds are allocated for 'future' media technologies. This is particularly remarkable because the BBC is a public service organisation that receives its annual operating budget directly from UK citizens, therefore putting the organisation under constant scrutiny.

The media sector, possibly more than any other, is directly subject to the new challenges presented by the digital age. As the digital divide continues to be bridged, and more people familiarise themselves with communicative technologies, they will become even more active with regard to what they consume – they may, as is the case of audiovisual production, even play a role in the creative process itself. The BBC has invested heavily in transforming itself into an organisation equipped for a precarious future, one in which audiences will be more active in choosing what, how and when they want to consume audiovisual content. If the BBC, as such a reputable and pioneering organisation, has embraced creative management and technology to such an extent, this should not be taken lightly. Consequently, this chapter suggests there is ample reason to support the increased use of technology in stimulating organisational creativity.

## REFERENCES

Amabile, T. (1996). *Creativity in context*. Boulder, CO: Westview.

Amabile, T. (1997). Motivating creativity in organisations: On doing what you love and loving what you do. *California Management Review*, *40*(1), 39–58.

Andriopoulos, C., & Dawson, P. (2009). *Managing change, creativity & innovation*. London: Sage Publications.

Asimakou, T. (2009). The knowledge dimension of innovation management. *Knowledge Management and Research & Practise*, 7(1), 82–90. doi:10.1057/kmrp.2008.34

BBC. (2006). *BBC annual report and accounts 2005/06*. BBC Publications.

BBC. (2008). *BBC annual report and accounts 2007/08*. BBC Publications.

BBC. (2009). *Commissioning*. Retrieved June 5, 2009, from http://www.bbc.co.uk/commissioning/tv/networ/wocc.shtml

Bergh, B. V., & Stuhlfaut, M. (2006). Is advertising creativity primarily an individual or a social process? *Mass Communication & Society*, *9*(4), 373–397. doi:10.1207/s15327825mcs0904_1

Bilton, C. (1999). Risky business: The independent production sector in Britain's creative industries. *International Journal of Cultural Policy*, *6*(1), 17–39. doi:10.1080/10286639909358110

Bilton, C. (2007). *Management of creativity: From creative industries to creative management*. Oxford: Blackwell Publishing.

Bilton, C., & Leary, R. (2002). What can managers do for creativity? Brokering creativity in the creative industries. *International Journal of Cultural Policy*, *8*(1), 49–64. doi:10.1080/10286630290032431

Born, G. (2004). *Uncertain vision: Birt, Dyke and the reinvention of the BBC*. London: Secker and Warburg.

Butler, T., & Murphy, C. (2007). Understanding the design of Information Technologies for Knowledge Management in organizations: A pragmatic perspective. *Information Systems Journal*, *17*, 143–163. doi:10.1111/j.1365-2575.2007.00237.x

Csikszentmihalyi, M. (1988). *Creativity*. New York: Harper Collins.

Csikszentmihalyi, M. (1997). *Creativity: The flow of psychology of the discovery and invention*. New York: Harper Collins.

Davis, H., & Scase, R. (2000). *Managing creativity*. London: Open University Press.

DCMS. (2007). *Staying ahead: The economic performance of the UK's creative industries*. Retrieved December 15, 2007, from http://www.culture.gov.uk/reference_library/publications/3672.aspx

Ekvall, G. (1996). Organisational climate for creativity and innovation. *European Journal of Work and Organizational Psychology*, *5*(1), 105–123. doi:10.1080/13594329608414845

Eysenck, H. (1995). *Genius: The natural history of creativity*. Cambridge, UK: Cambridge University Press. doi:10.1017/CBO9780511752247

Florida, R. (2002). *The rise of the creative class: And how it's transforming work, leisure, community and everyday Life*. New York: Basic Books.

Florida, R. (2004). *Creative cities*. New York: Basic Books.

Gardner, H. (1993). *Creating Minds*. New York: Basic.

Henry, J. (2001). *Creativity and perception in management*. London: Sage Publications.

Henry, J. (2006). *Creative management and development* (3rd ed.). London: Sage Publications.

Hesmondhalgh, D. (2007). *The cultural industries* (2nd ed.). London: Sage Publications.

Howkins, J. (2001). *The creative economy: How people make money from ideas*. London: Penguin.

Jackson, J. M., & Williams, K. D. (1985). Social loafing on difficult tasks. *Journal of Personality and Social Psychology*, *49*, 937–942. doi:10.1037/0022-3514.49.4.937

Jeffcutt, P., & Pratt, A. (2002). Managing creativity in the cultural industries. *Creativity and Innovation Management*, *11*(4), 225–233. doi:10.1111/1467-8691.00254

Leadbeater, C. (2009). *We think*. London: Profile Books.

Moultrie, J., & Young, A. (2009). Exploratory study of organizational creativity in creative organizations. *Creativity and Innovation Management*, *18*(4), 299–314. doi:10.1111/j.1467-8691.2009.00536.x

Nelson, S. (2007). *Speech given to BBC Vision in-house multiplatform day*. Retrieved June 15, 2009, from http://www.bbc.co.uk/pressoffice/speeches/stories/nelson_multiplatform.shtml

Nonaka, I., & Takeushi, H. (1995). *The knowledge creating company: How the Japanese companies create dynamics*. Oxford: Oxford University Press.

Osborn, A. (1963). *Applied imagination: Principles and procedures of creative problem solving* (3rd rev. ed.). New York: Charles Scribner's Sons.

Runco, M. (2007). *Creativity: Theories and themes: Research, development and practice*. San Diego: Elsevier Academic Press.

Rushton, K. (2008). Extra job losses loom at BBC. *Broadcast Magazine*. Retrieved June 19th, 2009, from http://www.broadcastnow.co.uk/news/multi-platform/news/extra-job-losses-loom-at-bbc/1924060.article

Sonnenburg, S. (2004). Creativity in communication: A theoretical framework for collaborative product creation. *Creativity and Innovation Management, 13*(4), 254–262. doi:10.1111/j.0963-1690.2004.00314.x

Stein, M. (1975). Stimulating creativity: *Vol. 2. Group procedures*. London: Academic Press.

Sweney, M. (2009). Project canvas start-up costed at 24 million. *The Guardian*. Retrieved August 8, 2009, from http://www.guardian.co.uk/media/2009/feb/26/project-canvas-strartup-costs

Webster, F. (2006). *Theories of the information society*. London: Routledge.

West, M., & Sacramento, C. (2008). Flourishing in teams: Developing creativity and innovation. In Henry, J. (Ed.), *Creative management and development*. London: Sage Publications.

Xu, F., & Rickards, T. (2007). Creative management: A predicted development from research into creativity and management. *Creativity and Innovation Management, 16*(3), 216–228. doi:10.1111/j.1467-8691.2007.00445.x

## ADDITIONAL READING

Andrew, P. (2002). Managing Creativity in the cultural industries [Editorial, and Special Guest Editor with Paul Jeffcutt]. *Creativity and Innovation Management, 11*(4), 225–233. doi:10.1111/1467-8691.00254

Arbor, I., & Bjerke, B. (1997). *Methodology for Creating Business Knowledge*. London: Sage.

Born, G. (2002). 'Reflexivity and Ambivalence: Culture, Creativity and Government in the BBC', *Cultural Values. Journal for Cultural Research, 6*(1-2), 65–90. doi:10.1080/1362517022019000

Born, G., & Prosser, T. (2001). 'Culture and Consumerism: Citizenship, Public Service Broadcasting and the BBC's Fair trading Obligations'. *The Modern Law Review, 64*(5), 657–687. doi:10.1111/1468-2230.00345

Chris, B. (2000). *Disorganised Creativity* (*Vol. 45*, pp. 17–39). Boekmancahier – Amsterdam.

De Bono Edward. (1990). *Lateral Thinking: A Textbook of Creativity*. London: Penguin Group.

DiMaggio, P. (2001). Making Sense of the Contemporary Firm and Prefiguring Its Future. In DiMaggio, P. (Ed.), *The Twenty-First-Century Firm: Changing Economic Organisation in International Perspective* (pp. 3–30). Princeton: Princeton University Press.

Georgina, B. (2000). Inside Television: Television Studies and the sociology of culture. *Screen, 41*(4), 404–424.

Getzels Jacob and Csikszentmihalyi Mihaly. (1976). *The Creative Vision: A longitudinal study of problem finding in Art*. New York: Wiley.

Greg, D. (2005). *Inside Story*. London: Harper Perennial.

Harris, M., & Wegg-Prosser, V. (2007). Post Bureaucracy and the Politics of Forgetting: The Management of Change at the BBC, 1991 – 2002. *Journal of Organizational Change Management*, *20*(3), 290–303. doi:10.1108/09534810710740146

Hesmondhalgh, D. (2007). *The Cultural Industries* (2nd ed.). London: Sage Publications.

Iosifidis, P. (2007). *Public Television in the Digital Era: Technological Strategies and New Challenges for Europe*. London: Palgrave Macmillan.

Küng-Shankleman, L. (2000). *Inside the BBC and CNN: Managing Media Organisations*. London: Routledge.

Mark, R. (2004). Everyone has Creative Potential. In Robert, S., Elena, G., & Singer, J. (Eds.), *Creativity: From Potential to Realisation* (pp. 21–30). Washington: American Psychology Association.

Phil, R. (2004), 'Public Service Content: The Conditions for Creativity' in Tambini Damian and Cowling Jamie (eds.), *From Public Service Broadcasting to Public Service Communications,* Institute for Public Policy Research, London.

Simon, D., Ana, L., & Stephen, P. (2008), 'No Third Way for Economic Organization? Networks and Quasi-Markets in Broadcasting' *Centre for Business Research*, University of Cambridge, Working Paper No. 360, March 2008.

Teresa, A. (2008). How to Kill Creativity. In Jane, H. (Ed.), *Creative Management and Development*. London: Sage Publications.

Throsby, D. (2000). *Economics and Culture*. Cambridge University Press.

Tracey, M. (1998). *The Decline and Fall of Public Service Broadcasting*. Oxford: Oxford University Press.

Hutton Will, O'Keeffe Aine, Turner Natali, (2005), The Work Foundation, *The Tipping Point, How much is Broadcast Creativity at Risk?*, (July 2005).

## KEY TERMS AND DEFINITIONS

**British Broadcasting Corporation:** A UK public service broadcaster. The oldest and largest Public Service Broadcaster in the world. Established in 1922, it is responsible mainly for broadcasting television and radio stations and broadcasts one of the most popular internet sites in the world.

**Creativity:** An implemented idea that is both relevant and original to the domain of the creative act.

**Creative Management:** A management paradigm that is based on creating the correct organisational environment in order to stimulate creativity of employees.

**Commissioning:** In contemporary audiovisual production, commissioning occurs when broadcasting organisations outsource content from independent producers but at the same time, are involved in the creative process.

**Competitive Advantage:** An attribute an organisation has that allows the organisation to outperform competitors.

**Knowledge Management:** A management paradigm that is based on techniques and tools that allow organisations to use, store and generate organisational knowledge, both tacit and explicit.

**Management:** The process of planning, organising and leading an organisation in order for that organisation to fulfil its goals and objectives.

**Public Service Broadcasting:** A form of broadcasting that is owned by, and intended to benefit the public of a particular country. Public Service Broadcasting needs to be objective from both commercial and state objectives.

## ENDNOTES

1. Citing Isaksen et al. (2000-2001, p. 172), Runco defines organisational climate as, "the recurring patterns of behaviour, at-

titudes and feelings that characterize life in the organisation…" (p. 164).

2 Michael Grade eventually left the BBC in November 2006 to take up a post at ITV plc.

3 According to the BBC 'future' is a more appropriate preface to 'media' than the more frequently used 'new'

# Chapter 15
# Sustainable Blended-Learning in HEI:
## Developing and Implementing Multi-Level Interventions

**Paula Peres**
*School of Accounting and Administration of Oporto, Portugal*

**Sandra Ribeiro**
*School of Accounting and Administration of Oporto, Portugal*

**Célia Tavares**
*School of Accounting and Administration of Oporto, Portugal*

**Luciana Oliveira**
*School of Accounting and Administration of Oporto, Portugal*

**Manuel Silva**
*School of Accounting and Administration of Oporto, Portugal*

## ABSTRACT

*This chapter aims to demonstrate how PAOL - Unit for Innovation in Education, a project from ISCAP - School of Accounting and Administration of Oporto - Institute Polytechnic of Oporto, Portugal - prompted new educational initiatives and new learning scenarios at a Higher Education Institution. Furthermore, it will demonstrate PAOL's lines of intervention through an extensive analysis based on the 6 years of experience that this unit has in the educational technology field; a project that began small but that, due to the force of innovation, has progressively conquered new adepts. Therefore the unit described in this chapter relates all these factors, as a whole, capable of attaining changes that influence mentalities and methodologies, overcoming cultural and technical barriers. This case study can serve as a catalyst, potentiating the creation of new multi-faceted projects in the scope of web technologies in higher education teaching-learning processes.*

DOI: 10.4018/978-1-60960-519-3.ch015

## INTRODUCTION

The success of any e-learning institutional project relies on the harmonization of several variables, interests and motivations. PAOL, ISCAP's Unit for Innovation in Education, described in this chapter, tries to relate all these factors as a whole, capable of attaining innovations that influence mentalities and methodologies, overcoming cultural and technical barriers.

One of PAOL's main goals is to facilitate the integration of Web Technologies at ISCAP (School of Accounting and Administration of Oporto – Porto Polytechnic Institute – Portugal) using them to complement face-to-face learning. This process of integrating Web Technologies in Higher Education Institutions occurs in a very specific educational reality – the Portuguese -demanding a profound knowledge of it in terms of the organizational structure of e-learning practices. "Background: A Global Vision of E-Learning Practices At Portuguese Higher Education" introduces, thus, the Portuguese learning structure, followed by a brief description of today's e-learning practices.

Given the difficulties in operationalizing the many approaches mentioned in the literature, "An Innovative Framework to Promote the Use of E-Learning Technologies as an Institutional Project" presents a new structure that emerged from our experience and may be of help to institutions in the development of e-learning projects, namely in the development of e-learning projects that aim at involving the entire learning community, adding value to the institution. This structure is described based on 6 years of experience, which have allowed the promotion of coherent and objective paths towards a sustainable, flexible, innovative and technology-supported learning process. The description of the elements taken into account is based on the concepts underlined by the methodological and pedagogical principles and grouped into 3 interventional vectors: institutional, technical and pedagogical dimension.

The institutional dimension includes the management of Human Resources, knowledge creation and knowledge sharing. The sharing dynamics, both within and outside the institution, is presented as an essential vector for the success of any institutional project.

The technical dimension shows that the adoption of Web Technologies begins by naturally transposing traditional learning practices into e-learning environments. The first stage of this process began with the use of a Learning Management System, where the main concern focused on enabling the community to use it. The analysis presented in "Technical Dimension" of "An Innovative Framework to Promote the Use of E-Learning Technologies as an Institutional Project," follows Rogers' innovation line to better portrait the different stages and results.

The pedagogical dimension of the Project supports the belief that the success of an e-learning Project is related to its usage in the learning process. This dimension appears as fundamental in the sustainable progress of an e-learning Project, being the one that demands the most significant cultural changes and which requires more time to be implemented. Cultural changes are naturally difficult to occur and when there are no practical models to support the new approaches, many of the stakeholders in the process prefer, despite the critique, traditional teaching and learning methods.

Promoting the educational dimension of PAOL was, thus, the part that required greater effort, in particular by seeking to develop innovative projects that would motivate and drive the community to (accept) change. The creativity of the projects developed appears as one of the motivating forces for participation and, therefore, for the community's adhesion. The description of the projects in "Pedagogical Dimension" may prove useful for future reuse in different contexts. Among many others, we have developed the "Digital Portfolios" and "Supporting Design for E-Learning" projects and have supported or cooperated with other preexistent projects in the institution.

In each of the institutional, technical and pedagogical vectors analyzed, the descriptions of the inter-relationships that make up the systemic view of PAOL's system are included.

For a project to be sustainable and enable the achievement of a consistent and evolving framework, a continuous process of evaluation and self-evaluation is mandatory. Thus "Evaluation" describes the successive evaluations carried out over the 6 years, which results contributed to consecutive improvements to the structure now described.

Any innovation project faces constraints, not always predictable, that must be described to assist new projects when developing a risk plan. Thus, "Promoting Innovation: Contextual Risks and Recommendations" presents a brief description of the risks associated with e-learning project and includes a list of recommendations that may serve to ease the development of certain stages in different projects.

Considering the maturity level achieved in e-learning, it is crucial to consider the future and plan new actions directed towards a broader audience that extends beyond the boundaries of the institution and reach international publics and to broaden the Unit's objectives. "Future Research Directions" presents the intended projects including: internationalization through projects such as Supporting Design for b-learning, employability, accessibility and blended-learning courses for lifelong learning.

To conclude, this chapter presents a general conclusion of a history that draws on innovative ways to integrate sustainable e-learning technologies in a Higher Education Institution. It summarizes the creative strategies and techniques used to motivate the community to be willingly part of a project, which only then becomes a true school project.

## Background: A Global Vision of E-Learning Practices At Portuguese Higher Education

Nowadays, the social and technological evolution is reflected on the education systems and justifies the growth of the presence of higher institutions in the internet. The level of web technology adopted by an institution may influence, as Sauter decribed it (2003), the prestige, quality and number of students enrolled. In today's society, learning is a natural act. It is a continuous action during the entire life, its scope being more than traditional formal learning added by updated trainings (Sauter, 2003). Today's socio-cultural pressures enhance the importance of promoting learning strategies that allow students to learn, to be, to criticize and to produce knowledge taking into account society's reality.

This scenario led higher education institutions to explore the information communication technologies in order to find new flexible and sustainable paths leading to lifelong learning. The acquaintance of students with web technologies is more explored everyday in order to promote learning. Today's education systems should accomplish the changing processes and give response to the emergent demands.

In Portugal, under support of different Ministries, there is a public and a private network of education institutions that are organized according to Figure 1.

The higher education system includes public, private and cooperative institutions. There are 14 public university institutions and 15 public polytechnic institutions. There are also 13 private and cooperative universities and 2 private polytechnic institutions. Moreover, there are 116 private or cooperative schools of higher education, polytechnic or mixed (Comissão Europeia, 2007).

E-Learning in Portuguese higher education is being promoted by a large number of specialists. However, as recognized by Carvalho (2006), the

*Figure 1. Portuguese Educational Structure*

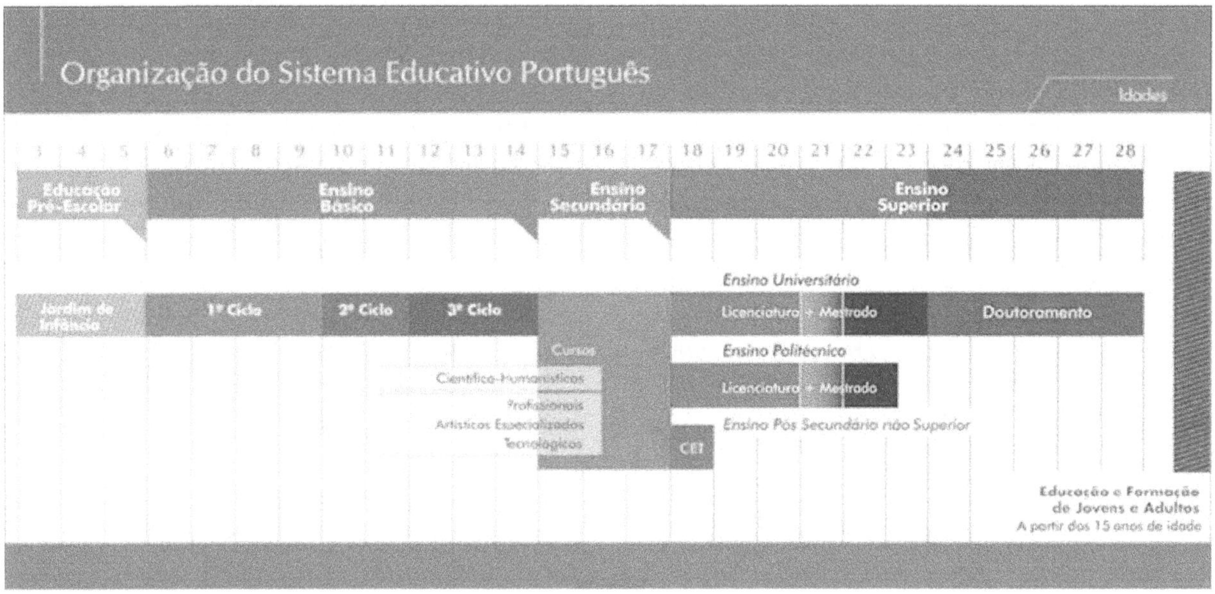


culture and system problems such as the national norms, teachers and students' habits, the lack of infrastructures and the accreditation mechanisms, among other factors, have been blocking the systematization of this specific type of learning environment. Students, parents and society are also used to the traditional school model and even when they criticize it, they do not have other references or enough experience in a different school system and, consequently, still prefer or understand the old model better (Moran, 2005).

Through a global analysis of today's e-learning practices in public higher education institutions (HEI) in Portugal, a study conducted by PAOL in 2007 (PAOL, 2008), it was possible to observe a growing concern from the institutions that belong to this system.

Usually, the institutions management decisions resulted in an increase in the use of the Internet mainly as a complementary way to promote communication between students and teachers (Isla, 2004). In general, observations and studies of Portuguese schools revealed a weak use of the numerous didactical-pedagogical possibilities offered by cyberspace (Silva, 2005). Due to this, many institutions decided to create specialized units in order to ensure sustainable web technology integration, as is the case of Oporto University (GATIUP, 2010), CEMED - Multimedia and e-learning center (CEMED, 2010) and PAOL – Unit for Innovation in Education (PAOL, 2010), among others.

Nevertheless, according to Nova & Alves (2003), the majority of web-technology integration processes are using the same methodologies as those used in a face-to-face environment. In this context, the quality of the training offers is questioned and the potential creativity of the Internet, applied to knowledge construction, is not being fully used. Due to the lack of knowledge or due to the existent passivity, we have to assume that there is some degree of resistance from teachers regarding the use of e-learning systems, mostly from those used to lecture sessions where students have a passive attitude, acting as information recipients.

Learning practices in the classroom are still partly oriented to improve teacher's presentations. For this purpose, e-learning tools contribute only as a new expression of old procedures and not to promote and develop the expected and needed innovation made possible by digital education (Nova &Alves, 2003). According to Dias (2004), studies reveal a focus on information and the transmission process, suggesting a focus concerning content, its organization and the way it is presented, not taking into account the interaction process that e-learning environments can support. Almeida (2003) enhances the need for change when she states that "changing the environment where education takes place means changing education itself". It is important to reflect on the type of training we want and the practices we should use to achieve it.

## Context of PAOL's Unit

PAOL is a project that has been developed at ISCAP. ISCAP is a higher education polytechnic institute that belongs to IPP (Porto Polytechnic Institute), in Portugal. Currently, it has a population of around 4,000 students and 230 teachers. It offers graduate and master programs in Accounting, International Commerce, Marketing, Business Communication and Administrative Assistance and Translation.

To ensure a higher success rate and support extra-curricular activities ISCAP developed several internal clusters, like PAOL – Unit for Innovation in Education, Multimedia Language Center (CML), Research Center and Services Rendered in Information Systems and Technologies (CEISE/STI) and Centre for Intercultural Studies (CEI), clusters that engage in and promote students' activities and meet both higher education and student demands.

## An Innovative Framework to Promote the Use of E-Learning Technologies as an Institutional Project

More and more frequently, institutions offer e-learning options, so students have references in regards to quality and its appropriateness to their own interests. In the beginning of 2003, aware of this and to support the use of web technologies in the learning-teaching process, ISCAP launched a project named PAOL.

The project is currently constituted by a Direction Board, a Coordinating Unit and specialized staff aims at providing support to the implementation of computer-assisted teaching and learning, being constituted by human resources and material resources, brought together to stimulate the use of information and communication technologies both by teachers and students in an harmonious and integrated approach.

In the construction of an institutional e-learning project, it is important to consider all aspects that affect its progress and determine the success of its implementation. A first basic description may be that of the daily tasks a project like this has to develop.

Possibly, like any other such unit, PAOL:

- Promotes different training courses in order to improve the use of the e-learning platform;
- Develops manuals to support training courses;
- Gives technical support to build resources and makes them available online;
- Promotes constant updates concerning Information and Communication Technologies (ICT) in education, through participation and follow-up actions in various projects and events;
- Creates e-learning contents and Learning Objects repositories;
- Implements innovations in learning/teaching strategies at ISCAP;
- Carries out scientific research in e/b-learninga and develops and cooperates with research teams in b-learning projects;
- Promotes e/b-learning events;
- Shares and discloses information on best practices as to the integration of technology in education;

- Seeks to involve the entire community in the school's e-learning project;
- Cooperates with other projects in ISCAP;
- Designs and endorses external courses;
- Imparts the project internally and externally, namely by newsletters;

The implementation, development and management of tasks such as the describe is nuclear for the success and growth of any project, but, as Kan (2005) refers, there is a need for the institution/project to develop specific strategies for implementing e-Learning. According to the same author, institutions should consider students as customers of education and training in a competitive market. Khan's model (2005), called "e-learning platform", was developed from the critical factors of the development of an e-learning experience and is based on eight dimensions: institutional, pedagogical, technological, interface design, evaluation, management, student support, and ethics. The combination of positive responses to all these dimensions, according to the author, provides sustainable paths to success.

The management of eLearning environments is composed, according to Ramos (2001), by 4 different stages: design, planning, implementation and evaluation procedures. This model, like most in the literature, focus on various stages, especially the need for evaluation of the work. Evaluation is seen as a process of analysis of the product created. A more formal definition is to think of evaluation as 'providing information to make decisions about the product or process' (Phillips: 2000). Conscious that initial solutions are not always a complete success, the enthusiasm for eLearning linked with reflections and the conclusions obtained from the procedural evaluation, leads to a new solution. The cycle then continues and new variants and approaches emerge (Meredith, 2002).

The experience accumulated over the last 6 years has enabled the creation of the following structure that integrates, consolidates and lists the elements identified by Khan, as well as other elements prompted by b-learning contextual practice in higher education and that have been validated in successive Evaluation cycles (Figure 2):

This structure and the associations that emerged from the operationalization of concepts and the result of a critical reflection will be described next. An innovative intervention strategy for the mobilization of management, technical and pedagogical competencies will be presented.

*Figure 2. Framework to help the use of e-Learning Technologies*

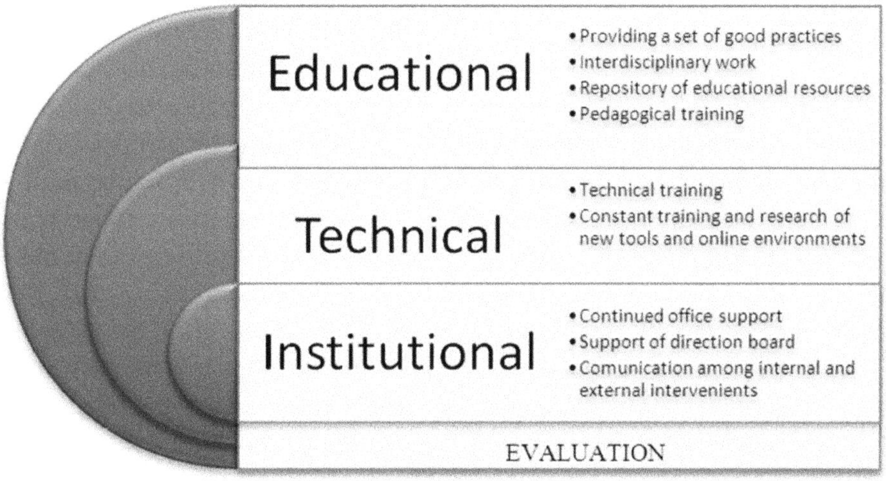

## Institutional Dimension

### Human Resources: The Project's Team

The clear politics supporting the use of educational technology at the school has had a positive impact on the allocation of institutional resources (both human and physical), on the coordination of efforts as well as on the sharing of practices and knowledge. The formulation of institutional politics and their adoption by HEI, is pointed out by several authors (McNaught and Kennedy 2000; Sherry 1998b; Garrison and Kanuka 2004) and recognized in the implementation process of this on-line support project, as part of the essential conditions.

With the development of the project, the creation of our own explicit and tacit knowledge has gained increasing importance and plays now an essential role in PAOL's activities. This knowledge is acquired through various channels including the knowledge accumulated by its team, composed of Professors, PhD students and teachers with Master's degree in b-learning, who seek training through participation in various courses to update technical and pedagogical knowledge, as well as through the presence or organization of national and international conferences, congresses and seminars.

### Enhancing Dynamic Knowledge Sharing

The promotion of a dynamic knowledge-sharing environment is another factor that contributes to the sustainability and continuous improvement of this project. We promote knowledge-sharing using several tools and approaches, such as the internal training courses that cover the different online applications used (LMS - Moodle, e-portfolio platform - ELGG, Survey Platform - LimeSurvey, Learning Design Platform - LAMS, etc.), the creation of a Portuguese Moodle user's manual, that PAOL has also developed and made available (for free) to the Moodle community, the conferences promoted in order to share our experience and learn with others that work in the same field.

An important factor that led to further external visibility comes from the fact that we've been promoting, on an annual basis, a round table discussion, covering all the major aspects and activities occurred during the year, that respect the use of Moodle and all other tools.

It is our belief that by sharing both good and bad experiences we are able to overcome obstacles and move forward as a whole. Very often, more personalized/individual conversations alert us to problems that must be solved or, on the other hand, reveal a practice worth noting and sharing.

The project has also gained visibility through the participation in both national and international events, as well as by providing external training. The results obtained are reflected on the integration of new innovative elements in a community that grows in knowledge generated through practice.

## Technical Dimension

### MOODLE as the Cornerstone

In its origin, PAOL had, as the main motivating element, the emerging need to develop sustainable solutions for technology integration as support for "regular" classes taught at the institution. This propelling element still remains as one of the main levers, now extended across the various areas of institutional knowledge and the inclusion of educational and pedagogical skills.

In the emergent technological context that we nowadays live, it is important to come forward with answers that correspond to present-day students' needs. This led to the implementation of the online platform known as WebCT at ISCAP in 2003, in an effort to promote e/b-learning. However, it soon became clear that other platforms could better satisfy the needs of our institution. Consequently, ISCAP changed its approach with the adoption of the open source platform Moodle in the year 2004.

The open source LMS Moodle was adopted, thus allowing students a broader online access to the subjects taught in class as well as new learning activities (individual and in group). In parallel, a website was developed to disseminate the main objectives of the project, activities and the national and international events about computer-assisted teaching and technology-enhanced learning. The site was equally used to give support to users (online tutorials, manuals, etc.) and to promote reflection on the intrinsic concepts of the Project and on computer-assisted teaching subjects, such as e-learning, blended-learning, learning objects, LMS, among others.

In Portugal, Moodle's usage has been increasing rapidly, thus justifying the opening of an open space to share experiences. Moodle has interesting characteristics, as it is an LMS that offers a set of activities adaptable to multiple subjects. It has several activities that can be used differently as educational instruments. It is possible to insert texts and the necessary documentation for the subjects, to suggest assignments, promote discussion forums and to create term glossaries in a cooperative form. The assessment and self-evaluation of knowledge is done through the pages of lessons and questionnaires.

In order to create a dynamic use of Moodle, PAOL has been promoting several trainings sessions on a continuous base, as well as in other educative technological areas, to motivate teachers in the usage of such instruments, helping them to manage Moodle and generate more creative educative offers for their students.

PAOL went through several stages over the years of its existence, especially due to the target audience of the project (students and teachers), trying to teach, motivate and disseminate the projects principles and the use of this kind of educative approach.

The use of Web technologies follows a similar trend to that of Roger's (1995) diffusion of innovation theory, as it developed due to the enthusiasm of some innovators and grew steadily in adepts, having now 166 teachers who use Moodle on a regular basis (see Figure 3). Thus a relation will be established between PAOL's development stages and Rogers' theory of the innovation adoption curve.

Teacher's role in this process can be classified into different types, having, for instance, teachers that are keener to participate and integrate technology, teachers that only allow themselves to participate if the experience is proven successful by others and teachers that are reluctant to participate in any form. It is important to analyse the

*Figure 3. The adoption of MOODLE in ISCAP*

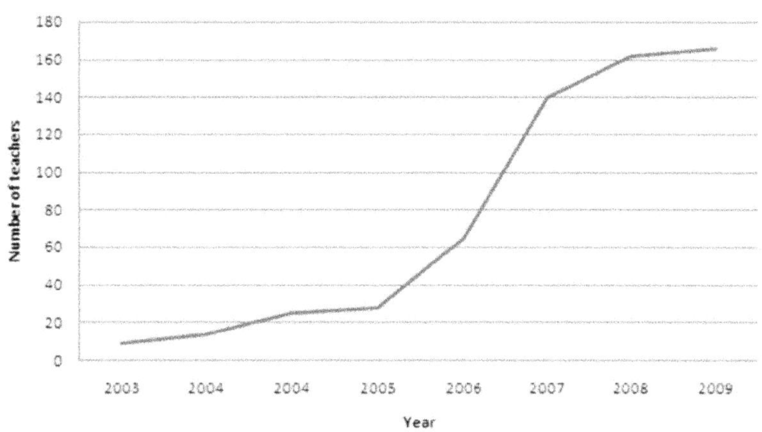

impact of these roles in the institutional development of b-learning so that we can develop new solutions and approaches that better meet the expectations of the institution.

When we analyse how these roles create an impact on the educational culture of the school, it is possible to verify that the integration of the innovation curve has reached a horizontal status, or what could be considered a comfort zone. This comfort zone is mainly occupied by the late majority, classified by Rogers as the teachers that were more reluctant to adopt Moodle, but ended up by accepting it because they were impelled either by their students or their pears. The level of uncertainty that classifies this group has, in fact, transformed this integration of technology into a very slow process where very basic practices are dominant.

This late majority group has been growing mainly in the past two years, when teachers began using Moodle to store and distribute their course contents. Of course, this represented a great step towards increasing student access to information and it became clear that almost all teachers were willing to use an LMS as part of their teaching process. This can be seen when analysing PAOL's annual reports on the use of Moodle at ISCAP, summed up in Figure 4.

As it is possible to verify, the number of online files/resources made available to students has nearly tripled from about 2000, in 2007, to approximately 6000 in 2009, but there isn't a significant growth of the use of other Moodle tools. This is mainly due to the Moodle adoption process carried out by the late majority. Unwilling to take risks with technology, this group has settled in a comfort zone, safely using technology to reproduce the traditional classroom teaching methodologies. This comfort zone is, therefore, characterized by a stagnation of practices of making contents available online and reusing them from one year to another, over and over. We believe that this tendency will allow us to perceive that the number of online contents will also stagnate, if not in the next school year, in a few years time.

This is the "sad truth about e-learning", as described by Dalziel (2005), in the majority of educational institutions, since the ambition to enhance technology-based education through digital reusable contents has lead to the content distribution habits and to rhetorical pedagogies. In fact, there is neither accommodating learning activities (Britain, S., 2004) nor students' opportunity to perceive the different phases of their learning process (Vogel, M., & Oliver, M., 2006).

*Figure 4. The use of MOODLE Activities and Resources at ISCAP in the last three years (Oliveira L. et al., 2009)*

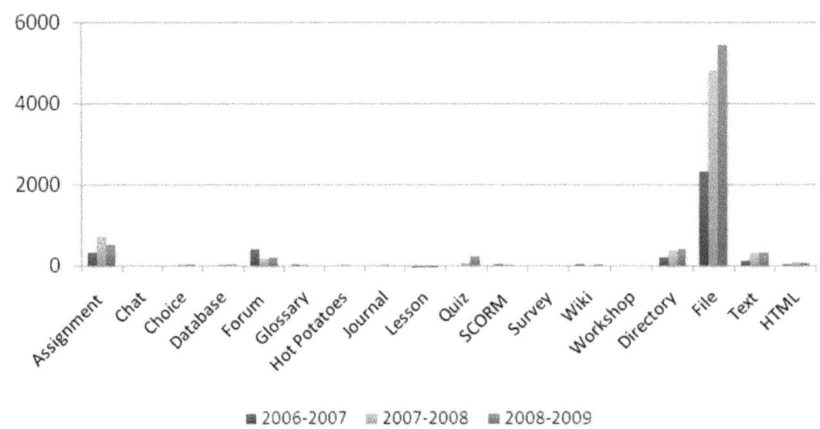

There is a notorious institutional tendency for the development of content-centred environments, and the integration of technology in the teaching-learning process has not been accompanied by a shift in the pedagogical methodologies that would to truly transport this process into Moodle and, in fact, foment the teaching and learning opportunities (Oliveira L., et al., 2009).

As shown by our experience, there are a lot of opportunities in terms of integrating and anchoring technologies to support HEI develop their educative environments and offers. The use of a technological LMS constitutes an innovation opportunity in the teaching/learning process, and can be seen as an interesting leverage for the creation of new educational contexts. At ISCAP the b-learning model that PAOL initiated has, in fact, been maturing the school's educational culture, according to the teacher's willingness or availability to integrate technology in the teaching-learning process.

## Pedagogical Dimension

The quality of the educative offers is a growing criterion used by the students in their choice of higher education institutions. Information and communication technologies are constantly evolving, providing an increasing amount of job offers, fact for which the students are profoundly aware. For this reason they demand an increasing effort by the teachers and the education institutions and look forward to explore/access these technologies at an educational level.

These modifications must, therefore, have an impact on the organizational culture of schools, its pedagogical practices and the definition of the roles of its stakeholders. Thus, and in order to cope with the social pressure, HEI need to act proactively in order to provide suitable educational offers, adapted to the new requirements and with the capacity to follow change.

As ICT practices evolve, and in order to fully take advantages of its positive influence in the learning and teaching process, teacher trainers focus, not only on the technology, but on the pedagogical strategies that can be used with a specific technological tool or application. Training often begins by explaining technological aspects of the tool or application. Nevertheless pedagogical suggestions are brainstormed and discussed, hoping to avoid the use of technology void of clear and specific objectives.

Even so, and despite the several activities available in the platform, we can see that, very often, its usage tends to correspond to a reproduction of the classroom rather than the use of innovation in the teaching/learning process. To fight this phenomenon, there is a continuous need for technological and pedagogical support which walks hand in hand with the emergent need to develop new mechanisms with potential to enrich online learning environments and thereby education itself, by supporting teachers and providing them with adequate training, tools and resources.

It is our view that, although the users present positive satisfaction levels in terms of technical support, there is a need to emphasize the pedagogical dimension, in order to extract meaningful and measurable advantages in learning.

Thus, it will be necessary for PAOL to accommodate new support mechanisms and new initiatives that, not only facilitate the improvement of innovators skills and activities, but also reinforce the global need for the readjustment of pedagogical methodologies amongst the first cycle late majority group.

As Salmon (2002) states, the success of online learning resides on the balanced arrangement of the old but still relevant teaching-learning concepts and the implementation of innovations that take full advantage of technology's potential (Oliveira L., et al., 2009). Therefore, it is necessary to instigate a new teaching and learning culture (Cox et al., 2003), where technology acts as a facilitator of fundamental changes germinating and diversifying new processes and contexts (Coghlan, M., 2003), promotes collaborative and personalized learning opportunities and generates higher levels of student performance and motivation (Means, B. & Golan, S., 1998).

## Promoting Creativity and Innovation

The pedagogical innovation and the temporal flexibility of the support for individual and collaborative study of our students are the central vectors of this project that converge to improve the quality of the existing educational offers. Along with these vectors, PAOL explores technologies as a pedagogical resource to contribute to the innovation of the teaching/learning process and to help students prepare their professional life seeking, at the same time, to promote their engagement in the study of the subjects. For this, it was important to motivate students by simplifying their access to knowledge, creating Internet using habits in study and working activities, making diversified material to backup the available online activities and resources and by providing different forms of learning, according to the various individual learning styles.

In order to promote the innovative introduction of pedagogical methods and approaches that are more creative and motivational, PAOL implemented two educational tools, ELGG and LAMS. Those tools enable the use of e-portfolios in class (ELGG) and the design of learning paths in the hope to potentiate a more active and self-aware approach to b-learning. These pedagogical tools were used in the educational projects described below.

### ELLG: Digital Portfolios

An activity we support is related to the creation of digital portfolios, more commonly known as e-portfolios using the ELGG platform. ePortfolios, considering the definition of the National Learning Infrastructure Initiative, quoted by Barret & Carney (2005), is defined as "a collection of authentic and diverse evidence, drawn from a larger archive representing what a person or organization has learned over time on which the person or organization has reflected, and designed for presentation to one or more audiences for a particular rhetorical purpose" (p. 1).

With ELGG we achieve exactly that, because it allowed the exploration of several educational dimensions, namely the student's reflective process and the analysis of their individual learning paths. They were used to reflect the *modus operandi* of their authors, thus allowing the collection of evidence that may contribute to the construction of more appropriate learning strategies by the teacher.

PAOL, implemented an e-portfolios project in ISCAP during the winter semester 2007-2008, in the course of Electronic Tools Applied to Translation, taught in the 3rd year in the Administrative Assistance and Translation degree. Students were expected to develop an e-portfolio in ELGG platform during the first and the second semester and were also expected to collaborate in answering the assessment tools that were applied to them. In order for them to be able to do that, PAOL promoted several trainings in ELGG and gave an ELGG manual for its usage.

This project expected to show how the learning styles (Felder & Soloman) and Multiple intelligences (Gardner, 2006) could be reflected in students' e-portfolios. It was a very complex process and we used several assessment tools, including grids of observation, the multiple intelligence test (MIT), the index of learning styles (ILS), a questionnaire and interviews, as shown in Figure 5.

With the development of this case study, we were able to understand more deeply the importance of the learning styles and multiple intelligences, but mostly we understood how they can be translated in the work students do (Tavares et al., 2008).

Several students were really committed to the task of looking at the e-Portfolio as a way of show-

*Figure 5. Assessment of learning styles and multiple intelligences through e-portfolios*

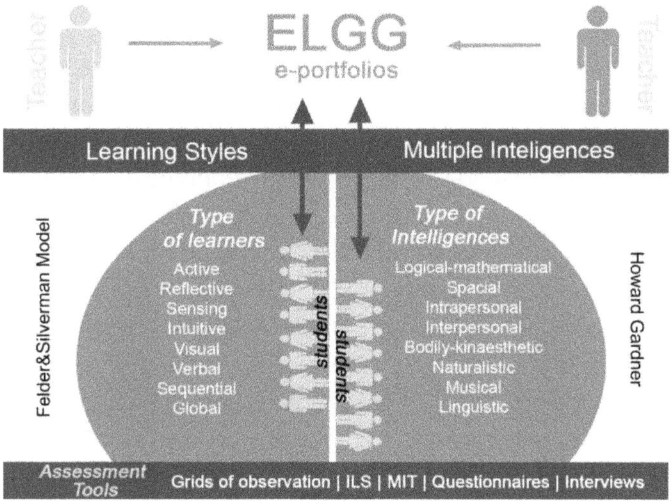


ing their personal and academic achievements. We had the opportunity to confirm that several assumptions constructed upon observing students' work were, in fact, translated in the results that the students achieved in the tests (ILS, MIT). For example, students that had their files very well organized in the ePortfolio, generally matched the ILS for having Sequential characteristics; students that were often commenting their colleagues blogs were very commonly to be found as students with a high Interpersonal intelligence in the MIT (Tavares et al., 2008).

The e-Portfolio revealed to be an excellent way of observing the work preferences of the students and therefore it could be important for teachers in the process of developing more adequate learning strategies for their students, according to their needs and motivations. It is a tool that has the potential to promote individual and adapted learning strategies to different learning styles, promote students' motivation and consequently adherence to the different educational activities planned.

## SD4L: Supporting Design for E-Learning

Despite the investment that organizations and governments have been making in teacher technical training, most teachers still lack essential pedagogical skills. Therefore, is hasn't been possible to take full advantage of educational technologies and to truly enhance education.

Teachers tend to build, adjust and implement teaching processes without having documented them. As a result there are very few collections of documented strategies that teachers can (re) use to satisfy their educational needs.

One of the activities that PAOL has been supporting is the introduction of Learning Design inspired technologies, such as LAMS, the Australian Learning Activity Management System. LAMS in an open source authoring tool and learning environment inspired by the IMS LD[1] (IMS, 2008) specification, which can be integrated into most of the wide spread LMS. This system was developed by the Macquarie E-Learning Centre Of Excellence (MELCOE) at Macquarie University, and has become the world's leading design software (Dalziel, 2006). PAOL has been developing the Supporting Design for Learning project in order to sustain teachers in the design of rich online pedagogical experiences, where the student is the central element.

This project aims to instigate a new teaching-learning culture by encouraging share and reuse habits amongst teachers. The existence of documented strategies facilitates their comprehension, their reuse and instigates the development of active learning scenarios by giving teachers concrete examples. This will not only allow the economy of time and effort inherent to its construction, but also allow augmenting the collections of documented strategies that teachers can use to satisfy their educational needs.

In order to carry this out we have assembled what we considered to be the essential technological infrastructure, using three components: a new learning environment – the LAMS – that was integrated with the pre-existent MOODLE platform and a small support website/repository as shown in Figure 5.

The use of MOODLE and LAMS as a built-in learning environment allowed teachers to access LAMS authoring environment through MOODLE and to add to their course pages the sequences of activities created in LAMS. In the same way, students only had to access the corresponding MOODLE course page in order to complete the sequence of tasks that the teacher had made available. However, adding an authoring and learning environment to MOODLE was not the only accomplishment of bringing these two systems together. Teachers also had the opportunity to monitor students' progress in real time, being able to open or close learning paths and to move students along the different tasks. Thus, LAMS integration with MOODLE allows teachers the benefit of orchestrating activities in different types of learning flows, which is not provided by MOODLE alone.

LAMS allows the creation and orchestration of activities in linear, conditional and branched learning flows, offering teachers a growing set of instruments: sequence management tools, activities and combined split-screen activities (Mason, 2001).

LAMS is composed of three main environments: the authoring environment – where activity sequences are constructed – the run-time environment – from which students have access to activities – and the monitoring environment – from which teacher can control the learning flow.

According to Britain (2004), the evaluation of Learning Design software may be considered in two perspectives: the level of conformance with the IMS LD (IMS, 2008) specification (conformance level) or its capability to design the learning

*Figure 6. Moodle/LAMS Technological infrastructure (Oliveira L. et al., 2009)*

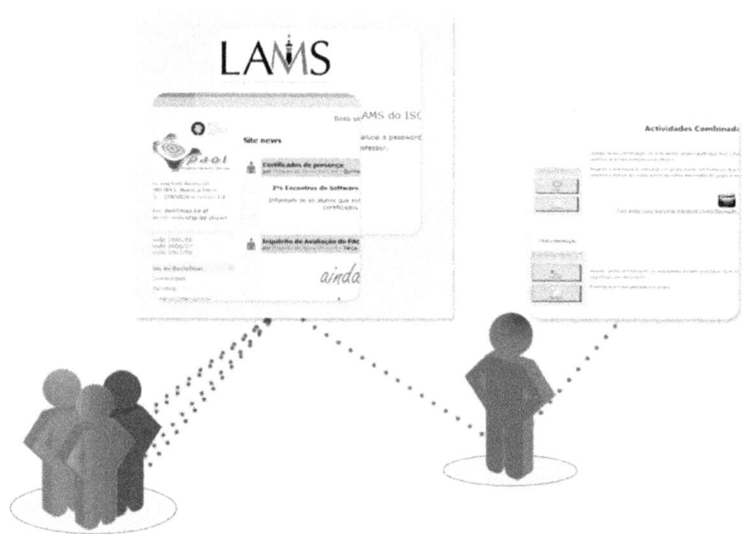

contexts/scenarios that have been motivating the construction of that same specification (educational potential).

When we consider the advantage of LAMS we cannot refer to high levels of conformance, but there are significant benefits for developing learning structures. The tool has a great pedagogical planning potential and offers an accessible interface that allows for a quick global comprehension and edition of learning sequences (Dalziel, 2006; Masterman, 2005).

LAMS construction philosophy also meets one of the biggest digital learning trends announced by IMS GLC in late 2008: "The future of digital learning revolves more around context than content. (...) Multi-media and online activities are the learning tools of tomorrow because you can tell so much more of a story with it than you can with the simple written word. Reading about science can be tedious and boring, but to actually watch and experience it can be amazing" (Siragusa & Dixon, 2005). Indeed, active learning grants higher levels of student motivation and assessment (Ghiglione, 2009).

In this project we have also developed the SD4L – Supporting Design for Learning Portal, where teacher can find a range of useful information and resources about the use of LAMS and about the design of learning paths. The majority of these resources consist of interactive tutorials about tools and strategies that where built on work developed by the LAMS Community. One of the main features of this portal is also the teaching-learning strategies repository that has been growing due to the contribution of several teachers. In this section of the portal, teachers can share and download almost ready-to-implement learning scenarios that were built by their colleagues, thus allowing them to spare time and effort throw the establishment of cyclical processes of reuse, as shown in Figure 7.

Within the implementation of this project, we concluded that it is essential to document pedagogical planning and strategies in order to provide teachers with concrete examples, leading to an economy of the time and effort inherent the development of these environments. The creation of a collection of documented pedagogical strategies will not only offer teachers tools and tips that can satisfy their educational needs, but will also contribute to the perfection of pedagogic strategies by its repeated usage in different educational contexts.

*Figure 7. Cyclical processes of reuse*

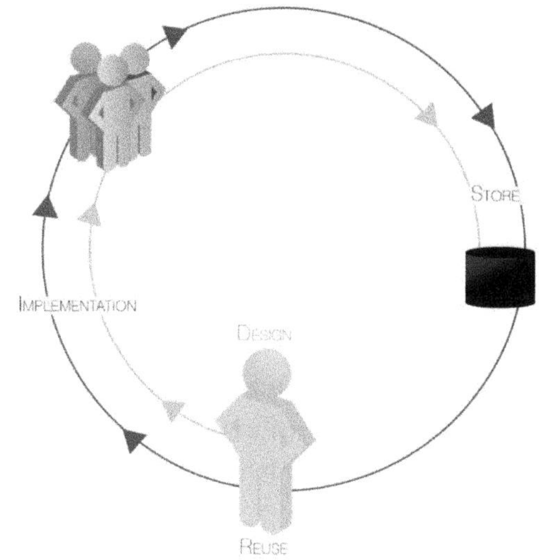


## Cooperation with others Projects in ISCAP

For a project to assume a school dimension it is important to involve and be involved in the entire community and the various existing projects. Thus, PAOL has established a partnership with the School's Psychology Office and, in order to fill gaps in students' prior knowledge, several other projects covering different areas that range from math to foreign languages.

In terms of Math, for example, PAOL has supported a project called "Matactiva", which intends to help first-year students in a self-learning process in order to fulfil the lack of some important mathematical concepts. PAOL has also supported the creation of a languages centres network - RECLES.pt, using Moodle platform.

## Evaluation

### Self-Evaluation Cycles

The management of e-Learning environments is composed of four different phases: conception, planning, execution and procedural evaluation (Ramos et al., 2002). This model, like most models presented in the literature focuses on various phases, especially on the necessity to evaluate the work carried out.

Therefore, procedural evaluation is an ongoing activity that hopefully leads to new ways of approaching the design, development, implementation and reflection on all teaching and learning activities carried out within the institution. As Ramboll (2004) states, evaluation could be an incentive to invest in pedagogical aspects and serves the interest of ICT support.

Based on these premises and on the perception that constant changes in ICT place new and different challenges, PAOL carries out an annual inquiry to validate the project's acceptance level and the users' satisfaction, covering all aspects of its activity. The inquiries focus on:

- Technology use (as an added value to the institution and the teaching-learning processes);
- Staff engagement and avalability (to integrate ICT into their practices);
- Improvements to the project's organizational structure;
- Analysis of Projects' main activities;
- Use of available technical means;
- Monitoring of the project's development process;
- Evaluating the workflow in order to improve it.

PAOL's self-evaluation process is carried out using a questionnaire that is adapted every year and distributed to both teachers and students. The analysis of the answers obtained allows the classification of the efficiency and effectiveness of the e-learning project implementation and development, in the institutional, technical and educational dimensions.

The questionnaire has been distributed, in recent years, in an electronic format and is meant to gather information from both students and teachers of both sexes and all ages in all courses at ISCAP. Given the incremental facility to obtain qualitative data through Moodle, the design of the questionnaire has shifted towards the inclusion of mainly qualitative issues, to include reflections on the PAOL's pedagogical approach.

Nevertheless, the structure of the questionnaire is designed to obtain the information the Project considers essential to evaluate its evolution and consequently, obtain a global perspective of its implementation and dissemination in the school. Both students' and teachers' questionnaire focus on:

- Users' profile (age, gender, etc.);
- Internet Access;
- Evaluation of "PAOL";
- Users' needs and expectations;
- LMS usage;
- Evaluation of the LMS Moodle.

The questionnaire also tries to understand Internet access available to teachers and students, users' experiences using Internet and computers, as well as the existence of eventual technical problems these might have had. Perceptions are gathered according to the degree of importance or Yes/No questions. Each question usually offers five choices, or, in terms of activities, a 4-point Linket scale is used.

Currently, the elaborated questionnaires tend to cover a lot more aspects than Moodle and Internet access. The questions asked address issues like types of resources to use in their classes, the tools made available for them and that weren't used before, like ELGG, Limesurvey, LAMS, or about other services that we have, such as the "Teachers Place", were an access is granted to all essential tools and resources to create multimedia contents

We also try to understand if there is a use of any of those tools (Moodle, ELGG, LAMS) to perform any type of online assessment. Furthermore, we try to collect their opinions about the services offered and their importance for the overall workflow and teaching-learning process. In order to do that, and because it is our perception that a closed questionnaire may not provide all the necessary information and, in certain situations, may appear to be limited, we use open-ended questions.

Data is then collected and analyzed and members of the project seek, by using both qualitative and quantitative analysis methods, to draw out key perceptions from learners and teachers for further analysis, group discussion and processes revision and update.

Trying to frame and compare the main results obtained in the context and supported by Meredith's (2002) heuristic integration model, we can claim that, although ISCAP is a centennial school with a strong traditional culture of teaching and learning based on expository methods, currently the majority of ISCAP's teachers have accepted the integration of technologies rather well and have considered the project as "a good achievement and with potential that needs to be further developed", as stated by a teacher in one of the questionnaires.

In consonance with Ramboll's Report (2004), experience demonstrates the existence of many opportunities in terms of integrating and anchoring technologies. The execution of evaluation cycles permitted the analysis of the benefits and problems, which have obviously led to changes in the implementation of new cycles.

Taking into account that the management of the learning environment is made up of 4 phases: conception, planning, execution and evaluation (Ramos et al, 2001) and, as in any cyclical implementation, after verifying that the initial solution was not a complete success (Meredith 2002), there is a need to adjust in order to reach a full institutional integration of the technologies. Therefore, every year, based on the results obtained, PAOL readjusts some of the processes and methodologies used. This occurs as, in each new cycle, the need to increment improvements, basically at the pedagogical support level, is emphasized.

However, this type of evaluation is generally not focused on the teaching and learning process having, therefore, few consequences in terms of learning improvements and teachers' pedagogical and professional development. Nevertheless, for the teachers that have already adhered to the project, this evaluation process helps us to foresee needs and implement strategies that motivate them to find new forms of pedagogical enrichment based on the use of technological resources.

## Promoting Innovation: Contextual Risks and Recommendations

In a project regarding the integration of Web technologies in a HEI, it is impossible to avoid some constraints that may influence the success of a school project. In this sense, it is important to develop a risk management plan, in particular to contend with the possible resistance to change by students and teachers. Lack of awareness of the added value due to the lack examples of good practices and the technological difficulties encountered by some teachers are some of the risks to be taken into account. To contend with these risks, users' manuals of platforms in use were made available. Furthermore, PAOL offers constant training and ongoing support, through the existence of an office where teachers and students can personally contact and place their questions and get technical and teaching support.

On a technical level, it is important to have a team responsible for the smooth operation of the web-based platforms in use and their integration with other existent information systems in the school. At an institutional level, it is important to ensure good communication and support from the school's Management, in terms of support for initiatives and as leverage for the necessary structural and cultural changes.

The reduced contribution of pedagogic performance throughout a teacher's career, insignificantly valued compared to investigative research, constrains the personal investment in the development of his/her teaching activity. This becomes an inhibiting factor, also highlighted in other HE systems by several authors (Hannan and Silver, 2000, Haywood et al., 2000). The workload constitutes a discouragement, as there is neither valorisation nor recognition for well-implemented pedagogy in HEI (Pereira, 2004). In this sense, it is once again important to emphasize the facilitator support offered by the Institution in order to minimize the above mentioned problems.

## Future Research Directions

In the future, it is foreseeable the growth of the Innovation in Education Unit (PAOL) due to the developments in the area of HEI internationalization, increasing employability, access and lifelong learning sought through projects such as SD4bL, employability, accessibility and b-learning courses, briefly described below.

In order to optimize the project's intervention in pedagogy in technology mediated education, PAOL intends to develop the project *SD4L (Supporting Design for b-Learning Project)* that includes the participation of foreign collaborators and sustains the idea of reusability as fundamental concept to e-learning sustainability. This innovative project, both in national and international educational contexts, aims to foster interdisciplinary and multi-level exchange of teaching strategies as well as knowledge in order reduce the time and effort that teachers need in the educational planning process.

This project aims to deliver:

1. A new teaching-learning and that operates within an integrated multilateral, interdisciplinary and multi-level Web-based technology;
2. A web-based pedagogical planner capable of sustaining the new model developed and the educational planning needs identified;
3. A digital library, constituted by learning strategies, that serves as a support tool for the conception of new learning contexts.

The first step for this project is to investigate and develop tools to optimize educational planning, while promoting the exchange and reuse of teaching strategies to maximize the sharing and saving time and effort in developing them in order to encourage creativity and innovation in learning strategies that motivate students and enhance their participation in learning activities. The second step is then to create a model adjusted for different levels of education, areas of knowledge and cultures to support the integration of web technologies in students' learning process. Subsequently, technical solutions that provide adequate support are essential.

The *e-portfolios project* that was developed at ISCAP has reached a point where it is important to move onto the next step, i.e., go beyond the focus on learning and bind e-portfolios to the labour market, providing a unique environment suitable for companies to explore students' digital identity, assess their ability to reflect on the contents, thus creating an environment that enhances the knowledge, skills and attitudes learned/developed and their growth potential.

This project will be part of the strategy of internationalization, aiming at enabling students to develop their personal learning space, by illustrating it with content, ideas and organizing their work in order to develop a profile to be presented in a global market.

The objective is that both the human and technical elements provide a contribution to stimulate teachers and students to use it, through the integration of information technology and communication to potentiate the skills and the abilities of our students in the labour market.

Another objective of this project is to investigate the contribution of the different methodological approaches in the construction of knowledge in different subjects and develop research projects that contribute to the improvement and innovation in educational practices.

This project will, in all its phases, assess the technological needs and the level of relationship with ICT teachers and students and define training courses for teachers and students, in order to create a greater awareness about all the possibilities available in the tool in use (ELGG).

The next phase will be the establishment of a set of parameters that should be a part of the student's e-portfolios, in order to represent the most important facts of their education/training and their professional life, if already existent. With this, we seek to promote the search and access to these e-portfolios by several companies (that we intend to combine in a database) in order to promote future job selection.

At the same time, we also seek to develop conditions to follow the innovation trends in the labour market, not only in terms of changing work methods, but also in terms of training requirements and the relationship between employer and the HEI.

A final aim is to meet European guidelines defined for e-Portfolios implementation and use. This project also has, as its concern, the promotion of the accessibility of online courses to the entire student population. Thus, in the future we will design new Moodle adaptations in order to create an *interactive multimodal learning environment* to encourage students' learning. Moodle allows the implementations of several strategies according to the courses' contextual specificity, incorporating cooperation, communication and coordination processes, conceived to be configured and adapted by the final users (Orlikowski & Hofman, 1996), therefore permitting the creation of specific applications, as each teacher can create his/her own "space" in order to implement a more personal pedagogical approach.

In future, we could attend to the needs and characteristics of *lifelong learning*, as blended-learning scenarios seem to provide both the necessary flexibility and the support for higher education courses and for the development of the skills for technology and time management.

## CONCLUSION

Whenever one reflects on the integration of web technology in the b-learning process, the importance of responding to the questions about "what to do?", "when to do?" and "how to do?" appears, because implementing a b-learning environment in a HEI requires cultural, structural and procedural changes.

Teachers need, in fact, assistance in the integration of web technologies in the teaching-learning process, mainly in finding which tools best suit their teaching needs and which pedagogical strategies best accomplish the learning goals that are posed to students. So, we think that the success of technology-enhanced learning relies deeply in the harmonious combination of the correct technologies with the most efficient pedagogies, allowing for the implementation of innovative, authentic and diversified teaching and learning opportunities, requiring the need to work on three levels: institutional, technical and pedagogical.

Therefore, it is our understanding that the basic level considers the institutional aspects and supports the progress of the project. This level includes the aspects related to management and ethical aspects. Without the clear support of the school's Board and a Management team, it is difficult to be successful in an e-learning institutional project and reach the school's full dimension. The help offered by the Management team encloses the creation of an office to support trainings and projects where specialized staff offers teachers and students continuous support in the integration of web technologies in face-to-face environments. Besides helping the internal communication, the Management team should also give support to the teams activities and development. At the same time, such a Management should create the conditions for the spreading of a new and more modern institutional image, thus reinforcing the competitiveness of the HEI.

The second level of the proposed structure refers to the technical dimension. In an e-learning project, it is important to select the web platform to be used. It is also important to give training and technical support to all teachers and students. We underline the constant technical evolution and the need for decision-making about its adoption. The first step for teachers that are interested in the use of e-learning methodologies is to obtain technical training on the web environment. Usually, teachers start by asking for initial training on the LMS and only after they reflect on how to use it improving the teaching-learning process.

The last level of the proposed model systematizes the educational perspective. After acknowledging how to make a file available online, teachers look for much more in order to use the LMS to improve learning. For example, the technical creation of a MOODLE forum, is easy, however the major difficulty is to determine it's use in order to engage students, to create motivation and improve learning. In this context, it is important to offer a good set of strategies, pedagogical practice manuals or an e-repository of digital contents, to fulfil that need.

These and other tools intend to reduce teachers' efforts on contents production and try to lead their effort towards a more pedagogical planning of the unit. Thus, the training on this level occurs after the technical training. In this phase, interdisciplinary work is essential, not only to inspire and promote the reuse of e-learning strategies but also to promote the soft skills, vital in today's society.

## REFERENCES


Almeida, M. (2003). Educação, ambientes virtuais e interactividade. In Silva, M. (Ed.), *Educação On-line. Sao Paolo: Edições Loyola.*

Barret, H., & Carney, J. (2005). Conflicting paradigms and competing purposes in electronic portfolio development. Educational assessment. *LEA Journal*. Retrieved from http://eletronicportfolio.org/portfolios/LEAJournal-BarrettCarney.pdf

Berge. (2004). Obstacles faced at various stages of capability regarding distance education in institutions of higher education: Survey results. Retrieved on January 22, 2010, from http://emoderators.com/barriers/hghred_stgs.shtml

Britain, S. (2004). A review of learning design: Concept, specifications and tools. *A report for the JISC E-learning Pedagogy Programme, JISC.* Retrieved January 2010, from http://www.jisc.ac.uk/uploaded_documents/ACF83C.doc

Carvalho, V. (2006). E-learning: A experiência do Instituto Superior de Engenharia do Porto. In Carvalho, V. (Ed.), *e-learning e formação avançada: casos de sucesso no Ensino Superior da Europa e America Latina* (pp. 73–88). Portugal: Edições Politema.

CEMED. (2010). *Centro Multimédia e de Ensino a Distância da Universidade de Aveiro*. Retrieved January 2010, from http://www.cemed.ua.pt/

Coghlan, M. (2003). Should using the internet change the way we teach? Paper presented at the Educause 2003, Adelaide, Austalasia. Retrieved from http://www.caudit.edu.au/educauseaustralasia/2003/EDUCAUSE/PDF/AUTHOR/ED030040.PDF

Comissão Europeia. (2007). *Estrutura dos Sistemas de Ensino, Formação Profissional e Ensino para Adultos na Europa*. Unidade Portuguesa da Rede Eurydice.

Cox, M., Abbott, C., Webb, M., Blakeley, B., Beauchamp, T., & Rhodes, V. (2003). ICT and attainment: A review of the research literature. Coventry, UK: Becta (ICT in Schools Research and Evaluation Series).

Dalziel, J. (2005). *From reusable e-learning content to reusable learning designs: Lessons from LAMS*. Retrieved January 2010 from http://rams.ramp.org.au/download/attachments/9469955/dalziel_reusable.pdf

Dalziel, J. (2006). *Future directions for LAMS and learning design*. Paper presented at the 1st International LAMS Conference. Retrieved January 2010 from http://saturn.melcoe.mq.edu.au/presentations/FutureofLAMS.LAMSConf.ppt

Dias, P. (2004). Processos de aprendizagem colaborativa nas comunidades onLine. In Dias, A., & Gomes, M. J. (Eds.), *E-Learning para E-formadores* (pp. 19–31). TecMinho Edições.

Felder, R., & Soloman, B. (2010). Learning styles and strategies. Retrieved January 2010, from http://www4.ncsu.edu/unity/lockers/users/f/felder/public/ILSdir/styles.htm

Gardner, H. (2006). *Multiple intelligences–new horizons*. New York: Basic Books.

Garrison, R. & Kanuka. (2004). Blended learning: Uncovering its transformative potential in higher education. *The Internet and Higher Education*, *7*, 95–105. doi:10.1016/j.iheduc.2004.02.001

GATIUP. (2010). *GATIUP - Gabinete de Apoio para as Novas Tecnologias na Educação da Universidade do Porto*. Retrieved January 2010, from http://sigarra.up.pt/up/web_base.gera_pagina?p_pagina=18377

Ghiglione, E. (2009). *LAMS 2. User documentation*. Retrieved January 2010, from http://wiki.lamsfoundation.org/display/lamsdocs/Home;jsessionid=6890123183F3AC8FCFBA1B51E0D5CA62

Hannan, A., & Silver, H. (2000). *Innovating in higher education-teaching, learning and institutional cultures*. Buckingham, UK: The Society for Research into Higher Education/Open University Press.

IMS. G.L.C. (2008). Seven trends to watch in digital learning. Retrieved January 2010, from http://www.imsglobal.org/articles/7trendsinDigitalLearning.html

IPP. (2007). *Self evaluation report by Instituto Politécnico do Porto*.

IPP. (2010). *Instituto Politécnico do Porto. Official website*. Retrieved January 2010, from http://www.ipp.pt

ISCAP. (2010). *Instituto Superior de Contabilidade e Administracao do Porto*. Retrieved December 2009, from http://www.iscap.ipp.pt/

Isla, J. (2004). Collaborative learning at Monterrey Tech-Virtual University. In Duffy, T., & Jamie, K. (Eds.), *Learner-centered theory and practice-cases from higher education* (pp. 297–319).

Khan, B. (2005). *Managing e-learning strategies. Design, delivery, implementation and evaluation*. Science Publishing.

Mason, J. (2001). *Qualitative researching*. London: Sage Publications.

Masterman, L., & Lee, S. (2005). *Evaluation of practitioner trial of LAMS: Final report to JISC*. Oxford University Computing Services.

McNaught, C. & Kennedy, P. (2000). Staff development at RMIT: Bottom-up work service by top-down investment and policy. *ALT-J*, 4-18.

Means, B., & Golan, S. (1998). *Transforming teaching and learning with multimedia technology*. Menlo Park, CA: SRI International.

Meredith, S. (2002). Models of e-learning: Technology promise vs learner needs literature. *The International Journal of Management Education*.

Moran, J. (2005). A pedagogia e a didática da educação On-line. In Vidigal da Silva, R. (Ed.), *Educação, Aprendizagem e Tecnologia. Um paradigma para professores do século XXI*. Portugal: Edições Sílabo.

Nova, C., & Alves, L. (2003). Estação online: a ciberescrita, as imagens e a EAD. In Silva, M. (Ed.), *Educação On-line. Sao Paolo: Edições Loyola*.

Oliveira, L., Albuquerque, A., Ribeiro, S., Peres, P., & Carvalho, C. V. (2009). A sustainable approach to the reuse of student-centered learning scenarios. *Proceedings of EDULEARN09 - International Conference on Education and New Learning Technologies*, Barcelona, Spain, 6-8 July.

Orlikowski, W., & Hofman, D. (1996). *An improvisational model of change management: The case of groupware technologies*. Massachusetts Institute of Technology, Sloan School of Management Working Paper #3869, Cambridge.

PAOL. (2008). *Uma Reflexão - Quarta Mesa Redonda*. Instituto Superior de Contabilidade e Administração do Porto.

PAOL. (2010). *Projecto de Apoio On-line*. Retrieved January 2010, from http://www.iscap.ipp.pt/paol.

Pereira, C. (2004). *Educação em Ciência e Tecnologia para o Séc. XXI. Necessidade de mudar de paradigma*.

Phillips, R. (2000). *Handbook for learning-centered evaluation of computer-facilitated learning projects in higher education*. Murdoch University.

Ramboll. (2004). *Studies in the context of the e-learning initiative*. Virtual Models of European Universities (Lot 1). Draft Final Report to the EU Commission, DG Education and Culture.

Ramos, F., Moreira, A., & Santos, A. (2001). Towards a reference framework for e-learning management. *Proceedings of DLA 2001 - Distance Learning Administration Conference, U.S.A.*

Rogers, E. (1995). *Diffusion of innovations*. The Free Press.

Salmon, G. (2002). *e-Learning works: Learning from the past, present and future*. WORLD DIDAC 2002, Zurich. Retrieved from http://www.atimod.com/research/presentations/ZURICH-SALMON1.htm

Sauter, V. (2003). Web design studio: A preliminary experiment in facilitating faculty use of internet. In Aggarvil, A. (Ed.), *Web-based education: Learning from experience* (pp. 131–154). Information Science Publishing.

Sherry, L. (1998b). Diffusion of the Internet within a graduate school of education. Doctoral dissertation of Philosophy - Educational Leadership and Innovation, University of Colorado.

Silva, R. (2005). Gestão da aprendizagem e do conhecimento. In Silva, R., & Silva, A. (Eds.), *Educação, Aprendizagem e Tecnologia: um paradigma para professores do século XXI* (pp. 43–65). Edições Sílabo.

Siragusa, L., & Dixon, K. (2005). *Closing the gap between pedagogical theory and online instructional design: A bridge too far: Methods and technologies for learning*. Southampton, UK: WIT Press.

Tavares, C., Moreira, A., Silva, M., & Albuquerque, A. (2008). *The use of an e-Porfolio as a way of understanding students learning preferences*. Paper presented at the ICERI 2008 - International Conference of Education, Research and Innovation.

Vogel, M., & Oliver, M. (2006). *Design for learning in virtual learning environments-insider perspectives*. London: Centre for Excellence in Learning Technology.

## ENDNOTE

1 The IMS Learning Design Specification was developed by the University of Netherland and offers a generic notational system that allows the description of the main components of the educational scenario, such as "roles", "methods", "learning-objectives", "contents", "prerequisites" and "metadata". For more information, please see: http://www.imsglobal.org/learningdesign/

# Section 5
# Conclusion

# Chapter 16
# The Opportunities and Challenges of Technology Driven Creative Collaborations

**Diego E. Uribe Larach**
*State University of New York, Buffalo State, USA*

**John F. Cabra**
*State University of New York, Buffalo State, USA*

## ABSTRACT

*The onset of the 21st century is marked by deep psychological and sociological transformations affecting every scale of human endeavor, ranging from individual to crowd behavior. Deep and central to these transformations is the penetration of digital communication and computer technology into modern day life. Above all, this new and evolving technological landscape has opened exciting new possibilities to drive creative behavior, organizational creativity and innovation through computer-mediated interactions. Such opportunities are met with equal challenges that need to be addressed in order to harness the full potential of massively distributed creative collaborations. This chapter will elaborate on the underlying trends that give rise to these opportunities and challenges and to what extent these trends will govern creativity and innovation in areas of organizational life such as business, education, science and design in the next 10 to 30 years.*

## INTRODUCTION

We have taken the task of writing this chapter with deep emotional and cognitive dissonance. It is almost a strip of dark comedy to write about technology and innovation in a medium that has remained unchanged for centuries and that is on the verge of becoming obsolete. There are indications that several formats of printed media are already losing ground against digital vehicles of information (http://www.youtube.com/watch?v=6ILQrUrEWe8, Retrieved October 26, 2009). Moreover, we know we are targeting

DOI: 10.4018/978-1-60960-519-3.ch016

a shrinking demographic unlike the vast audience of not-so-distant future teachers, managers, scientists, designers and engineers who should read this book but may miss the chance to do so because they were *connected* elsewhere. As media guru Marshall McLuhan (1964) succinctly stated, the medium is the message—printed books are living remnants of the 20th century persona, society and knowledge.

Take a minute to think about a typical day in your life. You are unlikely to find a day when you had no interaction with some sort of digital technology. On this day, you may have worked on your laptop computer, purchased something online, taken a picture with your digital camera and then uploaded it to a social media site. You certainly called somebody using your smartphone, listened to music on your mp3 player, interacted with a digital vending kiosk, or played a videogame. In all likelihood you did all of these things and more on that same day using only the smartphone you carry. Although these technologies are in their infant stages with no more than a few decades of existence, their influence and pervasiveness in modern day life have been remarkable. As of 2008, the average U.S. American between the ages 18 to 55 have consumed between 500 to 570 minutes of digital information per day through means such as TV, mobile phones and computer technologies (Council for Research Excellence, 2008). This is the equivalent of 8 ½ hours of our daily lives. It is yet to be seen and difficult to imagine the extent to which we will continue to integrate these technologies in a seamless and frictionless way. The thought of what lies ahead is evocatively exciting and frightening as we ponder the integration of technologies such as intra-body nanotechnology, sub-dermal radio frequency identification tags—just to name a few—with our psychological, sociological and even physiological being.

On one hand this integration is exciting because it has democratized computer-mediated networks. Whereby networks flourished only in select geographical locations, now they are free, accessible and ubiquitous to anyone who holds an internet connection or a mobile phone service.[1] Networks feed and breathe in a cloud.[2] With this phenomenon, clouds condense the time it takes to articulate interactions, lift the flow of information, expand the frequency of interactions and exacerbate contagion properties of such networks (Christakis & Fowler, 2009). The impact of this phenomenon has led to the creation of a myriad of outlets for content sharing and human expression. The explosion of this massively distributed sharing of content and ideas may turn out to be the greatest driver of creativity and innovation of the 21st century.

On the other hand this integration can be frightening because it can lead to misunderstandings or destructive behaviors such us flaming[3]. Worse, it can lead to unproductive speculations that humans will merge with computers[4] and consequently delineate biology as outdated, the end of human history as we know it. These technologies have also been predicted to facilitate communication at a dizzying speed, trade knowledge of our cultures and band memories and resources in never before seen ways. The information these systems churn are overwhelming and has been blamed for being the leading producer of information anxiety (Wurman, 1989). To be clear, the scope of this chapter is not about making destructive speculations, as we disagree with the naysayers. Our focus leans more on opportunities and challenges more closely and reasonably marked.

Although our reach to share content and ideas has been considerably expanded through computer mediated networks, the state of the art in communication technology still lacks qualities that complete the gestalt of human communication. According to Perkins (2008), 88% of what we communicate (message), of how we derive meaning and how we check for understanding is determined by body language and inflections in our tone of voice. That is, human communication depends on non-verbal feedback loops that guide and modulate communication responses. In most computer-mediated exchanges, these feedback loops are absent, which at times may hinder the overall quality of the exchange. Consequently, one of the biggest challenges in the domain of computer mediated creative collaboration is to find ways to convey nonverbal cues to remote counterparts. Addressing this gap is going to be crucial to harness the full power of distributed creative collaborations.

In the following pages, technology trends (such as the ones described above) are developed in the following sections within a context of the opportunities and challenges these trends pose for teams and organizations engaged in creative collaborations. Finally, a list of technology areas that have the potential to change the face of organizational creativity and innovation are listed and briefly elaborated. It must be understood that any particular technology as mentioned throughout this chapter is subject to obsolescence by the time this book goes to print.

## UNDERSTANDING CREATIVITY, INNOVATION & TECHNOLOGY

### About Creativity

We have transitioned from the information economy to the creative economy (Cooper, Hofheinz, & Purdy, 2009) and yet from a layperson's view there is little understanding of what creativity is, the dynamics that govern creative behavior, the processes that promote it, or the attributes that comprise a creative output. Generally audiences give constructive views of creativity. This hasn't always been the case. Historically, it was not uncommon to have participants say that creativity is "just for artists or scientists", "creativity equates to being eccentric", "to be creative, one has to be a bit crazy", "creativity is frivolous", etc. These views have changed based on sixty years of scientific study. Creativity seems to be now viewed in a more valuable way; that it is necessary in today's ever changing business environment.

Creativity is a set of abilities that can be nurtured over time (Parnes, 1987; Torrance, 1987). More closely, the creative act results in and is characterized by something that is novel and useful (Amabile, 1988: Boden, 2007; Puccio, Murdock, & Mance, 2007; Stein, 1974). Creativity is not a simple construct, but a systemic and multifaceted phenomenon (Murdock & Puccio, 1993; Woodman & Schoenfeldt, 1990) as shown in Figure 1. In this figure is a model that depicts how creativity works in an organization.

The three fundamental ingredients to creativity are the person (people), the process and the environment (Puccio et al. 2007). The ingredients can be thought of as follows:

*Figure 1. Creativity: A systems view*

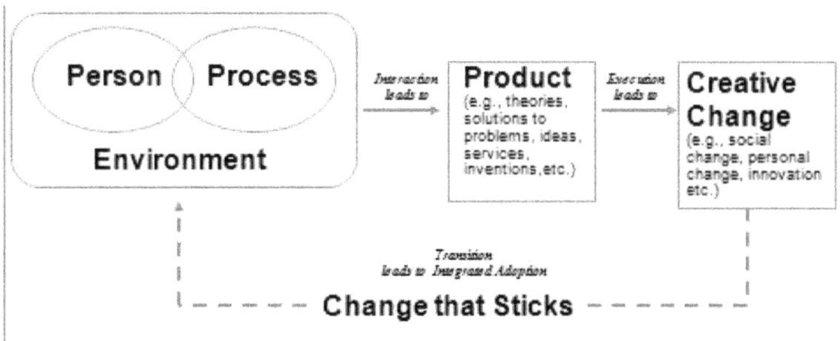


Adapted from Puccio, Murdock, & Mance, 2007

- **Person:** Qualities of the individual that predispose him or her to be creative, such as personality traits, domain relevant skills, problem-solving preferences and motivation.
- **Process:** Refers to the stages of thinking an individual or team goes through to creatively solve a problem.
- **Environment:** Relates to the climate, culture and physical attributes found in an organization that impacts the person and the process.

The interaction of these three factors, that is when people have the right skills, motivation and a supportive environment, leads to creative products (Puccio et al. 2007). In general, creativity leads to change when creative outputs like products, services, theories and/or processes are adopted by individuals, teams, organizations, and society. Specifically, innovation is an example of a creative change and is typically viewed as the successful adoption of a new product or service into the marketplace. This systems view of creativity therefore serves as reference to better plot the impact of technology trends in harnessing organizational creativity and innovation.

For the purpose of discussion and analysis, creativity will be conceptualized as an output that is both novel and useful. The creative process will be considered as a series of heuristics, deployed consciously or unconsciously, that maximize the probability of achieving a novel and useful outcomes.

## About Innovation

If creativity involves the production of outputs that are both novel and useful, then what's the relationship between creativity and innovation? Although intimately related, creativity and innovation are different constructs. Innovation can be defined as the deliberate introduction and adoption of new and useful products, processes, concepts, and organizational schemas to a relevant social unit such as an organization, a market, a field of study, and/or a nation (West & Rickards, 1999). All innovations stem from creative outputs yet not all creative outputs become innovations. That is, all creativity is a potential or unrealized innovation. In light of this definition, creativity influences innovation in two fundamental ways. First, the greater the number of creative outputs that an individual or social unit produces, the higher the probability that one of these creative outputs will successfully be commercialized (Stevens & Burley, 1997). Second, the stronger the skills related to an individual's (or group's) creativity, the higher the probability that the individual or group will overcome the myriad of obstacles and challenges that abound in the adoption phases of the innovation process. Consequently, creativity and the creative process are tightly woven into innovation, from conceptualization of novelty that is useful, to introduction and mass adoption.

## About Technology

Computer technologies are plentiful in modern life and thus it is not uncommon to perceive digital technology as an all-encompassing technology. However, technology is a much more distinct concept. As defined by the Merrian-Webster dictionary, technology is: (a) the practical application of knowledge, especially in a particular area; (b) a capability given by the practical application of knowledge; (c) a manner of accomplishing a task especially using technical processes, methods, or knowledge; and (d) the specialized aspects of a particular field of endeavor (Retrieved November 19th, 2009 from: http://www.merriam-webster.com/dictionary/technology). McLuhan (1964) stated that technology is simply an extension of man. Malafouris (2009) went further arguing that technology changes the formation of the human mind and consequently technology, which comprises material culture, is not a mirror image of the human mind but a tangible part of it. Technological extensions emerge from our physical/biological, emotional, spiritual, psychological and sociological nature. In the same way that clothing is an extension of our bodies, computers are extensions of our brains and cognitive capabilities. We assert therefore that the internet and the web are extensions of our social being (i.e., the need to reach out, communicate, share information and connect with others).

Another aspect of technology is that it is conditioned by properties of novelty, temporality and relevance. That means that once a technology, meaning any extension of ourselves, is culturally accepted and seamlessly integrated in our lifestyle, it ceases to be viewed as something novel; it becomes a part of the socio-technical landscape of the times. In this regard, we currently don't think of silver cutlery or even refrigerators in terms of technology, although anthropologically they are. As Alan Kay asserted, technology is everything that was invented after we were born (Tapscott, 2009). This assertion creates a problematic implication for the discussion and topics elaborated in this chapter, for there is a distinct sociological division between two generations, those who were born before the digital revolution and those whom were born after the digital revolution, the latter being referred to as the Net Generation (Tapscott, 2009) or Digital Natives (Prensky, 2001), and the former as Digital immigrants. In discussing a communication landscape that affects computer-mediated creativity in organizations, it should be noted that for a significant population who already is present in different domains of organizational life, computer and communication technologies, including the internet, are not technologies per se but rather they form part of the socio-technical vicissitudes of life in the 21st century reality. The technology trends discussed in the following pages, touch on mainly extensions of computer, communication and digital technologies.

## CHARTING THE SOCIO-TECHNICAL LANDSCAPE OF THE 21ST CENTURY

### Trends, Opportunities & Challenges in Harnessing Creativity & Innovation

#### The 3 W's: Whatever—Wherever—Whenever

Moore's Law, which predicts an exponential increase in a microprocessor's power, while simultaneously decreasing its size, has made it possible for us to live in a ubiquitous and omnipresent reality, a *now-reality* (Tapscott, 2009). In the now-reality the constructs of need and satisfaction collapse in the blink of an eye. Whether it is through smartphones, netbooks or any other networked device, we carry the world in our pockets and therefore can access this world of information wherever we are and whenever we want. From a sociological perspective we now expect and demand that information, communication and transactions be instantaneous. Actually, Digital Natives, those born in the age of digital technologies, regard email technology as a slow and archaic form of communication when compared to their preferred vehicles of Instant Messaging and Text Messaging services (Palfrey & Gasser, 2008; Tapscott, 2009). As these forms of technology continue to improve, they will become more ubiquitous. As organizations become more internationalized, organizational members will be expected to interface with overseas partners. As such, we anticipate organizations will engage in live and remote creative collaborations.

**Opportunities**

Face-to-face collaborations often involve the use of flipchart paper and Post-It notes, which are cumbersome and sluggish. We predict these platforms will fade. There are platforms now that can mind-map and visually tag data into more meaningful themes in a matter of minutes (Puccio & Cabra,

2009). Now let's take this further. Imagine that these software programs can be executed to create endless iterations of digitized prototypes, models, or simulations. It now exists. A software program called the Data Visualizer, in concert with another program called Catai, can perform thousands of iterative checks in seconds to examine the interface possibilities among component widgets (Schrage, 2006). The implication here suggests that reliance on non-digital technology (analog technology) so often used in creativity meetings, such as flipcharts, whiteboards, and post-its will no longer be practical. Comparatively speaking, the physical space it consumes slows the creative process. Analog technology appears to prime groups to clean up and inventory information instead of comprehensively expanding and connecting information.

Platforms range from simple instant messaging boards, collaborative document editing forms, and wikis, to sophisticated web based collaborative concept mapping programs, virtual worlds and tangible media. Instantly these platforms can relay information to its users to power divergent phases of the creative process either for continuing antecedent synchronous work or for live-remote collaborations. In short, ubiquitous technologies create a seamless and timeless interface that connect synchronous, asynchronous, remote and live creative collaborations. Creativity therefore has the potential to emerge and be harnessed whenever, wherever, and in whatever context.

One application of these ubiquitous technologies is the possibility to instantly capture insights while immersed in the context of discovery. For example, in design thinking, ethnographic observation is used as a source of inspiration (Brown, 2008; Kelley & Littman, 2001). If individuals could stream their images or video footages in real time into digital platforms, then they certainly would be able to capitalize on the moment to build a shared understanding and make sense of the data (Tassoul & Buijs, 2007), on demand and while out in the field. Now the innovation lab has become mobile as it can be staged whenever, wherever (creative environment), however (creative process), whatever (creative product) and with whomever (the creative person).

## Challenges

We live in a permanently connected and networked world. In the same way we can access indiscriminate information, whenever we desire it, and from mostly any location, the network can now do the same with us. We are immersed in a flood of requests, pushed by people asking for information or wanting interaction. Of course participation in the network is optional. Yet objectively speaking, the cost of nonparticipation outstrips the loss of opportunity. In a study of employee behaviors and use of email technology in an organization in the U.K., Jackson, Dawson and Wilson (2003) found that employees interrupted their workflow to check email every five minutes, while only taking 6 six seconds to reply to it. Interruptions such as these pose a major challenge to creative collaboration. First, given this permanent connected status, people expect an immediate response to their streams of communication. As such, individuals are given little time to think through their reply. This comes at the expense of prerequisite incubation time needed to formulate novel associations that feed the creative process (Sio & Ormerod, 2009). Accordingly, one would expect that a group who deliberately marginalizes technological interruptions would outperform a group who does not in idea fluency and quality output. This is important because incubation, which is a period of pre-conscious, fringe-conscious, off-conscious and even unconscious mental activity that occurs while the individual is detached from the task of problem solving, is a key contributor to the creative process (Wallas, 1921; Parnes, 1991; Puccio et al. 2007).

Csikszentmihalyi (1990) asserted that flow, which is a state of deep uninterrupted focus and creative immersion, depends on one's ability engage completely in a task. Neuroscience studies have shown that the associational cortices of the brain fire wildly while we are at rest (from deliberate conscious efforts). The brain keeps crafting associations even when we are not deliberately working on a problem (Andreasen, 2006). These studies suggest that incubation plays a crucial role in creative insight (e.g. Kekule, Newton, Einstein, Poincare, etc., see Runco, 2007), a view that would not sit well with individuals unable to delay their want of a technological "fix". Second, in the same fashion that networks help spread innovations within communities, a security breach on the other hand could result in a leak of top secret information related to a new product offering prior to market introduction (Christakis & Fowler, 2009). Worse, let's take the case of Google and the Republic of China. Google in 2010 discovered that the accounts of US, Chinese and European Gmail users, recognized as advocates of human rights in China, appeared to be the principal targets of attacks.

The 3W landscape offers a fertile ground to extend creative efforts of teams and organizations over space and time. Yet we have become subject to such a saturation of stimuli and request for attention that consequently little space and time is made available for deep creative cognition and incubation processes. Organizations wanting to take advantage of ubiquitous technology must strike a balance between the demands of efficiency to streamline the creative process and effectiveness to allot the time for groups or individuals to engage without distraction in deep processes conducive to creativity.

## The Hyper-Reality

The vision of a seamless digital landscape has been the inspiration and fascination of many movie directors, who have anticipated some of the most incredible immersive technologies decades in advance of their commercialization in real life. For example, in George Lucas' 1977 Star Wars movie, R2-D2 relays a 3D holographic message from Princess Leia to young Luke Skywalker; that scene planted the seeds for Cisco's telepresence technology into the collective unconscious and into the R&D labs of major players in the computer and communication technology business. Today telepresence, when coupled with Musion's holographic technology, is now available as a commercial product to organizations worldwide providing an enhanced and immersive remote collaboration experience. Telepresence, Virtual Worlds, Virtual Reality, Augmented Reality, and Tangible Digital Media (technologies that we have seen in many of Hollywood blockbusters) are all byproducts of the seamless and frictionless integration of what we hold to be part of the natural world and that of the digital world. This gives rise to a *Hyper-Reality*.

Hyper-reality, as we coin it, blends layers of digital information into our field of perception to heighten our experience of life. There are two kinds of technology that come to mind. One technology invites us to get immersed into digital realms such as Virtual Worlds. This technology recreates vivid and immersive realities for us to explore and are detached from our physical reality. Another technology brings the digital world into our physical reality such as Augmented Reality or Tangible Media devices. This new socio-technical landscape, which is made out of a rich blend between matter and bytes, offers significant opportunities to harness creativity and innovation.

### Opportunities within Virtual Worlds

The Hyper-Reality trend has an effect on every element of the systems view of creativity. For the creative person, Virtual Worlds (VW) such as Second Life (SL) offer opportunities to engage in pro-creative behaviors that otherwise would have been deemed as too risky to exhibit in real life. Second Life enables its users, called residents, to interface with each other through an avatar, a three dimensional depiction of a user himself. For example, it is unlikely that individuals working on a team would agree to dress up in costumes for a strategic planning meeting as a means to add a layer of playfulness; playfulness and humor have been associated with environments conducive to creativity and creative output (Ekvall & Tångeberg-Andersson, 1986; Brown, 2009). However, because avatar customization is a common practice in VW's, individuals meeting in such a virtual venue can easily experiment with their appearance by bringing that element of heightened playfulness to the online interaction. On a deeper level, individuals interacting in a VW may be more willing to share their ideas and opinions than if interacting face-to-face. It is not uncommon for some individuals to withdraw from a creative collaboration in fear of ridicule or muting effects stemming from highly extroverted members of a team (VanGundy, 1983). Ho and Mcleod's (2009) study of individuals' willingness to express their opinion showed that subjects were more inclined to express their opinions and points of view when interacting through computer mediated interfaces than in face-to-face interaction. This was especially so when the context was emotionally laden. One noteworthy aspect of this study involved maintaining the anonymity of the subjects. Most creative collaborations held in organizations are not anonymous, notwithstanding crowd sourcing practices. However, we hypothesize that virtual representations would provide a moderately safe haven for collaborators to express themselves free of worry and irreverence than compared to a face-to-face modality. For introverts, it would appear to level the playing field thus producing a more balanced contribution of ideas per team member.

As it relates to the creative environment, we discuss four platforms, namely Virtual Worlds, Augmented Reality, Virtual Reality and Tangible Media. Virtual worlds such as SL and Teleplace enable creative collaboration in a stimulating platform. For example, Uribe and Cabra (2010) facilitated a creative problem session in SL involving members of a Fortune 500 company. A survey of user experience suggested participants greatly enjoyed their experience. They found the virtual environment imaginative, playful and inspiring and concluded that their experience helped produce novel and useful associations during ideation. Such experiences are quite difficult to design in real life. A setting like Epcot or the Guggenheim museum is ideal as they provide an enriching experience. However, they are settings that cannot be accessed on demand unless they were planned as a costly offsite retreat. Within a virtual simulator it is possible to design specific settings that fit each stage of the creative process. A set design for a clarification phase of the creative process may be filled with facts related to context and questions to prompt participants to ask many questions and engage in deep inquiry. A set design for an ideation phase may contain imaginative visuals and fantastic scenery that occasions participants to use their imagination to craft novel and useful options to address a challenge.

Virtual Worlds allow an individual to customize environment experiences and share them in real-time with peers. For example, in SL collaborations one individual may set the virtual environment to simulate a sunset, with warm ambient light, while another individual might participate in the simulation as if it were midday, with a brighter ambient light. Thus each individual sets the environment to his or her preference to maximize creative engagement or to fit environment to person.

From a creative process perspective, the hyper-reality offers exciting opportunities to leverage creative discovery. First, and because of the detailed level of realism that virtual simulations offer, Virtual Worlds offer one of the most immersive platforms among remote synchronous collaboration technologies. As reported in Uribe and Cabra (in press), individuals feel like they are in close proximity to each other. In terms of specific activities and sub processes within the creative process, we see two core areas where the use of virtual world venues and the manipulation of digital data will increase creative effectiveness. First, virtual platforms allow us to enhance ideation initiatives. Our ideas now can be expressed digitally and in 3D, which add a new layer to the traditional mechanisms of ideation comprising the use of scribbled post it's and quick sketches. 3D models convey a more vivid picture of an idea than a 2D image. As such, it easier for team members to fully grasp a concept and then build on it thus further driving ideation. In addition, ideas can be expressed in a myriad of digital formats like video footage, pictures, and sounds for richer creative expression. Second, these technologies grant us an unprecedented ability to accurately prototype products and experiences (Puccio, Cabra, Fox, & Cahen, 2010).

Not too long ago, CAD software (Computer Aided Design) was highly expensive and required considerable technical knowledge to use. Today, CAD technology has been simplified and is massively accessible as a result of the development and distribution of open source software. In fact, Second Life can be viewed as a massive and collaborative CAD environment. Everything that exists in Second Life, from buildings, to clothes and objects has been designed and created by its users. From a product perspective, virtual models can be manipulated with our mouse in moving concepts from abstract conceptualization to concrete materialization. Moreover, with the availability of 3D printers we can now bring those models to reality at a very low cost in comparison to doing so in a fully equipped design shop. It is worth noting that prototyping is not limited to physical products. In 2006, the Starwood hotel chain prototyped their renewed concept for the Aloft hotel brand in Second Life. They built a virtual replica of their hotel concept and with that, invited people through an avatar presence to experience the architecture, the design check-in procedures, information accessibility points and security. This simulation yielded a large reserve of feedback that was later screened and incorporated into the hotel design. These capabilities, we imagine, will strengthen the refinement and verification phases of the creative process.

From a creative environment perspective, we suppose that hyper-realities will make work more rigorous and in the process transform the workspace, physical or virtual, into a dynamic place where organizational members will be invigorated to move back and forth among challenging and equally stimulating projects. Environments will be tailored to accommodate an organizational member's content preferences. Environments will be situational and iterative, learning in sync with users ongoing acquired heuristics. Overall, environments will engage us in a holistic fashion tapping into the many regions of the brain related with the human senses.

**Opportunities within Virtual Reality**

Virtual reality (VR) is a computer simulated setting. An example is the cockpit simulator, which is used extensively to train personnel in the military and in the airline industry. Virtual reality is now used for all kinds of settings or for producing various simulated experiences. Heim (1993) listed seven areas of VR comprising: simulation, interaction, artificiality, immersion, full-body immersion, network communication, and telepresence.

Bullinger, Müller-Spahn and Rößler (1996) identified potential benefits of VR. One immediate benefit is the promotion of flow. It would appear that when groups are using VR, the simulated experience promotes an increase in concentrated attention. This happens because the rich and dynamic setting marginalizes cognitive distractions. Another benefit is its flexibility. In computer simulated settings groups can intuitively and iteratively change the elements of virtual prototypes. It would also appear, that when individuals are permitted to use their visual, motor, and acoustic sense, they tend to remember more and likely to build more knowledge structures in less time. Because VR is visual, groups are able to improve their communication and understanding of concept proposals. Their interaction is enhanced is more impressionable.

**Opportunities within Augmented Reality**

While virtual reality requires donned equipment that facilitates the creation of a simulation not of the real world *per se* or in real time, Augmented Reality does not. Augmented Reality (AR) is a technology that permits live, direct or indirect examination of a tangible setting comprising the real-world. Reality can be augmented with audio, and someday, with a sense of touch. For example, imagine a football trading card, which is a physical element. Then, the football player emerges as a 3D object. It is also interactive as you can study the performance statistic of this football player, page to the next informational bit, and so on. AR, in short, merges 3D objects into video (real and virtual) in real time.

Streitz, Geißler Holmer, Konomi, Müller-Tomfelde, Reischl, Rexroth, Seitz and Steinmetz (1999) maintain that the real world should be more of a line to information than what is experienced in the present. They assert that virtual worlds should augment our awareness and interactions with the physical world. Nissan uses AR for design simulations. Vehicles are streamed into video as 3D objects that are then maneuvered across physical terrains. This capacity permits visualizing prototypes and changing them by component, color, or shape based on insights that were generated through the maneuvers, cutting down production time, costs, and time-to-market.

**Opportunities within Tangible Media**

Tangible Media is a technology that makes information tangible. For example, blocks that look like LEGOs can be joined, enlarged, colors changed through hand gestures. These blocks can be copied and the process used to construct some sort of prototypes can be replayed for analysis and further insight. These forms are not limited to blocks. Tangible landscapes or surfaces permit simulations of natural climate phenomena such as hurricanes. Typically TM is operated on a table top. Back, Matsumoto and Dunnigan (2009) designed a Tangible Media technology to simulate the use of Post-Its™ called Post-Bits. This media form provides groups with the ability to flex, move, sort, stack e-paper in the same way as seen in the real world. These Post-Bits were also designed to capture digital information. They explain how this form of tangible media can be used for stages of the creative process to tackle all kinds of challenges.

**Challenges**

The Hyper-Reality suffers from one major drawback, a lack of *intuitiveness*. Software and user interfaces of a virtual world, virtual reality and augmented reality technologies don't offer yet an instinctive and stress-free experience. It is not uncommon for people working in computer mediated or computer aided creative collaboration to lose time and energy struggling with the technology instead of struggling to overcome their blocks to creativity. Cumbersome user interfaces coupled with unstable technological platforms distract people, disrupt creative flow and undermine creative effectiveness. On one hand, platforms that are intuitive and simple (like wikis and chat clients), lack immersive quality. On the other hand, rich and engaging platforms lack intuitiveness as they are inundated with many features. Generally speaking digital mediums are still short of fluidity compared to flipchart paper, markers, and Post-Its. Although multi-touch technologies were designed to naturalize our interaction with digital media, that is to manipulate digital media with everyday gestures as we do with analog media, users even now prefer analog mediums as these pose no learning challenges (Brown & Katz, 2009).

In order to make the hyper-reality a truly engaging experience, human communications need to function holistically. Face-to-face interaction is richer and far more complex than a virtual avatar representation that we commonly find in computer-mediated interactions. They lack the affective cues, body language and non verbal feedback loops that frame and complete the gestalt of human interpersonal dynamics. Currently, avatars cannot frown, laugh or show bewilderment. This same deficit exists in text messaging, emailing or any other similar computer-mediated exchanges. We have all been there. What we intended to convey in an email was misunderstood, taken out of context, or misread by the receiver. The feedback loops that in any other case would have permitted the sender of an email to clarify his message were not available. A similar problem occurs in some computer mediated collaborations. This challenge has fueled a new area of research and development broadly referred to as *Affective Computing.* For example, de la Torre (2005) developed machine-learning algorithms that automatically characterize and classify human behavior from multimodal data using video recording, biofeedback mechanisms, and physical activity data inspectors. These measurement tools permit affective and emotional information to be streamlined into virtual world avatar expression and other emotion type visualizations that will give our distant collaborators cues to our emotional reactions in real time. We anticipate that systems such as the ones already described will allow us to customize the level of emotional expression that we opt to convey to our collaborators with a higher degree of accuracy than say, what is found in face-to-face interactions. Because creative collaboration mostly involves groups of people coming together to design solutions or solve complex problems, effective management of group dynamics and emotional states will play a critical role in leveraging group creative competencies (Amabile et al., 2005; Zhou & George, 2003).

If we really want to harness the power of remote computer-mediated creative collaborations, our ability to express and modulate emotion through computer interfaces needs to be dramatically enhanced for optimal individual and group creative performance. In addition, and especially for long term collaboration projects, research shows that the opportunity to meet collaborators face-to-face prior to the computer mediated interaction enhances the subsequent effectiveness of remote computer mediated collaborations (Hill et al., 2009).

Hyper-Reality provides a fertile landscape to leverage individual and group creativity. The ability to participate in immersive and inspiring environments that can transform our ideas into aesthetically pleasing sketches or 3D prototypes, we expect will streamline the innovation processes. However, to exploit the full potential of these technologies a frictionless experience is needed. Additionally, the more we embed affective computing components, the richer and more powerful the creative experience will be. The holistic attributes of the creativity phenomena therefore call upon the use of both cognitive and emotional faculties (Amabile et al. 2005; Puccio et al. 2007) and therefore, we must engage and be engaged in the Hyper-Reality but in a holistic way.

## The Superorganism

Most of our efforts in understanding the nature of our species have been dedicated to unveiling the factors that make each one of us unique rather than to understanding the structure, dynamics and influence of the complex social networks in which we participate. The latter has been far more complex and technically challenging to study (Christakis & Fowler, 2009). However, with affordable communication technologies, researchers have been able to assess the influence that social networks exert on individuals. Findings from several research efforts in the field of social networks have shifted our perspective from the archetype of the *homo-economicus* (self-centered individual that maximizes his own gain), to that of a networked individual, who shares, influences and is predisposed considerably by his peers (Christakis & Fowler, 2009).

Technology has leveraged a human trait, and this is our tendency to embed ourselves in social structures. For that reason, networks have expanded exponentially to change the social landscape from a few geographically concentrated networks to a massively distributed global social fabric, the Superorganism. Today's creative discoveries are byproducts of a distributed collective effort rather than that of the individual genius. With the aid of computer mediated social networks, individuals are constantly sharing ideas, perceptions and opinions. They comprise a very rich melting pot that fuels creativity and innovation in varied domains. As the visionary thinker Alvin Toffler predicted in his book the Third Wave (1980) we are transitioning to an age where boundaries that were once established in the industrial revolution between producers and consumers are fading away to gradually give rise to the socio-economical archetype of the *Prosumer*. Thus a new social contract has been established in which it is not enough for people to be passive recipients of goods and services, but where there is a participatory dialogue, articulation and co-creation of value between different individuals and sectors of society (Brown & Katz, 2009).

One of the clearest and earliest manifestations of this new socio-economical order is open source collaborative projects. Open source phenomena emerged from the software development discipline. Later, software development evolved and began including a process of remote collaborations. Potential users and developers were give access to program code and freedom to independently modify software. Examples of open source products and services are Linux based operating systems (e.g. Ubuntu), Wikipedia online encyclopedia, and the Mozilla Firefox web browser. Open source collaboration has transformed and revolutionized collaborative creative work. Creativity has moved beyond just R&D labs and design teams. Open source collaboration now functions organically to foster creative talent in all corners of the globe and by the masses. For example, we would be hard pressed to name the original creators of the Mozilla Firefox web browser. In recent years, the open source phenomenon has moved outside of the software industry into domains of education, design and manufacturing thus giving rise to a closely related practice, that of *crowd-sourcing*, which involves reaching out to the wisdom (and talent) of crowds by inviting *consumers* to co-create with the organization (Howe, 2008). Comet Skateboards recently mined a community of skaters worldwide to collaboratively design a long-board skateboard. Within a few weeks Comet was able to complete a few stages of innovation namely, receive hundreds of idea proposals, arrive at a consensus, and test pilot 14 prototypes (http://www.cometskateboards.com/ride/FSM/indexpand.php, Retrieved January 11th, 2010). In the coming years we are going to see many more examples like the one described above in very diverse fields and with different degrees of mining.

**Opportunities**

Sourcing practices foster creativity because its participants come from different cultures, domains and backgrounds. They reflect many personalities, cognitive styles, and various approaches to creative problem-solving. When managed appropriately, diversity can serve as a rich resource for novel and valuable ideas. When managed effectively, diverse groups can neutralize toxic personalities that prime dysfunctional group dynamics. People who gravitate toward open or crowd sourcing initiatives are passionate and intrinsically motivated, as these endeavors typically do not offer remuneration or even public recognition. As research has shown, intrinsic motivation is a powerful predictor of creative effectiveness (Amabile, 1985; Hennessey, 1999).

From a process perspective, it is evident that having many people with an array of backgrounds is conducive to generating many and varied ideas (Gryskiewicz, 1999). It is not uncommon for open source and crowd-sourced projects to have hundreds of collaborators. If each collaborator were only to submit one idea, the project would still benefit from hundreds if not thousands of other ideas. From a divergent thinking perspective, the more ideas, the greater the probability of finding a few very innovative ideas as quantity yields quality of ideas (Parnes & Noller, 1972; Puccio et al., 2007). Additionally, a robust pool of ideas greatly increases the chances of producing novel associations because they stem from combining parts of individual ideas within that pool (Ward 2007). Open sourcing is a great mechanism for peer review. Feedback loops allow for improvement iterations. As development responsibilities are shared, the work load is distributed in a way that fits tasks with collaborators' skills and expertise.

From a product perspective, open sourcing increases the quality of ideas because contributors are constantly manipulating and incrementally improving them. As a consequence of having many contributors, products are introduced to the marketplace faster. Open sourced products drive competitor costs down because they are free. They also are more stable because they are not vulnerable to vendor influences (i.e., selling the product to another organization without consent, changing the features of a product or a service offering). Along this line, open sourced products are flexible as they can be self-developed to accommodate personal needs.

From person and environment perspective, open sourcing creates a sense of belonging. In a study by Cabra, Talbot and Joniak (2005), sense of belonging was determined to be conducive to a creative climate. Open sourcing excites because collaborators have a chance to unlock and create new knowledge that is produced by other highly talented collaborators. They find the work intellectually stimulating. Open sourcing also presents an opportunity for collaborators to make a name for themselves as their work helps to increase their professional status within the community (Lakhani & Wolf, 2005).

## Challenges

The Superorganism is not free of problems. One challenge is the management of people. Another is the management of voluminous data. Large data sets overwhelm, complicate and are hard to distill. Wurman (1989) asserted that "information anxiety is produced by the ever-widening gap between what we understand and what we think we should understand. *Information anxiety* is the black hole between data and knowledge. It happens when information doesn't tell us what we want or need to know." (p. 34). Within this context, ideas run the risk of getting lost. Therefore, the ability to mine, screen, and quickly select data we surmise will be a critical creativity skill of the 21st century (Puccio & Cabra, 2009).

Quite often collaborators drift apart or find it difficult to keep a project on track. Collaborators are independent and sensitive to projects that contribute to their personal developmental needs. Additionally, open source projects are subject to holes attributed to collaborative inertia. For example, Sugarlab's open source Sugar Learning Environment is a operating system that nurtures creative collaboration among young learners. Unfortunately, the dearth of system and user support has precluded a wider dissemination into school systems worldwide. Such a problem is mitigated in the case of crowd-sourcing initiatives, as it is the host organization that controls to a great extent the direction and objective of such endeavors. Nonetheless, in this kind of creative collaboration the host organization must ensure that the invitation to the community is genuine and truly participatory. In other words, when engaging in crowd-sourcing practices, an organization needs to be aware that some control both in the process and final product has to be relinquished to the community of collaborators.

In these more porous structures, organizational life will be deeply affected by openness and mass collaboration. For example, leadership will be less and less determined by power structure but by merits of ideas and contributions to the network; as soon as an individual's idea usefulness or contribution declines, leadership will migrate to other network nodes on the rise (Hammel, 2009). This notion of leadership by merit is still quite new and only few organizations (e.g. Design firm IDEO) have adopted such loose and fluid leadership practices. The scope and extent to which individuals exercise creative leadership, which involves in part the role of building an environment that nurtures the emergence of novelty that is useful (Puccio et al., 2007), will also shift. Fulfilling this role is already made difficult for leaders working in close and relatively small organizations; it will certainly prove even more challenging in a remote collaborative environment that is also expansively distributed and dauntingly massive. Another organizational challenge includes the effective regulation of organizational openness. Uzzi and Spiro's (2005) study on networks suggest that organizations are better served when their structures are balanced with strong and weak network ties. They posit that a strong network alone would likely produce familiar ideas, while weak ties would likely produce more remote associations since weak ties are based on flatter and more diverse hierarchies of human resources. Moreover, too weak ties may lead to distrust among members of an organization to freely express themselves creatively. Conversely, too many strong ties would suggest a closed structured network, which would isolate the organization from environmental influences and likely trap it in old habits. The balance between strong ties and weak ties grant enough structure for trust to develop among members of the organization and at the same time make the organization more open and connected to other networks.

The biggest challenge to the collaborative model is moving projects from conceptualization to materialization. Creative collaboration may be efficient in the divergent phases of the creative process, yet in the convergent phases, where decisions need to be made and action items need to be executed may prove to be quite challenging for two specific reasons. First, it is difficult to get hundreds of collaborators to arrive at a consensus around a short-list of ideas. Second, idea ownership is distributed unevenly and in so many places, that consequently momentum is likely to lose steam. Crowd-sourcing practices may tackle the above shortcomings as the locus of decision making control rests on the host organization. But when this arrangement does not exist, organizations will need to develop and encourage resolve as exposure implies opening part of their intellectual property to the public and vulnerability to security breaches.

The Superorganism is by far the most important and pervasive of the discussed trends. The massive sharing and presence in social network engines as seen in Facebook, Linkedin and Twitter are examples of how we have decided to take our social entrenchment to a larger scale with the support of computer and communication technologies. As discussed above, this new socio-technical landscape opens vast collaborative creative possibilities that will be a core driver of technical and social innovation. Citizens of the 21st century are no longer bound by the scope of their geographic location and therefore they can now collaborate in creative endeavors locally and globally.

The Superorganism is shaping a new economy and social order. The open source phenomenon described previously is far more interesting for its sociological and economical implications than its technical feats. It is unprecedented for individuals to collaborate in the absence of their traditional views of profit, power and recognition so engrained in our current socioeconomically system. Despite the lack of these drivers, open source communities and projects have been able to develop and produce products as innovative as those developed by for-profit organizations. It is yet to be seen if the open source phenomenon yields a new 21st century economical and innovation model that challenges the 250 years of the industrial profit-driven model. However, the business arena is not the only ground where the Superorganism is shaping a new society. Increasingly, institutes of higher learning are posting their course material online. Take MIT's open courseware (OCW). Everything from course contract, to downloadable podcasts of class lectures and slides have been shared. Universities that place more energy into protecting intellectual property than into sharing it are being outpaced by other academic institutes. Sharing encourages pedagogical improvements. Since the start of MIT's initiative, the OCW site has received 40 million visits. 30 million comprised visitors from around the world. Educators around the world struggle with low budgets and cut programs. Students hungry for information often don't have access to it. OCW fills these gaps. An ancillary benefit of course is great public relations.

Universities and research centers have a greater chance of tackling the challenges of the 21st century by articulating, actively sharing, and networking with other research centers, organizations and individuals than by protecting themselves in silos. In this vein, we are seeing the emergence of the first open sourced universities and programs as well as open sourced curriculums to be freely used and modified by the teaching-learning community.

Open source creativity is expected to move outside the software industry into areas such as manufacturing, services, health and education. It also anticipated that organizational life will be deeply affected by this phenomenon. For example, leadership will not be determined by a power structure but by merits one's contributions to the network; as soon as an idea loses its appeal, leadership will then migrate to other network areas in search of new and improved ideas. Crowd sourcing practices we expect will be commonplace and as a result, the once clear-cut boundaries between producers and consumers, firms and customers will blur.

As exciting as these visions are, one sobering obstacle remains and reminds how easily we can get carried away with the next technological introduction. As Cerf and Schultz (2007) compassionately asserted, "Let me pause to point out the obvious. For about 2 billion people today, the Internet is NOT a reality; it is yet a distant dream or perhaps entirely unknown. Electricity is still unknown or unavailable or unreliable. Sanitation, housing, food, water and education are in limited supply. For those parts of our global society, the Internet and its benefits seem far beyond reach. We must not give up pursuit, however idealistic, of access for all." (p. 2)

## REFERENCES


Amabile, T. M. (1985). Motivation and creativity: Effects of motivational orientation on creative writers. *Journal of Personality and Social Psychology*, *48*(2), 393–399. doi:10.1037/0022-3514.48.2.393

Amabile, T. M., Barsade, S. G., Mueller, J. S., & Staw, B. M. (2005). Affect and creativity at work. *Administrative Science Quarterly*, *50*, 367–403. doi:10.2189/asqu.2005.50.3.367

Andreasen, N. C. (2006). *The creative brain: The science of genius*. New York: Plume.

Back, M., Matsumoto, T., & Dunnigan, A. (2009). Protying a tangible tool for design: Multimedia e-paper sticky notes. *Artificial Intelligence for Engineering Design, Analysis and Manufacturing*, *23*, 263–274. doi:10.1017/S0890060409000237

Brown, S. (2009). *Play: How it shapes the brain, opens the imagination, and invigorates the soul*. New York: Avery.

Brown, T., & Katz, B. (2009). *Change by design: How design thinking transforms organizations and inspires innovation*. New York: Harper Business.

Bullinger, H. J., Müller-Spahn, F., & Rößler, A. (1996). Encouraging creativity–support of mental processes by virtual experience. In *Virtual Reality World: IDG conferences and seminars*. Retrieved February 3, 2010, from http://www.creativethinkingnetwork.com/ dmdocuments/Encouraging%20 Creativity.pdf

Cabra, J. F., Talbot, R. J., & Joniak, A. J. (2005). Exploratory study of creative climate: A case from selected Colombian companies and its implications on organizational development. *Cuadernos de Administración*, *18*(29), 53–86.

Cerf, V., & Schutz, C. (2002). *Visions 2020: Transforming education and training through advanced technologies.* Washington, DC: U.S. Department of Commerce. Retrieved March 8, 2007, from http://www.technology.gov/reports/TechPolicy/2020Visions.pdf

Christakis, N. A., & Fowler, J. H. (2009). *Connected: The surprising power of our social networks and how they shape our lives*. New York: Little, Brown and Co./ Hachette Book Group.

Cooper, T., Hofheinz, P., & Purdy, M. (2009). *Skills for the future*. Retrieved November 17th, 2009 from http://www.accenture.com/NR/rdonlyres/55C4D829-CB2B-4553-9BC3-10F8D-BC4987C/0/SkillsfortheFuture.pdf.

Council for Research Excellence. (2008). *Ground breaking study of video viewing finds younger boomers consume more video media than any other group.* Retrieved from http://www.researchexcellence.com/vcm_brief.pdf

Csikszentmihalyi, M. (1990). *Flow: The psychology of optimal experience*. New York: Harper & Row.

De la Torre, F., Gross, R., & Baker, S. (2005). Representational oriented component analysis (ROCA) for face recognition with one sample image per training class. IEEE *Conference on Computer Vision and Pattern Recognition*.

Ekvall, G., & Tångeberg-Andersson, Y. (1986). Working climate and creativity: A study of an innovative newspaper office. *The Journal of Creative Behavior*, *20*(3), 215–225.

Florida, R. L. (2008). *Who's your city? How the creative economy is making where to live the most important decision of your life*. New York: Basic Books.

Gryskiewicz, S. S. (1999). *Positive turbulence*. San Francisco: Jossey Bass.

Hammel, G. (2009). Empowering natural leaders in 'Facebook generation' ways. Retrieved from http://blogs.wsj.com/management/2009/05/18/empowering-natural-leaders-in-facebook-generation-ways/

Heim, M. (1993). *The metaphysics of virtual reality*. Oxford: Oxford University Press.

Hennessey, B. A. (1999). Intrinsic motivation, affect and creativity. In Russ, S. (Ed.), *Affect, creative experience, and psychological adjustment* (pp. 77–90). Philadelphia: Brunner/Mazel.

Hill, N. S., Bartol, K. M., Tesluk, P. E., & Langa, G. A. (2009). Organizational context and face-to-face interaction: Influences on the development of trust and collaborative behaviors in computer-mediated groups. *Organizational Behavior and Human Decision Processes*, *108*(2), 187–190. doi:10.1016/j.obhdp.2008.10.002

Ho, S. S., & McLeod, D. M. (2008). Social-psychological influences on opinion expression in face-to-face and computer-mediated communication. *Communication Research*, *35*(2), 190–207.

Howe, J. (2008). *Crowdsourcing: Why the power of the crowd is driving the future of business*. New York: Crown Business.

Jackson, T., Dawson, R., & Wilson, D. (2003). Reducing the effect of email interruptions on employees. *International Journal of Information Management*, *23*, 55–65. doi:10.1016/S0268-4012(02)00068-3

Kelley, T., & Littman, J. (2001). *The art of innovation: Lessons in creativity from IDEO, America's leading design firm*. New York: Currency/Doubleday.

Kurzweil, R. (1999). *The age of spiritual machines: When computers exceed human intelligence*. New York: Viking.

Lakhani, K. R., & Wolf, R. (2005). Why hackers do what they do: Understanding motivation and effort in free/open source software projects. In Feller, J., Fitzgerald, B., Hissam, S., & Lakhani, K. (Eds.), *Perspectives on free and open source software*. Cambridge, MA: MIT Press.

Malafouris, L. (2009). Neuroarchaeology: Exploring the links between neural and cultural plasticity. *Progress in Brain Research, 178*, 253–261. doi:10.1016/S0079-6123(09)17818-4

McLuhan, M. (1964). *Understanding media: The extensions of man*. New York: McGraw-Hill.

Palfrey, J., & Gasser, U. (2008). *Born digital: Understanding the first generation of digital natives*. New York: Basic Books.

Parnes, S. J. (1987). The creative studies project. In Isaksen, S. G. (Ed.), *Frontiers of creativity research: Beyond the basics* (pp. 156–188). Buffalo, NY: Bearly Limited.

Parnes, S. J. (1991). Creative problem solving and visionizing. In Parnes, S. J. (Ed.), *Source book for creative problem-solving: A fifty year digest of proven innovation processes*. Buffalo, NY: Creativity Education Foundation.

Parnes, S. J., & Noller, R. B. (1972). Applied creativity: The creative studies project: Part 2–Results of the two-year program. *The Journal of Creative Behavior, 6*, 164–186.

Perkins, P. S. (2008). *The art and science of communication: Tools for effective communication in the workplace*. Hoboken, N.J.: John Wiley & Sons.

Prensky, M. (2001). Digital natives, digital immigrants part 2: Do they really think differently? *On The Horizon - The Strategic Planning Resource for Education Professionals, 9*(6), 1-6.

Puccio, G. J., & Cabra, J. F. (2009). Creative problem solving: Past, present and future. In T. Rickards, H. A. Runco. & S. Moger (Eds.), *The Routledge companion to creativity*, (pp. 327-337). Oxford: Routledge.

Puccio, G. J., Cabra, J. F., Fox, M. J., & Cahen, H. (2010). Creative problem-solving, history and future. *Artificial Intelligence for Engineering Design, Analysis and Manufacturing, 24*, 153–159. doi:10.1017/S0890060410000028

Puccio, G. J., Murdock, M., & Mance, M. (2007). *Creative leadership: Skills that drive change*. Thousand Oaks, CA: SAGE Publications.

Runco, M. A. (2007). *Creativity theories and themes: Research, development and practice*. Burlington, MA: Elsevier Academic Press.

Sio, U. N., & Ormerod, T. C. (2009). Does incubation enhance problem solving? A meta-analytic review. *Psychological Bulletin, 135*(1), 94–120. doi:10.1037/a0014212

Stevens, G. A., & Burley, J. (1997). 3000 raw ideas equals 1 commercial success. *Research Technology Management, 40*(3), 16–27.

Streitz, N. A., Geißler, J., Holmer, T., Konomi, S., Müller-Tomfelde, C., Reischl, W., et al. (1999). i-LAND: An interactive landscape for creativity and innovation. In *Proceedings of the ACM Conference on human Factors in Computing Systems*, (pp. 120-127). Pittsburg, PA: ACM Press.

Tapscott, D. (2009). *Grown up digital: How the net generation is changing your world*. New York: McGraw-Hill.

Tassoul, M., & Buijs, J. (2007). Clustering: An essential step from diverging to converging. *Creativity and Innovation Management, 16*(1), 16–26. doi:10.1111/j.1467-8691.2007.00413.x

Taylor, C. (2009). The creative industries, governance and economic development: A UK perspective. *GeoJournal Library, 98*, 153–166. doi:10.1007/978-1-4020-9949-6_10

Toffler, A. (1980). *The third wave*. New York: Morrow.

Torrance, E. P. (1987). Teaching for creativity. In Isaksen, S. G. (Ed.), *Frontiers of creativity research: Beyond the basics*. Buffalo, NY: Bearly Limited.

Uribe, D. E., & Cabra, J. F. (2010). Creative problem-solving in Second Life: An action research study. *Creativity and Innovation Management, 19*(2), 167–179. doi:10.1111/j.1467-8691.2010.00550.x

Uzzi, B., & Spiro, J. (2005). Collaboration and creativity: The small world problem. *American Journal of Sociology, 111*(2), 447. doi:10.1086/432782

VanGundy, A. B. (1983). Brainwriting for new product ideas: An alternative to Brainstorming. *Journal of Consumer Marketing, 1*(2), 67–74. doi:10.1108/eb008097

Ward, T. B. (2007). Creative cognition as a window on creativity. *Methods (San Diego, Calif.), 42*, 28–37. doi:10.1016/j.ymeth.2006.12.002

West, M. A., & Rickards, T. (1999). Innovation. In M.A. Runco & S.R. Pritzker (Eds.), *Encyclopedia of Creativity, 2*, 35-43. San Diego: Academic Press.

Wurman, R. S. (1989). *Information anxiety*. New York, NY: Doubleday.

Zhou, J., & George, J. M. (2003). Awakening employee creativity: The role of leader emotional intelligence. *The Leadership Quarterly, 14*, 545–568. doi:10.1016/S1048-9843(03)00051-1

## KEY TERMS AND DEFINITIONS

**Creativity:** The production and realization of ideas that are both novel and useful.

**Innovation:** The successful introduction and sustained adoption of creative outputs by a social unit.

**Technology:** An extension of human abilities that result in the practical application of knowledge.

**Socio-Technical Landscape:** The technological milieu that informs the perception of what's the ordinary reality in any given age and culture.

**Communication:** Process by which we attribute and share the meaning of our human experience.

**Virtual Worlds:** A synchronous ongoing network of people, represented as avatars, whose interactions are hosted and facilitated by networked computers.

**Augmented Reality:** Live direct or indirect view of a physical real-world environment whose objects and properties are enhanced by computer-generated imagery.

**Tangible Media:** User interface in which digital content is manipulated through direct physical interaction (i.e. touch).

## ENDNOTES

1 Although we do not reject Florida's (2008) assertion that physical place continues to determine much of a network's structure and properties, we maintain that physical space does not discriminate who receives access.

2 A Cloud is a metaphor for Internet space mediated computing.

3 Flaming is the deliberate act of insulting or creating dissent within a group through posting or writing messages on internet bulletin boards and message groups.

4 Ray Kurzweil,the father of term Singularity, predicted a moment in time when the first "super-intelligent" machine will arise, capable of improving on its own source code without human input.

# Compilation of References


Abrami, P. C., & Barrett, H. (2005). Directions for research and development on electronic portfolios. *Canadian Journal of Learning and Technology, 31*(3). Retrieved on February 5, 2006, from http://www.cjlt.ca/content/vol31.3/abrami.html

Ackermann, F. (1996). Participants perceptions on the role of facilitators using group decision support systems. *Group Decision and Negotiation, 5*(1), 93 519.

Adair, J. (1990). *The challenges of innovation.* England: The Talbot Adair Press.

Adams, R. (2006). Innovation measurement: A review. *International Journal of Management Reviews, 8*(1), 21–47. doi:10.1111/j.1468-2370.2006.00119.x

Adelman, L., Gualtieri, J., & Stanford, S. (1995). Examining the effect of causal focus on the option generation process: An experiment using protocol analysis. *Organizational Behavior and Human Decision Processes, 61*, 54–66. doi:10.1006/obhd.1995.1005

Agres, A., de Vreede, G. J., & Briggs, R. O. (2005). A tale of two cities: Case studies of GSS transition in two organizations. *Group Decision and Negotiation, 14*(4), 256–266. doi:10.1007/s10726-005-0315-6

Ahmed, P. K. (1998). Culture and climate for innovation. *European Journal of Innovation Management, 1*(1), 30–43. doi:10.1108/14601069810199131

Ahmed, P., Loh, A., & Zairi, M. (1999). Cultures for continuous improvement and learning. *Total Quality Management, 10*(4/5), 426–434.

Ahuja, G. (2000). Collaboration networks, structural holes, and innovation: A longitudinal study. *Administrative Science Quarterly, 45*(3), 425–455. doi:10.2307/2667105

Ajzen, I. (1991). The theory of planned behavior. *Organizational Behavior and Human Decision Processes, 50*(2), 179–211. doi:10.1016/0749-5978(91)90020-T

Alferes, V. R. (2002). *Programas e rotinas complementares do SPSS.* Retrieved January 20, 2010, from http://www.fpce.uc.pt/nucleos/niips/

Allard, S., Levine, K. J., & Tenopir, C. (2008). Design engineers and technical professionals at work: Observing information usage. *Journal of the American Society for Information Science and Technology, 60*, 443–454. doi:10.1002/asi.21004

Almeida, M. (2003). Educação, ambientes virtuais e interactividade . In Silva, M. (Ed.), *Educação On-line. Sao Paolo: Edições Loyola.*

Amabile, T. M. (1996). *Creativity in context*. Boulder, CO: Westview Press.

Amabile, T. M., Barsade, S. G., Mueller, J. S., & Staw, B. M. (2005). Affect and creativity at work. *Administrative Science Quarterly, 50*(3), 367–403. doi:10.2189/asqu.2005.50.3.367

Amabile, T. M. (1983). The social psychology of creativity: A componential conceptualization. *Journal of Personality and Social Psychology, 45*(2), 357–376. doi:10.1037/0022-3514.45.2.357

Amabile, T. M., Conti, R., Coon, H., Lazenby, J., & Herron, M. (1996). Assessing the work environment for creativity. *Academy of Management Journal, 39*(5), 1154–1184. doi:10.2307/256995

Amabile, T. (1997). Motivating creativity in organisations: On doing what you love and loving what you do. *California Management Review, 40*(1), 39–58.

Amabile, T. M. (1985). Motivation and creativity: Effects of motivational orientation on creative writers. *Journal of Personality and Social Psychology, 48*(2), 393–399. doi:10.1037/0022-3514.48.2.393

Amabile, T. (1988). A model of creativity and innovation in organizations . In Staw, B., & Cummings, L. (Eds.), *Research in organizational behaviour* (*Vol. 10*, pp. 123–167). Greenwich, CT: JAI Press.

Amaral, L. (1994). *Um referencial para o planeamento de sistemas de informação*. Unpublished doctoral dissertation, University of Minho, Portugal.

Ambrosini, V., & Bowman, C. (2001). Tacit knowledge: Some suggestions for operationalization. *Journal of Management Studies, 38*(6), 811–829. doi:10.1111/1467-6486.00260

Anderson, N. R., & West, M. A. (1998). Measuring climate for work group innovation: Development and validation of the team climate inventory. *Journal of Organizational Behavior, 19*, 235–258. doi:10.1002/(SICI)1099-1379(199805)19:3<235::AID-JOB837>3.0.CO;2-C

Anderson, N., & Gasteiger, R. M. (2007). Helping creativity and innovation thrive in organizations: Functional and dysfunctional perspectives . In Langan-Fox, J., Cooper, C. L., & Klimoski, R. J. (Eds.), *Research companion to the dysfunctional workplace: Management challenges and symptoms* (pp. 422–440). Cheltenham, UK: Edward Elgar Publishing.

Anderson, J. V. (1993). Mind mapping: A tool for creative thinking. *Business Horizons, 36*(1), 41-46. Retrieved October 9, 2008, from http://www.sciencedirect.com/science/article/B6W45-4FN1XTN-7/2/a433a975929052093f-e8aee1bf11b2e8

Andreasen, N. C. (2006). *The creative brain: The science of genius*. New York: Plume.

Andrews, F. M. (1975). Social and psychological factors that influence the creative process . In Taylor, I. A., & Getzels, W. (Eds.), *Perspectives in creativity* (pp. 117–145). Chicago: Aldine.

Andriopoulos, C., & Dawson, P. (2009). *Managing change, creativity & innovation*. London: Sage Publications.

Angle, H. L. (1989). Psychology and organizational innovation . In Van deVen, A. H., Angle, H. L., & Poole, M. S. (Eds.), *Research on the management of innovation: The Minnesota studies* (pp. 135–170). New York: Harper & Row.

Antunes, P., Ho, T., & Carriço, L. (1999). *A GDSS agenda builder for inexperienced facilitators.* (pp. 1-15). Paper presented at the 10th Euro GDSS Workshop, Copenhagen, Denmark.

Arabie, P., & Hubert, L. J. (1994). Cluster analysis in marketing research. In R.P. Bagozzi (Ed.), *Advanced methods of marketing research,* (pp.160-189). Basil: Blackwell.

Argyris, C. (1982). *Reasoning, learning and action: Individual and organizational*. San Francisco: Jossey-Bass.

Argyris, C. (1999). *On organizational learning*. Oxford: Blackwell Publishers.

Argyris, C., & Schön, D. (1978). *Organizational learning: A theory of action perspective*. Reading, MA: Addison-Wesley.

Arrow, K. (1962). The economic implications of learning by doing. *The Review of Economic Studies*, *29*, 155–173. doi:10.2307/2295952

Arthur, W. (1996). Increasing returns and the new world of business. *Harvard Business Review*, *74*, 100–109.

Arundel, A., Bordoy, C. & Kanerva, M. (2008). *Neglected innovators: How do innovative organizations that do not perform R&D innovate?* Pro Inno Europe Report: Innometrics, March 31.

Ash, J. (1998). Managing knowledge gives power. *Communication World*, *3*(15), 23–26.

Asimakou, T. (2009). The knowledge dimension of innovation management. *Knowledge Management and Research & Practise*, *7*(1), 82–90. doi:10.1057/kmrp.2008.34

Assink, M. (2006). Inhibitors of disruptive innovation capability: A conceptual model. *European Journal of Innovation Management*, *9*(2), 215. doi:10.1108/14601060610663587

Atkinson, R., & Andes, S. (2009). *The Atlantic Century: Benchmarking EU & US innovation and competitiveness*. Washington: The Information and Innovation Technology Foundation.

Back, M., Matsumoto, T., & Dunnigan, A. (2009). Protying a tangible tool for design: Multimedia e-paper sticky notes. *Artificial Intelligence for Engineering Design, Analysis and Manufacturing*, *23*, 263–274. doi:10.1017/S0890060409000237

Badaracco, J. L. (1991). *The knowledge link: How firms compete through strategic alliances*. Boston: Harvard Business School Press.

Baer, J. (1997). Gender differences in the effects of anticipated evaluation on creativity. *Creativity Research Journal*, *10*, 25–31. doi:10.1207/s15326934crj1001_3

Baer, J., & Kaufman, J. C. (2006). Creativity research in English-speaking countries . In Kaufman, J. C., & Sternberg, R. J. (Eds.), *The international handbook of creativity* (pp. 10–38). N.Y.: Cambridge University Press.

Baetz, W. G. (2003). *Organizational learning practices.* Doctoral dissertation, University of Albany, State University of New York. ProQuest Information and Learning Company.

Baker, F. B., & Hubert, L. J. (1976). A graph-theoretic approach to goodness-of-fit in complete-link hierarchical clustering. *Journal of the American Statistical Association*, *71*, 870–878. doi:10.2307/2286853

Balkundi, P., & Kilduff, M. (2005). The ties that lead: A social network approach to leadership. *The Leadership Quarterly*, *16*(6), 941–961. doi:10.1016/j.leaqua.2005.09.004

Bandow, R. (2009). *Balance scorecard and benchmarking*. Berlin: Fraunhofer-Gesellschaft.

Bandura, A. (2001). Social cognitive theory: An agentic perspective. *Annual Review of Psychology*, *52*(1), 1–26. doi:10.1146/annurev.psych.52.1.1

Baracskai, Z. (1998). *Profi problémamegoldó (Master of Problem Solving)*. Nyíregyháza: Szabolcs-Szatmár-Bereg megyei Könyvtárak Egyesülés.

Barret, H., & Carney, J. (2005). Conflicting paradigms and competing purposes in electronic portfolio development. Educational assessment. *LEA Journal*. Retrieved from http://eletronicportfolio.org/portfolios/LEAJournal-BarrettCarney.pdf

Barrett, F. D. (1978). Creativity techniques: Yesterday, today, and tomorrow. *Advanced Management Journal*, *43*(1), 25-35. Retrieved October 9, 2008, from http://search.ebscohost.com/login.aspx?direct=true&db=buh&AN=4606291&site=ehost-live

Barron, F., & Harrington, D. (1981). Creativity, intelligence, and personality. *Annual Review of Psychology*, *32*, 439–476. doi:10.1146/annurev.ps.32.020181.002255

Barron, F. B., & Harrington, D. M. (1981). Creativity, intelligence, and personality. *Annual Review of Psychology*, *32*, 439–476. doi:10.1146/annurev.ps.32.020181.002255

Barron, F., & Harrington, D. M. (1981). Creativity, intelligence, and personality. *Annual Review of Psychology*, *32*, 439–476. doi:10.1146/annurev.ps.32.020181.002255

Baruch, Y. (1999). Response rate in academic studies: A comparative analysis. *Human Relations*, *52*, 421–438. doi:10.1177/001872679905200401

Basadur, M. S. (1997). Organizational development interventions for enhancing creativity in the workplace. *The Journal of Creative Behavior*, *31*(1), 59–73.

Basadur, M. (2004). Leading others to think innovatively together: *Creative* leadership. *The Leadership Quarterly*, *15*(1), 103–121. doi:10.1016/j.leaqua.2003.12.007

Basadur, M., & Gelade, G. (2006). The role of knowledge management in the innovation process. *Creativity and Innovation Management*, *15*(1), 45–61. doi:10.1111/j.1467-8691.2006.00368.x

Basadur, M. S. (1997). Organizational development interventions for enhancing creativity in the workplace. *The Journal of Creative Behavior*, *31*(1), 59–73.

Basadur, M. S. (1999). *Simplex: A flight to creativity*. Buffalo, NY: The Creative Education Foundation.

Basadur, M. S., & Hausdorf, P. A. (1996). Measuring divergent thinking attitudes related to creative problem solving and innovation management. *Creativity Research Journal*, *9*(1), 21–32. doi:10.1207/s15326934crj0901_3

Basadur, M. S., Pringle, P. F., & Kirkland, D. (2002). Crossing cultures: Effects on the divergent thinking of Spanish-speaking South American Managers. *Creativity Research Journal*, *14*(3-4), 395–408. doi:10.1207/S15326934CRJ1434_10

Basadur, M. S., Pringle, P. F., Speranzini, G., & Bacot, M. (2000). Collaborative problem solving through creativity in problem definition: Expanding the pie. *Creativity and Innovation Management*, *9*(1), 54–76. doi:10.1111/1467-8691.00157

Basadur, M. & Paton, B. (1993). Using creativity to boost profits in recessionary times. *Business Strategies*, Jan-Feb., 14-19.

Basadur, M. S. (2000). *The economic, social and psychological outcomes of implementing a deliberate process of organizational creativity*. Working paper n° 100. McMaster University, Management of Innovation and New Technology Research Center.

Bassie, L. J. (1997). Harnessing the power of intellectual capital. *Training & Development*, *12*(51), 25–30.

Bates, R., & Khasawneh, S. (2005). Organizational learning culture, learning transfer climate and perceived innovation in Jordanian organizations. *International Journal of Training and Development*, *9*(2), 96–109. doi:10.1111/j.1468-2419.2005.00224.x

Batey, M., Chamorro-Premuzik, Y., & Furnham, A. (2009). Intelligence and personality as predictors of divergent thinking: The role of general, fluid and crystallized intelligence. *Thinking Skills and Creativity*, *4*(1), 60–69. doi:10.1016/j.tsc.2009.01.002

Baughman, W. A., & Mumford, M. D. (1995). Process-analytic models of creative capacities: Operations influencing the combination-and-reorganization process. *Creativity Research Journal*, *8*, 37–62. doi:10.1207/s15326934crj0801_4

BBC. (2006). *BBC annual report and accounts 2005/06*. BBC Publications.

BBC. (2008). *BBC annual report and accounts 2007/08*. BBC Publications.

BBC. (2009). *Commissioning*. Retrieved June 5, 2009, from http://www.bbc.co.uk/commissioning/tv/networ/wocc.shtml

Bell, M. A. (1982). Phases in group problem-solving. *Small Group Behavior*, *13*, 475–495. doi:10.1177/104649648201300405

Benson, J., & Clark, F. (1982). A guide for instrument development and validation. *The American Journal of Occupational Therapy.*, *36*(12), 789–800.

Bentley, T. (2008). Learning beyond the classroom. *Educational Management Administration & Leadership*, *28*(3), 353–364. doi:10.1177/0263211X000283008

Berge. (2004). Obstacles faced at various stages of capability regarding distance education in institutions of higher education: Survey results. Retrieved on January 22, 2010, from http://emoderators.com/barriers/hghred_stgs.shtml

Bergh, B. V., & Stuhlfaut, M. (2006). Is advertising creativity primarily an individual or a social process? *Mass Communication & Society*, *9*(4), 373–397. doi:10.1207/s15327825mcs0904_1

Bessant, J., & Caffyn, S. (1997). High-involvement innovation through continuous improvement. *International Journal of Technology Management*, *14*(1), 7–28. doi:10.1504/IJTM.1997.001705

Bichteler, J., & Ward, D. (1989). Information-seeking behavior of geoscientists. *Special Libraries*, *79*, 169–178.

Bilton, C. (1999). Risky business: The independent production sector in Britain's creative industries. *International Journal of Cultural Policy*, *6*(1), 17–39. doi:10.1080/10286639909358110

Bilton, C. (2007). *Management of creativity: From creative industries to creative management*. Oxford: Blackwell Publishing.

Bilton, C., & Leary, R. (2002). What can managers do for creativity? Brokering creativity in the creative industries. *International Journal of Cultural Policy*, *8*(1), 49–64. doi:10.1080/10286630290032431

Black, P. (1998). *Testing: Friend or foe? Theory and practice of assessment and testing*. London: RoutledgeFalmer.

Blackler, F., Reed, M., & Whitaker, A. (1993). Editorial introduction: Knowledge workers and contemporary organizations. *Journal of Management Studies*, (November): 851–862. doi:10.1111/j.1467-6486.1993.tb00469.x

Blake, P. (1998). The knowledge management expansion. *Information Today*, *1*(15), 12–13.

Boden, M. A. (Ed.). (2004). *The creative mind*. London: Routledge.

Borghini, S. (2005). Organizational creativity: Breaking equilibrium and order to innovate. *Journal of Knowledge Management*, *9*(4), 19–33. doi:10.1108/13673270510610305

Born, G. (2004). *Uncertain vision: Birt, Dyke and the reinvention of the BBC*. London: Secker and Warburg.

Bostrom, R. P., & Nagasundaram, M. (1998). Research in creativity and GSS. *Proceedings of the 31st Hawaii International Conference on System Sciences*. Los Alamitos: IEEE Computer Society Press.

Bouthillier, F. & Shearer, K. (2002). Understanding knowledge management and information management: The need for an empirical perspective. *Information Research, 8*(1).

Bragg, A. & Bragg, M. (2005). *Developing new business idea: The fast-track to creating viable new businesses for executives and entrepreneurs*. Edinburg: Pearson Education Limited.

Braybrooke, D., & Lindblom, C. E. (1963). *A strategy of decision*. New York: The Free Press.

Brennan, A., & Dooley, L. (2005). Networked creativity: A structured management framework for stimulating innovation. *Technovation*, *25*, 1388–1399. doi:10.1016/j.technovation.2004.08.001

Bridgstock, R. (2009). The graduate attributes we've overlooked: Enhancing graduate employability through career management skills. *Higher Education Research & Development*, *28*(1), 27–39. doi:10.1080/07294360802444347

Brief, A. P., & Downey, H. K. (1983). Cognitive and organizational structures: A conceptual analysis of implicit organizing theories. *Human Relations*, *36*, 1065–1090. doi:10.1177/001872678303601201

Briggs, R. O., Adkins, M., Mittleman, D., Kruse, J., Miller, S., & Nunamaker, J. F. Jr. (1999). A technology transition model derived from field investigation of GSS use aboard the USS Coronado. *Journal of Management Information Systems*, *15*(3), 151–195.

Briggs, R. O., de Vreede, G. J., & Nunamaker, J. F. Jr. (2003). Collaboration engineering with thinkLets to pursue sustained success with group support systems. *Journal of Management Information Systems*, *19*(4), 31–63.

Briggs, R. O., Kolfschoten, G. L., Lukosch, S. G., de Vreede, G. J., & Dean, D. (2010). *Facilitator in a box: CACE and PSS for rapid development of collaborative applications for high-value tasks.* (pp. 1-10). Paper presented at the Hawaii International Conference on System Science. Waikoloa, HI.

Brigthman, H. J., Elrod, R., & Ramakrishna, H. (1988). Matching problem diagnostic tools to managers' decision styles: A contingency approach. *Omega*, *16*, 1–9. doi:10.1016/0305-0483(88)90002-3

Britain, S. (2004). A review of learning design: Concept, specifications and tools. *A report for the JISC E-learning Pedagogy Programme, JISC*. Retrieved January 2010, from http://www.jisc.ac.uk/uploaded_documents/ACF83C.doc

Brown, J. S., & Duguid, P. (1991). Organizational learning and communities-of-practice: Toward a unified view of working, learning, and innovation. *Organization Science*, *2*, 40–57. doi:10.1287/orsc.2.1.40

Brown, A. (1998). *Organisational culture* (2nd ed.). Harlow, UK: Prentice Hall.

Brown, J. S., & Duguid, P. (1999). *The social life of information*. Boston: Harvard Business School Press.

Brown, S. (2009). *Play: How it shapes the brain, opens the imagination, and invigorates the soul*. New York: Avery.

Brown, T., & Katz, B. (2009). *Change by design: How design thinking transforms organizations and inspires innovation*. New York: Harper Business.

Brown, J. S., & Gray, E. S. (2004). Introduction . In Conner, M. L., & Clawson, J. G. (Eds.), *Creating a learning culture: Strategy, technology, and practice* (pp. 1–16). Cambridge, U.K.: Cambridge University Press.

Bull, G., & Kajder, S. (2005). Digital storytelling in the language arts classroom. *Learning and Leading with Technology*, *32*(4), 46–49.

Bullinger, H. J., Müller-Spahn, F., & Rößler, A. (1996). Encouraging creativity–support of mental processes by virtual experience. In *Virtual Reality World: IDG conferences and seminars*. Retrieved February 3, 2010, from http://www.creativethinkingnetwork.com/ dmdocuments/Encouraging%20Creativity.pdf

Burt, R. S. (2000). The network structure of social capital. *Research in Organizational Behavior*, *22*, 345–423. doi:10.1016/S0191-3085(00)22009-1

Burt, R. S. (2004). Structural holes and good ideas. *American Journal of Sociology*, *110*(2), 349–399. doi:10.1086/421787

Bush, T. (2002). Authenticity–reliability, validity and triangulation . In Coleman, M., & Briggs, A. (Eds.), *Research methods in educational leadership and management*. London: Paul Chapman.

Butler, T., & Murphy, C. (2007). Understanding the design of Information Technologies for Knowledge Management in organizations: A pragmatic perspective. *Information Systems Journal*, *17*, 143–163. doi:10.1111/j.1365-2575.2007.00237.x

Bygrave, W. D., & Hofer, C. W. (1991). Theorizing about entrepreneurship. *Entrepreneurship: Theory & Practice*, *16*(2), 13–22.

Byrne, B. M. (1994). *Structural equation modeling with EQS and EQS/Windows: Basic concepts, applications and programming*. Thousand Oaks, CA: Sage.

Cabra, J. F., Talbot, R. J., & Joniak, A. J. (2005). Exploratory study of creative climate: A case from selected Colombian companies and its implications on organizational development. *Cuadernos de Administración*, *18*(29), 53–86.

Calinski, R. B., & Harabasz, J. (1974). A dendrite method for cluster analysis. *Communication Statistics*, *3*, 1–27. doi:10.1080/03610927408827101

Campbell, T., & Cairns, H. (1994). Developing and measuring the learning organization from buzz words to behaviours. *Industrial & Commercial Training*, *26*(7), 10–15. doi:10.1108/00197859410064583

Cardoso, L.M. (2007). *Gerir conhecimento e gerar competitividade. Estudo empírico sobre a gestão do conhecimento e seu impacto no desempenho organizacional.* Penafiel: Editorial Novembro.

Carnegie, R., & Butlin, M. (1993). *Managing the innovative enterprise: Australian companies competing against the world's best.* Melbourne: Business Council of Australia.

Carneiro, A. (2000). How does knowledge management influence innovation and competitiveness? *Journal of Knowledge Management*, *4*(2), 87–98. doi:10.1108/13673270010372242

Carson, P. O., & Carson, E. D. (1993). Managing creativity enhancement through goal-setting and feedback. *The Journal of Applied Psychology*, *27*, 36–45.

Carvalho, V. (2006). E-learning: A experiência do Instituto Superior de Engenharia do Porto . In Carvalho, V. (Ed.), *e-learning e formação avançada: casos de sucesso no Ensino Superior da Europa e America Latina* (pp. 73–88). Portugal: Edições Politema.

Carvalho, Lourenço, and Dimas. (2010). Cultura y eficacia: El papel de la orientación cultural para el aprendizaje en la determinación de la eficacia del grupo/equipo. *Ciencia Y Trabajo, 12*(36), 306-311.

Cebon, P., Newton, P., & Noble, P. (1999). *Innovation in organizations: Towards a framework for indicator development.* Melbourne Business School Working Paper #99-9, September.

CEMED. (2010). *Centro Multimédia e de Ensino a Distância da Universidade de Aveiro.* Retrieved January 2010, from http://www.cemed.ua.pt/

Cerf, V., & Schutz, C. (2002). *Visions 2020: Transforming education and training through advanced technologies.* Washington, DC: U.S. Department of Commerce. Retrieved March 8, 2007, from http://www.technology.gov/reports/TechPolicy/2020Visions.pdf

Chen, H. L., Cannon, D., Gabrio, J., Leifer, L., Toye, G., & Bailey, T. (2005). Using Wikis and weblogs to support reflective learning in an introductory engineering design course . In Gero, J. S., & Lindermann, U. (Eds.), *Human Behaviour in Design 05* (pp. 95–105). Key Centre of Design Computing and Cognition, University of Sydney.

Cheng, Y. T., & Van de Ven, A. H. (1996). Learning the innovation journey: Order out of chaos? *Organization Science*, *7*(6), 593–614. doi:10.1287/orsc.7.6.593

Chesbrough, H. (2003). *Open innovation: The new imperative for creating and profiting from technology*. HBS Press.

Chia, R. (1998). From complexity science to complex thinking: Organization as simple location. *Organization*, *5*(3), 341–369. doi:10.1177/135050849853003

Choo, C. W. (1998). *The knowing organization: How organizations use information to construct meaning, create knowledge, and make decisions.* New York: Oxford University Press.

Christakis, N. A., & Fowler, J. H. (2009). *Connected: The surprising power of our social networks and how they shape our lives.* New York: Little, Brown and Co./ Hachette Book Group.

Chua, C. K., Leong, K. F., & Lim, C. S. (2003). *Rapid prototyping: Principals and applications* (2nd ed.). Singapore: World Scientific Publishing.

Churchill, G. A. (1979). A paradigm for developing better measures of marketing constructs. *JMR, Journal of Marketing Research*, *15*, 64–73. doi:10.2307/3150876

Clark, K. B., & Fujimoto, T. (1991). *Product development performance: Strategy, organization, and management in the world auto industry*. Boston: Harvard University Press.

Clawson, V. K., Bostrom, R., & Anson, R. (1993). The role of the facilitator in computer-supported meetings. *Small Group Research*, *24*(4), 547–565. doi:10.1177/1046496493244007

Clement, J. (1989). Learning via model construction and criticism: Protocol evidence on sources of creativity in science . In Glover, J., Ronning, R., & Reynolds, C. (Eds.), *Handbook of creativity: Assessment, theory and research* (pp. 341–381). NY: Plenum.

Coenen, L., Moodyson, J., & Asheim, B. T. (2004). Nodes, networks and proximities: On the knowledge dynamics of the Medicon Valley biotech cluster. *European Planning Studies*, *12*(7), 1003–1018.

Coghlan, M. (2003). Should using the internet change the way we teach? Paper presented at the Educause 2003, Adelaide, Austalasia. Retrieved from http://www.caudit.edu.au/educauseaustralasia/2003/EDUCAUSE/PDF/AUTHOR/ED030040.PDF

Cohen, M. D., & Bacdayan, P. (1994). Organizational routines are stored as procedural memory: Evidence from a laboratory study. *Organization Science*, *5*, 554–568. doi:10.1287/orsc.5.4.554

Cohen, W. M. (1991). Individual learning and organizational routine: Emerging connections. *Organization Science*, *2*, 135–139. doi:10.1287/orsc.2.1.135

Cohen, W. M., & Levinthal, D. A. (1990). Absorptive capacity: A new perspective on learning and innovation. *Administrative Science Quarterly*, *35*, 128–152. doi:10.2307/2393553

Cohen, W. M., & Levinthal, D. A. (1990). Absorptive capacity: A new perspective on learning and innovation. *Administrative Science Quarterly*, *35*, 128–152. doi:10.2307/2393553

Cole-Gomolski, B. (1997). Chase uses new apps to ID best customers. *Computerworld*, *35*(31), 49–50.

Cole-Gomolski, B. (1997a). Users loath to share their know-how. *Computerworld*, *46*(31), 6.

Cole-Gomolski, B. (1998). Vendors cram knowledge-ware market. *Computerworld*, *5*(31), 55–56.

Coleman, J.S. (1988). Social capital in the creation of human capital. *American Journal of Sociology, 94*(Supplement: Organizations and Institutions: Sociological and Economic Approaches to the Analysis of Social Structure), 95-120.

Collins, J. C., & Porras, J. I. (1994). *Built to last: Successful habits of visionary companies*. New York: HarperBusiness.

Collins, A. M., & Loftus, E. F. (1975). A spreading-activation theory of semantic processing. *Psychological Review*, *82*(6), 407–428. doi:10.1037/0033-295X.82.6.407

Comissão Europeia. (2007). *Estrutura dos Sistemas de Ensino, Formação Profissional e Ensino para Adultos na Europa*. Unidade Portuguesa da Rede Eurydice.

Commins, E. J. (2004). *The role of the Enterprise Ireland Technologist in facilitating Innovation Management in Irish SMEs*. Unpublished Master of Science Dissertation in Technology Management. Dublin: University College Dublin.

Conner, M. L., & Clawson, J. G. (2004). Afterword . In Conner, M. L., & Clawson, J. G. (Eds.), *Creating a learning culture: Strategy, technology, and practice* (pp. 327–338). Cambridge, U.K.: Cambridge University Press.

Cook, P. (1998). *Best practice creativity*. Aldershot, UK: Gower Publishing.

Cook, S. D. N., & Yanow, D. (1993). Culture and organizational learning. *Journal of Management Inquiry*, *2*(4), 373–390. doi:10.1177/105649269324010

Cook, S. D., & Brown, J. S. (1999). Bridging epistemologies: The generative dance organizational knowledge and organizational knowing. *Organization Science*, *10*(4), 381–400. doi:10.1287/orsc.10.4.381

Cooke, P., Heidenreich, M., & Braczyk, H. J. (2004). *Regional innovation systems*. London: Routledge.

Cooper, P. (2005). A study of innovators' experience of new product innovation in organisations. *R & D Management*, *35*(5), 525–533. doi:10.1111/j.1467-9310.2005.00409.x

Cooper, T., Hofheinz, P., & Purdy, M. (2009). *Skills for the future*. Retrieved November 17th, 2009 from http://www.accenture.com/NR/rdonlyres/55C4D829-CB2B-4553-9BC3-10F8DBC4987C/0/SkillsfortheFuture.pdf.

Corti, E., & lo Storto, C. (2000). Knowledge creation in small firms during product innovation: An empirical analysis of cause-effect relationships among its determinants. *Enterprise & Innovation Management Studies*, *1*(3), 245–263. doi:10.1080/14632440010023190

Corti, E. & lo Storto, C. (1999). Factors to sustain the creation of knowledge, competences and routines in small firms during product development. *Piccola Impresa/Small Business, 2*, 33-57.

Corti, E., & lo Storto, C. (1997). Technical problem-solving and learning processes during product innovation: Some empirical evidence for the development of a theoretical framework of knowledge creation in small firms. *Proceedings of the PICMET'97*, Portland (Oregon), July.

Costa, P. T., & McCrae, R. R. (1993). *Revised NEO personality inventory and NEO five-factor inventory, professional manual*. Odessa, FL: Psychological Assessment Resources.

Council for Research Excellence. (2008). *Groundbreaking study of video viewing finds younger boomers consume more video media than any other group.* Retrieved from http://www.researchexcellence.com/vcm_brief.pdf

Covin, J. G., & Slevin, D. P. (1988). The influence of organization structure on the utility of an entrepreneurial top management style. *Journal of Management Studies*, *25*, 217–234. doi:10.1111/j.1467-6486.1988.tb00033.x

Cox, M., Abbott, C., Webb, M., Blakeley, B., Beauchamp, T., & Rhodes, V. (2003). ICT and attainment: A review of the research literature. Coventry, UK: Becta (ICT in Schools Research and Evaluation Series).

Craft, A. (2005). *Creativity in schools: Tensions and dilemmas, Oxford: Routledge Falmer*. London: QCA. doi:10.4324/9780203357965

Crooks, T. J., Kane, M. T., & Cohen, A. S. (1996). Threats to the valid use of assessment. *Assessment in Education*, *3*(3), 265–285. doi:10.1080/0969594960030302

Csikszentmihalyi, M. (1997). *Creativity: The flow of psychology of the discovery and invention*. New York: Harper Collins.

Csikszentmihalyi, M. (1990). *Flow: The psychology of optimal experience*. New York: Harper & Row.

Csikszentmihalyi, M. (1992). Motivation and creativity . In Albert, R. S. (Ed.), *Genius and eminence* (pp. 19–33). Oxford: Pergamon Press.

Csikszentmihalyi, M. (1994). The domain of creativity . In Feldman, D. H., Csikszentmihalyi, M., & Gardner, H. (Eds.), *Changing the world: A framework for the study of creativity*. London: Praeger.

Csíkszentmihályi, M. (2006). A systems perspective on creativity . In Henry, J. (Ed.), *Creative management and development* (3rd ed., pp. 3–17). London: Sage.

Cyert, R. M., & March, J. G. (1963). *A behavioral theory of the firm*. Englewood Cliff, NJ: Prentice Hall.

Czaja, R., & Blair, J. (1996). *Designing surveys: A guide to decisions and procedures*. Thousand Oaks, CA: Pine Forge Press.

Daft, R. L., & Lengel, R. H. (1986). Organizational information requirements, media richness and structural design. *Management Science*, *32*, 554–571. doi:10.1287/mnsc.32.5.554

Daft, R. L., & Macintosh, N. B. (1981). A tentative explanation into the amount and equivocality of information processing in organizational work units. *Administrative Science Quarterly*, *26*, 207–224. doi:10.2307/2392469

Daft, R. L., & Weick, K. E. (1984). Toward a model of organizations as interpretation systems. *Academy of Management Review*, *9*, 284–295. doi:10.2307/258441

Dalal, S. (2008). *The innovation boot camp*. Orange, CA: The Institute for Effective Innovation.

Dalziel, J. (2005). *From reusable e-learning content to reusable learning designs: Lessons from LAMS*. Retrieved January 2010 from http://rams.ramp.org.au/download/attachments/9469955/dalziel_reusable.pdf

Dalziel, J. (2006). *Future directions for LAMS and learning design*. Paper presented at the 1st International LAMS Conference. Retrieved January 2010 from http://saturn.melcoe.mq.edu.au/presentations/FutureofLAMS.LAMSConf.ppt

Darroch, J. (2005). Knowledge management, innovation and firm performance. *Journal of Knowledge Management*, *9*(3), 101–115. doi:10.1108/13673270510602809

Dasgupta, S. (1996). *Technology and creativity*. New York: Oxford University Press.

Davenport, T. H. (1993). *Process innovation: Reengineering work through information technology*. New York: Harvard Business Press.

Davenport, T. H., Long, D. W. D., & Beers, M. C. (1998). Building successful Knowledge Management projects. *Sloan Management Review*, *39*(2), 45–57.

Davenport, T. H., & Prusak, L. (1997). *Information ecology*. Oxford University Press.

Davenport, T., Eccles, R. G., & Prusak, L. (1992). Information politics. *Sloan Management Review*, (Fall): 53–65.

Davenport, T., Hammer, M., & Metsisto, T. (1989). How executives can shape their company's information systems. *Harvard Business Review*, (March-April): 130–134.

Davenport, T. H., & Prusak, L. (1998). *Working knowledge: How organizations manage what they know*. Boston: Harvard Business School Press.

Davenport, T. H., De Long, D. W., & Beers, M. C. (1999). Successful knowledge management projects . In Cortada, J. W., & Woods, J. A. (Eds.), *The knowledge management yearbook 1999-2000* (pp. 89–107). Boston: Butterworth-Heinemann.

Davis, G. A., & Manski, M. E. (1966). An instructional method for increasing originality. *Psychonomic Science*, *6*, 73–74.

Davis, H., & Scase, R. (2000). *Managing creativity*. London: Open University Press.

Dawson, S. (2008). A study of the relationship between student social networks and sense of community. *Journal of Educational Technology & Society*, *11*(3), 224–238.

Dawson, S. (2009). 'Seeing' the learning community: An exploration of the development of a resource for monitoring online student networking. *British Journal of Educational Technology*.

DCMS. (2007). *Staying ahead: The economic performance of the UK's creative industries*. Retrieved December 15, 2007, from http://www.culture.gov.uk/reference_library/publications/3672.aspx

de Bono, E. (1971). *Lateral thinking for management: A handbook*. London: McGraw-Hill.

de Bono, E. (1973). *Po: Beyond yes and no*. London: Penguin Books.

de Bono, E. (1990). *Six thinking hats*. London: Penguin Books.

de Bono, E. (1993). *Teach your child how to think*. London: Penguin Books.

de Bono, E. (1994). *Parallel thinking: From socratic to De Bono thinking*. London: Viking, Penguin Group.

De Bono, E. (1993). *Serious creativity: Using the power of lateral thinking to create new ideas*. HarperBusiness.

De Dreu, C. K. W., & West, M. A. (2001). Minority dissent and team innovation: The importance of participation in decision making. *The Journal of Applied Psychology*, *86*(6), 1191–1201. doi:10.1037/0021-9010.86.6.1191

De la Fuente, A., & Ciccone, A. (2002). *Human capital in a global and knowledge-based economy*. Brussels: European Commission Directorate-General for Employment and Social Affairs.

De la Torre, F., Gross, R., & Baker, S. (2005). Representational oriented component analysis (ROCA) for face recognition with one sample image per training class. IEEE *Conference on Computer Vision and Pattern Recognition*.

De Long, D. (1999). *Building the knowledge-based organization: How culture drives knowledge behaviors*. Center for Business Innovation: Ernst & Young LLP.

de Vreede, G. J., Davison, R., & Briggs, R. O. (2003a). How a silver bullet may lose its shine-learning from failures with group support systems. *Communications of the ACM*, *46*(8), 96–101. doi:10.1145/859670.859676

de Vreede, G. J., & de Bruijn, J. A. (1999). Exploring the boundaries of successful GSS application: Supporting inter-organizational policy networks. *Database*, *30*(3-4), 111–131.

de Vreede, G. J., & Briggs, R. O. (2005). *Collaboration engineering: Designing repeatable processes for high-value collaborative tasks.* Paper presented at the Hawaii International Conference on System Science, Kauai HI.

de Vreede, G. J., Vogel, D. R., Kolfschoten, G. L., & Wien, J. S. (2003). Fifteen years of GSS in the field: A comparison across time and national boundaries. *Proceedings of the 36th Hawaii International Conference on System Sciences*. IEEE Computer Society Press.

Deal, T., & Kennedy, A. (1982). *Corporate cultures*. Reading, MA: Addison-Wesley.

Dennis, A. R., Aronson, J. E., Heninger, W. G., & Walker, E. D. II. (1999). Structuring time and task in electronic brainstorming. *Management Information Systems Quarterly*, *23*(1), 95–108. doi:10.2307/249411

Dennis, A. R., & Wixom, B. H. (2001). Investigating the moderators of the group support systems use with meta-analysis. *Journal of Management Information Systems*, *18*(3), 235–257.

Dennis, A. R., Wixom, B. H., & Vandenberg, R. J. (2001). Understanding fit and appropriation effects in group support systems via meta-analysis. *Management Information Systems Quarterly*, *25*(2), 167–183. doi:10.2307/3250928

Dery, D. (1983). Decision-making, problem-solving, and organizational learning. *Omega*, *11*, 321–328. doi:10.1016/0305-0483(83)90025-7

Dewey, J. (1910). *How we think*. New York: Heath. doi:10.1037/10903-000

Dias, P. (2004). Processos de aprendizagem colaborativa nas comunidades onLine . In Dias, A., & Gomes, M. J. (Eds.), *E-Learning para E-formadores* (pp. 19–31). TecMinho Edições.

DiBella, A., Nevis, E., & Gould, J. (1996). Organizational learning style as a core capability . In Moingeon, B., & Edmondson, A. (Eds.), *Organizational learning and competitive advantage* (pp. 38–55). London: Sage.

Dickson, G. W., Limayem, M., Lee Partridge, J., & DeSanctis, G. (1996). Facilitating computer supported meetings: A cumulative analysis in a multiple criteria task environment. *Group Decision and Negotiation*, *5*(1), 51–72.

Diehl, M., & Stroebe, W. (1987). Productivity loss in brainstorming groups: Toward the solution of a riddle. *Journal of Personality and Social Psychology*, *53*, 497–509. doi:10.1037/0022-3514.53.3.497

Diehl, M., & Stroebe, W. (1991). Productivity loss in idea-generating groups: Tracking down the blocking effects. *Journal of Personality and Social Psychology*, *61*(3), 31–63. doi:10.1037/0022-3514.61.3.392

Dillion, J. T. (1982). Problem finding and solving. *The Journal of Creative Behavior*, *16*, 97–111.

DiMattia, S., & Oder, N. (1997). Knowledge management: Hope, hype, or harbinger? *Library Journal*, *15*(122), 33–35.

Dodgson, M. (1993). Organizational learning: A review of some literatures. *Organization Studies*, *14*, 375–394. doi:10.1177/017084069301400303

Domínguez, M. S. (2001). Importancia de la gestión del conocimiento: Homo sapiens versus maquina sapiens. *Proceedings of the 2nd Conference of the Portuguese Association of Information Systems*. Évora: Portuguese Association of Information Systems.

Donnellon, A. (1996). *Team talk: The power of language in team dynamics*. Harvard Business School Press.

Dorenbosch, L., Van Engen, M. L., & Verhagen, M. (2005). On-the-job innovation: The impact of job design and human management through production ownership. *Creativity and Innovation Management*, *14*(2), 129–141. doi:10.1111/j.1476-8691.2005.00333.x

Dörfler, V. (2010). The fitness of innovation in an evolutionary framework . In Jolly, A. (Ed.), *The innovation handbook: How to develop, manage and protect your most profitable ideas*. London: Kogan Page.

Dörfler, V. (2004). *Descriptive model of learning capability.* Paper presented at the 3rd Annual Conference on Information Science Technology and Management, Alexandria, Egypt.

Dowd, E. T. (1989). The self and creativity. Several constructs in search of a theory . In Glover, J. A., Ronning, R. R., & Reynolds, C. R. (Eds.), *Handbook of creativity* (pp. 233–241). New York: Plenum Press.

Drazin, R., & Van de Ven, A. H. (1985). Alternate form of fit in contingency theory. *Administrative Science Quarterly*, *30*, 514–539. doi:10.2307/2392695

Drenth, P. J. (1998). Research in work and organizational psychology: Principles and methods. In P.J. Drenth, H. Thierry & C. J. de Wolff (Eds.), *Handbook of work and organizational psychology, (vol. 1), Introduction to work and organizational psychology (2nd ed.),* (pp. 11-46). Hove, East Sussex, UK: Psychology Press.

Duncan, R., & Weiss, A. (1979). Organizational learning: Implications for organizational design. *Research in Organizational Behavior*, *1*, 75–123.

Duncan, R. B. (1972). Characteristics of organizational environments and perceived environmental uncertainty. *Administrative Science Quarterly*, *17*, 313–327. doi:10.2307/2392145

Duncker, K. (1945). *On problem solving. Psychological Monographs Series, 58(5)*. California: American Psychological Association.

Dyer, J. H., & Singh, H. (1998). The relational view: Cooperative strategy and sources of interorganizational competitive advantage. *Academy of Management Review*, *23*(4), 660–679. doi:10.2307/259056

Easterby-Smith, M., Lyles, M., & Tsang, E. (2008). Interorganizational knowledge transfer: Current themes and future prospects. *Journal of Management Studies*, *45*(4), 678–690. doi:10.1111/j.1467-6486.2008.00773.x

Edmonds, E., Candy, L., Jones, R., & Soufi, B. (1994). Support for collaborative design: Agents and emergence. *Communications of the ACM*, *37*(7), 41–47. doi:10.1145/176789.176793

Edmonds, G., Branch, R., & Mukherjee, P. (1994). A conceptual framework for comparing instructional design models. *Educational Technology Research and Development*, *42*(4), 55–72. doi:10.1007/BF02298055

Edmondson, A., & Moingeon, B. (1998). From organizational learning to the learning organization. *Management Learning*, *29*(1), 5–20. doi:10.1177/1350507698291001

Edmonson, A. (1999). Psychological safety and learning behavior in work teams. *Administrative Science Quarterly*, *44*(2), 350–383. doi:10.2307/2666999

Edquist, C. (1997). *Systems of innovation: Technologies, institutions and organizations*. London: Printer.

Egan, T. M., Yang, B., & Bartlett, K. R. (2004). The effects of organizational learning culture and job motivation to transfer learning and turnover intention. *Human Resource Development Quarterly*, *15*(3), 279–301. doi:10.1002/hrdq.1104

Einstein, A. (1956/1984). *The world as I see it*. New York: Kensington Publishing.

Ekvall, G. (1996). Organisational climate for creativity and innovation. *European Journal of Work and Organizational Psychology*, *5*(1), 105–123. doi:10.1080/13594329608414845

Ekvall, G., & Tångeberg-Andersson, Y. (1986). Working climate and creativity: A study of an innovative newspaper office. *The Journal of Creative Behavior*, *20*(3), 215–225.

Encyclopædia Britannica Online. (2010). *Engineering*. Retrieved April 01, 2010, from www.britannica.com/EBchecked/topic/187549/engineering

Ericsson, K. A. (1996). The acquisition of expert performance: An introduction to some of the issues. In Ericsson, K. A. (Ed.), *The road to excellence. The acquisition of expert performance in the arts and sciences, sports, and games* (pp. 1–50). Mahwah, NJ: Erlbaum.

European Commission. (1998). *A guide to Technology Management and innovation for organizations*. TEAM-GUIDE.

European Commission. (1996). *Innovation Management Tools–a review of selected methodologies (MINT)*. EIMS publication No 30.

European Commission. (2004). *Final report of the expert group, "Education for Entrepreneurship". Making progress in promoting entrepreneurial attitudes and skills through primary and secondary education*. Brussels: European Commission. Retrieved February 8, 2008, from http://ec.europa.eu/enterprise/policies/sme/files/support_measures/training_education/doc/entrepreneurship_education_final_en.pdf

Eysenck, H. (1995). *Genius: The natural history of creativity*. Cambridge, UK: Cambridge University Press. doi:10.1017/CBO9780511752247

Fallon, K.K. (2000, April 3). *Design firm principals unite to demand better technology: Leading architects examine practice, question impact of CAD tools on architecture, engineering, and construction*. [Press release]. New York: Business Wire.

Faltin, G. (2001). Creating a culture of innovative entrepreneurship. *Journal of International Business and Economy*, *2*(1), 123–140.

Fasko, D. (2000-2001). Education and creativity. *Creativity Research Journal*, *13*(3&4), 317–327.

Federal Ministry of Economics and Technology (2007, August). Gründungsideen entwickeln - Weniger Glück als Verstand. *GründerZeiten Nr. 39*.

Federal Ministry of Education and Science (Ed.). (2005). *Wissens- und technologieorientiertes Gründungsgeschehen. Kienbaum-Bestandsaufnahme für eine Weiterentwicklung von EXIST.* Bonn, Berlin: Federal Ministry of Education and Science. Retrieved July 21, 2006, from www.bmbf.de/pub/wissens_technologieorientiertes_gruendungsgeschehen.pdf

Feist, G. J., & Barron, F. X. (2003). Predicting creativity from early to late adulthood: Intellect, potential, and personality. *Journal of Research in Personality*, *37*(2), 62–88. doi:10.1016/S0092-6566(02)00536-6

Felder, R., & Soloman, B. (2010). Learning styles and strategies. Retrieved January 2010, from http://www4.ncsu.edu/unity/lockers/users/f/felder/public/ILSdir/styles.htm

Feldman, M. S. (2000). Organizational routines as a source of continuous change. *Organization Science*, *6*, 611–629. doi:10.1287/orsc.11.6.611.12529

Feyerabend, P. K. (1987). *Farewell to reason*. New York: Verso.

Feyerabend, P. K. (1993). *Against method* (3rd ed.). London: Verso.

Fidel, R., & Green, M. (2004). The many faces of accessibility: Engineers' perception of information sources. *Information Processing and Management: an International Journal*, *40*, 563–581. doi:10.1016/S0306-4573(03)00003-7

Finerty, P. (1997). Improving customer care through knowledge management. *Coastal Management*, *9*(71), 33.

Fink, A., & Kosecoff, J. (1985). *How to conduct surveys: A step-by-step guide*. Newbury Park: Sage.

Finke, R. A. Ward., T.B. & Smith, S.M. (1992). *Creative cognition: Theory, research, and applications.* Cambridge, MA: MIT Press.

Fiol, C. (1994). Consensus, diversity, and learning in organizations. *Organization Science*, *5*, 403–420. doi:10.1287/orsc.5.3.403

Fiol, C., & Lyles, M. (1985). Organizational learning. *Academy of Management Review, 10*(4), 803–813. doi:10.2307/258048

Fjermestad, J., & Hiltz, S. R. (1999). An assessment of group support systems experimental research: Methodology and results. *Journal of Management Information Systems, 15*(3), 7–149.

Fjermestad, J., & Hiltz, S. R. (2001). A descriptive evaluation of group support systems case and field studies. *Journal of Management Information Systems, 17*(3), 115–159.

Fleck, J. (1994). Learning by trying: The implementation of configurational technology. *Research Policy, 23*, 637–652. doi:10.1016/0048-7333(94)90014-0

Florida, R., & Tinagli, I. (2004). *Europe in the Creative Age*. London: Demos.

Florida, R. (2003). *The rise of the creative class: And how it's transforming work, leisure, community and everyday life. North Melbourne*. Pluto Press.

Florida, R. (2004). *Creative cities*. New York: Basic Books.

Florida, R. L. (2008). *Who's your city? How the creative economy is making where to live the most important decision of your life*. New York: Basic Books.

Fontana, L. A., Dede, C., White, C. S., & Cates, W. M. (1993). *Multimedia: A gateway to higher-order thinking skills*. Center for Interactive Educational Technology: George Mason University, VA.

Freire, P. (2001). *Gestão do conhecimento e conhecimento da gestão*. Retrieved January 20, 2010, from http//:www.kmol.online.pt

Frost & Sullivan. (2007). *Meetings around the world: The impact of collaboration on business Performance*. Retrieved on March 30, 2010, from http://www.verizonbusiness.com/us/resources/conferencing/impactcollab.pdf

Fry, L. W., & Slocum, J. W. (1984). Technology, structure, and work group effectiveness: A test of a contingency model. *Academy of Management Journal, 27*, 221–246. doi:10.2307/255923

Furnham, A., Crump, J., Batey, M., & Chamorro-Premuzik, T. (2009). Personality and ability predictors of the "consequences" test of divergent thinking in a large non-student sample. *Personality and Individual Differences, 46*(4), 536–540. doi:10.1016/j.paid.2008.12.007

Gabor, D. (1970). *Innovations: Scientific, technological and social*. Oxford: Oxford University Press.

Gagne, R. M., & Dick, W. (1983). Instructional psychology. *Annual Review of Psychology, 34*(1), 261–295. doi:10.1146/annurev.ps.34.020183.001401

Galbraith, J. R. (1973). *Designing complex organizations*. Reading, MA: Addison-Wesley.

Gallupe, R. B., Dennis, A. R., Cooper, W. H., Valacich, J. S., Bastianutti, L. M., & Nunamaker, J. F. Jr. (1992). Electronic brainstorming and group size. *Academy of Management Journal, 35*(2), 350–369. doi:10.2307/256377

Gardner, H. (1993). *Creating minds*. New York: Basic Books.

Gardner, H. (2006). *Multiple intelligences–new horizons*. New York: Basic Books.

Gardner, H. (1998). *Extraordinary minds*. London: Phoenix.

Garfield, M. J., Taylor, N. J., Dennis, A. R., & Satzinger, J. W. (2001). Research report: Modifying paradigms-individual differences, creativity techniques, and exposure to ideas in group idea generation. *Information Systems Research, 12*(3), 322–333. doi:10.1287/isre.12.3.322.9710

Garner, W. R. (1962). *Uncertainty and structure as psychological concepts*. New York: John Wiley & Sons.

Garrison, R. & Kanuka. (2004). Blended learning: Uncovering its transformative potential in higher education. *The Internet and Higher Education, 7*, 95–105. doi:10.1016/j.iheduc.2004.02.001

Garyd, R., & Kumaraswamy, A. (1995). Technological and organizational designs for realizing economies of substitution. *Strategic Management Journal, 16*, 93–109. doi:10.1002/smj.4250160919

GATIUP. (2010). *GATIUP - Gabinete de Apoio para as Novas Tecnologias na Educação da Universidade do Porto*. Retrieved January 2010, from http://sigarra.up.pt/up/web_base.gera_pagina?p_pagina=18377

Gentry, W. A., Mondore, S. P., & Cox, B. D. (2007). A study of managerial derailment characteristics and personality references. *Journal of Management Development*, *26*(9), 857–873. doi:10.1108/02621710710819348

George, J. (2007). Creativity in organizations. *The Academy of Management Annals*, *1*(1), 439–477. doi:10.1080/078559814

Georgsdottir, A. S., & Getz, I. (2004). How flexibility facilitates innovation and ways to manage it in organizations. *Creativity and Innovation Management*, *13*(3), 166–175. doi:10.1111/j.0963-1690.2004.00306.x

Gersick, C. J. G., & Hackman, J. R. (1990). Habitual routines in task-performing groups. *Organizational Behavior and Human Decision Processes*, *47*, 65–97. doi:10.1016/0749-5978(90)90047-D

Gerstenfeld, A., & Berger, P. (1980). An analysis of utilization differences for scientific and technical information. *Management Science*, *26*, 165–179. doi:10.1287/mnsc.26.2.165

Geschka, H. (2007). Kreativitätstechniken . In Köhler, R., Küpper, H.-U., & Pfingsten, A. (Eds.), *Handwörterbuch der Betriebswirtschaft* (6th ed., pp. 992–1005). Stuttgart: Schäffer-Poeschel Verlag.

Gettys, C. F., Pliske, R. M., Manning, C., & Casey, J. T. (1987). An evaluation of human act generation performance. *Organizational Behavior and Human Decision Processes*, *39*(1), 23–51. doi:10.1016/0749-5978(87)90044-6

Getzels, F., & Csikszentmyhalyi, M. (1976). *The creative vision: A longitudinal study of problem solving in art*. New York: Wiley.

Getzels, J. W., & Csikszentmihalyi, M. (1975). From problem solving to problem finding . In Taylor, I. A., & Getzels, J. W. (Eds.), *Perspectives in creativity* (pp. 90–116). Chicago: Aldine.

Ghiglione, E. (2009). *LAMS 2. User documentation*. Retrieved January 2010, from http://wiki.lamsfoundation.org/display/lamsdocs/Home;jsessionid=6890123183F3AC8FCFBA1B51E0D5CA62

Ghiglione, R., & Matalon, B. (1997). *O inquérito, teoria e prática* (3ª ed.). Oeiras: Celta Editora.

Gibson, C. B., & Gibbs, J. L. (2006). Unpacking the concept of virtuality: The effects of geographic dispersion, electronic dependence, dynamic structure, and national diversity on team innovation. *Administrative Science Quarterly*, *51*(3), 451–495.

Gieski, J.F. (1999). *Continuous improvement and global innovation management trial project*. Euro-Australian Cooperation Centre Newsletter, No. 2.

Gipps, C. V. (1994). *Beyond testing: Towards a theory of educational assessment*. London: The Falmer Press.

Gladwell, M. (2005). *Blink: The power of thinking without thinking*. London: Penguin Books.

Glaser, B. G., & Strauss, A. (1967). *The discovery of Grounded Theory: Strategies for qualitative research*. New York: Aldine.

Glaser, R. (1986). On the nature of expertise . In Klix, F., & Hagendorf, H. (Eds.), *Human memory and cognitive capabilities: Mechanisms and performances* (pp. 915–928). Amsterdam: Elsevier Science Publishing.

Gloet, M., & Terziovski, M. (2004). Exploring the relationship between knowledge management practices and innovation performance. *Journal of Manufacturing Technology Management*, *15*(5), 402–409. doi:10.1108/17410380410540390

Glover, J. A., Ronning, R. R., & Reynolds, C. R. (1989). *Handbook of creativity*. New York: Plenum Press.

Glynn, M. A. (1996). Innovative genius: A framework for relating individual and organizational intelligences to innovation. *Academy of Management Review*, *21*(4), 1081–1111. doi:10.2307/259165

Goldberg, J. C. (1980). Counseling the adult learner: A selective review of the literature. *Adult Education Quarterly*, *30*(2), 67–81. doi:10.1177/074171368003000201

Gomes, A. D. (2000). *Cultura Organizacional: Comunicação e Identidade*. Coimbra: Quarteto Editora.

Gomes, A. D. (1992). Formas de organização e técnicas de gestão: Análise comparativa e níveis de análise. *Revista Portuguesa de Gestão*, *2*, 29–47.

Google Wave. (2010). *Homepage information*. Retrieved May 9, 2010, from http://wave.google.com

Gough, H. G. (1957). *California psychological inventory manual*. Consulting Psychologists Press.

Granovetter, M. (1973). The strength of weak ties. *American Journal of Sociology*, *78*(6), 1360–1380. doi:10.1086/225469

Granovetter, M. (1985). Economic action and social structure: The problem of embeddedness. *American Journal of Sociology*, *91*(3), 481–510. doi:10.1086/228311

Granovetter, M. (2005). The impact of social structure on economic outcomes. *The Journal of Economic Perspectives*, *19*(1), 33–50. doi:10.1257/0895330053147958

Grant, R. M. (1996). Prospering in dynamically-competitive environments: Organizational capability as knowledge integration. *Organization Science*, *7*, 375–387. doi:10.1287/orsc.7.4.375

Greene, P. G., Katz, J. A., & Johannisson, B. (2004). Entrepreneurship education. *Academy of Management Learning & Education*, *3*(3), 238–241.

Gresov, C. (1989). Exploring fit and misfit with multiple contingencies. *Administrative Science Quarterly*, *34*, 431–453. doi:10.2307/2393152

Griffiths-Hemans, J., & Grover, R. (2006). Setting the stage for creative new products: Investigating the idea fruition process. *Journal of the Academy of Marketing Science*, *34*(1), 27–39. doi:10.1177/0092070305281777

Gryskiewicz, S. S. (1999). *Positive turbulence*. San Francisco: Jossey Bass.

Guildford, J. (1950). Creativity. *The American Psychologist*, *5*, 444–454. doi:10.1037/h0063487

Guilford, J. P. (1977). *Way beyond the IQ*. Buffalo, NY: Bearly Limited.

Guilford, J. P. (1968). *Intelligence, creativity and their educational implications*. San Diego: Knapp.

Guilford, J. P. (1988). Some changes in the structure-of-intellect model. *Educational and Psychological Measurement*, *48*(1), 1–4. doi:10.1177/001316448804800102

Häcker, H., & Stapf, K. H. (Eds.). (1998). *Dorsch. Psychologisches Wörterbuch* (13th ed.). Bern, Göttingen, Toronto, Seattle: Huber.

Hadamard, J. (1954). *The Psychology of invention in the mathematical field*. New York: Dover Publications.

Hagardon, A. (1999). Group cognition and creativity in organizations. *Research on Managing Groups and Teams*, *2*, 137–155.

Hagardon, A. (2000). *Building an innovation factory. Harvard Business Review*. May-June.

Hage, J. (1999). Organizational innovation and organizational change. *Annual Review of Sociology*, *25*, 597–622. doi:10.1146/annurev.soc.25.1.597

Hall, S. C. (1986). Thoughts on emerging technology. *Information Technology Quarterly*, *5*(1), 13–20.

Hamel, G., & Prahalad, C. (1994). *Competing for the future*. Boston: Harvard Business School Press.

Hammel, G. (2009). Empowering natural leaders in 'Facebook generation' ways. Retrieved from http://blogs.wsj.com/management/2009/05/18/empowering-natural-leaders-in-facebook-generation-ways/

Hannan, A., & Silver, H. (2000). *Innovating in higher education-teaching, learning and institutional cultures*. Buckingham, UK: The Society for Research into Higher Education/Open University Press.

Harter, J., Schmidt, F., & Keyes, L. (2003). *Well-being in the workplace and its relationship with business outcomes: A review of the Gallup studies*. Washington, D.C.: American Psychological Association.

Hauser, J., Tellis, G. J., & Griffin, A. (2006). Research on innovation: A review and agenda for Marketing Science. *Marketing Science, 25*(6), 687–717. doi:10.1287/mksc.1050.0144

Hayes, J. R. (1978). *Cognitive psychology. Thinking and creating*. Homewood, IL: Dorsey Press.

Healy, J. M. (1998). *Failure to connect: How computers affect our children's minds–for better and worse*. New York: Simon & Schuster.

Hearn, G. (2006). Value creating ecologies: Understanding next generation business systems. *Foresight, 8*(1), 55–65. doi:10.1108/14636680610647147

Hearn, G., & Bridgstock, R. S. (2010). Education for the creative economy: Innovation, transdisciplinarity and networks . In Peters, M., & Araya, D. (Eds.), *Education in the creative economy*. New York: Peter Lang.

Hedberg, B., Nystrom, P. C., & Starbuck, W. H. (1976). Camping on seesaws: Prescriptions for a self-designing organization. *Administrative Science Quarterly, 21*, 41–65. doi:10.2307/2391877

Hedberg, B. (1981). How organizations learn and unlearn . In Nystrom, P. C., & Starbuck, W. H. (Eds.), *Handbook of organizational design* (pp. 3–27). New York: Oxford University Press.

Heilman, K. M., Nadeau, S. E., & Beversdorf, D. O. (2003). Creative innovation: Possible brain mechanisms. *Neurocase, 9*(5), 369–379. doi:10.1076/neur.9.5.369.16553

Heilman, K. M. (2005). *Creativity and the brain*. New York: Taylor & Francis group.

Heim, M. (1993). *The metaphysics of virtual reality*. Oxford: Oxford University Press.

Heller, K. A. (2004). Identification of gifted and talented students. *Psychological Science, 46*(3), 302–323.

Hellstrom, C., & Hellstrom, T. (2002). Highways, alleys and by-lanes: Charting the pathways for ideas and innovation in organizations. *Creativity and Innovation Management, 11*(2), 107–114. doi:10.1111/1467-8691.00242

Hender, J. M., Dean, D. L., Rodgers, T. L., & Nunamaker, J. F. Jr. (2002). An examination of the impact of stimuli type and GSS structure on creativity: Brainstorming versus non-brainstorming techniques in a GSS environment. *Journal of Management Information Systems, 18*(4), 59–85.

Hender, J. M., Rodgers, T. L., Dean, D. L., & Nunamaker, J. F., Jr. (2001). Improving group creativity: Brainstorming versus non-brainstorming techniques in a GSS environment. *Proceedings of the 36th Hawaii International Conference on System Sciences*. IEEE Computer Society Press.

Henderson, A. J. (2003). *The e-learning question and answer book: A survival guide for trainers and business managers*. Broadway, NY: American Management Association.

Hennessey, B. A. (1999). Intrinsic motivation, affect and creativity . In Russ, S. (Ed.), *Affect, creative experience, and psychological adjustment* (pp. 77–90). Philadelphia: Brunner/Mazel.

Henry, J. (2001). *Creativity and perception in management*. London: Sage Publications.

Henry, J. (2006). *Creative management and development* (3rd ed.). London: Sage Publications.

Henry, C., Hill, F. & Leitch, C. (2005b). Entrepreneurship education and training: Can entrepreneurship be taught? Part 2. *Education + Training, 47*(3), 158-169.

Henry, C., Hill, F., & Leitch, C. (2005a). Entrepreneurship education and training: Can entrepreneurship be taught? Part 1. *Education+ Training, 47*(2), 98–111.

Herring, S. R., Jones, B. R., & Bailey, B. P. (2009). Idea generation techniques among creative professionals. *Proceedings of the 42nd Hawaii International Conference on System Sciences*. Los Alamitos, CA: IEEE Computer Society Press.

Herrington, J., & Oliver, R. (1998). *Using situated learning and multimedia to promote higher-order thinking*. Western Australia: Edith Cowan University.

Herstatt, C., Verworn, B., & Nagahira, A. (2004). Reducing project related uncertainty in the "fuzzy front end" of innovation — a comparison of German and Japanese product innovation projects. *International Journal of Product Development, 1*(1), 43–65. doi:10.1504/IJPD.2004.004890

Hertzum, H., & Pejtersen, A. M. (2000). The information-seeking practices of engineers: Searching for documents as well as for people. *Information Processing & Management, 36*, 761–778. doi:10.1016/S0306-4573(00)00011-X

Hesmondhalgh, D. (2007). *The cultural industries* (2nd ed.). London: Sage Publications.

Hewstone, M. (1989). *Causal attribution: From cognitive processes to collective beliefs*. Oxford: Blackwell.

Hibbard, J. (1997). Knowing what we know. *Information Week, 653*, 49–54.

Higgins, J. M. (1996). Innovate or evaporate: Creative techniques for strategists. *Long Range Planning, 29*(3), 370–380. doi:10.1016/0024-6301(96)00023-4

Higgins, J. M. (1994). *101 creative problem solving techniques: The handbook of new ideas for business*. WInter Park, FL: New Management Publishing Company.

Hildrum, J. M. (2009). Sharing tacit knowledge online: A case study of e-learning in Cisco's network of systems integrator partner firms. *Industry and Innovation, 16*(2), 197–218. doi:10.1080/13662710902764360

Hill, M., & Hill, A. (2000). *Investigação por questionário*. Lisboa: Sílabo.

Hill, R. (1996). A measure of the learning organization. *Industrial & Commercial Training, 28*(1), 19–25. doi:10.1108/00197859610105440

Hill, N. S., Bartol, K. M., Tesluk, P. E., & Langa, G. A. (2009). Organizational context and face-to-face interaction: Influences on the development of trust and collaborative behaviors in computer-mediated groups. *Organizational Behavior and Human Decision Processes, 108*(2), 187–190. doi:10.1016/j.obhdp.2008.10.002

Ho, S. S., & McLeod, D. M. (2008). Social-psychological influences on opinion expression in face-to-face and computer-mediated communication. *Communication Research, 35*(2), 190–207.

Hockey, J. (2004). Practical creativity. *Medical Device Technology, 15*(1), 20–23.

Hollingworth, R. W., & McLoughlin, C. (2001). Developing science students' metacognitive problem solving skills online. *Australian Journal of Educational Technology, 17*(1), 50–63.

Holzbaur, U. (2007). *Entwicklungsmanagement. Mit hervorragenden Produkten zum Markterfolg*. Berlin, Heidelberg, New York: Springer.

Horton, G. (2006, July). *Idea engineering: Teaching students how to generate ideas*. Paper presented at the 9th International Conference on Engineering Education, San Juan, Puerto Rico.

Howard-Jones, P. (2002). A dual-state model of creative cognition for supporting strategies that foster creativity in the classroom. *International Journal of Technology and Design Education, 12*(3), 215–226. doi:10.1023/A:1020243429353

Howe, J. (2008). *Crowdsourcing: Why the power of the crowd is driving the future of business*. New York: Crown Business.

Howell, J. M., & Boies, K. (2004). Champions of technological innovation: The influence of contextual knowledge, role orientation, idea generation, and idea promotion on champion emergence. *The Leadership Quarterly, 15*(1), 123–143. doi:10.1016/j.leaqua.2003.12.008

Howkins, J. (2001). *The creative economy: How people make money from ideas*. London: Penguin.

Huang, J., & Li, Y. (2009). The mediating effect of knowledge management on social interaction and innovation performance. *International Journal of Manpower, 30*(3), 285–301. doi:10.1108/01437720910956772

Huber, G. P. (1991). Organizational learning: The contributing processes and a review of the literature. *Organization Science*, *2*, 88–115. doi:10.1287/orsc.2.1.88

Huberman, A. M., & Miles, M. B. (2002). *The qualitative researcher's companion*. Thousand Oaks, CA: Sage.

Hughes, C. E., & Stapleton, C. B. (2005). *The shared imagination: Creative collaboration in augmented virtuality*. Paper presented at HCI International, Las Vegas, NV, July 2005.

Huhtala, H., & Parzefall, R. (2007). A review of employee well-being and innovativeness: An opportunity for a mutual benefit. *Creativity and Innovation Management*, *16*(3), 299–306. doi:10.1111/j.1467-8691.2007.00442.x

Humble, J., & Jones, G. (1989). Creating a climate for innovation. *Long Range Planning*, *22*(4), 46–51. doi:10.1016/0024-6301(89)90081-2

Hunter, S. T., Friedrich, T. L., Bedell, K. E., & Mumford, M. D. (2006). Creative thought in real world innovation. *Serbian Journal of Management*, *1*, 29–39.

Hurley, R. F., & Hult, G. T. (1998). Innovation, market orientation, and organizational learning: An integration and empirical examination. *Journal of Marketing*, *62*, 42–54. doi:10.2307/1251742

Hwang, G. J., Hsiao, C. L., & Tseng, J. C. R. (2003). A computer-assisted approach to diagnosing student learning problems in science courses. *Journal of Information Science and Engineering*, *19*(2), 229–248.

IMS. G.L.C. (2008). Seven trends to watch in digital learning. Retrieved January 2010, from http://www.imsglobal.org/articles/7trendsinDigitalLearning.html

Inouye, A. S., Mitchell, W. J., & Blumenthal, M. S. (2003). *Beyond productivity: Information technology, innovation, and creativity*. Washington, D.C.: National Academies Press.

IPP. (2007). *Selfevaluation report by Instituto Politécnico do Porto*.

IPP. (2010). *Instituto Politécnico do Porto. Official website*. Retrieved January 2010, from http://www.ipp.pt

Isaksen, S. G., Puccio, G. J., & Treffinger, D. J. (1993). An ecological approach to creativity research: Profiling for creative problem solving. *The Journal of Creative Behavior*, *27*(3), 149–170.

ISCAP. (2010). *Instituto Superior de Contabilidade e Administracao do Porto*. Retrieved December 2009, from http://www.iscap.ipp.pt/

Isla, J. (2004). Collaborative learning at Monterrey Tech-Virtual University . In Duffy, T., & Jamie, K. (Eds.), *Learner-centered theory and practice-cases from higher education* (pp. 297–319).

Itami, H. (1987). *Mobilizing invisible assets*. Cambridge, MA: Harvard University Press.

Jackson, J. M., & Williams, K. D. (1985). Social loafing on difficult tasks. *Journal of Personality and Social Psychology*, *49*, 937–942. doi:10.1037/0022-3514.49.4.937

Jackson, T., Dawson, R., & Wilson, D. (2003). Reducing the effect of email interruptions on employees. *International Journal of Information Management*, *23*, 55–65. doi:10.1016/S0268-4012(02)00068-3

Janik, A. (1988). Tacit knowledge, working life and scientific method . In Göranzon, B., & Josefson, I. (Eds.), *Knowledge, skill and artificial intelligence* (pp. 53–63). London, Berlin: Springer & Verlag.

Jannson, D. G., & Smith, S. M. (1991). Design fixation. *Design Studies*, *12*(1), 3–11. doi:10.1016/0142-694X(91)90003-F

Janssen, O. (2003). Innovative behavior and job involvement at the price of conflict and less satisfactory relations with co-workers. *Journal of Occupational and Organizational Psychology*, *76*(3), 347–364. doi:10.1348/096317903769647210

Jeffcutt, P., & Pratt, A. (2002). Managing creativity in the cultural industries. *Creativity and Innovation Management*, *11*(4), 225–233. doi:10.1111/1467-8691.00254

Jelinek, M. (1979). *Institutionalizing innovation*. New York: Praeger.

Johnson-Laird, P. N. (1983). *Mental modes*. Cambridge, MA: Harvard University Press.

Jonassen, D. H. (2000). *Computers as mindtools for schools: Engaging critical thinking* (2nd ed.). Upper Saddle River, NJ: Merrill/Prentice Hall.

Jonassen, D. H., Peck, K. L., & Wilson, B. G. (1999). *Learning with technology–a constructivist perspective*. Upper Saddle River, NJ: Merrill/Prentice Hall.

Jonassen, D. H., & Carr, C. S. (2000). Mindtools: Affording multiple knowledge representations for learning . In Lajoie, S. P. (Ed.), *Computers as cognitive tools, vol. two: No more walls* (pp. 165–196). Mahwah, NJ: Lawrence Erlbaum.

Jones, S. (1996). *Developing a learning culture: Empowering people to deliver quality, innovation and long-term success*. London: McGraw-Hill.

Jones, C. & English, J. (2004). A contemporary approach to entrepreneurship education. *Education + Training, 46*(8/9), 416-423.

Jong, J. P. J., & Kemp, R. (2003). Determinants of co-workers' innovative behaviour: An investigation into knowledge intensive services. *International Journal of Innovation Management, 7*(2), 189–212. doi:10.1142/S1363919603000787

Jong, J. P., & Hartog, D. N. (2007). How leaders influence employees' innovative behaviour. *European Journal of Innovation Management, 10*(1), 41–64. doi:10.1108/14601060710720546

Ju, T. L., Li, C. Y., & Lee, T. S. (2006). A contingency model for knowledge management capability and innovation. *Industrial Management & Data Systems, 106*(6), 855–877. doi:10.1108/02635570610671524

Judge, W., Fryxell, G., & Dooley, R. (1997). The new task of R & D management: Creating goal directed communities for innovation. *California Management Review, 39*, 72–84.

Kandemir, D., & Hult, G. T. (2005). A conceptualization of an organizational learning culture in international join ventures. *Industrial Marketing Management, 34*, 430–439. doi:10.1016/j.indmarman.2004.10.002

Kanter, R. M. (1983). *The change masters*. London: Routledge.

Kao, J. (1989). *Entrepreneurship, creativity, & organization*. New York: Prentice-Hall.

Karau, S., & Williams, K. (1993). Social loafing: A meta- analytic review and theoretical integration. *Journal of Personality and Social Psychology, 65*, 681–706. doi:10.1037/0022-3514.65.4.681

Kasperson, C. J. (1978). Psychology of the scientist: 37. Scientific creativity: A relationship with information channels. *Psychological Reports, 42*, 691–694.

Katz, J. A. (2007). Education and training in entrepreneurship . In Baum, J. R., Frese, M., & Baron, R. (Eds.), *The psychology of entrepreneurship* (pp. 209–235). Mahwah, NJ/ London: Lawrence Erlbaum Associates.

Kaufman, J. C., & Sternberg, R. J. (Eds.). (2006). *The international handbook of creativity*. New York: Cambridge University Press.

Kearney, M., & Treagust, D. F. (2001). Constructivism as a referent in the design and development of a computer program using interactive digital video to enhance learning in physics. *Australian Journal of Educational Technology, 17*(1), 64–79.

Kelley, T., & Littman, J. (2001). *The art of innovation: Lessons in creativity from IDEO, America's leading design firm*. New York: Currency/Doubleday.

Khan, B. (2005). *Managing e-learning strategies. Design, delivery, implementation and evaluation*. Science Publishing.

Kiesler, S., & Sproull, L. (1982). Managerial response to changing environments: Perspectives on problem sensing from social cognition. *Administrative Science Quarterly, 27*, 548–570. doi:10.2307/2392530

Kirby, D.A. (2004). Entrepreneurship education: Can business schools meet the challenge? *Education + Training, 46*(8/9), 510-519.

Kirton, M. J. (2003). *Adaption-Innovation: in the context of diversity and change*. London: Routledge.

Kline, R. B. (2005). *Principles and practice of structural equation modelling* (2nd ed.). New York: The Guilford Press.

Kline, S. J., & Rosenberg, N. (1986). An overview of innovation . In Landau, R., & Rosenberg, N. (Eds.), *The positive sum game* (pp. 275–305). Washington, DC: National Academy Press.

Knights, D., Murray, F., & Hugh, W. (1993). Networking as knowledge work: A study of strategy development in the financial services industry. *Journal of Management Studies, 30*(6), 975–995. doi:10.1111/j.1467-6486.1993.tb00475.x

Knoll, S. W., & Horton, G. (2010) Changing the perspective: Improving generate thinkLets for Ideation. *Proceedings of the 43rd Hawaii International Conference on System Sciences*. Los Alamitos, CA: IEEE Computer Society Press.

Koestler, A. (1964). *The act of creation*. New York: Macmillan.

Kogut, B., & Zander, U. (1993). Knowledge of the firm and the evolutionary theory of the multinational corporation. *Journal of International Business Studies, 24*(4), 625–646. doi:10.1057/palgrave.jibs.8490248

Kogut, B., & Zander, U. (1996). What firms do? Coordination, identity, and learning. *Organization Science, 7*(5), 502–518. doi:10.1287/orsc.7.5.502

Kolb, D. A. (1974). Learning and problem solving on management and the learning process . In Kolb, D. A., Rubin, I. M., & McIntyre, J. M. (Eds.), *Organizational psychology: A book of readings* (pp. 27–42). Englewood Cliffs, NJ: Prentice-Hall.

Kolfschoten, G.L., Briggs, R.O., & Vreede, G.J., de, Jacobs, P.H.M. & Appelman, J.H. (2006). Conceptual foundation of the thinkLet concept for collaboration engineering. *International Journal of Human Computer Science, 64*(7), 611–621. doi:10.1016/j.ijhcs.2006.02.002

Kolfschoten, G. L., Briggs, R. O., & de Vreede, G. J. (2009). A technology for pattern-based process design and its application to collaboration engineering . In Rummler, S., & Ng, K. B. (Eds.), *Collaborative technologies and applications for interactive information design: Emerging trends in user experiences* (pp. 1–18). Information Science Reference.

Kolfschoten, G. L., & Veen, W. (2005). *Tool support for GSS session design.* Paper presented at the Hawaii International Conference on System Sciences, Kauai HI.

Kolfschoten, G. L., Niederman, F., de Vreede, G. J., & Briggs, R. O. (2008). *Roles in collaboration Support and the effect on sustained collaboration support.* (pp. 1-10). Paper presented at the Hawaii International Conference on System Science, Waikoloa HI.

Kontoghiorghes, C., Awbrey, S. M., & Feurig, P. L. (2005). Examining the relationship between learning organization characteristics and change adaptation, innovation, and organizational performance. *Human Resource Development Quarterly, 16*(2), 185–211. doi:10.1002/hrdq.1133

Koper, R., & Bennett, S. (2008). Learning design: Concepts. In H.H. Adelsberger, Kinshuk, J. M. Pawlowski & D. Sampson (Eds.), *Handbook on information technologies for education and training* (2nd ed., pp. 135-154). Heidelberg: Springer.

Kovács, I. (1993). Mudança tecnológica e organizacional: Análise de tendências na Indústria. *Socius Working Paper, 2*(94). 1-54.

Kubiszyn, T., & Borich, G. (2007). *Educational testing and measurement–classroom application and practice* (8th ed.). USA: John Wiley & Sons, Inc.

Kulicke, M., Stahlecker, T., Lo, V., & Wolf, B. (2006). *EXIST - Existenzgründungen aus Hochschulen. Bericht der wissenschaftlichen Begleitung zum Förderzeitraum 1998 bis 2005 (Kurzfassung)*. Berlin: Bundesministerium für Wirtschaft und Technologie. Retrieved September 7, 2007, from http://www1.isi.fraunhofer.de/p/download/EXIST/exist_kurzfassung.pdf

Kurzweil, R. (1999). *The age of spiritual machines: When computers exceed human intelligence*. New York: Viking.

Laberis, B. (1998). One big pile of knowledge. *Computerworld, 5*(32), 97.

Lakhani, K. R., & Wolf, R. (2005). Why hackers do what they do: Understanding motivation and effort in free/open source software projects . In Feller, J., Fitzgerald, B., Hissam, S., & Lakhani, K. (Eds.), *Perspectives on free and open source software*. Cambridge, MA: MIT Press.

Lam, A. (2004). *Organizational innovation*. MPRA Paper No. 11539.

Lambert, J. (2006). *Digital storytelling-capturing lives, creating community*. Berkeley, CA: Digital Diner Press.

Lambert, J. (2007). Digital storytelling–how digital media help preserve cultures. *The Futurist, 41*(2), 25.

Lange, A. (2009a). Entrepreneurs as exceptional: Indications from giftedness and expertise research. In M. Rebernik, B. Bradac & M. Rus (Eds.), *The winning products: Proceedings of the 29th Conference on Entrepreneurship and Innovation Maribor - PODIM, Maribor, 25th - 26th March 2009* (pp. 89-105). Maribor: IRP Institute for Entrepreneurship Research.

Lange, A. (2009b). Gender-specific beliefs about entrepreneurship: Specifications and implications. In J. Braet (Ed.), *The Proceedings of the 4th European Conference on Entrepreneurship and Innovation, The University of Antwerp, Belgium, 10-11 September 2009*, (pp. 273-280). Antwerp: University of Antwerp.

Lange, A. (2010). TPB-basierte Überzeugungen bei Studierenden mit unterschiedlichen Gründungsintentionen: Ausprägungen und Schlussfolgerungen. In A. Rese, D. Baier, M. Mißler-Behr & M. Kaiser (Eds.), *Entrepreneurship Education: Symposium des Brandenburgischen Instituts für Existenzgründung und Mittelstandsförderung (BIEM e. V.) "Gründung und Innovation" vom 11.-12. Juni 2009, Brandenburgische Technische Universität Cottbus* (pp. 83-109). Lohmar: Eul.

LaPlante, A. (1997). Sharing the wisdom. *Computerworld, 22*(41), 73.

Leadbeater, C. (2009). *We think*. London: Profile Books.

Leiponen, A. (2006). Managing knowledge for innovation: The case of business-to-business services. *Journal of Product Innovation Management, 23*, 238–258. doi:10.1111/j.1540-5885.2006.00196.x

Lentzy, J., & Weineck, G. (2009). *Linking findings in innovation management to entrepreneurship: Clustering potential entrepreneurs by innovator types*. Paper presented at the 1st International Conference on Strategic Innovation and Future Creation, March 2009, Malta.

Levitt, B., & March, J. G. (1988). Organizational learning. *Annual Review of Sociology, 14*, 319–340. doi:10.1146/annurev.so.14.080188.001535

Levitt, B., & March, J. (1988). Organizational learning. *Annual Review of Sociology, 14*, 319–340. doi:10.1146/annurev.so.14.080188.001535

Lim, C. P., & Tay, L. Y. (2003). Information and communication technologies (ICT) in an elementary school: Students' engagement in higher order thinking. *Journal of Educational Multimedia and Hypermedia, 12*(4), 425–451.

Litman, J. A. (2008). Interest and deprivation factors of epistemic curiosity. *Personality and Individual Differences, 44*(7), 1585–1595. doi:10.1016/j.paid.2008.01.014

Locke, E. A. (1999). *The essence of leadership: the four keys to leading successfully*. Boston: Lexington Books.

Lonergan, D. C., Scott, G. M., & Mumford, M. D. (2004). Evaluative aspects of creative thought: Effects of idea appraisal and revision standards. *Creativity Research Journal, 16*, 231–246. doi:10.1207/s15326934crj1602&3_7

López, S. P., Péon, J. M., & Órdas, C. J. (2004). Managing knowledge: The link between culture and organizational learning. *Journal of Knowledge Management, 8*(6), 93–104. doi:10.1108/13673270410567657

Louis, M. R. (1985). An investigator's guide to workplace culture . In Frost, P. J., Moore, F. L., Louis, M. R., Lundberg, C. C., & Martin, J. (Eds.), *Organizational culture* (pp. 73–93). Beverly Hills, CA: Sage.

Lu, Y., Tsang, E. W., & Peng, M. W. (2008). Knowledge management and innovation strategy in the Asia Pacific: Toward an institution-based view. *Asia Pacific Journal of Management, 25*, 361–374. doi:10.1007/s10490-008-9100-9

Lubart, T. I. (2001). Models of the creative process: Past, present, future. *Creativity Research Journal, 13*(3/4), 295–308. doi:10.1207/S15326934CRJ1334_07

Lubart, T. I. (2001). Models of the creative process: Past, present and future. *Creativity Research Journal, 13*(3/4), 295–308. doi:10.1207/S15326934CRJ1334_07

Luczkiw, E. (2008). Entrepreneurship education in an age of chaos, complexity and disruptive change . In Potter, J. (Ed.), *Entrepreneurship and higher education* (pp. 65–93). OECD Publishing. doi:10.1787/9789264044104-5-en

Lumsdaine, E., & Lumsdaine, M. (1995). *Creative problem solving: Thinking skills for a changing world.* McGraw-Hill.

Lundberg, C. C. (1985). On the feasibility of cultural intervention in organizations . In Frost, P. J., Moore, F. L., Louis, M. R., Lundberg, C. C., & Martin, J. (Eds.), *Organizational culture* (pp. 169–185). Beverly Hills, CA: Sage.

Lundvall, B., & Nielsen, P. (2007). Knowledge management and innovation performance. *International Journal of Manpower, 39*(3/4), 207–223. doi:10.1108/01437720710755218

Lundvall, B. A. (2001). Innovation policy in the globalising learning economy . In Archibuigi, D., & Lundvall, B. A. (Eds.), *The globalising learning economy*. Oxford: Oxford University Press.

Lyles, M. A., & Mitroff, I. I. (1980). Organisational problem formulation: An empirical study. *Administrative Science Quarterly, 25*, 102–119. doi:10.2307/2392229

Macaulay, L. A., Alabdulkarim, A., & Kolfschoten, G. L. (2006). *An analysis of the role of the facilitator and alternative scenarios for collaboration support.* Paper presented at the First HICSS Symposium on Case and Field Studies of Collaboration, Waikoloa HI.

Maccoby, M. (2003). The seventh rule: Create a learning culture. *Research Technology Management, 46*(3), 59–60.

Machin, S., & Vignoles, A. (2005). *What's the good of education? The economics of education in the UK.* Princeton: Princeton University Press.

Machlup, F., & Mansfield, U. (1983). *The study of information: Interdisciplinary messages*. New York: John Wiley & Sons.

MacKinnon, D. W. (1965). Personality and the realization of creative potential. *The American Psychologist, 20*(4), 273–281. doi:10.1037/h0022403

Mahler, J. (1997). Influences of organizational culture on learning in public agencies. *Journal of Public Administration: Research and Theory, 7*(4), 519–540.

Maidique, M. A., & Zirger, B. J. (1984). The new product learning cycle. *Research Policy, 14*, 299–313. doi:10.1016/0048-7333(85)90001-0

Majaro, S. (1988). *The creative gap–managing ideas for profit*. Longman.

Malafouris, L. (2009). Neuroarchaeology: Exploring the links between neural and cultural plasticity. *Progress in Brain Research, 178*, 253–261. doi:10.1016/S0079-6123(09)17818-4

Malhotra, Y. (1998). Tools @ work: Deciphering the knowledge management hype. *Journal for Quality and Participation, 4*(21), 58–60.

March, J. C., & Olsen, J. P. (1975). The uncertainty of the past: Organizational learning under ambiguity. *European Journal of Political Research, 3*, 147–171. doi:10.1111/j.1475-6765.1975.tb00521.x

March, J. G., & Simon, H. A. (1958). *Organizations*. New York: John Wiley and Sons.

Marples, D. L. (1961). The decisions of engineering design. *IEE Transactions on Engineering Management, 20*, 55–71. doi:10.1109/IRET-EM.1961.5007593

Marquardt, M. (1996). *Building the learning organization*. New York: McGraw-Hill.

Marsick, V. J., & Watkins, K. E. (2003). Demonstrating the value of an organization's learning culture: The dimensions of learning organization questionnaire. *Advances in Developing Human Resources, 5*(2), 132–151. doi:10.1177/1523422303005002002

Mårtensson, M. (2000). A critical review of knowledge management as a management tool. *Journal of Knowledge Management, 4*(3), 204–216. doi:10.1108/13673270010350002

Martin, J., & Meyerson, D. (1988). Organizational cultures and the denial, channeling and acknowledgement of ambiguity . In Pondy, L. R., Boland, R. J., & Howard, T. (Eds.), *Managing ambiguity and change* (pp. 93–125). New York: Wiley.

Martins, E. C., & Terblanche, F. (2003). Building organisational culture that stimulates creativity and innovation. *European Journal of Innovation Management, 6*(1), 64–74. doi:10.1108/14601060310456337

Martiny, M. (1998). Knowledge management at HP Consulting. *Organizational Dynamics, 2*(47), 71–77. doi:10.1016/S0090-2616(98)90025-0

Mason, J. (2001). *Qualitative researching*. London: Sage Publications.

Masterman, L., & Lee, S. (2005). *Evaluation of practitioner trial of LAMS: Final report to JISC*. Oxford University Computing Services.

Mayer, R. E. (1983). *Thinking, problem solving, and cognition*. San Francisco: Freeman.

Mayo, A. (1998). Memory bankers. *People Management, 2*(4), 34–38.

McAdam, R., & McClelland, J. (2002). Sources of new product ideas and creativity practices in the UK textile industry. *Technovation, 22*, 113–121. doi:10.1016/S0166-4972(01)00002-5

McAdam, R., & McClelland, J. (2002a). Individual and team-based idea generation within innovation management: Organizational and research agendas. *European Journal of Innovation Management, 5*(2), 86–97. doi:10.1108/14601060210428186

McAdam, R., Mason, B., & McCrory, J. (2007). Exploring the dichotomies within the tacit knowledge literature: Towards a process of tacit knowing in organizations. *Journal of Knowledge Management, 11*(2), 43–59. doi:10.1108/13673270710738906

McCall, M. W. (1977). Making sense with nonsense: Helping frames of reference clash . In Nystrom, P. C., & Starbuck, W. H. (Eds.), *Prescriptive model of organizations* (pp. 111–123). Amsterdam: North-Holland.

McEvily, S., & Chakravarthy, B. (2002). The persistence of knowledge-based advantage: An empirical test for product performance and technological knowledge. *Strategic Management Journal, 23*(4), 285–305. doi:10.1002/smj.223

McFadzean, E. (1998). The creativity continuum: Towards a classification of creative problem solving techniques. *Creativity and Innovation Management, 7*(3), 131. doi:10.1111/1467-8691.00101

McGill, M., & Slocum, J. Jr. (1993). Unlearning the organization. *Organizational Dynamics*, *22*(2), 67–79. doi:10.1016/0090-2616(93)90054-5

McKee, D. (1992). An organizational learning approach to product innovation. *Journal of Product Innovation Management*, *9*, 232–245. doi:10.1016/0737-6782(92)90033-9

McLean, L. (2005). Organizational culture's influence on creativity and innovation: A review of the literature and implications for human resource development. *Advances in Developing Human Resources*, *7*(2), 226–246. doi:10.1177/1523422305274528

McLuhan, M. (1964). *Understanding media: The extensions of man*. New York: McGraw-Hill.

McNaught, C. & Kennedy, P. (2000). Staff development at RMIT: Bottom-up work service by top-down investment and policy. *ALT-J*, 4-18.

McWilliam, E. (2008). *The creative workforce: How to launch young people into high flying futures*. Sydney: UNSW Press.

McWilliam, E., & Dawson, S. (2008). Teaching for creativity: Towards sustainable and replicable pedagogical practice. *Higher Education*, *56*(6), 633–643. doi:10.1007/s10734-008-9115-7

Means, B., & Golan, S. (1998). *Transforming teaching and learning with multimedia technology*. Menlo Park, CA: SRI International.

Mednick, S. A. (1962). The associative basis of the creative process. *Psychological Review*, *69*(3), 220–232. doi:10.1037/h0048850

Mehra, A., Dixon, A. L., Brass, D. J., & Robertson, B. (2006). The social network ties of group leaders: Implications for group performance and leader reputation. *Organization Science*, *17*(1), 64–79. doi:10.1287/orsc.1050.0158

Meindl, J. M., Stubbart, C., & Porac, J. F. (1996). Introduction . In Meindl, J. M., Stubbart, C., & Porac, J. F. (Eds.), *Cognition within and between organizations* (pp. ix–xxiii). Thousand Oaks, CA: Sage Publications.

Mendonça, S. (2008). Brave old world: Accounting for 'high-tech' knowledge in 'low-tech' industries. *Research Policy*, *38*, 470–482. doi:10.1016/j.respol.2008.10.018

Meredith, S. (2002). Models of e-learning: Technology promise vs learner needs literature. *The International Journal of Management Education*.

Merriam, S. B. (2001). Andragogy and self-directed learning: Pillars of adult learning theory. *New Directions for Adult and Continuing Education*, *89*(3).

Messick, S. (1989). Validity . In Linn, R. L. (Ed.), *Educational management* (3rd ed., pp. 13–103). London: Collier Macmillan.

Milligan, G. W. (1981). A Monte Carlo study of thirty internal criterion measures for cluster analysis. *Psychometrika*, *46*, 187–199. doi:10.1007/BF02293899

Milligan, G. W., & Cooper, M. C. (1985). An examination of procedures for determination of clusters in a data set. *Psychometrika*, *50*, 159–179. doi:10.1007/BF02294245

Milligan, G. W., & Cooper, M. C. (1988). A study of standardization of variables in cluster analysis. *Journal of Classification*, *5*, 181–204. doi:10.1007/BF01897163

Mintzberg, H. (1976). Planning on the left side and managing on the right side. *Harvard Business Review*, *54*, 49–58.

Mintzberg, H., Duru, R., & Theoret, A. (1976). The structure of unstructured decision process. *Administrative Science Quarterly*, *21*, 246–275. doi:10.2307/2392045

Mohr, L. B. (1982). *Explaining organization behavior*. San Francisco: Jossey-Bass.

Monteiro, I., & Sousa, F. (2008). A liderança inovadora na hotelaria algarvia. *Revista Portuguesa e Brasileira de Gestão*, *7*(2), 68–78.

Moran, J. (2005). A pedagogia e a didática da educação On-line . In Vigidal da Silva, R. (Ed.), *Educação, Aprendizagem e Tecnologia. Um paradigma para professores do século XXI*. Portugal: Edições Sílabo.

Moreira, P. (1997). *A Aprendizagem Organizacional, a Performance Organizacional e a Vantagem Competitiva Sustentável. Dissertação de Mestrado não publicada.* ISCTE - Instituto Superior das Ciências do Trabalho e da Empresa.

Morrison, D. G. (1967). Measurement problems in cluster analysis. *Management Science, 13*, B774–B780. doi:10.1287/mnsc.13.12.B775

Mostert, N. M. (2007). Diversity of the mind as the key to successful creativity at Unilever. *Creativity and Innovation Management, 16*(1), 93–100. doi:10.1111/j.1467-8691.2007.00422.x

Mouchiroud, C., & Lubart, T. (2006). Past, present, and future perspectives on creativity in France and French-speaking Switzerland. In J.C. Kaufman & R.J. Sternberg (Eds.), *The international handbook of creativity,* (pp. 96-123). NewYork: Cambridge University Press.

Moultrie, J., & Young, A. (2009). Exploratory study of organizational creativity in creative organizations. *Creativity and Innovation Management, 18*(4), 299–314. doi:10.1111/j.1467-8691.2009.00536.x

Mumford, M. D., & Baughman, W. A. (1996). Process-based measures of creative problem-solving skills: 1. Problem construction. *Creativity Research Journal, 9*, 63–76. doi:10.1207/s15326934crj0901_6

Mumford, M. D., Baughman, W. A., Maher, M. A., Costanza, D. P., & Supinski, E. P. (1997). Process-based measures of creative problem-solving skills 4: Category combination. *Creativity Research Journal, 10*, 59–71. doi:10.1207/s15326934crj1001_7

Mumford, M. D., Baughman, W. A., Supinski, E. P., & Maher, M. A. (1996). Process-based measures of creative problem-solving skills: 2. Information Encoding. *Creativity Research Journal, 9*, 77–88. doi:10.1207/s15326934crj0901_7

Mumford, M. D., Mobley, M. I., Uhlman, C. E., Reiter-Pamon, R., & Doares, L. (1991). Process analytic models of creative capacities. *Creativity Research Journal, 4*, 91–122. doi:10.1080/10400419109534380

Mumford, M. D., Supinski, E. P., Baughman, W. A., Costanza, D. P., & Threlfall, K. V. (1997). Process-based measures of creative problem-solving skills: 5. Overall prediction. *Creativity Research Journal, 10*, 77–85.

Mumford, M. D., Supinski, E. P., Threlfall, K. V., & Baughman, W. A. (1996). Process-based measures of creative problem-solving skills, 3: Category selection. *Creativity Research Journal, 9*, 395–406. doi:10.1207/s15326934crj0904_11

Mumford, M. D. (2000). Managing creative people: Strategies and tactics for innovation. *Human Resource Management Review, 10*(3), 313–351. doi:10.1016/S1053-4822(99)00043-1

Mumford, M. D., Mobley, M. I., Reiter-Palmon, R., Uhlman, C. E., & Doares, L. M. (1991). Process analytic models of creative capacities. *Creativity Research Journal, 4*(2), 91–122. doi:10.1080/10400419109534380

Mumford, M. D., Reiter-Palmon, R., & Redmond, M. R. (1994). Problem construction and cognition: Applying problem representations in ill-defined domains . In Runco, M. (Ed.), *Problem finding, problem solving, and creativity* (pp. 3–39). Norwood, NJ: Ablex.

Mumford, M. D., & Hunter, S. T. (2005). Innovation in organizations: A multi-level perspective on creativity . In Yammarino, F. J., & Dansereau, F. (Eds.), *Research in multi-level issues* (*Vol. 4*, pp. 11–74). Oxford, England: Elsevier.

Mumford, M.D., Lonergan, D.C. & Scott, G.M. (2002). Evaluating creative ideas: Processes, standards, and context. *Inquiry: Critical thinking across the disciplines, 22*, 21-30.

Munkvold, B. E., & Anson, R. (2001). *Organizational adoption and diffusion of electronic meeting systems: A case study.* (pp. 279–287). Paper presented at the Proceedings of ACM GROUP01, New York.

Murray, P. (2000). Designing for business benefits from knowledge management . In Despres, C., & Chauvel, D. (Eds.), *Horizons knowledge: The present and the promise of knowledge management.* Boston: Butterworth Heinemann.

Myers, I. B. (1962). *The Myers-Briggs type indicator*. Palo Alto, CA: Consulting Psychologists Press.

Nelson, R. R. (2004). The market economy and the scientific commons. *Research Policy*, *33*, 455–471. doi:10.1016/j.respol.2003.09.008

Nelson, R. R., & Winter, S. G. (1982). *An evolutionary theory of economic change*. Cambridge, MA: Harvard University Press.

Nelson, S. (2007). *Speech given to BBC Vision in-house multiplatform day*. Retrieved June 15, 2009, from http://www.bbc.co.uk/pressoffice/speeches/stories/nelson_multiplatform.shtml

Neo, K. T. K., & Neo, M. (2001). A constructivist learning experience: Reconstructing a web site using web based multimedia authoring tools. *Australian Journal of Educational Technology*, *17*(3), 330–350.

Nerney, C. (1997). Getting to know knowledge management. *New World (New Orleans, La.)*, *39*(14), 101.

Neto, E. D., & Figueiredo, A. D. (2001). The risks of systems and information technology: A multidimensional approach. *Proceedings of the 2nd Conference of the Portuguese Association of Information Systems*. Évora: Portuguese Association of Information Systems.

Nevis, E., DiBella, A., & Gould, J. (1995). Understanding organizations as learning systems. *Sloan Management Review*, *36*(2), 73–85.

Newell, A., & Simon, H. A. (1972). *Human problem solving*. Englewood Cliffs, NJ: Prentice Hall.

Newell, A., Shaw, J. C., & Simon, H. A. (1979). The processes of creative thinking . In Simon, H. A. (Ed.), *Models of thought* (pp. 144–174). New Haven: Yale University Press.

Nichols, P. D., Meyers, J. L., & Burling, K. S. (2009). A framework for evaluating and planning assessments intended to improve student achievement. *Educational Measurement: Issues and Practice*, *28*(3), 14–23. doi:10.1111/j.1745-3992.2009.00150.x

Nijstad, B. A., & Stroebe, W. (2006). How the group affects the mind: A cognitive model of ideation in groups. *Personality and Social Psychology Review*, *10*(3), 186–213. doi:10.1207/s15327957pspr1003_1

Ninan, A. (2005). *What are the roles of networks and clusters in the operation of an industry? The case of Queensland music*. Brisbane: Queensland University of Technology.

Nissen, M., Kamel, M., & Sengupta, K. (2000). Integrated analysis and design of knowledge systems and processes. *Information Resources Management Journal*, *13*(1), 24–43.

Nonaka, I. (1994). A dynamic theory of organizational knowledge creation. *Organization Science*, *5*, 14–36. doi:10.1287/orsc.5.1.14

Nonaka, I., & Konno, N. (1998). The concept of 'ba': Building a foundation for knowledge creation. *California Management Review*, *40*, 40–54.

Nonaka, I., & Takeuchi, H. (1995). *The knowledge-crating company: How Japanese companies create the dynamics of innovation*. New York: Oxford University Press.

Nova, C., & Alves, L. (2003). Estação online: a ciberescrita, as imagens e a EAD . In Silva, M. (Ed.), *Educação On-line. Sao Paolo: Edições Loyola.*

Nunamaker, J. F. Jr, Dennis, A., Valacich, J., Vogel, D., & George, J. F. (1991). Electronic meeting systems to support group work. *Communications of the ACM*, *34*(7), 40–61. doi:10.1145/105783.105793

Nunamaker, J. F. Jr, Briggs, R. O., Mittleman, D. D., Vogel, D., & Balthazard, P. A. (1997). Lessons from a dozen years of group support systems research: A discussion of lab and field findings. *Journal of Management Information Systems*, *13*(3), 163–207.

Nunnally, J. C. (1967). *Psychometric theory* (1st ed.). New York: McGraw Hill.

Nunnally, J. (1978). *Psychometric theory*. New York: Mc Graw-Hill.

Nystrom, P. C., & Starbuck, W. H. (1984). To avoid organizational crises. *Organizational Dynamics*, (Spring): 53–65. doi:10.1016/0090-2616(84)90011-1

Nystrom, H. (1990). *Technological and market innovation: Strategies for product and company development*. London: John Wiley & Sons.

O'Brien, J. A. (1993). *Management information systems: A managerial end user perspective*. Homewood, IL: Dow Jones Irwin.

O'Reilly, C. (1989). Corporations, culture and commitment: Motivation and social control in large organizations. *California Management Review*, *31*(4), 9–25.

Obstfeld, D. (2005). Social networks, the tertius iungens orientation, and involvement in innovation. *Administrative Science Quarterly*, *50*(1), 100–130.

OECD. (2005). *Oslo Manual-guidelines for collecting and interpreting innovation data* (3rd ed.). Paris.

Oke, A., Munshi, N., & Walumbwa, F. O. (2009). The influence of leadership on innovation process and activities. *Organizational Dynamics*, *38*(1), 64–72. doi:10.1016/j.orgdyn.2008.10.005

Oldham, G. R., & Cummings, A. (1996). Employee creativity: Personal and contextual factors at work. *Academy of Management Journal*, *39*(3), 607–655. doi:10.2307/256657

Oliveira, L., Albuquerque, A., Ribeiro, S., Peres, P., & Carvalho, C. V. (2009). A sustainable approach to the reuse of student-centered learning scenarios. *Proceedings of EDULEARN09 - International Conference on Education and New Learning Technologies*, Barcelona, Spain, 6-8 July.

Oliver, K., & Hannafin, M. (2000). Student management of web-based hypermedia resources during open-ended problem solving. *The Journal of Educational Research*, *94*(2), 75–92. doi:10.1080/00220670009598746

Olton, R. M. (1980). Experimental studies of incubation: Searching for the elusive. *The Journal of Creative Behavior*, *13*, 9–13.

Orlikowski, W. J. (2002). Knowing in practice: Enacting a collective capability in distributed organizing. *Organization Science*, *13*(3), 249–273. doi:10.1287/orsc.13.3.249.2776

Orlikowski, W., & Hofman, D. (1996). *An improvisational model of change management: The case of groupware technologies*. Massachusetts Institute of Technology, Sloan School of Management Working Paper #3869, Cambridge.

Osborn, A. F. (1963). *Applied imagination: Principles and procedures of creative thinking* (2nd ed.). New York: Scribner's & Sons, Inc.

Ostro, N. (1997). The corporate brain. *Chief Executive*, *123*, 58–62.

Otte, E., & Rousseau, R. (2002). Social network analysis: A powerful strategy, also for the information sciences. *Journal of Information Science*, *28*(6), 441–453. doi:10.1177/016555150202800601

Palfrey, J., & Gasser, U. (2008). *Born digital: Understanding the first generation of digital natives*. New York: Basic Books.

PAOL. (2008). *Uma Reflexão - Quarta Mesa Redonda*. Instituto Superior de Contabilidade e Administração do Porto.

PAOL. (2010). *Projecto de Apoio On-line*. Retrieved January 2010, from http://www.iscap.ipp.pt/paol.

Paquin, R. (2007). *Brokering industrial ecosystems: Creating regional institutional forms.* Paper presented at the Conference on the Institutional Foundations for Industry Self-Regulation, Boston, MA.

Parker, S. K. (1998). Enhancing role breadth self-efficacy: The role of job enrichment and other organizational interventions. *The Journal of Applied Psychology*, *83*(6), 835–852. doi:10.1037/0021-9010.83.6.835

Parnes, S. J., & Noller, R. B. (1972). Applied creativity: The creative studies project: Part 2–Results of the two-year program. *The Journal of Creative Behavior*, *6*, 164–186.

Parnes, S. J. (1987). The creative studies project . In Isaksen, S. G. (Ed.), *Frontiers of creativity research: Beyond the basics* (pp. 156–188). Buffalo, NY: Bearly Limited.

Parnes, S. J. (1991). Creative problem solving and visionizing . In Parnes, S. J. (Ed.), *Source book for creative problem-solving: A fifty year digest of proven innovation processes*. Buffalo, NY: Creativity Education Foundation.

Paulus, P. B. (1999). Group creativity . In Runco, M. A., & Pritzker, S. R. (Eds.), *Encyclopedia of creativity* (pp. 779–795). CA: Academic Press.

Paulus, P. (2008). Fostering creativity in groups and teams . In Zhou, J., & Shalley, C. (Eds.), *Handbook of organizational creativity*. New York: Lawrence Erlbaum Associates.

Pedler, M., Burgoyne, J., & Boydell, T. (1991). *The learning company*. Maidenhead: McGraw-Hill.

Pentland, B. T., & Rueter, H. H. (1994). Organizational routines as grammar of actions. *Administrative Science Quarterly*, *39*, 484–510. doi:10.2307/2393300

Pereira, C. (2004). *Educação em Ciência e Tecnologia para o Séc. XXI. Necessidade de mudar de paradigma.*

Perkins, P. S. (2008). *The art and science of communication: Tools for effective communication in the workplace*. Hoboken, N.J.: John Wiley & Sons.

Perrow, C. (1967). *Organizational analysis: A sociological view*. Belmont, CA: Wadsworth.

Perry, N. (1993). Scientific communication, innovation networks and organization structures. *Journal of Management Studies*, *30*(6), 957–973. doi:10.1111/j.1467-6486.1993.tb00474.x

Perry-Smith, J. E. (2006). Social yet creative: The role of social relationships in facilitating individual creativity. *Academy of Management Journal*, *49*(1), 85–101.

Perry-Smith, J. (2008). When being social facilitates creativity . In Zhou, J., & Shalley, C. (Eds.), *Handbook of organizational creativity*. New York: Lawrence Erlbaum Associates.

Peters, T., & Waterman, R. (1982). *In search of excellence: Lessons from Americas best run companies*. New York: Warner Books.

Pettigrew, A. M. (2000). Foreword . In Ashakanasy, N. M., Wilderom, C. P., & Peterson, M. F. (Eds.), *Handbook of organizational culture and climate* (pp. xiii–xv). Thousand Oaks: Sage.

Phillips, R. (2000). *Handbook for learning-centered evaluation of computer-facilitated learning projects in higher education*. Murdoch University.

Picken, J., & Dess, G. (1997). Out of (strategic) control. *Organizational Dynamics*, (Summer): 35–48. doi:10.1016/S0090-2616(97)90026-7

Pierce, J. L., & Delbecq, A. L. (1977). Organization structure, individual attitudes and innovation. *Academy of Management Review*, *2*(1), 27–37. doi:10.2307/257602

Pinchot, E. & Pinchot, G. (1996). Seeding a climate for innovation. *Executive Excellence, June,* 17-18.

Pink, D. H. (2005). *A whole new mind*. New York: Penguin.

Pisano, G. P. (1994). Knowledge, integration, and the locus of learning: An rmpirical snalysis of process development. *Strategic Management Journal*, *15*, 85–100. doi:10.1002/smj.4250150907

Pol, E., & Ville, S. (2009). Social innovation: Buzz word or enduring term? *Journal of Socio-Economics*, *38*, 878–885. doi:10.1016/j.socec.2009.02.011

Polanyi, M. (1958). *Personal knowledge: Toward a post-critical philosophy*. Chicago: University of Chicago Press.

Polanyi, M. (1966). *The tacit dimension*. London: Routdedge & Kegan Paul.

Polito, T., & Watson, K. (2002). Toward an interdisciplinary organizational learning framework. *The Journal of American Academy of Business*, *2*(1), 162–166.

Polya, G. (1945). *How to solve it: A new aspect of mathematical method*. Princeton University Press. reprinted in 2004

Popper, M., & Lipshitz, R. (2000). Organizational learning: Mechanisms, culture, and feasibility. *Management Learning, 31*(2), 181–196. doi:10.1177/1350507600312003

Post, B. Q. (1993). A business case framework for group support technology. *Journal of Management Information Systems, 9*(3), 7–26.

Potts, J., & Cunningham, S. (2008). Four models of the creative industries. *International Journal of Cultural Policy, 14*(3), 233–247. doi:10.1080/10286630802281780

Pounds, W. F. (1969). The process of problem finding. *Industrial Management Review, 11*, 1–19.

Powell, W., Koput, K., & Smith-Doerr, L. (1996). Interorganisational collaboration and locus of innovation: Networks of learing in biotechnology. *Administrative Science Quarterly, 41*, 116–145. doi:10.2307/2393988

Preiser, S. (2006). Creativity research in German-speaking countries . In Kaufman, J. C., & Sternberg, R. J. (Eds.), *The international handbook of creativity* (pp. 167–201). New York: Cambridge University Press.

Prensky, M. (2001). Digital natives, digital immigrants part 2: Do they really think differently? *On The Horizon - The Strategic Planning Resource for Education Professionals, 9*(6), 1-6.

Prietula, M., & Simon, H. (1998). The experts in your midst. *Harvard Business Review*, (January-February): 120–124.

Prusak, L. (1996). The knowledge advantage. *Planning Review, 2*(24), 6–8.

Puccio, G. J., Firestien, R. L., Coyle, C., & Masucci, C. (2006). A review of the effectiveness of CPS training: A focus on workplace issues. *Creativity and Innovation Management, 15*(1), 19–33. doi:10.1111/j.1467-8691.2006.00366.x

Puccio, G. J., Treffinger, D. J., & Talbot, R. J. (1995). Exploratory examination of relationships between creativity styles and creative products. *Creativity Research Journal, 8*(2), 157–172. doi:10.1207/s15326934crj0802_4

Puccio, G. J., Cabra, J. F., Fox, M. J., & Cahen, H. (2010). Creative problem-solving, history and future. *Artificial Intelligence for Engineering Design, Analysis and Manufacturing, 24*, 153–159. doi:10.1017/S0890060410000028

Puccio, G. J., Murdock, M., & Mance, M. (2007). *Creative leadership: Skills that drive change*. Thousand Oaks, CA: SAGE Publications.

Puccio, G. J., & Cabra, J. F. (2009). Creative problem solving: Past, present and future. In T. Rickards, H. A. Runco. & S. Moger (Eds.), *The Routledge companion to creativity,* (pp. 327-337). Oxford: Routledge.

Quinn, J. B. (1985). Managing innovation: Controlled chaos. *Harvard Business Review, 63*(3), 73–84.

Raaheim, K. (1974). *Problem solving and intelligence*. Oslo: Norwegian Universities Press.

Ramboll. (2004). *Studies in the context of the e-learning initiative*. Virtual Models of European Universities (Lot 1). Draft Final Report to the EU Commission, DG Education and Culture.

Ramos, F., Moreira, A., & Santos, A. (2001). Towards a reference framework for e-learning management. *Proceedings of DLA 2001 - Distance Learning Administration Conference, U.S.A.*

Rank, J., Pace, V. L., & Frese, M. (2004). Three avenues for future research on creativity, innovation, and initiative. *Applied Psychology: An International Review, 53*(4), 518–528. doi:10.1111/j.1464-0597.2004.00185.x

Ray, G., Barney, J. B., & Muhanna, W. A. (2004). Capabilities, business processes, and competitive advantage: choosing the dependent variable in empirical tests of the resource-based view. *Strategic Management Journal, 25*(1), 23–37. doi:10.1002/smj.366

Rebelo, T., & Almeida, N. (2008). Cultura de aprendizagem: O papel das lideranças intermédias. *Psychologica, 47*, 125–143.

Rebelo, T., Gomes, A. D., & Cardoso, L. (2002). Orientações culturais para a aprendizagem nas organizações: homogeneidade e/ou heterogeneidade. *Psychologica, 30*, 345–363.

Rebelo, T., Gomes, A. D., & Cardoso, L. (2005). Cultura de aprendizagem: A (bi)dimensionalidade da escala OCA. *Psychologica, 38*, 191–208.

Rebelo, T., & Gomes, A. (2009). Different types of organization, different cultural orientations towards learning: What factors explain this? In Fanti, K. A. (Ed.), *Applying psychological research to understand and promote the well-being of clinical and non-clinical populations* (pp. 175–186). Athens: ATINER.

Rebelo, T. (2001). *Organização, aprendizagem e cultura: Estudo sobre a homogeneidade/heterogeneidade da orientação cultural para a aprendizagem*. Unpublished master dissertation, University of Coimbra, Portugal.

Rebelo, T. (2006). Orientação cultural para a aprendizagem nas organizações: Condicionantes e consequentes. Unpublished doctoral dissertation, University of Coimbra, Portugal.

Rebelo, T., Gomes, A., & Cardoso, L. (2007). *Is learning culture a good bet? A cultural orientation and its impact on firm performance.* Paper presented at the 13th European Congress of Work and Organizational Psychology, Stockholm, Sweden.

Rebelo, T., Gomes, A. D., & Cardoso, L. (2003). Cultures d'apprentissage: L'échelle OCA. In C. Vandenberghe, N. Delobbe & G. Karnas (Eds). *Développement des compétences, investissement professionnel et bien-être des personnes – Actes du 12e Congrès de Psychologie du Travail et des Organisations: Vol 2. Dimensions individuelles et sociales de l'investissement professionnel,* (pp. 497-507). Louvain–la-Neuve: Presses Universitaires de Louvain.

Redway, K. (2003). *Make it happen! A step-by-step guide from creativity and innovation.* UK: Judy Pitkus.

Reeves, T. (1996). Rogue learning on the company reservation. *The Learning Organization, 3*(2), 20–29. doi:10.1108/09696479610113774

Reigeluth, C. M. (1996). A new paradigm of ISD? *Educational Technology, 36*(3), 13–20.

Rein, M., & Schön, D. (1977). Problem setting in policy research . In Weiss, C. (Ed.), *Using social research in public policy making* (pp. 235–251). Lexington, MA: D. C. Heath.

Reiter-Palmon, R., Herman, A. E., & Yammarino, F. J. (2008). Creativity and cognitive processes: Multi-level linkages between individual and team cognition . In Mumford, M. D., Hunter, S. T., & Bedell-Avers, K. E. (Eds.), *Multi-level issues in creativity and innovation* (pp. 203–267). Oxford: Elsevier.

Rhodes, R. H., Hill, C. E., Thompson, B. J., & Elliott, R. (1994). Client retrospective recall of resolved and unresolved misunderstanding events. *Journal of Counseling Psychology, 41*(4), 473–483. doi:10.1037/0022-0167.41.4.473

Rickards, T., & Freedman, B. (1978). Procedures for management in idea deficient situations: An examination of brainstorming approaches. *Journal of Management Studies, 15*, 43–55. doi:10.1111/j.1467-6486.1978.tb00908.x

Robertson, B. F., & Radcliff, D. F. (2009). Impact of CAD tools on creative problem solving in engineering design. *Computer Aided Design, 41*, 136–146. doi:10.1016/j.cad.2008.06.007

Robinson, A. (1998). *Corporate creativity: How innovation & improvement actually happen.* San Francisco: Berret Koehler Publishers.

Robinson, K. (2000). *Out of our minds: Learning to be creative*. Oxford: Capstone.

Rodriguez, E., Franco, T., & Santos, M. (2006). *Nature and antecedents of organizational commitment: Considerations for human resources management.* ISEG Working Papers, 12, 75-86.

Roe, A. (1976). Psychological approaches to creativity in science . In Rothenberg, A., & Hausman, C. R. (Eds.), *The creativity question* (pp. 165–175). Durham, NC: Duke University Press.

Roffe, I. (1999). Innovation and creativity in organizations: a review of the implications for training and development. *Journal of European Industrial Training, 23*(4), 224–237. doi:10.1108/03090599910272103

Rogers, E. (1962/2003). *Diffusion of innovations* (5th ed.). New York: Free Press.

Romano, N. C., Jr., Nunamaker, J. F., Jr., Briggs, R. O., & Mittleman, D. D. (1999). *Distributed GSS facilitation and participation: Field action research.* Paper presented at the Hawaii International Conference on System Sciences, Waikoloa. HI.

Rooney, D., Hearn, G., Mandeville, T., & Joseph, R. (2003). *Public policy in knowledge-based economies: Foundations and frameworks*. Cheltenham, UK: Edward Elgar.

Rooney, D., & Schneider, U. (2005). A model of the material, mental, historical and social character of knowledge. In Rooney, D., Hearn, G., & Ninan, A. (Eds.), *Handbook on the knowledge economy* (pp. 19–36). Cheltenham, UK: Edward Elgar.

Rosenberg, V. (1967). Factors affecting the preferences of industrial personnel for information gathering methods. *Information Storage and Retrieval, 3*, 119–127. doi:10.1016/0020-0271(67)90005-8

Rosenberg, N. (1982). Learning by using. In Rosenberg, N. (Ed.), *Inside the black box: Technology and economics* (pp. 120–140). Cambridge, UK: Cambridge University Press.

Rostan, S. M. (1997). A study of young artists: The development of artistic talent and creativity. *Creativity Research Journal, 10*, 175–192. doi:10.1207/s15326934crj1002&3_6

Ruggles, R. L. (1999). The state of the notion: Knowledge management in practice. In Rutkowski, M. L., & Stasko, S. R. (Eds.), *Knowledge management: People make the difference*. The Knowledge Chronicle, Walsh College.

Rumelt, R., Schendel, D. & Teece, D.J. (1992). Strategic management and economics. *Strategic Management Journal, Winter Special Issue, 12*, 5-29.

Runco, M. A., & Okuda, S. M. (1988). Problem-discovery, divergent thinking, and the creative process. *Journal of Youth and Adolescence, 17*, 211–220. doi:10.1007/BF01538162

Runco, M. A. (2004). Creativity. *Annual Review of Psychology, 55*, 657–687. doi:10.1146/annurev.psych.55.090902.141502

Runco, M., & Chand, I. (1995). Cognition and creativity. *Educational Psychology Review, 7*(3), 243–267. doi:10.1007/BF02213373

Runco, M. A. (2004). Creativity. *Annual Review of Psychology, 55*(1), 657–687. doi:10.1146/annurev.psych.55.090902.141502

Runco, M. A., & Okuda, S. M. (1988). Problem discovery, divergent thinking, and the creative process. *Journal of Youth and Adolescence, 17*(3), 211–220. doi:10.1007/BF01538162

Runco, M. A. (2007). *Creativity theories and themes: Research, development and practice*. Burlington, MA: Elsevier Academic Press.

Rushton, K. (2008). Extra job losses loom at BBC. *Broadcast Magazine.* Retrieved June 19th, 2009, from http://www.broadcastnow.co.uk/news/multi-platform/news/extra-job-losses-loom-at-bbc/1924060.article

Sadler-Smith, E., & Badger, B. (1998). Cognitive style, learning and innovation. *Technology Analysis and Strategic Management, 10*(2), 247–265. doi:10.1080/09537329808524314

Salaman, G. (2001). A response to Snell: The learning organization: Fact or fiction? *Human Relations, 54*(3), 343–359. doi:10.1177/0018726701543004

Salas, E., & Cannon-Bowers, J. A. (2001). The science of training: A decade of progress. *Annual Review of Psychology, 52*(1), 471–499. doi:10.1146/annurev.psych.52.1.471

Salmon, G. (2002). *e-Learning works: Learning from the past, present and future*. WORLD DIDAC 2002, Zurich. Retrieved from http://www.atimod.com/research/presentations/ZURICHSALMON1.htm

Salter, A., & Gann, D. (2003). Sources of ideas for engineering design. *Research Policy, 32*, 1309–1324. doi:10.1016/S0048-7333(02)00119-1

Santana, S. (1999). *Tecnologias de informação e da comunicação e pequenas e médias empresas: Uma abordagem centrada na aprendizagem organizacional*. Unpublished doctoral dissertation, University of Aveiro, Portugal.

Santanen, E. L., Briggs, R. O., & de Vreede, G. J. (2004). Causal relationships in creative problem solving: Comparing facilitation interventions for ideation. *Journal of Management Information Systems, 20*(4), 167–197.

Santanen, E. L. (2005). *Resolving ideation paradoxes: Seeing apples as oranges through the clarity of ThinkLets.* Paper presented at the Hawaii International Conference on System Sciences, Kauai HI.

Santanen, E. L., & de Vreede, G. J. (2004). *Creative approaches to measuring creativity: Comparing the effectiveness of four divergence Thinklets.* Paper presented at the Hawaiian International Conference on System Sciences, Waikoloa HI.

Santanen, E. L., Briggs, R. O., & de Vreede, G. J. (2002). Toward an understanding of creative solution generation. *Proceedings of the 35th Hawaii International Conference on System Sciences*. IEEE Computer Society Press.

Sarapuu, T., & Adojann, K. (1999). *Usage of educational web pages to develop students' higher-order thinking skills*. Science Didactics Department, University of Tartu.

Sarmento, A., Batista, J., Cardoso, L., Lousã, M., Rebelo, T., Babo, R., et al. (2000). Gestão do Conhecimento e Sistemas de Informação: Que relação? *Proceedings of the VIII Accounting and Auditing Congress*, Aveiro: ISCAA. CD-ROM Edition.

Sá-Soares, D. (1998). *Planeamento de sistemas de informação: Estudo das variáveis que condicionam a sua estratégia de execução*. Unpublished master thesis, University of Minho, Portugal.

Sauter, V. (2003). Web design studio: A preliminary experiment in facilitating faculty use of internet . In Aggarvil, A. (Ed.), *Web-based education: Learning from experience* (pp. 131–154). Information Science Publishing.

Sawyer, K. (2007). *Group genius: The creative power of collaboration*. New York: Basic Books.

Scarborough, H. (1993). Problem-solutions in the management of information systems expertise. *Journal of Management Studies, 30*(6), 939–955. doi:10.1111/j.1467-6486.1993.tb00473.x

Scarbrough, H. (2003). Knowledge management, HRM and the innovation process. *International Journal of Manpower, 24*(5), 501–516. doi:10.1108/01437720310491053

Schein, E. H. (1992). *Organizational culture and leadership* (2nd ed.). San Francisco: Jossey-Bass.

Schein, E. H. (1993). On dialogue, culture and organizational learning. *Organizational Dynamics*, (Winter): 40–51. doi:10.1016/0090-2616(93)90052-3

Schein, E. H. (1994). *Organizational and managerial culture as a facilitator or inhibitor of organizational learning*. M.I.T. - Sloan School of Management.

Schein, E. H. (1996). Culture: The missing concept in organization studies. *Administrative Science Quarterly, 41*(2), 229–241. doi:10.2307/2393715

Schein, E. H. (1997). *Three cultures of management: The key to organizational learning in the 21st century*. M.I.T. - Sloan School of Management.

Schein, E. H. (1996). Organizational learning as a cognitive re-definition: Coercive persuasion revisited. Retrieved from http:/learning.mit.edu/res/wp/10010.html

Scherer, J. (2007). *Kreativitätstechniken. In 10 Schritten Ideen finden, bewerten, umsetzen*. Offenbach, Germany: Gabal.

Schneider, B., Brief, A., & Guzzo, R. (1996). Creating a climate and culture for sustainable change. *Organizational Dynamics*, (Summer): 7–20. doi:10.1016/S0090-2616(96)90010-8

Schneiderman, B. (1998). Codex, memex, genex: The pursuit of transformational technologies. *International Journal of Human-Computer Interaction, 10*(2), 87–106. doi:10.1207/s15327590ijhc1002_1

Schneiderman, B. (2000). Creating creativity: User interfaces for supporting innovation. *Transactions on Computer-Human Interaction, 7*(1), 114–138. doi:10.1145/344949.345077

Schneiderman, B. (2002). Creativity support tools. *Communications of the ACM, 45*(10), 116–120.

Schneiderman, B., Hewett, T., Fischer, G., Jennings, P., Czerwinski, M., & Kules, B. (2006). Creativity support tools: Report from a U.S. National Science Foundation sponsored workshop. *International Journal of Human-Computer Interaction, 20*(2), 61–77. doi:10.1207/s15327590ijhc2002_1

Schneiderman, B. (1999) User interfaces for creativity support tools. *Proceedings of the 3rd Conference on Creativity & Cognition,* (pp. 15-22). New York: ACM Press.

Schoonhoven, C. B. (1981). Problems with contingency theory: Testing assumptions hidden within the language of contingency theory. *Administrative Science Quarterly, 26*, 349–377. doi:10.2307/2392512

Schrader, S., Riggs, W. M., & Smith, R. P. (1991). Problem solving in the management of technology and innovation: Choosing the uncertainty-ambiguity boundary. MIT Working Paper # 3345-91-BPS.

Schumpeter, J. A. (1934). *Invention and economic growth.* Cambridge: Harvard University Press.

Schumpeter, J. (1911/1934). *The theory of economic development.* Cambridge: Harvard University Press.

Schumpeter, J. (1939/1961). *Business cycles.* New York: MacGraw Hill.

Schumpeter, J. (1947/1976). *Capitalism, socialism and democracy* (5th ed.). London: George Allen & Unwin.

Schwartz, S. J., & Waterman, A. S. (2006). Changing interests: A longitudinal study of intrinsic motivation for personality salient activities. *Journal of Research in Personality, 40*(6), 1119–1136. doi:10.1016/j.jrp.2005.12.003

Schwenk, C., & Thomas, H. (1983). Formulating the mess: The role of decision aids in problem formulation. *Omega, 11*, 239–252. doi:10.1016/0305-0483(83)90049-X

Scott, G. M., Leritz, L. E., & Mumford, M. D. (2004). The effectiveness of creativity training: A quantitative review. *Creativity Research Journal, 16*, 361–388. doi:10.1080/10400410409534549

Scott, S., & Bruce, R. (1994). The influence of leadership, individual attributes, and climate on innovative behavior: A model of individual innovation in the workplace. *Academy of Management Journal, 37*, 580–607. doi:10.2307/256701

Scott, S. G., & Bruce, R. A. (1994). Determinants of innovative behavior: A path model of individual innovation in the workplace. *Academy of Management Journal, 37*(3), 580–607. doi:10.2307/256701

Seibert, S. E., Kraimer, M. L., & Liden, R. C. (2001). A social capital theory of career success. *Academy of Management Journal, 44*(2), 219–237. doi:10.2307/3069452

Senge, P. M. (1990). *The fifth discipline.* New York: Doubleday.

Senge, P., Kleiner, A., Roberts, C., Ross, R., Roth, G., & Smith, B. (1999). *The dance of change: The challenges to sustaining momentum in learning organizations.* New York: Doubleday Currency.

Shalley, C. E., Zhou, J., & Oldham, G. R. (2004). The effects of personal and contextual characteristics on creativity: Where should we go from here? *Journal of Management, 30*(6), 933–958. doi:10.1016/j.jm.2004.06.007

Shalley, C. (2008). Creating roles . In Zhou, J., & Shalley, C. (Eds.), *Handbook of organizational creativity.* New York: Lawrence Erlbaum Associates.

Shane, S., & Venkataraman, S. (2000). The promise of entrepreneurship as a field of research. *Academy of Management Review*, *25*(1), 217–226. doi:10.2307/259271

Shapiro, S. M. (2001). *24/7 Innovation: A blueprint for surviving and thriving in an age of change*. New York: McGraw-Hill.

Sharma, S., Durand, R. M., & Gur-Arie, O. (1981). Identification and analysis of moderator variables. *JMR, Journal of Marketing Research*, *18*, 291–300. doi:10.2307/3150970

Sherry, L. (1998b). Diffusion of the Internet within a graduate school of education. Doctoral dissertation of Philosophy - Educational Leadership and Innovation, University of Colorado.

Shneiderman, B. (2000). Creating creativity: User interfaces for supporting innovations. *ACM Trans. On Computer-Human Interaction, 7*(1), 114-138.

Siegle, D. (2008). Promoting creativity through curiosity and engagement: Wonder is not enough. *Parenting for High Potential,* 3.

Silber, K. (1998). The cognitive approach to training development: A practitioner's assessment. *Educational Technology Research and Development*, *46*(4), 58–72. doi:10.1007/BF02299674

Silva, R. (2005). Gestão da aprendizagem e do conhecimento . In Silva, R., & Silva, A. (Eds.), *Educação, Aprendizagem e Tecnologia: um paradigma para professores do século XXI* (pp. 43–65). Edições Sílabo.

Sim, E. W., Griffin, A., Price, R. L., & Vojak, B. A. (2007). Exploring differences between inventors, champions, implementers and innovators in creating and developing new products in large, mature firms. *Creativity and Innovation Management*, *16*(4), 422–436. doi:10.1111/j.1467-8691.2007.00457.x

Simmel, G. (1950). *The sociology of Georg Simmel* (Wolff, K., Trans.). Glencoe, IL: Free Press.

Simon, H. (1973). The structure of ill structured problems. *Artificial Intelligence*, *4*(3-4), 181–201. doi:10.1016/0004-3702(73)90011-8

Simon, H. A. (1979). Rational decision-making in business organizations. *The American Economic Review*, *69*, 493–513.

Simon, H. A. (1987). Making management systems: The role of intuition and emotion. *The Academy of Management Executive*, *1*, 57–64.

Simonton, D. K. (1988). *Scientific genius: A psychology of science*. New York: Cambridge University Press.

Sio, U. N., & Ormerod, T. C. (2009). Does incubation enhance problem solving? A meta-analytic review. *Psychological Bulletin*, *135*(1), 94–120. doi:10.1037/a0014212

Siragusa, L., & Dixon, K. (2005). *Closing the gap between pedagogical theory and online instructional design: A bridge too far: Methods and technologies for learning.* Southampton, UK: WIT Press.

Škerlavaj, M., Štemberger, M., Škrinjar, R., & Dimovski, V. (2007). Organizational learning culture–the missing link between business process change and organizational performance. *International Journal of Production Economics*, *106*, 346–367. doi:10.1016/j.ijpe.2006.07.009

Škerlavaj, M., Song, J.H. & Lee, Y. (2010). Organizational learning culture, innovative culture and innovations in South Korean firms. *Expert Systems and Applications.*

Skovvang, M., Elback, M. K., & Hertzum, M. (2006). Personometrics: Mapping and visualizing communication patterns in R&D projects. In F. Crestani and I. Ruthven (Eds.), *COLIS5: Proceedings of the Fifth International Conference on Conceptions of Library and Information Sciences (Glasgow, UK, June 5-8)*, (pp. 141-154).

Smilansky, J. (1984). Problem solving and the quality of invention: An empirical investigation. *Journal of Educational Psychology*, *76*, 377–386. doi:10.1037/0022-0663.76.3.377

Smith, G. F. (1989a). Defining managerial problems: A framework for prescriptive theorizing. *Management Science*, *35*, 963–981. doi:10.1287/mnsc.35.8.963

Smith, G. F. (1989b). Managerial problems identification. *OMEGA International Journal of Management Sciences*, *17*, 27–36. doi:10.1016/0305-0483(89)90017-0

Smith, G. F. (1998). Idea-generation techniques: A formulary of active ingredients. *The Journal of Creative Behavior*, *32*(2), 107–133.

Smith, G. J. W., & Carlsson, I. (2006). Creativity under the northern lights. Perspectives from Scandinavia . In Kaufman, J. C., & Sternberg, R. J. (Eds.), *The international handbook of creativity* (pp. 202–234). New York: Cambridge University Press.

Soliman, F., & Spooner, K. (2000). Strategies for implementing knowledge management: Role of human resources management. *Journal of Knowledge Management*, *4*(4), 337–345. doi:10.1108/13673270010379894

Sonnenburg, S. (2004). Creativity in communication: A theoretical framework for collaborative product creation. *Creativity and Innovation Management*, *13*(4), 254–262. doi:10.1111/j.0963-1690.2004.00314.x

Sousa, F. (2004). Innovación, creatividad y liderazgo organizacional. *Creatividad y Sociedad*, *5*, 11–17.

Sousa, F., & Andrade, C. (2007). Factores organizacionais e pessoais como determinantes da percepção do clima organizacional. *Inuaf-Studia*, *10*, 219–235.

Sousa, C. (1999). A gestão do conhecimento ou as virtudes da partilha. *Recursos Humanos Magazine*, *1*(4), 20–24.

Sousa, S. (1997). *Tecnologias de informação: O que são? Para que servem?*Lisboa: Editora de Informática.

Sousa, F. (2008). Still the elusive definition of creativity. *International Journal of Psychology: A Biopsychosocial Approach, 2*, 55-82.

Spearman, C. (1927). *The abilities of man*. New York: Macmillan.

Spence, W. R. (1994). *Innovation: The communication of change in ideas, practices and products*. London: Chapman & Hall.

Spender, J. C. (1996). Making knowledge the basis of a dynamic theory of the firm. *Strategic Management Journal*, *17*, 45–62.

Sta. Maria, R. F. (2003). Innovation and organizational learning culture in the Malaysian public sector. *Advances in Developing Human Resources*, *5*(2), 205–214. doi:10.1177/1523422303005002008

Stacey, R. D. (2001). *Complex responsive processes in organizations: Learning and knowledge creation*. London: Routledge.

Stake, R. E. (1995). *The art of case study research*. Thousand Oaks, CA: SAGE.

Stake, R. E. (2006). *Multiple case study analysis*. New York: The Guilford Press.

Stam, W., & Elfring, T. (2008). Entrepreneurial orientation and new venture performance: The moderating role of intra-and extra-industry social capital. *Academy of Management Journal*, *51*(1), 97–111.

Stein, M. I., & Heinze, S. J. (1960). *Creativity and the individual*. Illinois, USA: Free Press of Glencoe.

Stein, M. (1975). Stimulating creativity: *Vol. 2. Group procedures*. London: Academic Press.

Sternberg, R. J., O'Hara, L. A., & Lubart, T. I. (1997). Creativity as investment. *California Management Review*, *40*(1), 8–21.

Sternberg, R. J. (Ed.). (1999). *Handbook of creativity*. New York: Cambridge University Press.

Sternberg, R. J. (1998). Principles of teaching for successful intelligence. *Educational Psychologist*, *33*(2/3), 64–72.

Sternberg, R. J. (2006b). The nature of creativity. *Creativity Research Journal*, *18*(1), 87–98. doi:10.1207/s15326934crj1801_10

Sternberg, R. J., & Kaufman, J. C. (1998). Human abilities. *Annual Review of Psychology*, *49*(1), 479–502. doi:10.1146/annurev.psych.49.1.479

Sternberg, R. J. (2006a). Introduction . In Kaufman, J. C., & Sternberg, R. J. (Eds.), *The international handbook of creativity* (pp. 1–9). New York: Cambridge University Press.

Stevens, G. A., & Burley, J. (1997). 3000 raw ideas equals 1 commercial success. *Research Technology Management*, *40*(3), 16–27.

Stiggins, R. J. (2002). Assessment crisis: The absence of assessment for learning. *Kappan Professional Journal, 83*(10), 758-765. Retrieved on March 16, 2006, from http://electronicportfolios.org/afl/Stiggins-AssessmentCrisis.pdf

Stobart, G. (2005). Fairness in multicultural assessment systems. *Assessment in Education*, *12*(3), 275–287. doi:10.1080/09695940500337249

Stobart, G. (2006). The validity of formative assessment . In Gardner, J. (Ed.), *Assessment and learning* (pp. 133–146). London: Sage Publications.

Stoner, J., & Freeman, R. (1992). *Management*. Englewood Cliffs, NJ: Prentice Hall.

Strauss, A. L., & Corbin, J. M. (1998). *Basics of qualitative research: Techniques and procedures for developing Grounded Theory* (2nd ed.). Thousand Oaks, CA: Sage Publications.

Stream Work. (2010). *Features*. Retrieved May 9, 2010, from http://www.sapstreamwork.com/features/

Streitz, N. A., Geißler, J., Holmer, T., Konomi, S., Müller-Tomfelde, C., Reischl, W., et al. (1999). i-LAND: An interactive landscape for creativity and innovation. In *Proceedings of the ACM Conference on human Factors in Computing Systems,* (pp. 120-127). Pittsburg, PA: ACM Press.

Stricker, L. J., & Ross, J. (1964). An assessment of some structural properties of the Jungian personality typology. *Journal of Abnormal and Social Psychology*, *68*(1), 62–71. doi:10.1037/h0043580

Summers, I., & White, D. E. (1976). Creativity techniques: Toward improvement of the decision process. *Academy of Management Review*, *1*(2), 99–108. doi:10.2307/257490

Sveiby, K. E. (1997). *The new organizational wealth: Managing and measuring knowledge-based assets*. San Francisco: Berret-Koehler Publishers, Inc.

Sweller, J., van Merrienboer, J. J. G., & Paas, F. (1998). Cognitive architecture and instructional design. *Educational Psychology Review*, *10*(3), 251–296. doi:10.1023/A:1022193728205

Sweney, M. (2009). Project canvas start-up costed at 24 million. *The Guardian.* Retrieved August 8, 2009, from http://www.guardian.co.uk/media/2009/feb/26/project-canvas-strartup-costs

Symoens, J. (1998). Site server is a fine set of tools for web site building. *InfoWorld*, *4*(20), 128.

Taggar, S. (2001). Group composition, creative synergy, and group performance. *The Journal of Creative Behavior*, *35*(4), 261–286.

Takeuchi, H. (2001). Towards a universal management of the concept of knowledge . In Nonaka, I., & Teece, D. (Eds.), *Managing industrial knowledge: Creation, transfer and utilization* (pp. 315–329). London: Sage Publications.

Taminiau, Y., Smit, W., & Lange, A. (2009). Innovation in management consulting firms through informal knowledge sharing. *Journal of Knowledge Management*, *13*(1), 42–55. doi:10.1108/13673270910931152

Tapscott, D. (2009). *Grown up digital: How the net generation is changing your world.* New York: McGraw-Hill.

Tarricone, P., & Luca, J. (2002). Employees, teamwork and social interdependence–a formula for successful business? *Team Performance Management*, *8*(3/4), 54–59. doi:10.1108/13527590210433348

Tassoul, M., & Buijs, J. (2007). Clustering: An essential step from diverging to converging. *Creativity and Innovation Management*, *16*(1), 16–26. doi:10.1111/j.1467-8691.2007.00413.x

Tavares, C., Moreira, A., Silva, M., & Albuquerque, A. (2008). *The use of an e-Porfolio as a way of understanding students learning preferences*. Paper presented at the ICERI 2008 - International Conference of Education, Research and Innovation.

Taylor, J. (1970). Introducing social innovation. *The Journal of Applied Behavioral Science*, *6*(1), 69–77. doi:10.1177/002188637000600104

Taylor, C. (2009). The creative industries, governance and economic development: A UK perspective. *GeoJournal Library*, *98*, 153–166. doi:10.1007/978-1-4020-9949-6_10

Teare, R., & Rayner, C. (2002). Capturing organizational learning. *International Journal of Contemporary Hospitality Management*, *14*(7), 354–360. doi:10.1108/09596110210440657

Teece, D., Pisano, G., & Shuen, A. (1990). *Firm capabilities, resources, and the concept of strategy*. Center for Research on Management, University of California, Berkeley, CCC Working Paper 90-8.

Think Tank. (2010). *Group systems solutions*. Retrieved May 9, 2010, from http://www.groupsystems.com/solutions/

Thompson, M. A., & Kahnweiler, W. M. (2002). An exploratory investigation of learning culture theory and employee participation in decision making. *Human Resource Development Quarterly*, *13*(3), 271–288. doi:10.1002/hrdq.1031

Tiwana, A. (2008). Do bridging ties complement strong ties? An empirical examination of alliance ambidexterity. *Strategic Management Journal*, *29*(3), 251. doi:10.1002/smj.666

Toffler, A. (1970). *Future shock*. London: Bodley Head.

Toffler, A. (1980). *The third wave*. New York: Morrow.

Torrance, E. P., & Myers, R. E. (1970). *Creative learning and teaching*. New York: Dodd.

Torrance, E. P. (1963). *Creativity*. USA: American Educational Research Association of the National Education Association.

Torrance, E. P. (1987). Teaching for creativity . In Isaksen, S. G. (Ed.), *Frontiers of creativity research: Beyond the basics*. Buffalo, NY: Bearly Limited.

Treffinger, D.J. & Isaksen, S.G. (1992). *Creative problem solving: An introduction*. Center for Creative Learning: Sarasota.

Tsoukas, H. (1996). The firm as a distributed knowledge system: A constructionist approach. *Strategic Management Journal*, *17*, 11–25.

Tsoukas, H. (2003). Do we really understand tacit knowledge? In Easterby-Smith, M., & Lyles, M. A. (Eds.), *The Blackwell handbook of organizational learning and knowledge management* (pp. 411–427). Cambridge, MA: Blackwell Publishing.

Tulving, E., & Schachter, D. L. (1990). Priming and human memory systems. *Science*, *247*, 301–306. doi:10.1126/science.2296719

Tushman, M., & Moore, W. (1988). *Readings in management of innovation*. New York: Harper Collins Publishers.

Ulrich, K. T., & Eppinger, S. (1995). *Product design and development*. New York: McGraw-Hill.

Unsworth, K. L. (2005). Creative requirement: A neglected construct in the study of employee creativity? *Group & Organization Management*, *30*, 541–560. doi:10.1177/1059601104267607

Urabe, K. (1988). Innovation and the Japanese management system . In Urabe, K., Child, J., & Kagono, T. (Eds.), *Innovation and management international comparisons* (pp. 3–26). Berlin: Walter de Gruyter.

Urban, K. K. (2002). From creativity to responsible Createlligence® as future competence. *Journal of Educational Sciences*, *1*(2), 200-170. Retrieved December 27, 2008, from http://qspace.qu.edu.qa/handle/10576/8482

Uribe, D. E., & Cabra, J. F. (2010). Creative problem-solving in Second Life: An action research study. *Creativity and Innovation Management*, *19*(2), 167–179. doi:10.1111/j.1467-8691.2010.00550.x

Uzzi, B., & Spiro, J. (2005). Collaboration and creativity: The small world problem. *American Journal of Sociology*, *111*(2), 447. doi:10.1086/432782

Valacich, J. S., Dennis, A. R., & Connolly, T. (1994). Idea generation in computer based groups: A new ending to an old story. *Organizational Behavior and Human Decision Processes*, *57*(3), 448–467. doi:10.1006/obhd.1994.1024

Van de Ven, A. H., Polley, D. E., Garud, R., & Venkataraman, S. (1999). *The innovation journey*. New York: Oxford University Press.

Van de Ven, A. H. (1986). Central problems in the management of innovation. *Management Science*, *32*(5), 590–607. doi:10.1287/mnsc.32.5.590

Van de Ven, A., & Engleman, R. (2004). Central problems in managing corporate innovation and entrepreneurship. *Advances in Entrepreneurship . Firm Emergence and Growth*, *7*, 47–72. doi:10.1016/S1074-7540(04)07003-5

Van der Meer, H. (2007). Open innovation–the Dutch treat: Challenges in thinking in business models. *Creativity and Innovation Management*, *16*(2), 192–202. doi:10.1111/j.1467-8691.2007.00433.x

Van Der Zee, J.T.M. & De Jong, B. (1999). *Alignment is not enough: Integrating business and information technology management with the balanced business scorecard.*

Vandijck, E. (1998). Information and communication technology in the 21st century. *Revista Portuguesa de Gestão*, *2*, 15–22.

VanGundy, A. B. (1988). *Techniques of structured problem solving*. New York: Van Nostrand Reinhold Company Incorporated.

VanGundy, A. B. (2005). *101 activities for teaching creativity and problem solving*. San Francisco: Wiley & Sons, Inc.

VanGundy, A. B. (1983). Brainwriting for new product ideas: An alternative to Brainstorming. *Journal of Consumer Marketing*, *1*(2), 67–74. doi:10.1108/eb008097

Venkatraman, N. (1989). The concept of fit in strategy research: Toward verbal and statistical correspondence. *Academy of Management Review*, *14*, 423–444. doi:10.2307/258177

Vogel, M., & Oliver, M. (2006). *Design for learning in virtual learning environments-insider perspectives*. London: Centre for Excellence in Learning Technology.

von Bertalanffy, L. (1981). *A systems view of man*. Boulder, CO: Westview Press.

von Hippel, E. (1978). A customer-active paradigm for industrial product idea generation. *Research Policy*, *3*, 240–266. doi:10.1016/0048-7333(78)90019-7

Von Hippel, E. (1994). "Sticky information" and the locus of problem solving: Implications for innovation. *Management Science*, *40*(4), 429–439. doi:10.1287/mnsc.40.4.429

von Hippel, E., & Tyre, M. (1993). How learning by doing is done: Problem identification in novel process equipment. Massachusetts Institute of Technology, SSM Working Paper#BPS 3521-93.

Von Krogh, G., Ichijo, K., & Nonaka, I. (2000). *Enabling knowledge creation: How to unlock the mystery of tacit knowledge and release the power of innovation*. New York: Oxford University Press.

Von Seggern, M., & Jourdain, J. M. (1996). Technical communications in engineering and science: The practices within a government defense laboratory. *Special Libraries*, *87*, 98–119.

Wagner, C. (1993). Problem solving and diagnosis. *OMEGA International Journal of Management Sciences*, *21*, 645–656. doi:10.1016/0305-0483(93)90006-7

Wagner, R. K., & Sternberg, R. J. (1985). Practical intelligence in real world pursuits: The role of tacit knowledge. *Journal of Personality and Social Psychology*, *49*, 436–458. doi:10.1037/0022-3514.49.2.436

Wallas, G. (1926). *The art of thought*. New York: Harcourt Brace.

Walsh, J. P., & Ungson, G. R. (1991). Organizational memory. *Academy of Management Review*, *16*, 57–91. doi:10.2307/258607

Ward, T. B. (2006). Creative cognition as a window on creativity. *Methods (San Diego, Calif.)*, *42*(1), 28–37. doi:10.1016/j.ymeth.2006.12.002

Ward, T. B., Smith, S. M., & Finke, R. A. (2009). Creative cognition . In Sternberg, R. J. (Ed.), *Handbook of Creativity* (pp. 189–212). New York: Cambridge University Press.

Warr, A., & O'Neill, E. (2005). Understanding design as a social creative process. *Proceedings of the 5th conference on Creativity & Cognition,* (pp. 118-127). New York: ACM Press.

Wasserman, S., & Faust, K. (1994). *Social network analysis: Methods and applications*. Cambridge: Cambridge University Press.

Watson, S. (1998). Getting the 'aha'! *Computerworld*, *4*(32), 1–2.

Webster, F. (2006). *Theories of the information society*. London: Routledge.

Weick, K. E. (1995). *Sensemaking in organizations*. Thousand Oaks, CA: SAGE Publications.

Weick, K. (1995). *Sensemaking in organizations*. Thousand Oaks, CA: Sage.

Weick, K. E., & Westley, F. (1996). Organizational learning: Affirming an oxymoron . In Clegg, S. R., Hardy, C., & Nord, W. R. (Eds.), *Handbook of organization studies* (pp. 440–458). London: Sage.

Weisberg, R. W. (1993). *Creativity: Beyond the myth of genius*. New York: W. H. Freeman & Co.

Weisberg, R. W. (2006). Modes of expertise in creative thinking: Evidence from case studies . In Ericsson, K. A., Charness, N., Feltovich, P. J., & Hoffman, R. R. (Eds.), *The Cambridge handbook of expertise and expert performance* (pp. 761–787). New York: Cambridge University Press.

Wenger, E. (1999). *Communities of practice: Learning, meaning and identity*. Cambridge: Cambridge University Press.

Wenger, E. C., & Snyder, W. M. (2000). Communities of practice: The organisational frontier. *Harvard Business Review*, (January–February): 139–145.

Wertheimer, M. (1945). *Productive thinking*. New York: Harper.

West, M. A. (2002). Sparkling fountains or stagnant ponds: An integrative model of creativity and innovation implementation in work groups. *Applied Psychology: An International Review*, *51*(3), 355–424. doi:10.1111/1464-0597.00951

West, M. A., & Farr, J. L. (1989). Innovation at work: Psychological perspectives. *Social Behaviour*, *4*, 15–30.

West, M., & Sacramento, C. (2008). Flourishing in teams: Developing creativity and innovation . In Henry, J. (Ed.), *Creative management and development*. London: Sage Publications.

West, M. A. (2001). Management of creativity and innovation in organizations . In Smelser, N., & Baltes, P. (Eds.), *International encyclopedia of the social and behavioral sciences* (pp. 2895–2900). Oxford: Pergamon Press.

West, M. A. (1990). The social psychology of innovation in groups . In West, M. A., & Farr, J. L. (Eds.), *Innovation and creativity at work: Psychological and organizational strategies* (pp. 4–36). Chichester, UK: Wiley.

West, M. A., & Rickards, T. (1999). Innovation. In M.A. Runco & S.R. Pritzker (Eds.), *Encyclopedia of Creativity, 2*, 35-43. San Diego: Academic Press.

Wilderom, C. P. M., Glunk, U., & Maslowsky, R. (2000). Organizational culture as a predictor of organizational performance . In Ashakanasy, N. M., Wilderom, C. P., & Peterson, M. F. (Eds.), *Handbook of organizational culture and climate* (pp. 193–209). Thousand Oaks: Sage.

Williams, L. J., Ford, L. R., & Nguyen, N. (2002). Basic and advanced measurement models for confirmatory factor analysis . In Rogelberg, S. G. (Ed.), *Handbook of research methods in industrial and organizational psychology* (pp. 366–389). Oxford: Blackwell.

Winograd, T., & Flores, F. (1986). *Understanding computers and cognition: A new foundation for design*. Norwood, NJ: Ablex.

Winter, S. G. (1987). Knowledge and competence as strategic assets . In Teece, D. J. (Ed.), *The competitive challenge: Strategies for industrial innovation and renewal* (pp. 159–184). Cambridge, MA: Ballinger.

Wolfe, R. A. (1994). Organizational innovation: Review, critique and suggested research directions. *Journal of Management Studies, 31*, 405–431. doi:10.1111/j.1467-6486.1994.tb00624.x

Woodman, R. W., Sawyer, J. D., & Griffin, R. W. (1993). Toward a theory of organizational creativity. *Academy of Management Review, 18*(2), 293–321. doi:10.2307/258761

Wurman, R. S. (1989). *Information anxiety*. New York, NY: Doubleday.

Xu, F., & Rickards, T. (2007). Creative management: A predicted development from research into creativity and management. *Creativity and Innovation Management, 16*(3), 216–228. doi:10.1111/j.1467-8691.2007.00445.x

Yang, B. (2003). Identifying valid and reliable measures for dimensions of a learning culture. *Advances in Developing Human Resources, 5*(2), 152–162. doi:10.1177/1523422303005002003

Yeh, J.-C., Chang, J.-Y., & Qyang, Y.-J. (2000). Content and knowledge management in digital library and museum. *Journal of the American Society for Information Science American Society for Information Science, 4*(51), 371–379. doi:10.1002/(SICI)1097-4571(2000)51:4<371::AID-ASI6>3.0.CO;2-6

Yeung, A. K., Ulrich, D. O., Nason, S. W., & Glinow, M. A. (1999). *Organizational learning capability: Generating and generalizing ideas with impact*. Oxford: Oxford University Press.

Yusuf, S. (2009). From creativity to innovation. *Technology in Society, 31*(1), 1–8. doi:10.1016/j.techsoc.2008.10.007

Zacharis, N. Z. (2010). Innovative assessment for learning enhancement: Issues and practices. *Contemporary Issues in Education Research, 3*(1), 61–69.

Zaltman, G., Duncan, R., & Holbek, J. (1973). *Innovation & organizations*. New York: John Wiley & Sons.

Zampetakis, L., & Moustakis, V. (2006). Linking creativity with entrepreneurial intentions: A structural approach. *The International Entrepreneurship and Management Journal, 2*(3), 413–428. doi:10.1007/s11365-006-0006-z

Zander, U., & Kogut, B. (1995). Knowledge and the speed of the transfer and imitation of organizational capabilities: An empirical test. *Organization Science, 6*(1), 76–92. doi:10.1287/orsc.6.1.76

Zephram. (2010). *Home*. Retrieved May 9, 2010, from http://www.zephram.de/?lang=en

Zhou, J., & Shalley, C. (Eds.). (2008). *Handbook of organizational creativity*. New York: Lawrence Erlbaum Associates.

Zhou, J., & George, J. M. (2003). Awakening employee creativity: The role of leader emotional intelligence. *The Leadership Quarterly, 14*, 545–568. doi:10.1016/S1048-9843(03)00051-1

Zipperer, L. (1993). The creative professional and knowledge. *Special Libraries, 84*, 69–78.

Zizek, S. (2008). The prospects of radical politics today. *International Journal of Baudrillard Studies, 5*(1).

Zorrinho, J. (1995). *Gestão da informação: Condição para vencer*. IAPMEI, Ministry of Industry and Energy.

Zuboff, S. (1988). *In the age of the smart machine: The future of work and power*. New York: Basic Books.

Zwicky, F., & Wilson, A. (1965). *New methods of thought and procedure*. New York: Springer-Verlag.

# About the Contributors

**Anabela Mesquita** is Professora Adjunta at ISCAP (School of Accountancy and Administration of Porto). Since 2007 she is the Vice Dean of ISCAP. She is also Member of the Agoritmi Center at Universidade do Minho. Her research interests include Knowledge and Innovation Management, Impact of Information Systems in Organization, Life Long Learning at Higher Education level and e-learning. She has been involved in several European and National research projects. She has published numerous papers in various international journals and Conference Proceedings. She serves as Member of the Editorial Board and referee for the IGI Global. She also serves as Associate Editor of the *Information Resources Management Journal* and is the Editor in Chief of the *International Journal of Technology and Human Interaction*. She serves as referee of the *Journal of Cases of Information Technology*. She has also been evaluator and reviewer for European Commission projects.

* * *

**Astrid Lange**, born in 1980 in Bernau, Germany, studied Psychology (Diploma) at the University Potsdam. Since October 2007 she is employed as Scientific Assistant at the Brandenburg University of Technology Cottbus. She examines student's entrepreneurial intentions within the research project "Mobilization of female innovative entrepreneurial potential in the area of technology (MobI$^{UP}$)", public funded by the German Federal Ministry of Research and Technology and the European Social Fund. Her engagement in the EXIST-III - project "Development of a Workshop for Entrepreneurial and Team Competence for Diverse Disciplines and Start-up Teams", public funded by the German Federal Ministry of Economic Affairs and Technology and the European Social Fund, focuses on qualification of potential entrepreneurs. Her trainings include topics as creativity and idea generation, team building, and career management. Working on her PhD, she investigates the "subjective beliefs regarding entrepreneurship".

**Calvin Lee** is one of the co-founders of the consultancy firm Genuince which specializes in Enterprise Architecture and Business Process Modelling. He is always interested in finding ways to help organizations implement their strategy and projects in an integral manner. In this role he has worked for different governmental organizations and has been invited for guest lectures and written several articles. Since January 2009 he is the Managing Director of TeamSupport. A company that markets a fully web-based application to support brainstorm and group meetings anywhere and anytime. With this tool and our facilitation services we will introduce the world how simple it is to make organizations' collaboration efforts more creative, sustainable and of greater value.

**Célia Tavares** is a graduate in Specialized Translation and Interpretation from the School of Accounting and Administration of Oporto (ISCAP) and has a Master in Multimedia in Education by University of Aveiro (Portugal). Professionally works since 2005 in PAOL - Unit for Innovation in Education. PAOL is a unit held in ISCAP that is entirely dedicated to the development and research in the e/b-learning field. Along with this activity, she is also an Assistant Lecturer at ISCAP, teaching "Technical English" and "English for Accounting". The author has also several publications/communications in the e/b-learning area, as well as in the e-portfolios area.

**Cher Ping Lim** (PhD) is Professor of Education in the Department of Curriculum and Instruction, Hong Kong Institute of Education. He has been the chief investigator of several major research projects that include: (1) Effective Integration of IT in Singapore Schools: Pedagogical and Policy Implications, (2) Supporting E-discussions and e-sharing with New Technologies in Learning Communities, (3) Digital Curricular Literacies, and (4) 3D Multi-User Virtual Environments in Schools. He has also provided consultancy services to the World Bank, World Links, UNESCO, Inter-American Development Bank, World Bank, World Links, Microsoft, ACER Computers, schools and the Government of Barbados and Oman.

**Corrado lo Storto** is an Associate Professor in Economics and Engineering Management with the University of Naples Federico II (Italy) where he teaches Business Management and Project and Innovation Management. He holds a Laurea in Aeronautical Engineering, an MBA and a PhD in Science of Industrial Innovation. He is the Chair of the Chapter of the Technology Management Council in the Italian IEEE Section. His research and professional interests include technology transfer, innovation and knowledge management, technology planning, new product development, risk management and project valuation, and on these topics he has published several papers in international journals and conference proceedings.

**Diego E. Uribe Larach** is a creativity consultant and specialist with a Master's degree in Creativity from the International Center for Studies in Creativity, State University of New York, Buffalo State. He is the head of IDEMAX consulting and works as a trainer and facilitator of Creativity, Creative Problem-Solving (CPS), the LEGO® SERIOUS PLAY methodology, change leadership, and innovation methods for both profit and non-profit organizations such as: Xerox, IBM, Taca Airlines, Buffalo State College, the Western NY Green Economy initiative, Spark Labs, Unilever, Kraft Foods, Albright Knox Art Gallery, Canal 13, MBA UC Santiago, Chile, MBA UAI Santiago, Chile, and Universidad del Desarrollo, Santiago, Chile. Diego is also a Fulbright scholar and an adjunct professor at the International Center for Studies in Creativity and lecturer at the Master de Innovación, Universidad Adolfo Ibañez, Chile. His research work focuses on the opportunities and challenges of the marriage between digital technologies and creative behavior/expression.

**Duarte Gomes** is Associate Professor and coordinator of Work and Organizational Psychology Unit of the Faculty of Psychology and Education Sciences, University of Coimbra. He holds a PhD in Psychology, Work and Organizational Psychology - University of Coimbra, 1990. Pre-graduate and post-graduate teaching activity in work and organizational psychology area. National Coordinator of the European Master on Work, Organizational, and Personnel Psychology (WOP-P), supported by the European Commission through the Erasmus Mundus Programme. Organizational culture and leadership are his main research interests. Author of various books, book chapters and scientific papers.

**Fernando Cardoso de Sousa**, President of the Executive Board of APGICO, holds a PhD in Organizational Behaviour and is the director of the Marketing and Research Unit - GAIM (www.gaim.pt), the Portuguese Creativity and Innovation Association - APGICO (www.apgico.pt), and of the M.A. in Human Resource Management, and BsC. In Real Estate Management courses at the INUAF, Loule, Portugal.. Guest speaker in universities and companies in, he is author of books and articles on subjects like creativity, innovation, leadership, human resources, psychology and military sociology. His latest book in Portuguese is "Leading groups in complex problem solving". Consultant of companies in human resources and organizational behaviour, he was recently awarded the "Excellence in Innovation Award", at the 3rd International India Innovation Summit, in Mumbai (http://www.3is.biz/about_new.html).

**Graham Horton** studied Computer Science at the University of Erlangen, Germany, obtaining his Masters degree ("Diplom") in 1989. He obtained his PhD in Computer Science in 1991 and his "Habilitation" in 1998 at the same university, both in the field of simulation. Since 2001, he is Professor for Simulation at the Computer Science department of the University of Magdeburg. His research interests include state-space-based simulation of stochastic systems and computer-supported methods for ideation and innovation processes. He has been teaching the Idea Engineering course to students from a wide variety of study programs for five years. He is also one of the founders of the Zephram corporation, a consultancy which specializes in the early stages of the innovation process.

**Greg Hearn** is Research Professor in the Creative Industries Faculty at QUT. His work focuses on applications development and evaluation of new technologies and services in the creative industries. He has been involved in high level consulting and applied research examining new media and industry/organisational forms for more than two decades, with organisations including British Airways, Hewlett Packard, and Australian and international agencies. He was a consultant to the Broadband Services Expert Group, the national policy group that formulated Australia's foundational framework for the internet in 1994. In 2005 he was an invited member of a working party examining the role of creativity in the innovation economy for the Australian Prime Minister's Science Engineering and Innovation Council. He has authored or co-authored over 20 major research reports and a number of books including The knowledge economy handbook (2005: Edward Elgar) and Knowledge Policy: Challenges for the 21st Century (2008: Edward Elgar).

**Gwendolyn L. Kolfschoten** is an assistant Professor at Delft University of Technology in the Netherlands. She is an experienced facilitator of thinkLets-based Group Support System workshops having worked with numerous public and private organizations. Her research focuses on collaboration support, both from a technical and from a social perspective. She developed the first example of Computer Supported Collaboration Engineering (CACE) technology – an integrated support suite to assist collaboration engineers in process design, and she co-developed the Collaboration Engineering approach to collaboration process design. She has organized successful minitracks and tutorials at HICSS for the past four years. Her research has been presented at HICSS, CRIWG, AMCIS, EE and GDN conferences and has been published in the Journal of Management Information Systems, International Journal of Human-Computer Studies, Journal of the AIS, Computers and Education and Group Decision and Negotiation.

**Ileana Pardal Monteiro** holds a PhD in Organizational Psychology and is a board member of Apgico (www.apgico.pt). She is an assistant professor of the Management, Hotel and Tourism School (ESGHT) of the University of Algarve. She is the author of publications in the area of Organizational Behaviour, leadership, organizational creativity and innovation, creative problem solving and research methodology. She published recently the Portuguese book "Leading groups in complex problem solving". She worked in Human Resources Management in the Directorate-General of Public Administration.

**John F. Cabra** is an Assistant Professor at the International Center for Studies in Creativity. He was a former operations manager for American Airlines in Buffalo, New York. He has also held internal consulting positions for American Airlines in the fields of training and development, employee relations, and organizational development for its Miami, Caribbean, and Latin America division. Before joining American, John was a bilingual training and organizational development specialist for Fisher-Price Toys. As an external consultant, John designed and presented programs on leadership and creativity for organizations such as IBM, Cadbury Schweppes, DDI, Kraft Foods, United Airlines, ConAgra Foods, Greatbatch, The Consortium of College and University Media Centers, Chilean Travel and Tourism Association, Boehringer Maanheim, Quaker Oats, University of Rochester, Hyatt Aruba, Transitions Optical, and El Dorado Bogotá International Airport Authority. John has presented his work in 22 countries. John's research interests are organizational creativity, perceptions of creativity across cultures and the efficacy of 3D virtual platforms in the classroom. He has published articles in U.S. and Latin American business journals. He serves as an editorial board member of Creatividad y Sociedad, a creativity journal published at the University of Barcelona, Spain. He holds a PhD in Organizational Psychology from the University of Manchester, England, and a MS in Creativity and Change Leadership from the State University of New York, Buffalo State College.

**Joshua Fairchild** is a graduate student in the Industrial and Organizational Psychology doctoral program at The Pennsylvania State University. He received his bachelor's degree in psychology from the University of Connecticut, and has worked in institutional research at Southern Connecticut State University. His primary research areas include creativity and innovation management, leadership, and teamwork, and the integration of technology in these areas is a chief pursuit.

**Lee Yong Tay** is the Head of Department for Research and concurrent Head of Department for Information Communication Technology in Beacon Primary School, one of the Future Schools in Singapore. He has been actively involved in the integration of ICT into the school curriculum for the past decade. He obtained his PhD from Edith Cowan University, Western Australia. He also has a degree of Master of Arts in Instructional Design and Technology from the National Institute of Education, Singapore. His current research interest are in the areas of ICT integration into the curriculum, educational use of Multi-User Virtual Environment to re-engage academically at-risk students in academically related tasks and activities, and the use of ICT tools to facilitate the learning of higher-order thinking.

**Leonor Cardoso** is an Assistant Professor and member of Work and Organizational Psychology Unit of the Faculty of Psychology and Education Sciences – University of Coimbra. She holds a PhD in Psychology, Work and Organizational Psychology, University of Coimbra, 2004. Her professional activity relates to pre-graduate and post-graduate teaching activity in work and organizational psychology area. She is the Vice-Coordinator of the European Master on Work, Organizational and Personnel Psychology (WOP-P), supported by the European Commission through the Erasmus Mundus Programme. Knowledge management and its relationships with other organizational variables (e.g. organizational performance, total quality and human resources management, organizational cooperation and commitment) are her main research interests. She is the author of various book chapters and scientific papers.

**Lily Cushenbery** is a graduate student in Industrial and Organizational Psychology at the Pennsylvania State University. Her research areas include leader error recovery and malevolent creativity. Lily is also a research assistant at the International Center for the Study of Terrorism and has managed several research studies and applied projects. She has been an author for several articles and book chapters in the areas of leadership and innovation.

**Luciana Oliveira** is graduated in Assistance Management by the School of Accounting and Administration of Oporto (ISCAP) and has a Master in Multimedia by Faculty of Engineering of Oporto (Portugal). Professionally works since 2005 in PAOL - Unit for Innovation in Education. PAOL is a unit held in ISCAP that is entirely dedicated to the development and research in the e/b-learning field. She is the author of several publications/communications in the e/b-learning area, particularly in the Learning Design domain.

**Manuel Moreira da Silva** is a lecturer at School of Accounting and Administration of Porto - IPP, in the Department of Languages and Cultures. He has also taught at the School of Management and Industrial Studies (ESEIG) and has collaborated as invited teacher with New University of Lisbon (FSCH – UNL). He is the Coordinator of Multimedia Language Centre and a member of the executive board of PAOL, Unit for Innovation in Education. Since 2005, he is a researcher at INESC Porto, in its the Manufacturing Systems Unit (UESP), in the area of Enterprise Cooperation Networks, and, since 2007, a researcher with CINAMIL, the Military Academy Research Centre.

**Marc B. Stierand** is a Senior Lecturer at the Academy of Hotel & Facility Management at NHTV Breda University of Applied Sciences. His research is focused on two main areas: the first is concerned with the phenomenological modelling of lived experiences of creativity; and the second is concerned with managing creativity in an organizational context. One of his most recent studies includes the phenomenological modelling of culinary innovation for which he has interviewed 19 of the world's best chefs. Another focuses on the introduction of slowness as a notion in organizations. Marc has authored several journal papers and conference papers. He holds an honours degree in Hospitality Management from the University of Brighton, a masters degree in Business Economics from the International University of Applied Sciences Bad Honnef – Bonn, a postgraduate certificate in research methodology, and a PhD in Management Science both from the University of Strathclyde.

**Milan Radosavljevic** is director of postgraduate programmes and director of MSc Project Management in the School of Construction Management and Engineering, University of Reading. He sits on ARCOM Committee and holds editorship of ARCOM newsletter. He serves as Vice-Chairman of CIRIA Construction Process Advisory Panel and represents Reading on CIRIA Council. Milan regularly lectures at the Norwegian School of Management, Tallinn University of Technology, University of Zagreb and University of Ljubljana where he also holds visiting professorship. He has delivered numerous executive courses in the UK, Norway, Finland, Estonia, Mauritius, Slovenia and Croatia. Milan has got more than five years of hands on experience as production director in a precast concrete manufacturing company where he developed a novel short-term planning methodology that led to more than 50% increase in profitability. In the past he co-ordinated Scottish Demonstration Projects and was involved in several EU funded projects (ITC Euromaster, E-site, ITC@EDU).

**Natalya Sergeeva** is a PhD Candidate in the School of Construction Management and Engineering at the University of Reading, UK. She received her MSc with distinction in Construction Management and Economics from The Novosibirsk State University of Architecture and Civil Engineering (Sibstrin), Russia. The main topic of Natalya's MSc research was water pricing with a particular focus on Russian large water-supply company where she also worked for several years as assistant economist and accountant. Natalya's Doctoral research is on the influence of personal characteristics on employees' creative participation or willingness to put forward innovative and problem-solving ideas. As part of her research she is also working on the first of its kind idea radicality levels framework. Natalya regularly publishes in high profile conferences and is involved in teaching and student administration. Her other interest lies on the learning cutting edge topics such as open innovation, construction and R&D management, innovative behaviour.

**Nicholas Nicoli** is currently a lecturer at the Department of Communications, University of Nicosia, Cyprus. His research has focused on creativity in organisations, creative industries and television production. He has published on Cypriot digital switchover and public service broadcasting in Cyprus.

**Paula Peres** has a PhD in the e-learning area. Master in computer science and graduate in Math Computer. She is currently teaching on the computer scientific science area in the School of Accounting and Administration of Porto, ISCAP. She is developing many research activities concerning the Integration of the Information Technologies and Communication in higher education context. She has already participated in some initiatives and courses in blended learning mode and she uses these technologies as a support and to complement her learning practices. She has some books published in the information system area and she has also cooperated in e-learning editions.

**René Pellissier** holds a PhD in Systems Eng (University of Pretoria) and is Professor of Business Management in the Department of Business Management at the University of South Africa. She has held several executive positions in industry and lectured at most of Southern Africa's major universities. She has written/contributed to several books, e.g., 'The Little Statistics Book for (busy) business people', 'In search of the Quantum Organization: The IT circle of excellence', 'IT make or break issues', 'Business Research made Easy'. She was visiting professor in Information Systems in France and Professor of Information Management in the UK. She has been editor of several management journals and guest editor of an IT special issue. She is founder of Ymholiad Pty Ltd – International Research Institute in Leadership and Innovation. She has extensive consulting, teaching and research experience in: Statistics, Informational Engineering, Knowledge Management, Value Chain Management, Operations Management, Strategy, Business and Process Engineering and Scientific Research.

**Ruth Bridgstock** is Vice Chancellor's Research Fellow at Queensland University of Technology, and a key researcher within the Creative Workforce Research Program at the ARC Centre of Excellence for Creative Industries and Innovation. Ruth's research relates to lifelong individual career and organisational development in the post-compulsory education and business contexts of the 21$^{st}$ century. Her PhD research took a longitudinal approach to the investigation of individual and contextual predictors of graduate employability and early career success in creative sectors. Her current Fellowship project builds and tests theory about the development and deployment of capabilities for innovation in professionals working in the key sectors of science and technology, and the creative industries. The theory will then be used to inform curriculum development and pedagogic practice in university education.

**Samuel T. Hunter** is an Assistant Professor of Industrial and Organizational Psychology at the Pennsylvania State University. His primary areas of research include leadership and innovation management. Within these areas, Dr. Hunter has published over 35 journal articles, books, and book chapters in outlets such as the Leadership Quarterly, the Journal of Applied Psychology, and the Creativity Research Journal. Dr. Hunter has also received funding for his work from agencies such as the National Science Foundation and the Office of Naval Research.

**Sandra Ribeiro** is a PhD Student at Universidade de Aveiro in Multimedia in Education, where she is currently investigating Digital Storytelling in Higher Education. She has a Masters Degree in Multimedia in Education, which focuses on eLearning. Her research interests are eLearning and Multimedia in Higher Education in Language Learning and Student Motivation. Sandra has been a lecturer at School of Accounting and Administration of Porto - IPP, in the Department of Languages and Cultures since the year 2000. She is a member of PAOL – Unit for Innovation in Education, collaborating with the school's online support unit.

**Scott Cassidy** is a PhD candidate in the Industrial and Organizational Psychology program at the Pennsylvania State University. He received his BS, cum laude, in biology/pre-medical studies at James Madison University and his MA in Experimental Psychology at the College of William and Mary. His research interests include creativity and innovation management, as well as leadership and personnel selection. Within these and other areas, he has published several journal articles and book chapters in outlets such the *Journal of Personality and Social Psychology.*

**Shane Dawson** is the Director of Arts Instructional Support and Information Technology (ISIT) at the University of British Columbia, Vancouver, Canada. His role focuses on supporting and developing the diverse learning, teaching and technical requirements of the Faculty of Arts. Shane's research focuses on the application of quantitative data derived from student online activity to inform teaching and learning practice. In particular, Shane's research has demonstrated the use of student online communication data to provide lead indicators of student learning support, sense of community and course satisfaction. He is currently finalizing a large international project investigating the application of academic analytics and student social network analyses to benchmark student learning engagement. Shane is a lead developer of SNAPP a social network visualization tool for assisting teaching staff in better understanding, identifying and evaluating the impact of learning interventions to improve student engagement, academic performance and creative capacity.

**Siew Khiaw Lim** (M.InfoSys & MA) is currently the vice-principal of Beacon Primary School, one of the future schools under the FutureSchools@Singapore Program. She obtained her B.Sc. from the National University of Singapore and 2 masters in Management of Information Systems and Information Communication Technology in Education from the Central Queensland University, Australia and Institute of Education, University of London, respectively. She has been a secondary school mathematics teacher, senior IT instructor at the Education Technology Division - Ministry of Education (Singapore), and vice-principal of St. Anthony's Canossian Secondary School. Her areas of interest in research have been in professional development of teachers, electronic portfolio, and assessment.

**Stefan Werner Knoll** studied Computational Visualistics at the University of Magdeburg, obtaining his Masters degree ("Diplom") in 2006. He has been working at the Computer Science department of the University of Magdeburg since April 2007 as a PhD student in the research field of collaboration engineering and computer-supported methods for ideation. Since 2008, he is funded by a scholarship from the state of Saxony-Anhalt, Germany. The task of his PhD thesis is to analyze and define requirements for a generic group support system that uses the Collaboration Engineering approach.

**Teresa Rebelo** is Assistant Professor and member of Work and Organizational Psychology Unit of the Faculty of Psychology and Education Sciences – University of Coimbra. She holds a PhD in Psychology, Work and Organizational Psychology, University of Coimbra, 2006. Her professional activities are related to Pre-graduate and post-graduate teaching activity in work and organizational psychology area, namely in the European Master on Work, Organizational, and Personnel Psychology (WOP-P), supported by the European Commission through the Erasmus Mundus Programme. Organizational culture, organizational learning and corporate social responsibility are her main research interests. She is the author of various book chapters and scientific papers.

**Viktor Dörfler** is a Lecturer in the Department of Management Science at Strathclyde Business School. His research is focused on two interrelated areas: the first covers modeling of personal knowledge and knowledge increase in organizational context; and the second covers knowledge-based expert systems, in particular Doctus KBS (www.viktordorfler.com/doctus/), and related intelligent applications. His most recent research includes developing a typology of personal knowledge, modelling levels of personal knowledge, and examining the rationality of decision takers. Viktor has authored and coauthored two books, four book chapters, several journal papers, and a few dozen conference papers. He holds a PhD from the Strathclyde Business School; prior to which he obtained master's degrees in Mathematical Engineering, in International Business Relations, in Engineering Education, and an MBA.

# Index

## D



## E



## F

## K



## L



## M



## N



## O

## P



## Q



## R

www.ingramcontent.com/pod-product-compliance
Ingram Content Group UK Ltd.
Pitfield, Milton Keynes, MK11 3LW, UK
UKHW050952290926
15470UKWH00005B/1774

9 781609 605193